MW00446418

DICTIONARY OF COMPUTER SCIENCE, ENGINEERING, AND TECHNOLOGY

EDITOR-IN-CHIEF

Phillip A. Laplante

CRC Press

Boca Raton London New York Washington, D.C.

Library of Congress Cataloging-in-Publication Data

Dictionary of computer science, engineering, and technology / edited by Phillip Laplante.
p. cm.
ISBN 0-8493-2691-5 (alk. paper)
1. Computer science—Dictionaries. 2. Engineering—Dictionaries. 3. Technology—Dictionaries. I. Laplante, Phillip A.

QA76.15.D5258 2000
004'.03—dc21 00-052882

International Standard Book Number 0-8493-2691-5
Library of Congress Card Number 00-052882
Printed in the United States of America 1 2 3 4 5 6 7 8 9 0
Printed on acid-free paper

PREFACE

One can only appreciate the magnitude of effort required to develop a dictionary by actually experiencing it. Although I had written nine other books, I certainly did not know what I was getting into when in January of 1996 I agreed to serve as Editor-in-Chief of CRC Press's *Comprehensive Dictionary of Electrical Engineering*. Published 2 1/2 years later in 1999, I finally understood what I had gotten myself into. Unlike other books that I had written, creating that dictionary was more a test of will and stamina and an exercise in project management than in mere writing. And although I have managed organizations of up to 80 academics, nothing is more like "herding cats" than motivating an international collection of more than 100 distinguished engineers, scientists, and educators scattered around the globe, almost entirely via email. Yet, I think there is no other way to develop a dictionary. I still marvel at how Noah Webster must have managed to construct his *English Dictionary* without the benefits of modern communication.

But I learned many lessons from the *Electrical Engineering Dictionary* project that made the development of this dictionary far easier. For example, I put the development schedule on a much faster track, and I didn't burden the contributors with excessive formatting guidelines. Nevertheless, like the *Electrical Engineering Dictionary,* this dictionary, as much as it is an organizational achievement, is really the collaborative work of many brilliant and dedicated men and women. This is their dictionary and your dictionary. I hope you refer to it regularly and enjoy it.

Phillip A. Laplante, PE, Ph.D.
Editor-in-Chief
President
Pennsylvania Institute of Technology
Media, Pennsylvania

FOREWORD

How Was the Dictionary Constructed

As I knew this project would require a divide-and-conquer approach with fault-tolerance, I sought to partition the dictionary by defining areas that covered all aspects of Computer Science, Computer Engineering, and Computer Technology.

The partitions I selected were based largely on the Denning report on the Computer Science Curriculum (Peter Denning, et al, "Computing as a Discipline," *IEEE Computer Journal,* 22, 63–70, Feb. 1989), with some additions. The partitioning was made with intentional overlap, since I knew that many terms needed to be defined several different ways depending on usage, and I needed to ensure that every term would be defined at least once.

The resultant areas were:

- Algorithms, Data Structures, and Problems
- Artificial Intelligence
- Communications and Information Processing
- Computer Engineering (Processors)
- Computer Engineering (I/O and Storage)
- Computer Graphics
- Database Systems
- Numerical Computing
- Operating Systems
- Programming Languages
- Software Engineering
- Robotics and Synthetic Environments
- Computer Performance Analysis

While sufficient hardware terms were included to understand the basics of digital electronics, those interested in additional electrical engineering terms should consult CRC's *Comprehensive Dictionary of Electrical Engineering* which includes many more such terms.

Given the area editor structure, constructing the dictionary then consisted of the following steps:

1. Creating a terms list for each area
2. Defining terms in each area
3. Cross-checking terms within areas

4. Cross-checking terms across areas
5. Compiling and proofing the terms and definitions
6. Reviewing the compiled dictionary
7. Final proofreading

The first and most important task undertaken by the area editors was to develop a terms list to be defined. A terms list is a list of terms (without definitions), proper names (such as important historical figures or companies), or acronyms relating to their areas. What went into each terms list was left to the discretion of the area editor based on the recommendations of the contributing authors. However, lists were to include all technical terms that related to the area (and subareas). Technical terms of an historical nature were only included if it was noted in the definition that the term is "not used" in modern engineering or that the term is "historical" only. Although the number of terms in each list varied somewhat, each area's terms list consisted of approximately 700 items. The dictionary includes contributions from over 100 contributors from 17 countries.

Once the terms lists were created, they were merged and scrutinized for any obvious omissions. These missing terms were then picked up from other sources including CRC handbooks and dictionaries. These included *The Control Handbook, Electronics Handbook, Image Processing Handbook, Circuits and Filters Handbook,* and *The Electrical Engineering Handbook.* About 1000 terms were taken from the CRC handbooks. We also borrowed, with permission from IEEE, about 40 definitions that could not be found elsewhere or could not be improved upon. The process of developing and collecting the terms took 1 1/2 years.

Once all of the terms and their definitions were collected, the process of converting, merging, and editing began. This process took an additional six months.

Although authors were provided with a set of guidelines to write terms definitions, they were free to exercise their own judgment and to use their own style. As a result, the entries vary widely in content from short, one-sentence definitions to rather long dissertations. While I tried to provide some homogeneity in the process of editing, I neither wanted to tread on the feet of the experts and possibly corrupt the meaning of the definitions, nor did I want to interfere with the individual styles of the authors. As a result, I think the dictionary contains a diverse and rich exposition that collectively provides good insights into the areas intended to be covered by the dictionary. Moreover, I was pleased to find the resultant collection much more lively, personal, and user-friendly than typical dictionaries.

But despite the incredible support from my area editors, individual contributors, and staff at CRC Press, the final task of arbitrating conflicting definitions, rewording those that did not seem descriptive enough, and identifying missing ones was left to me. I hope that I have not failed you in my task.

How to Use the Dictionary

The dictionary is organized like a standard language dictionary except that not every word is defined (this would necessitate a complete embedding of an English dictionary). However, I tried to define most nonobvious technical terms used in the definition of another term. In some cases, more than one definition is given for a term. These are denoted (1), (2), (3), ..., etc. Multiple definitions were given in cases where the term has multiple distinct meanings in differing applications areas, in which case the definition is preceded by an identifier (e.g., "in robotics") to indicate the relevant area, except when the area is obvious or readily determined from the definition itself. When more than one equivalent but uniquely descriptive definition was available, to help increase understanding I included both. In most cases, the possible definitions are ordered according to their most likely usage. But I leave it to the reader to pick the definition that seems to fit your situation most closely.

The notations 1., 2., etc. are used to itemize certain elements of a definition and are not to be confused with multiple definitions. All noun terms are defined in the singular form, except for collective nouns or where the singular form is not used.

Acronym terms are listed by their expanded name. Under the acronym the reader is referred to that term. For example, if you look up "RISC" you will find "See reduced instruction set computer," where the definition can be found. The only exceptions are in the cases where the expanded acronym might not make sense, or where the acronym itself has become a word (such as "Fortran" or "Basic").

Finally, I tried to avoid proprietary names and trade names where possible. Some crept in because of their importance, however.

Acknowledgments

A project of this scope literally requires hundreds of participants. I would like to take this moment to thank these participants both collectively and individually. I thank, in no particular order:

- The editorial board members and contributors.
- Ron Powers, CRC President of Book Publishing, for conceiving this dictionary and believing in me.
- Nora Konopka, Acquisitions Editor from CRC, for providing incredible editorial support.
- The many other people at CRC who provided all kinds of support in the development and production of this book.
- My wife Nancy for converting, typing, and/or entering many of the terms.
- Susan Fox for providing excellent copy editing of the final manuscript.

This achievement is as much theirs as it is mine. Please accept my apologies if anyone was left out — this was not intentional and will be remedied in future printings of this dictionary.

Finally, thank you to my wife Nancy and children Christopher and Charlotte for their incredible patience and endurance while I literally spent hundreds of hours giving birth to this dictionary.

How to Report Errors/Omissions

Because of the magnitude of this undertaking and because I attempted to develop new definitions completely from scratch, I have surely omitted (though not deliberately) many terms. In addition, some definitions are possibly incomplete, weak, or even incorrect. I want to evolve and improve this dictionary in subsequent printings and editions. You are encouraged to participate in this collaborative, global process. Please send any suggested corrections, improvements, or new terms (along with suggested definitions) to me at p.laplante@ieee.org or plaplante@pit.edu. If your submission is incorporated, you will be recognized as a contributor in future editions of the dictionary.

EDITOR-IN-CHIEF

Phil Laplante is President of the Pennsylvania Institute of Technology, a two-year, private college that focuses on technology training and re-training. Prior to this, he was the founding dean of the BCC/NJIT Technology and Engineering Center in Southern New Jersey. He was also Professor of Computer Science and Chair of the Mathematics, Computer Science, and Physics Department at Fairleigh Dickinson University, New Jersey. In addition to his academic career, Dr. Laplante spent almost eight years as a software engineer designing avionics systems, a microwave CAD engineer, a software systems test engineer, and a consultant. He has written dozens of articles for journals, newsletters, magazines, and conferences, mostly on real-time computing and image processing. He has authored 13 other technical books and co-founded the journal *Real-Time Imaging,* as well as two book series including the CRC Press Series on Image Processing.

Dr. Laplante received his B.S., M. Eng., and Ph.D. in Computer Science, Electrical Engineering, and Computer Science, respectively, from Stevens Institute of Technology and an MBA from the University of Colorado. He is a senior member of IEEE and a member of numerous other professional societies, program committees, and boards. He is a licensed professional engineer in New Jersey and Pennsylvania and actively consults to high-tech start-ups, established technology firms, investors, and venture capitalists on technology and business strategy. Dr. Laplante is married with two children and resides in Pennsylvania.

AREAS AND AREA EDITORS

- **Algorithms, Data Structures and Problems**: *Paul E. Black, NIST*
- **Artificial Intelligence**: *Evelyn Rosengarten, Aurora Pharmaceutical*
- **Communications and Information Processing**: *Ling Guan, University of Sydney*
  - **Information Processing**: *Marco Gori, University of Florence (Associate Editor)*
  - **Communications**: *Ian Oppermann, University of Sydney (Associate Editor)*
- **Computer Engineering (Processors)**: *Janusz Zalewski, University of Central Florida*
- **Computer Engineering (I/O and Storage)**: *Amos Omondi, Flinders University*
- **Computer Graphics**: *Robert Fisher, University of Edinburgh*
- **Computer Performance Analysis**: *Eugene Veklerov, Lawrence Berkeley Laboratory*
- **Database Systems**: *Colm O'Riordan, Information Technology Centre, National University of Ireland, Galway*
- **Numerical Computing**: *Amos Omondi, Flinders University*
- **Operating Systems**: *Janusz Zalewski, University of Central Florida*
- **Programming Languages**: *Joseph Newcomer, Joseph M. Newcomer Co.*
- **Robotics and Synthetic Environments:** *John Draper, Oak Ridge National Laboratories*
- **Software Engineering**: *Paolo Nesi, University of Florence*

Trade and Service Marks

- The Graphics Interchange Format is the copyright property of CompuServe Incorporated.
- GIF (sm) is a service mark property of CompuServe Incorporated.
- UNIX is a registered trademark of AT&T's UNIX System Laboratories.

Companies

- AT&T Bell Laboratories
- Digital Equipment Corporation
- Hewlett-Packard Corporation
- IBM Corporation
- Inmos Corporation
- IEEE
- Intel Corporation
- Interactive Corporation
- Lotus Development Corporation
- Microsoft Corporation
- Motorola
- Santa Cruz Operation (SCO)
- Unimation, Inc.

CONTRIBUTORS

Amy W. Apon
University of Arkansas
Fayetteville, Arkansas

Anthony Ashbrook
University of Edinburgh
Edinburgh, Scotland

Carter Bays
University of South Carolina
Columbia, South Carolina

Pierfrancesco Bellini
University of Florence
Firenze, Italy

Lars Bengtsson
Halmstad University
Halmstad, Sweden

Edoardo Biagioni
University of Hawaii
Honolulu, Hawaii

Paul E. Black
NIST
Gaithersburg, Maryland

Joachim Charzinski
SIMENS AG/Information and
Communications Networks
Muenchen, Germany

Frederick Dahlgren
Chalmers University of Technology
Gothenburg, Sweden

Susan R. Dickey
Berkeley, California

John Draper
Oak Ridge National Laboratories
Oak Ridge, Tennessee

Stefan Edelkamp
Albert-Ludwigs-Universitat Freiburg
Freiburg, Germany

Mustafa Ege
Hacettepe University
Ankara, Turkey

Peter Fenwick
University of Auckland
Auckland, New Zealand

Fabrizio Fioravanti
University of Florence
Florence, Italy

Bob Fisher
University of Edinburgh
Edinburgh, Scotland

Leonard Franken
KPN Research
The Netherlands

Joseph L. Ganley
Oak Hill, Virginia

Marco Gori
University Degli Studi di Siena
Siena, Italy

Ling Guan
University of Sydney
Sydney, NSW, Australia

Haldun Hadimioglu
Polytechnic University
New York, New York

Joshua Hale
University of Glasgow
Glasgow, Scotland

John Harauz
John Harauz, Engineering Standards
Toronto, Ontario, Canada

Igor Hawryszkiewyc
University of Technology
Sydney NSW, Australia

Brian Henderson-Sellers
University of Technology
Sydney, NSW, Australia

Art S. Kagel
Bloomberg L.P.
East Brunswick, New Jersey

Christopher Lee Kuszmaul
NASA
Moffett Field, California

Philip Laplante
Pennsylvania Institute of Technology
Media, Pennsylvania

Kim Skak Larsen
University of Southern Denmark
Odense, Denmark

Shikharesh Majumdar
Carleton University
Ottawa, Canada

S. Manoharan
University of Auckland
Auckland, New Zealand

Conrad Martinez
University Politecnica de Catalunya
Barcelona, Spain

Eric McKenzie
University of Edinburgh
Edinburgh, Scotland

Jon Meddes
University of Edinburgh
Edinburgh, Scotland

Veena Mendiratta
Lucent Technologies
Naperville, Illinois

Mart Molle
University of California
Riverside, California

David Monniaux
LIENS
Paris, France

Paolo Nesi
University of Florence
Firenze, Italy

Joseph Newcomer
Joseph M. Newcomer Co.
Pittsburgh, Pennsylvania

Amos Omondi
Nanyang Technological University
Singapore

Ian Oppermann
Camperdown, NSW, Australia

Colm O'Riordan
National University of Ireland
Galway, Ireland

John Patterson
University of Glasgow
Glasgow, Scotland

Craig Robertson
University of Edinburgh
Edinburgh, Scotland

Patrick Rogers
Winchester, Virginia

Evelyn Rosengarten
Hanover Direct, Inc.
Weehawken, New Jersey

Robert S. Seiner
The Data Administration Newsletter
Pittsburgh, Pennsylvania

Rajeev Shorey
IBM Solutions Research Center
New Delhi, India

Sandeep Kumar Shukla
Intel Corporation
Santa Clara, California

Ian Sommerville
Lancaster University
Lancaster, United Kingdom

Eugene Veklerov
Lawrence Berkeley Laboratory
Berkeley, California

Chris Verhoef
University of Amsterdam
Amsterdam, The Netherlands

Gordon Watson
University of Edinburgh
Edinburgh, Scotland

Naofel Werghi
University of Edinburgh
Edinburgh, Scotland

Hau-San Wong
University of Sydney
Sydney, NSW, Australia

Vladimír Zábrodsky
CKD
Blansko, The Czech Republic

Janusz Zalewski
University of Central Florida
Orlando, Florida

John Zaleski
Coatesville, Pennsylvania

Special Terms

$/tps a benchmark rating — dollar cost for a system divided by its tps rating. *See* tps rating.

0-address machine *See* zero-address machine.

0-ary function a function with no arguments, also known as a constant function. *See also* unary function, binary function, N-ary function.

10base2 a type of coaxial cable used to connect nodes on an Ethernet network. The 10 refers to the transfer rate used on standard Ethernet, 10 megabits per second. The base means that the network uses baseband communication rather than broadband communications and the 2 stands for the maximum length of cable segment, 185 meters (almost 200). This type of cable is also called "thin" Ethernet because it is a smaller diameter cable than the 10base5 cables.

10base5 a type of coaxial cable used to connect nodes on an Ethernet network. The 10 refers to the transfer rate used on standard Ethernet, 10 megabits per second. The base means that the network uses baseband communication rather than broadband communications and the 5 stands for the maximum length of cable segment of approximately 500 meters. This type of cable is also called "thick" Ethernet because it is a larger diameter cable than the 10base2 cables.

10baseT a type of coaxial cable used to connect nodes on an Ethernet network. The 10 refers to the transfer rate used on standard Ethernet, 10 megabits per second. The base means that the network uses baseband communication rather than broadband communication and the T stands for twisted (wire) cable.

1NF *See* first normal form.

2 phase commit a commit procedure used in distributed databases. The first phase involves force-writing logs at all participating sites. The second phase involves either committing or undoing the transactions depending on the outcome of the first phase.

2D two-dimensional.

2NF *See* second normal form.

3 phase commit a commit procedure used in distributed databases. The first phase involves force-writing logs at all participating sites. The second phase involves a precommit or undoing transactions at each site (depending on whether the force-writing succeeded at all sites or not). The third phase involves committing all transactions. The 3 phase commit protocol is superior to the 2 phase commit in handling failures.

3D three-dimensional.

3D television stereoscopic television.

3GL *See* third generation language.

3NF *See* third normal form.

4GL *See* fourth generation language.

4NF *See* fourth normal form.

5GL *See* fifth generation language.

5NF *See* fifth normal form.

80/20 rule in networking, an empirical observation about the locality of traffic,which states that 80% of the traffic generated on a local area network is sent to another host in the same work group, and 20% is sent to an outside destination. Recently, because of the introduction of client-server applications, centralized servers and the popularity of both the Internet and corporate intranets, the traffic locality rule has been reversed so that about 80% of the traffic is sent to an outside destination.

0-8493-2691-5/01/$0.00+$.50

A* search *See* best first search.

A/D *See* analog-to-digital converter.

abdominal simulator a virtual reality simulation of the human abdomen, used for training and surgery planning in telemedicine.

abnormal event any external or program-generated event that makes further normal program execution impossible or undesirable, resulting in a system interrupt. Examples of *abnormal events* include: system detection of power failure, attempt to divide by 0, attempt to execute privileged instruction without privileged status, memory parity error.

abort to terminate the attempt to complete the transaction, usually because there is a deadlock or because completing the transaction would result in a system state that is not compatible with "correct" behavior, as defined by a consistency model, such as sequential consistency. Synonym for *rollback*.

absolute address an address within an instruction which directly indicates a location in the program's address space. Programming languages may or may not support the use of *absolute addresses,* depending upon their design. *See also* relative addressing.

absolute addressing an addressing mode where the address of the instruction operand in memory is a part of the instruction so that no calculation of an effective address by the CPU is necessary.

For example, in the Motorola M68000 architecture instruction: ADD 5000,D1, a 16-bit word operand, stored in memory at the word address 5000, is added to the lower word in register D1. The address "5000" is an example of using the *absolute addressing* mode. *See also* addressing mode.

absolute encoder an optical device mounted to the shaft of a motor consisting of a disc with a pattern and light sources and detectors. The combination of light detectors receiving light depends on the position of the rotor and the pattern employed (typically the Gray code). Thus, absolute position information is obtained. The higher the resolution required, the larger the number of detectors needed. *See also* encoder.

absolute error the difference between the actual result and computed result. Sometimes confusingly used for the magnitude of the difference.

absolute form for time in temporal logic systems, when time is referenced to a general system clock and the value is expressed in seconds (milliseconds). *See* relative form for time.

absolute input device an input device that reports its actual position, rather than relative movements. A data tablet or Polhemus tracker operates in this way. *See also* relative input device.

absolute path *See* full path.

absolute performance guarantee an approximation algorithm with an *absolute performance guarantee* is guaranteed to return a feasible solution whose value differs additively from the optimal value by a bounded amount.

absolute position the position of an object in space as referenced to its environment; different from relative position, which expresses position with regard to some other object in the environment.

abstract class a class defined such that no objects of the class can ever be created. The value of an abstract class is that other classes, objects of which can be created, can be derived classes or subclasses of the abstract class. Also known in some languages as pure classes.

abstract data type (ADT) a data type which hides its implementation. It may be a user-defined type or a base type in the language. The behavior of an abstract data type should be

0-8493-2691-5/01/$0.00+$.50

entirely deducible from the specification of its methods, and not require understanding the details of its implementation. Instances of the type are created and inspected only by using calls to the access functions. This allows the implementation of the type to be changed without requiring any changes outside the module in which it is defined. ADTs are central to object-oriented and object-based programming. A classic example of an ADT is a stack data type for which functions might be provided to create an empty stack, to flush a stack, to push values onto a stack, and to pop values from a stack.

abstract interpretation a partial execution of a program which gains information about its semantics (e.g., control structure, flow of information) without performing all the calculations. Abstract interpretation is typically used by compilers to analyze programs in order to decide whether certain optimizations or transformations are applicable. The objects manipulated by the program (typically values and functions) are represented by points in some domain. Each abstract domain point represents some set of real values.

abstraction a technique by which the domain of discourse of an algorithm is brought closer to the problem statement than the realization of the algorithm in code. The goal of many programming languages is to provide means by which the programmer can write algorithms without making the details visible. A word processor is an example of an *abstraction. See also* abstract data type, implementation.

abstract operation a generic operation allowed for objects or systems, which partially performs the whole calculation of the algorithm that should be associated with it. This concept is in some cases stressed by associating with the abstract operations only their definitions.

abstract syntax a general representation of a programming language or language constructs is typically specified by an *abstract syntax* in terms of categories such as "statement", "expression", and "identifier". The syntax is abstract in two senses: it is independent of machine-oriented structures and encodings and also independent of the source syntax (concrete syntax) of the language being compiled (though it will often be very similar). A parse tree is similar to a syntax tree but it also contains features such as parentheses which are syntactically significant but which are implicit in the structure of the syntax tree. In some cases, a description of a data structure is independent of machine-oriented structures and encodings.

abstract system a system defined by a functional design. It is not a physical system to be found in the real world, but a conceptual system behaving as specified in the functional design. It is in the understanding of the abstract system that it is possible to reason about the specified behavior and verify if it will satisfy the functional requirements.

A-buffering an antialiased extension to Z-buffering. An A-buffer identifies visible segments within a sub-pixel area which are represented with bit masks and area sampled for pixel intensity. The technique employs logical operations on the bit masks and thus avoids floating point geometry calculations.

accelerated test a test in which the applied stress level is chosen to exceed that stated in the reference conditions in order to shorten the time duration required to observe the stress response of the item, or to magnify the response in a given time duration.

acceleration change in the velocity of an object.

accelerometer sensor used to measure changes in object velocity.

accept a designation applied to the various types of automata to indicate that, having processed a sequence of input tokens in accordance with the rules for a specific automaton, the automaton is left in one of the designated accepting states.

acceptance criteria the list of requirements that a software system must satisfy before customers take delivery.

acceptance testing formal testing conducted to determine whether a system satisfies its acceptance criteria and thus whether the customer should accept the system.

accepting state any state in an automaton which is designated as a state that is entered upon processing the last symbol of an input string when the input string constitutes a valid sentence in the grammar recognized by the automaton.

access control a means of allowing access to an object based on the type of access sought, the accessor's privileges, and the owner's policy.

access control list a list of items associated with a file or other object; the list contains the identities of users that are permitted access to the associated file; there is information (usually in the form of a set of bits) about the types of access (such as read, write, or delete) permitted to the user.

access control matrix a tabular representation of the modes of access permitted from active entities (programs or processes) to passive entities (objects, files, or devices). A typical format associates a row with an active entity or subject and a column with an object; the modes of access permitted from that active entity to the associated passive entity are listed in the table entry.

accessibility refers to the ease with which a human may enter an environment and ranges from high (for example, an office building) to low (for example, a human blood vessel). Two aspects of the work environment, variability and *accessibility,* determine the applicability of autonomous robots, humans, and teleoperators. Accessibility determines whether a human is applicable: if a human cannot enter an environment or the environment is harmful, an autonomous robot or a teleoperator is the best choice. Where *accessibility* is low but variability is high, teleoperators are best. *See also* complexity.

access level equivalent to visibility inside classes. In the C++ language three types of access levels are defined: private, protected, and public. The access level is used to enforce encapsulation by identifying the visibility of attributes and classes with respect to clients.

access line a communication line that connects a user's terminal equipment to a switching node.

access matrix a matrix with a row for each subject and a column for each subject and each object. Records authorized access modes (such as read, write, execute) subjects may be to objects. Represents the current access authorization policy, but never implemented directly as an array because it would be too sparse.

access mechanism a circuit board or an integrated chip that allows a given part of a computer system to access another part. This is typically performed by using a specific access protocol.

access method a method in which a file is accessed. The access methods possible depend on the file organization.

access mode a file's complete set of access permissions.

access path (**1**) a pair (name, index) which allows the selection of a value from an object. Note that for composite objects the index may not be a simple value such as an integer but may include such concepts as field selection.

(**2**) a designation in a machine instruction of the computation of an operand location, usually as a (name, index) pair, where the name designates a storage class.

access protocol a set of rules that establishes communication among different parts. These can involve both hardware and software specifications.

access right permission to perform an operation on an object, usually specified as the type of operation that is permitted, such as read, write, or delete. *Access rights* can be included in access control lists, capability lists, or in an overall access control matrix.

access time the total time needed to retrieve data from memory. For a disk drive this is the sum of the time to position the read/write head over the desired track and the time until the desired data rotates under the head.

access transparency a system property by which the complications involved in providing access to something (e.g., data) are not apparent to the accessor.

accident an event characterized by the unwanted release of energy or toxic material, possibly resulting in death, injury, or environmental damage. An *accident* may be due to a critical failure of an item, but not necessarily (e.g., it may be due to an external cause). The term mishap covers other types of unacceptable events, e.g., large financial loss.

accommodation (**1**) the ability of a robot to respond to forces and adjust itself to compensate for positioning errors; may be active, using sensors, a control loop, and actuators, or passive, taking advantage of compliance.

(**2**) changes made to the shape of the lens of the human eye with the goal of focusing images on the retina.

account (**1**) an interface and set of resources available to a user on a computer system.

(**2**) a unique data collection point for financial or measurement data. For instance, the AXE45 can be an *account* for a public telephony switch, and as such uniquely identifiable.

accumulation error image error caused mainly by the discretization process.

accumulator historically, the single register on a machine in which a computation would produce a result. Currently, any register in the CPU (processor) that stores one of the operands prior to the execution of an operation, and into which the result of the operation is stored. An *accumulator* serves as an implicit source and destination of many of the processor instructions. For example, register A of the Intel 8085 is an *accumulator*. Generally, however, usage of this term is considered archaic. *See also* CPU and processor.

accuracy (**1**) in a floating-point number representation, this is the number of significant bits contained. (Note that this is different from conventional usage, although it is related.) Also referred to as effective accuracy.

(**2**) the degree of precision with which a robot is able to move to a position in space.

(**3**) the capability of a software or system to provide the right or agreed results or effects with the needed degree of precision. *See* precision.

achromatic light without color. The quantity of light is the only attribute associated with *achromatic* light. In physical terms, this is the intensity or luminance or in the psychological sense it is the perceived intensity in which case the term brightness is used. In the YIQ or YUV representations, this is the Y component. In the HSV representations, it is the V (value) component. In the HSL representations, it is the L (lightness or intensity) component.

achromatic visual field that portion of the visual field within which color vision is not possible.

ACID an acronym meaning atomic, consistent, isolated, and durable.

Ackermann's function a function of two parameters whose value grows very fast. It is defined as $A(1, j) = 2^j$, $A(i, 1) = A(i-1, 2)$, and $A(i, j) = A(i-1, A(i, j-1))$.

acknowledge (**1**) a signal which indicates that some operation, such as a data transfer, has been successfully completed.

(**2**) to detect the successful completion of an operation and produce a signal indicating the success.

acoustic memory a form of circulating memory in which information is encoded in acoustic waves. Now obsolete.

acoustic tracker device using a sound emitter attached to the target to sense its position; used in sets of at least three trackers to allow triangulation of the emitter.

acquirer an organization that acquires or procures a system, product, or service from a supplier. The *acquirer* could be one of the following: buyer, customer, owner, user, purchaser.

acquisition (1) the process of obtaining a system, software product, or software service.

(2) in digital communications systems, the process of acquiring synchronism with the received signal. There are several levels of *acquisitions* and for a given communication system several of them have to be performed in the process of setting up a communication link: frequency, phase, spreading code, symbol, frame, etc.

(3) in vision processing, the process by which a scene (physical phenomenon) is converted into a suitable format that allows for its storage or retrieval. *See also* synchronization.

action an operation performed by a system or object.

action expression the expression specifying the rule behind the action to be performed.

action semantics a variation of denotational semantics where low-level details are hidden by use of modularized sets of operators and combinators.

action state the state associated with the action. Typically, the state of the system in which the action is performed.

activation the event in which a latent fault gives rise to a failure in response to a trigger. Same as manifestation of a fault.

activation function in an artificial neural network, a function that maps the net output of a neuron to a smaller set of values. This set is usually [0, 1]. The original idea was to approximate the way neurons fired, and the activation function took on the value 0 until the input became large and the value jumped to 1. Typical functions are the sigmoid function or singularity functions like the step or ramp.

activation record a record containing all of the information associated with an activation or call of a procedure or function. This information includes: the return address of the caller, the procedure's parameters and local variables, and the frame pointer of the caller.

active agent a program sent to another computer for execution on behalf of the sending computer.

active class in object-based or oriented programming, a class that generates objects with the autonomous behavior. An object instantiated by a pure *active class* does not receive requests from any other objects.

active contour a deformable template matching method that, by minimizing the energy function associated with a specific model (i.e., a specific characterization of the shape of an object), deforms the model in conformation to salient image features. Also called a snake.

active database a database within which the DBMS monitors the database and triggers certain actions to occur.

active device a device that can convert energy from a DC bias source to a signal at an RF frequency. Active devices are required in oscillators and amplifiers.

active error handling the action of reporting and managing unexpected program errors.

active logic digital logic which operates all of the time in the active, dissipative region of the electronic amplifiers from which it is constructed. The output of such a gate is determined primarily by the gate and not by the load. Such logic is generally faster than saturated logic, but it dissipates much more energy. *See* saturated logic.

active neuron a neuron with a non-zero output. Most neurons have an activation threshold. The output of such a neuron has zero output until this threshold is reached.

active object an object instantiated from an active class. An *active object* presents a given degree of autonomy and is typically responsible for the activities of other objects. In object-oriented systems, active objects produce asynchronous actions without the stimulation of other objects. Typically, in software applications there exist only few active objects managing all the other application objects. These can be pure passive or may present a certain degree of autonomy. In some texts, the non-pure active objects are considered active objects as well. A high number of active objects identify a high degree of parallelism for the application producing problems of synchronization.

active quality assurance a quality assurance group that plays an active role in reviews, testing, and other quality related issues, as opposed to passive quality assurance. The difference is often the amount of money that is spent on the quality assurance group. If it is not enough it is passive; otherwise it can be active.

active redundancy the adoption of redundant components working simultaneously to allow or prevent the recovering from failure.

active rule a rule that states a condition and an associated action. If the condition is satisfied, the rule is triggered and the specified action occurs.

active transaction a transaction which is pending, and not yet committed or aborted.

activity (**1**) in a software lifecycle, a heterogeneous collection of tasks which, together, define a large-scale goal to be achieved.

(**2**) in state modeling of a class, any operation which takes a significant time to execute.

activity diagram the diagram reporting the activities of software components. A type of flow chart to describe the behavior of class methods or of the whole class. This is typical of UML notation.

activity packet *See* template.

actual parameter the value or reference computed at the time of a function call of a function. Some language definitions refer to this as an argument. *See also* parameter.

actuator (**1**) a transducer which converts electric, hydraulic, or pneumatic energy to effective motion. For example, in robots, *actuators* set the manipulator in motion through actuation of the joints. Industrial robots are equipped with motors which are typically electric, hydraulic, or pneumatic. *See* industrial robot.

(**2**) in computers, a device, usually mechanical in nature, that is controlled by a computer, e.g., a printer paper mechanism or a disk drive head positioning mechanism.

actuator subsystem components of a robotic system including actuators, means of transmitting actuator forces to links, and means for providing actuators with power.

Ada a programming language developed by the U.S. Department of Defense to support writing reliable applications. Named after Augusta Ada Byron, Countess Lovelace (1815–1852), often called the world's first programmer because she prepared algorithms for Charle's Babbages Analytical Engine, an early mechanical calculator.

adaptability the capability of the software product to be adapted for different specified environments without applying actions or means other than those provided for this purpose for the software considered. *Adaptability* includes the scalability of internal capacity (e.g., screen fields, tables, transaction volumes, report formats). If the software has to be adapted by the end user, *adaptability* corresponds to suitability for individualization and may affect operability.

adaptation process that allows humans to adjust to changing conditions (e.g., illumination) or relationships (e.g., control order) in their environment; for example, *adaptation* allows a user to see in low lighting, and to perform teleoperation when the line of gaze from camera to end-effector is not the same as the normal one from head to hands.

adaptive algorithm an algorithm whose properties are adjusted continuously during execution with the objective of optimizing some criterion.

adaptive coding a coding scheme that adapts itself in some fashion to its input or output.

adaptive control advanced control scheme in which a sensor or sensors monitor a task or process and adjust robot parameters to optimally meet the needs of the mission.

adaptive forward differencing an efficient way to evaluate parametric functions describing curves or surfaces. Each value of the function is determined as the sum of the previous value and a difference term. The distance between points at which the function is evaluated is adapted to the flatness of the function. The value can be a vector as well as a scalar, and this is useful for calculating B-splines.

adaptive fuzzy system fuzzy inference system that can be trained on a data set through the same learning techniques used for neural networks. *Adaptive fuzzy systems* are able to incorporate domain knowledge about the target system given from human experts in the form of fuzzy rules and numerical data in the form of input-output data sets of the system to be modeled. *See also* neural network, fuzzy inference system.

adaptive logic network tree-structured network whose leaves are the inputs and whose root is the output. The first hidden layer consists of linear threshold units and the remaining layers are elementary logic gates, usually AND and OR gates. Each linear threshold unit is trained to fit input data in those regions of the input space where it is active (i.e., where it contributes to the overall network function).

adaptive maintenance the maintenance actions performed on a system in order to modify its behavior for satisfying new needs or environmental features. Classical adaptive maintenance tasks could be the porting, the addition of new functionalities, etc.

adaptive partitioning an approach for parallel job scheduling in which the size of the partition allocated to a job is adaptable to system load and is computed before starting job execution on the system. Both user input and current system load may be used in arriving at the partition size to be allocated to a job.

adaptive predictor a digital filter whose coefficients can be varied, according to some error minimization algorithm, such that it can predict the value of a signal, say N sampling time intervals, into the future. The *adaptive predictor* is useful in many interference cancellation applications.

adaptive resonance theory (ART) network a clustering network developed to allow the learning of new information without destroying what has already been learned. Each cluster is represented by a prototype and learning is achieved by comparing a new input pattern with each prototype. If a prototype is found that is acceptably close to that input, the new pattern is added to that prototype's cluster and the prototype is adjusted so as to move closer to the new input. If no prototype is acceptable, the pattern becomes a new prototype around which a new cluster may develop.

adaptive sampling a method of reducing aliasing artifacts when rendering by adapting the sampling rate in response to the local characteristic of the object being rendered. This technique is often useful to reduce the jagged edges at the edges of objects (or jaggies).

adaptive sorting algorithm a sorting algorithm that can take advantage of existing order in the input, reducing its requirements for computational resources as a function of the disorder in the input.

adaptive subdivision a paradigm for representing data in a hierarchical manner by repeatedly dividing and classifying it until no further definition is necessary, given an error tolerance. *See* octree and quadtree.

adder a logic circuit used for adding binary numbers.

add instruction a machine instruction that causes two numeric operands to be added together. The operands may be from machine registers, memory, or from the instruction itself, and the result may be placed in a machine register or in memory.

additive adder an adder in which addition is not realized as a subtraction, i.e., $A + B$ is not realized as $A - (-B)$.

additive color model colors are defined as a sum of contributions from primary colors. The most commonly used *additive color model* is the Red-Green-Blue model.

additive normalization a procedure used in the computation of mathematical functions, such as sine, cosine, etc. To compute $f(x)$, two variables, say X and Y, are maintained such that as X is reduced to some known constant (usually 0 or 1), Y tends to $f(x)$. The reduction of X is by the addition of a sequence of constants.

additive rule given sets A, B, and C, if $A \rightarrow C$ and $A \rightarrow B \models A \rightarrow BC$.

address a unique identifier for the place where information is stored (as opposed to the contents actually stored there). Most storage devices may be regarded by the user as a linear array, such as bytes or words in RAM or sectors on a disk. The *address* is then just an ordinal number of the physical or logical position. In some disks, the *address* may be compound, consisting of the cylinder or track and the sector within that cylinder.

In more complex systems, the *address* may be a "name" which is more relevant to the user but which must be translated by the underlying software or hardware.

addressable unit a specification of the smallest value that can be "named" by a computer implementation. Although most modern architectures have chosen the byte of 8 bits as the basic addressable unit, historically the addressable unit has been as many bits as the architect has chosen. Many machines, and even modern machines used as digital signal processors, have only words as addressable units, with lengths such as 16, 32, 64, and even 80 bits.

address aliasing *See* cache aliasing.

address bus the set of wires or tracks on a backplane, printed circuit board, or integrated circuit to carry binary address signals between different parts of a computer. The number of bits of address bus (the width of the bus) determines the maximum size of memory that can be addressed. Modern microchips have 32 address lines, thus 4 gigabytes of main memory can be accessed.

address-calculation sort a sort algorithm which uses knowledge of the domain of the items to calculate the position of each item in the sorted array. *See also* radix sort, bucket sort.

address decoder logic which decodes an address.

1. A partial decoder responds to a small range of addresses and is used when recognizing particular device addresses on an I/O address bus, or when recognizing that addresses belong to a particular memory module.

2. A full decoder takes N bits and asserts one of 2^N outputs, and it is used within memories (often within RAM chips themselves).

address error an exception (error interrupt) caused by a program's attempt to access unaligned words or long words on a processor that does not accommodate such requests. The *address error* is detected within the CPU. This contrasts with problems that arise in accessing the memory itself, where a logic circuit external to the CPU itself must detect and signal the error to cause the CPU to process the exception. Such external problems are called bus errors. *See* bus error.

address field the portion of a program instruction word that holds an address.

address generation interlock (AGI) a mechanism to stall the pipeline for one cycle when an address used in one machine cycle is being calculated or loaded in the previous cycle. Address generation interlocks cause the CPU to be delayed for a cycle. (AGIs on the Pentium are even more important to remove, because two execution time slots are lost).

addressing (**1**) in processors, a mechanism to refer to a device or storage location by an identifying number, character, or group of characters which may contain a piece of data or a program step.

(**2**) In networks, the process of identifying a network component, for instance, the unique address of a node on a local area network.

addressing fault an error that halts the mapper when it cannot locate a referenced object in main memory.

addressing mode a form of specifying the address (location) of an operand in an instruction. Some of the addressing modes, found in most processors, are: direct or register direct, where the operand is in a CPU register; register indirect (or simply indirect), where a CPU register contains the address of the operand in memory; and immediate, where the operand is a part of the instruction. *See also* central processing unit.

addressing range numbers that define the number of memory locations addressable by the CPU. For a processor with one address space, the range is determined by the number of signal lines on the address bus of the CPU.

address locking a mechanism to protect a specific memory address so that it can be accessed exclusively by a single processor.

address map a table that associates a base address in main memory with an object (or page) number.

address mapping the translation of virtual addresses into real (i.e., physical) addresses for memory access. *See* virtual memory.

address register a register used primarily to hold the address of a location in memory. The location can contain an operand or an executable instruction.

address size prefix a part of a machine instruction that provides information as to the length or size of the address fields in the instruction.

address space (**1**) an area of memory which is seen or used by a program and generally managed as a continuous range of addresses. Many computers use separate address spaces for code and data; some have other address spaces for system. An *address space* is usually subject to protection, with references to a space checked for valid addresses and access (such as read only).

The physical *address space* of a computer (2^{32} bytes, and up to 2^{64} bytes) is often larger than the installed memory. Some parts of the address range (often at extreme addresses) may be reserved for input-output device addresses. *See also* byte, memory, memory mapped I/O, and processor.

(**2**) the range of potential addresses returned by a hash function.

address spoofing any enemy computer's impersonation of a trusted host's network address.

address translation *See* address mapping.

ad hoc query a query in which data is retrieved from a database using a query of the form "return all data from a given set in the database that matches these conditions."

adjacency graph a graph in which each node represents an object, component, or feature in an image. An edge between two nodes indicates two components that are touching or connected in the image.

adjacency-matrix representation a representation of a directed graph with n vertices using an $n \times n$ matrix, where the entry at (i, j) is 1 if there is an edge from vertex i to vertex j; otherwise the entry is 0. A weighted graph may be represented using the weight as the entry. An undirected graph may be represented using the same entry in (i, j) and (j, i) or using an upper triangular matrix.

admissibility condition in artificial intelligence and optimization theory, the necessity that the heuristic measure never overestimates the cost of the remaining search path, thus ensuring that an optimal solution will be found.

ADT *See* abstract data type.

advanced manufacturing system (AMS) modern manufacturing system combining computer assisted design, automation, robotics, and humans in a manner that maximizes product quality, optimizes production costs, and increases the flexibility of the means of production.

adversary the input sequence can be thought of as being generated by an adversary that uses information about the past moves of the on-line algorithm to choose inputs that maximize the ratio between the cost to the algorithm and the optimal cost.

aerodynamic head *See* disk head.

afferent filter a filter acting on information coming in to a receiving system.

affine function a geometric image transformation including one or more translations, rotations, scales, and shears that is represented by a 4×4 matrix allowing multiple geometric transformations in one transform step. Affine transformations are purely linear and do not include perspective or warping transformations. *Affine functions* are functions of several variables which can be defined as

$$f(x_1, x_2, \ldots, x_n) = c_0 + c_1 x_1 + c_2 x_2 + \cdots + c_n x_n$$

where c_i are known constants.

affine map *See* affine function.

affine transform *See* function.

affordance information present in the environment and available for processing, and which is related to invariants within that environment.

after image a copy of a database object after it is updated.

agent (1) a computational entity that acts on behalf of other entities in an autonomous fashion.

(2) in the client-server model, the part of the system that performs information preparation and exchange on behalf of a client or server. Especially in the phrase "intelligent agent" it implies some kind of automatic process which can communicate with other agents to perform some collective task on behalf of one or more humans.

agent-based system an application whose components are agents. *See* agent.

aggregate class a class that contains an aggregation of other objects.

aggregate function a function supported in SQL (not expressible in the relational algebra) that applies mathematical functions to a collection of values. Examples include average, count, sum, etc.

aggregate traffic in an analytic model that has multiple classes of service, or multiple chains of customers, the aggregate traffic is the combined traffic from one or more classes of service. In this case, the performance metrics obtained are for the combination of classes. A class may be aggregated (combined) with another when no performance metrics are desired for the class, or when individual class measurements are difficult to obtain.

aggregation a whole-part relationship in which the whole object has emergent and resultant properties and the relationship is asymmetric and irreflexive at the instance level.

AGI *See* address generation interlock.

Aiken, Howard Hathaway (1900–1973) Born: Hoboken, New Jersey.

Aiken is best know as the inventor of the Mark I and Mark II computers. While not commercially successful, these machines were significant in the development of the modern computer. The Mark I was essentially a mechanical computer. The Mark II was an electronic computer. Unlike UNIVAC (see Eckert, John), these machines had a stored memory. Aiken was a professor of mathematics at Harvard University, Cambridge, Massachusetts. He was given the assignment to develop these computers by the Navy. Among his colleagues in this project were three IBM scientists and Grace Hopper. It was while working on the Mark I that Grace Hopper pulled the first "bug" (a moth) from a computer.

AIP cube autonomy, interaction, presence; three axes of a space describing key characteristics of synthetic environments.

air gap *See* magnetic recording air gap.

algebra *See* Boolean algebra.

algebraic language (**1**) a language in which the traditional algebraic operators (add, subtract, multiply, divide, and, depending on the language, a few others) are available.

(**2**) a language in which the operator precedence follows the traditional algebraic hierarchy (for example, multiply binds operands more closely than add).

algebraic method a specification method that sees the system and its parts as a type. The type is defined with a set of operators that are specified with a logical language asserting how the elements of the type are changed. Typical algebraic methods are Z and VDM.

algebraic reconstruction the process of reconstructing an image $\mathbf{x}$ from a noise-corrupted and blurred image $\mathbf{y}$. An arbitrary image is selected as the initial condition of an iterative algorithm for solving a set of linear equations. A set of linear constraints is specified. In each iteration, one constraint is applied to a linear equation. The constraints are repeated in a cyclic fashion until convergence is reached. The linear constraints are vectors in a vector space with specified basis images for the type of problem to be solved.

algebraic surface a surface defined by the set of points for which an algebraic function is equal to a constant value. For an algebraic function $f(\mathbf{p})$ at point p and a constant value c the surface S can be formally defined as $S = \{\mathbf{p} : f(\mathbf{p}) = c\}$.

Algol acronym for algorithmic language, a family of languages starting with Algol-58 (1958), Algol-60 (the best-known member of the family, 1960), and Algol-68 (1968). Historically important because Algol-60 is the prototype for most modern language designs, including SIMULA, Pascal, Ada, C, and C++.

algorithm (**1**) a systematic and precise, step-by-step procedure (such as a recipe, a program, or set of programs) for solving certain kinds of problems or accomplishing a task, for instance converting a particular kind of input data to a particular kind of output data, or controlling a machine tool. An *algorithm* may be expressed in ordinary language, in a programming language, or in machine code. An *algorithm* transforms some initial data to another form, which is its result. An *algorithm* can be executed by a machine.

(**2**) in image processing, algorithms can be either sequential, parallel, or ordered. In sequential algorithms, pixels are scanned and processed in a particular raster-scan order. As a given pixel is processed, all previously scanned pixels have updated (processed) values, while all pixels not yet scanned have old (unprocessed) values. The algorithm's result will, in general, depend on the order of scanning. In a parallel algorithm, each pixel is processed independently of any changes in the others, and its new value is written in a new image, such that the algorithm's result does not depend on the order of pixel processing. In an ordered algorithm, pixels are put in an ordered queue, where priority depends on some value attached to each pixel. At each time step, the first pixel in the queue is taken out of it and processed, leading to a possible modification of priority of pixels in the queue. By default, an *algorithm* is usually considered as parallel, unless stated otherwise.

algorithm animation the process of abstracting the data, operations, and semantics of computer programs and then creating animated graphical views of those abstractions. This technique is frequently used for better showing the algorithm's evolution especially for teaching.

algorithmic state machine (ASM) a sequential logic circuit whose design is directly specified by the algorithm for the task the machine is to accomplish.

alias a relation between two names used in a program that may refer to the same memory location.

aliasing (1) an issue in languages that support call-by-name, call-by-reference, and call-by-value-result. A location is said to be aliased when two or more means exist to access or modify it from the body of a function. This can have serious implications on understanding the behavior of a program, and can often result in incorrect code when code optimizations are performed when the key assumptions that allow the optimization are violated in the presence of aliasing. *See also* side effect.

(2) in computer graphics, the effect seen when a continuous shape (such as a line or curve) must be rendered on a device which only supports quantized representations (a display that renders in pixels, for example).

(3) in image processing, image distortion introduced by the subsampling process.

alignment the requirement that a datum (or block of data) be mapped at an address with certain characteristics, usually that the address modulo the size of the datum or block be zero. For example, the address of a naturally aligned long word is a multiple of four.

alignment shift the first step in the addition of two floating-point numbers is the alignment of the radix points. This is done by shifting one of the significands and correspondingly adjusting its exponent; the shifting is known as alignment.

all-definition-use-path-coverage variable naming convention where each occurrence of a variable in the source code is labeled either as a definition of the variable or as a use of the variable. All paths between the definition of a variable and the use of that definition are now identified. A test case is selected and executed for each such path.

allocate to create a block of storage of a given size in some memory, which is not to be used for any other purpose until expressly freed. The block of storage will be allocated from the heap (also known as the free storage pool), and remain allocated until the storage is either explicitly de-allocated, implicitly de-allocated, or garbage collected.

allocation the act of allocating. *See* allocate.

allocation schema the definition of the allocation of fragments to the different sites.

allocation unit on certain versions of DOS, a group of bytes of fixed size (e.g., 2,048 bytes) that is locked out during low-level formatting if it contains a bad block.

all-or-none fallacy belief that robots are either fully autonomous or not really robots, when the truth is that robotic devices exist on a continuum from fully autonomous to manually controlled, with many gradations of human responsibility and autonomy in between.

all pairs shortest path an algorithm to find the shortest paths between all pairs of vertices in a weighted, directed graph. *See also* Dijkstra's algorithm, Floyd-Warshall algorithm.

alphabet synonym for the set of terminal symbols in a grammar.

alpha-beta bounds the conventional name for the bounds on a depth-first minimax procedure that are used to prune away redundant subtrees in two-person games.

alpha blending a technique for computing the color of a pixel when multiple structures contribute to the pixel (e.g., at a region boundary where we want to avoid aliasing problems arising from partial pixel coverage transparency).

alpha channel (1) the collection of alpha values associated with an image where each alpha value represents the coverage of each pixel in the image. The alpha values are used in the process of alpha blending.

(2) a grayscale image associated with the color channels of an image that dictates the opacity/transparency of the corresponding color channel pixels. If the color channels are multiplied by the alpha channel when stored, the image is referred to as premultiplied; otherwise it is known as unpremultiplied.

alpha-channel compositing *See* alpha blending.

alpha motion a code motion characterized by moving a computation to the head of a branching construct such that it is executed once, before any branch is taken. *Contrast with* omega motion.

alphanumeric mode relates to alphabetic characters, digits, and other characters such as punctuation marks. Alphanumeric is a mode of operation of a graphic terminal or other input/output device. The graphics terminal should toggle between graphic and alphanumeric data.

alpha shapes a technique that allows one to represent the concept of shape applied to a collection of points in space.

alpha test the unit, module, or component test phase performed inside the company that produces the software. *Compare with* beta test.

alteration authorization authorization to alter the definition of a relation.

alternating path a path with alternating free and matched edges. *See also* augmenting path.

alternating Turing machine a nondeterministic Turing machine having universal states, from which the machine accepts only if all possible moves out of that state lead to acceptance. *See also* model of computation, nondeterministic Turing machine, oracle Turing machine, probabilistic Turing machine, universal Turing machine.

alternation a model of computation proposed by Chandra, Stockmeyere, and Kozen, which has two kinds of states, AND and OR. The definition of accepting computation is adjusted accordingly. *See also* time/space complexity, Turing machine.

ambient some computer languages or object models support the notion of an object inheriting properties not from a superclass but from an "ambient context" in which the object is embedded.

ambient lighting a global (artificial) illumination level representing infinite diffuse reflections from all surfaces within a scene ensuring that all surfaces are visible (lit) particularly those without direct illumination. *Ambient lighting* is usually treated as a constant in local shading functions but is simulated directly in radiosity calculations.

ambiguity in artificial intelligence, the presence of more than one meaning or possibility.

ambiguous grammar a grammar that can generate the same sentence by more than one sequence of grammatical rules, or alternatively, a grammar which can parse a given sentence by two or more reduction sequences.

Amdahl's law states that the speedup factor of a multiprocessor system is given by

$$S(n) = \frac{n}{1 + (n-1)f}$$

where there are n processors and f is the fraction of computational that must be performed sequentially (by one processor alone). The remaining part of the computation is assumed to be divided into n equal parts each executed by a separate processor but simultaneously. The speedup factor tends to $1/f$ as $n \to \infty$ which demonstrates that under the assumptions given, the maximum speedup is constrained by the serial fraction.

American National Standards Institute (ANSI) (1) the U.S. government body responsible for approving U.S. standards in many areas, including computers and communications. ANSI is a member of ISO. ANSI sells ANSI and ISO (international) standards.

(2) informally, refers to the standard ANSI-X3.4, which defined the 8-bit extension to the 7-bit ASCII character set standard. *See also* ASCII, International Standards Organization (ISO).

American standard code for information interchange (ASCII) a binary code comprised of seven digits, originally used to transmit telegraph signal information.

Originally defined as a 7-bit standard, it has been extended in various ways providing some limited international support. (See ASCII Code Chart on page 17).

amortized cost the cost of an operation considered to be spread over a sequence of many operations.

amortized worst case *See* amortized cost.

AMS *See* advanced manufacturing system.

anaglyph a stereoscopic picture consisting of two images of the same object, taken from slightly different angles, in two complementary colors. When viewed through colored spectacles, the images merge to produce a stereoscopic sensation.

analog control Control scheme in which a continuous, as opposed to digital, input signal is mapped onto system actions.

analog-to-digital (A/D) conversion a method by which a continuously varying signal (voltage) is sampled at regularly occurring intervals. Each sample is quantized to a discrete value by comparisons to preestablished reference levels. These quantized samples are then formatted to the required digital output (e.g., binary pulse code words). The A/D converter is "clocked" to provide updated outputs at regular intervals. In order not to lose any baseband information, sampling must occur at a rate higher than twice the highest incoming signal frequency component.

analysis activity in which the problem domain is understood and in which there should be no constraints related to any possible solution. In some cases, a model of the analysis is produced in some formal or semiformal model.

analysis-by-synthesis coding refers to the class of source coding algorithms where the coding is based on parametric synthetization of the source signal at the encoder. The synthesized signal is analyzed, and the parameters that give the "best" result are chosen and then transmitted (in coded form). Based on the received parameters, the speech is re-synthesized at the receiver.

analysis method a notation and heuristics for creating models of customer needs and constraints; for instance, the unified modeling process using UML and use case driven analysis. *See* analysis.

analysis reuse *See* reuse analysis.

analysis time the capability of a software system or component to be diagnosed for deficiencies or causes of failures, or for the parts to be modified to be identified.

analytical benchmarking the quantification of how effectively each machine in a heterogeneous computing environment can perform different categories of computation. The fact that a given high-performance machine can achieve near-peak performance for only a relatively small set of code types is the underlying motivation for heterogeneous computing. Although some general frameworks and tools for the process of analytical benchmarking have been proposed, more research is needed to achieve complete and effective automation of this process.

anatomical model virtual model of the human anatomy, used for training or planning.

ancestor in object-oriented programming, a class from which the current class has been specialized. It is another term for super-class.

anchored instruction instructional design methodology in which learning takes place in a meaningful context (the context serves as the

ASCII Code Chart

Hex	Char	Hex	Char	Hex	Char	Hex	Char
00	nul	20	sp	40	@	60	`
01	soh	21	!	41	A	61	a
02	stx	22	"	42	B	62	a
03	etx	23	#	43	C	63	c
04	eot	24	$	44	D	64	d
05	enq	25	%	45	E	65	e
06	ack	26	&	46	F	66	f
07	bel	27	'	47	G	67	g
08	bs	28	(	48	H	68	h
09	ht	29	)	49	I	69	i
0A	lf	2A	*	4A	J	6A	j
0B	vt	2B		4B	K	6B	k
0C	ff	2C	,	4C	L	6C	l
0D	cr	2D	-	4D	M	6D	m
0E	so	2E	.	4E	N	6E	n
0F	si	2F	/	4F	O	6F	o
10	dle	30	0	50	P	70	p
11	dc1	31	1	51	Q	71	q
12	dc2	32	2	52	R	72	r
13	dc3	33	3	53	S	73	s
14	dc4	34	4	54	T	74	t
15	nak	35	5	55	U	75	u
16	syn	36	6	56	V	76	v
17	etb	37	7	57	W	77	w
18	can	38	8	58	X	78	x
19	em	39	9	59	Y	79	y
1A	sub	3A	:	5A	Z	7A	z
1B	esc	3B	;	5B	[	7B	{
1C	fs	3C	<	5C	\	7C	|
1D	gs	3D	=	5D	]	7D	}
1E	rs	3E	>	5E	^	7E	~
1F	us	3F	?	5F	_	7F	

"anchor" for knowledge or skill acquisition); a potential application for virtual reality (VR) in education because of the ability of VR to provide such context.

anchor record the first record in each block of the data file.

AND The Boolean operator that implements the conjunction of two predicates. The truth table for $\wedge \equiv X$ and Y is

X	Y	$X \wedge Y$
F	F	F
F	T	F
T	F	F
T	T	T

n-ary ands can be obtained as conjunction of binary ands.

and *See* AND.

AND gate a device which implements the Boolean AND operation. *See* AND.

AND/OR tree a tree that enables the expression of the decomposition of a problem into subproblems, thus enabling alternate solutions to subproblems through the use of AND/OR mode labeling schemes.

animation (1) a medium that provides the illusion of a moving scene using a sequence of still images.

(2) techniques used in the production of animated films. In computer graphics this primarily concerns controlling the motion of computer models and the camera.

(3) the process by which the behavior defined by a formal specification is examined and validated against the informal requirements.

(4) graphic representation with movements of complex algorithms or for simple visualization of moving drawn object. It can be 2D or 3D.

anisotropic diffusion a process of progressive image smoothing as a function of a time variable t, such that the degree and orientation of smoothing at a point vary according to certain parameters measured at that point (e.g., gray-level gradient, curvature, etc.) in order to smooth image noise while preserving crisp edges. The progressively smoothed image $I(x, y, t)$ (where x, y are spatial coordinates and t is time) satisfies the differential equation

$$\partial I / \partial t = div(c \nabla I) \ ,$$

where the diffusion factor c is a decreasing function of the spatial gradient ∇I. When c is constant, this reduces to the heat diffusion equation

$$\partial I / \partial t = c \Delta I \ .$$

Other mathematical formulations have been given, where edge-preserving smoothing is realized by a selective diffusion in the direction perpendicular to the gradient. *See also* multiresolution analysis, mathematical morphology.

anisotropic filtering image filtering that produces different amounts of filtering (e.g., smoothing filtering in different directions at each pixel in an image). Two uses of anisotropic filtering in graphics are to: (1) produce textures with different spatial frequency distributions in different directions, and (2) to reduce aliasing effects along edges without blurring the edges as much. Anisotropic filtering can be done in either the image or the frequency domains.

annul bit a bit that is used to reduce the effect of pipeline breaks by executing the instruction after a branch instruction. The annul bit in a branch allows us to ignore the delay-slot instruction if the branch goes the wrong way. With the annul bit not set, the delayed instruction is executed. If it is set, the delayed instruction is annulled.

anomaly any detected difference between the expected values and functionalities with respect to verified old versions or documents.

ANSI *See* American National Standards Institute.

ANSI X3J11 ANSI standard definition for the C language.

anthropometry study of human dimensions, particularly as related to designing appropriately for the size of users.

anthropomorphic having shape or size similar to that of humans.

anthropomorphic design a design, particularly of a robotic manipulator, with link lengths, joints, and motions similar to humans.

anthropomorphic manipulator a manipulator which consists of two shoulder joints, one for rotation about a vertical axis and one for elevation out of the horizontal plane, an elbow joint with axis parallel to the shoulder elevation joint, and two or three wrist joints at the end of the manipulator (see figure below). An anthropomorphic manipulator is sometimes called a jointed, elbow, or articulated manipulator.

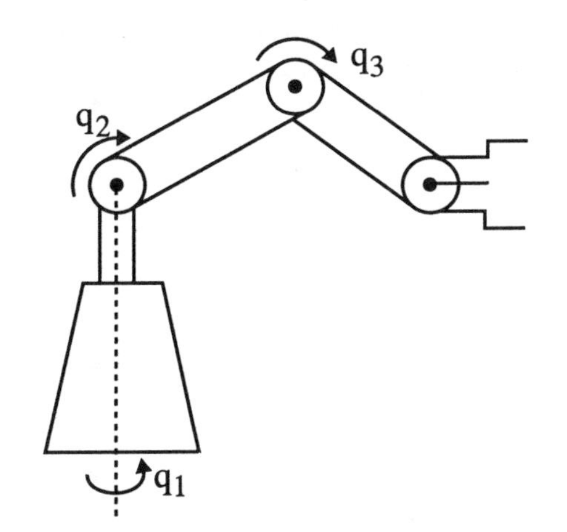

An anthropomorphic manipulator.

anthropomorphic robot robot similar in size, shape, and movements to a human.

anthropomorphic teleoperator teleoperated manipulator having the approximate size and shape of a human; an anthropomorphic robot that is teleoperated.

anthropomorphism design philosophy that robotic systems function best when approximating human size and shape.

anthropomorphizing Attributing human traits or motivations to animals or inanimate objects; a logical error.

antialiasing a method of reducing or preventing aliasing artifacts when rendering by using color information to simulate higher screen resolutions. In one technique, blurred pixels are introduced by filtering the image, or individual elements, to remove spatial frequencies that are greater than the pixel sample rate by convolution. If high frequencies remain they may cause other visual artifacts such as Moiré patterns. An alternative and often preferable technique is supersampling, where many samples per pixel are estimated and combined.

antialiasing filter typically, a filter which provides a prefiltering operation to ensure that the frequency components of a signal above the Nyquist frequency are sufficiently attenuated so that, when aliased, they will cause a negligible distortion to the sampled signal. *See also* aliasing, Nyquist frequency.

antichain a subset of mutually incomparable elements in a partially ordered set.

anticipatory paging the transfer of a page from auxiliary storage prior to the moment of need for this page.

antidependence a data dependence relation between two statements or operations where the first uses a data element and the second subsequently overwrites it with a new value.

antidependency a potential conflict between two instructions when the second instruction alters an operand which is read by the first instruction. For correct results, the first instruction must read the operand before the second alters it. Also called a write-after-read hazard.

aperiodic pertaining to a process or task executing in a non-repeatable fashion, at some unbounded rate.

aperiodic convolution the convolution of two sequences. *See* convolution.

aperiodic task A task that is generally started as a response to an external/internal event (i.e., a recovery task started on a failure event) that is not periodic. If the aperiodic task is under execution, nothing can be said of its next occurrence.

aperture problem given a sequence of images over time, we would like to infer the motion (optical flow) field. Based on local image information (i.e., based on the values of those pixels falling within some aperture), only the component of motion along the gray-level gradient can be inferred; the fact that the component of motion perpendicular to the gray-level gradient can only be known by resorting to global methods is known as the aperture problem. *See* optical flow, optical flux.

API *See* application program interface.

APL (**1**) acronym: a programming language. An abstract notation developed by Ken Iverson of IBM T.J. Watson Research, which he used in a formal description of the IBM /360 computer architecture.

(**2**) an implementation of the abstract notation of (**1**) that embodies many constructs of programming languages such as functions, assignment, and goto.

a posteriori probability *See* posterior statistics.

apparent concurrency within an interval of time more than one process executes on a computer, although at the instruction level, instructions from only one process run at any single point in time. *See also* concurrency.

appearance-based recognition recognizing objects based on views, generally using properties such as surface reflectance patterns; often in contrast with model-based recognition.

applet a small application. An applet often executes as a component of a larger application that has responsibility for managing the environment in which it executes. Examples of applets include small modules written in languages like Java and VBScript.

application (**1**) a program, or suite of programs, designed to solve a particular problem. When a suite of programs defines an application, they may be written in different programming languages.

(**2**) a term used to distinguish programs that run outside the operating system, hence with lower privilege than the operating system.

application context the set the application features that establish the description of the environmental aspects in which the application is used or is supposed to be used.

application family a generic classification of application systems. One purpose is to allow several abstract systems, possibly defined using different design languages, to be composed in one application. A second purpose is to factor out the support systems, which are generic and evolve independently from applications. The concept supports heterogeneous applications, not easily covered by a single system description expressed in one of the design languages.

application gateway a relay and filtering program that operates layer seven of the network stack.

application generator a program that takes as input a specification of the required product. This specification can be in a high level language (a 4GL program). The product of the generator can only be usually modified by the generator by changing the input specification. Application generators for some problem domains are typically complex applications.

application program software written to solve specific problems such as payroll preparation, inventory, word processing, and so on.

application program interface (API) the interface presented to writers of an application by the underlying operating system. The degree to which the API is made visible in a programming language influences how portable the programming language is to other systems. An API is defined at source code level and provides a level of abstraction between the application and the kernel (or other privileged utilities) to ensure the portability of the code. An API can also provide an interface between a high level language and lower level utilities and services which were written without consideration for the calling conventions supported by compiled languages. In this case, the API's main task may be the translation of parameter lists from one format to another and the interpretation of call-by-value and call-by-reference arguments in one or both directions.

Many languages abstract greatly from the API, providing operations such as "file open", "read data", "create new object", and leave it as an exercise for the writer of the compiler and run-time system to map these general operations to the underlying API. In many cases, the value representation of the language is unsuited to passing values to the underlying API (for example, most FORTRAN programs could not easily create the sequence of values required by some operating systems), and a suitable layer of subroutine library must provide language-specific access to the API.

Examples of APIs include OpenGL, Java3D, Allegro.

application semantics the part of the software that is not the user interface.

application software a complete, self-contained program that performs a specific function directly for the user.

application-specific integrated circuit (ASIC) in the broadest sense, integrated circuits that are designed for a specific application. The term is used to describe VLSI circuits of standard form which can be configured either in manufacture or on site to meet the specific needs of the application. The configuration process is generally so inexpensive that such circuits are economical to produce in extremely small runs.

application system the application part of a system instance implementation. An application system defines the application (the behavior) a customer wants to buy in terms of implementation code. It is normally a partial implementation lacking the necessary support to execute. An application system is used to produce the concrete systems that actually execute and is expressed using some high-level programming language.

applicative order evaluation an execution order in which the arguments in a function call are evaluated before the body of the function.

applicative-order reduction a reduction strategy in which the argument must be reduced to evaluated form before a function application can be reduced. Also called call-by-value reduction.

applied software measurement the practical set of soft factors, productivity data, quality data, project size, and risk information needed to manage software projects and to assess software products.

appraisal *See* assessment.

approval cycle the process of gaining funding or management approval to introduce a new tool or method into an organization.

approximate coding a process, defined with respect to exact coding, which deals with irreversible and information-lossy processing of two-level pictures to improve compression ratio with significant degradation of picture quality. Exact coding schemes depend on the ability to predict the color of a pixel or the progression of a contour from line-to-line. Irreversible processing techniques try to reduce prediction errors by maintaining the continuity of the contours from line-to-line. With predictive coding, the number of pixels can be changed to reduce those having nonzero prediction error. With block coding, the compression efficiency can be improved by increasing the probability of occurrence of the all zero block. The third approximate block coding scheme is pattern matching. In this scheme the identification codes of the repeated patterns are transmitted to the receiver. A library of patterns is maintained for continuous checking. *See* exact coding.

approximate reasoning an inference procedure used to derive conclusions from a set of fuzzy if-then rules and some conditions (facts). The most used approximate reasoning methods are based on the generalized modus ponens. *See* generalized modus ponens, fuzzy if-then rule, linguistic variable, imprecise computation.

approximation algorithm for solving an optimization problem. An algorithm that runs in time polynomial in the length of the input and outputs a feasible solution that is guaranteed to be nearly optimal in some well-defined sense called the performance guarantee.

a priori probability *See* prior statistics.

APT *See* automatic programming of tools.

arc a connection between two nodes of a graph. If the arc has a specified direction of traversal, it is known as a directed arc. Also referred to as an edge of the graph.

architectural design the process for defining the modules and their use relationships of a system. Large systems are divided into smaller subsystems and modules. The term has been initially used for processors. More recently, it has also been used for defining complex systems, e.g., software architecture, network architecture.

architectural visualization use of virtual reality to simulate, examine, and interact with building designs before design finalization or construction.

architecture the software architecture of a program or computing system is the structure or structures of the system, which comprise software components, the externally visible properties of those components, and the relationships among them. In object oriented modeling, the *architecture* is at high level the general system decomposition into the main sub-systems. In some cases it may include the system description in terms of class hierarchy.

archive (**1**) generally, a collection of objects, such as text files, stored as a unit; in a Unix environment, often called an archive library or simply a library, a collection of executable functions (in object code format) stored in a single file suitable for linking into an executable program; often containing a group of related functions that provide a set of reusable programming utilities.

(**2**) a set of inference rules that apply to functional dependencies.

archive library *See* archive.

argument (**1**) the formal parameter to a function.

(**2**) the actual parameter to a function. Some languages distinguish between a formal parameter value, as specified in a function definition, and the actual parameter value, as determined at a call site, by calling the former a "parameter" and the latter an "argument". Other languages make no such distinction. *See also* parameter.

(**3**) a piece of data given to a hardware operator block.

arithmetic and logic unit (ALU) a combinational logic circuit that can perform basic arithmetic and logical operations on n-bit binary operands.

arithmetic coding a method (due to Elias, Pasco, Rissanen, and others) for lossless data compression. This incremental coding algorithm works efficiently for long block lengths and achieves an average length within one bit of the entropy for the block. The name comes from the fact that the method utilizes the structures of binary expansions of the real numbers in the unit interval.

arithmetic error the error that is generated because a machine operation is only an approximation of the true mathematical operation or because a machine operator propagates onto its output errors in the operands; that is, it is the sum of the generated error and the propagated error.

arithmetic instruction a machine instruction that performs computation, such as addition or multiplication.

arithmetic-logic unit *See* arithmetic-logic unit.

arithmetic operation any of the following operations and combinations thereof: addition, subtraction, multiplication, division.

arithmetic operator an operator which performs a numeric computation. In most algebraic languages, the traditional infix operators such as "+" (add), "−" (subtract), "*" (multiply), and "/" (divide) are in common usage to indicate operations on individual values, but other operators such as truncating integer division (Pascal's " " operator), modulo (C's "%" operator) and exponentiation (FORTRAN's "**" operator) are provided. These operators may be extended to other arithmetic data types, such as complex numbers, arrays, and the like, or to represent more complex computations such as polar coordinate arithmetic or vector arithmetic. In a more general sense, an arithmetic operator may be represented syntactically as a function call but nonetheless be an arithmetic operator, for example $\sin(x)$.

ARM (**1**) acronym, the Annotated Reference Manual (for C++).

(**2**) acronym, «Advanced RISC Machine», a computer architecture.

arm a part of a robot. A robot is composed of an arm (or mainframe) and a wrist plus a tool. For many industrial robots the arm subassembly can move with three degrees of freedom. Hence, the arm subassembly is the positioning mechanism. *See* industrial robot.

Armstrong's axioms a set of inference rules that applies to functional dependencies.

Armstrong's inference rules *See* Armstrong's axioms.

ARQ *See* automatic repeat request.

array a data structure which is typically organized so that it represents an ordered set of values that can be accessed by supplying one or more values which uniquely identify one of the values of the set. These values are called subscripts. The dimension of an array is the number of subscript values required to uniquely identify a member. An array with one dimension is sometimes called a vector. A partial specification of an array (for an n-dimensional array, specifying no more than n-1 subscripts) is referred to as a slice. An array may be homogeneous or heterogeneous depending on the programming language. The nature of the subscript is language-dependent, and although most languages support the use of a subrange of integers as subscripts, other data types are possible. *See also* associative array.

array bounds checking a form of execution semantics in which the indices of the array access are checked to see that they are valid for the implementation of the array. In many implementations, the programmer can disable array bounds checking to improve performance.

array processor an array of processor elements operating in lockstep in response to a single instruction and performing computations on data that are distributed across the processor elements.

articulation vertex a vertex whose deletion disconnects a graph into two or more connected components. Also called cut vertex.

artifact (**1**) an error or aberration in a signal that is the result of aliasing, a quantization error, some form of noise, or the distorting effects of some type of processing. *See also* outlier.

(**2**) a classifiable visual error, e.g., a loss of resolution when zooming into an image or incorrect depth sorting due to the painter's algorithm.

artificial intelligence the study of computer techniques that emulate aspects of human intelligence, such as speech recognition, logical inference, and ability to reason from partial information.

artificial life the attempt to understand the emergence of life by re-creating possible lifeforms as simulation programs on computers and studying their behaviors. In other words, rather than using a descriptive approach as in biological disciplines, a synthetic approach is adopted to understand those aspects of the life phenomenon that are independent of the media in which it is implemented. According to Langton, the artificial life approach allows us to understand not only the current existing lifeforms ("life-as-we-know-it"), but also other lifeforms which are feasible but not yet created in nature ("life-as-it-could-be"). Evolutionary computation is usually adopted to simulate the evolutionary aspect of life emergence. Examples of artificial life programs include Tierra, Swarm, and Boids.

artificial muscle actuator designed to work in a fashion similar to human muscles, that is, by linear extension and contraction.

artificial neural network a complex nonlinear modeling technique based on a model of a human neuron. A neural net is used to predict outputs (dependent variables) from a set of inputs (independent variables) by taking linear combinations of the inputs and then making nonlinear transformations of the linear combinations using an activation function. In particular, artificial neural networks have been designed and used for performing pattern recognition operations. *See also* pattern recognition, perceptron.

artificial neuron an elementary analog of a biological neuron with weighted inputs, an internal threshold, and a single output. When the activation of the neuron equals or exceeds the threshold, the output takes the value $+1$, which is an analog of the firing of a biological neuron. When the activation is less than the threshold, the output takes on the value 0 (in the binary case) or -1 (in the bipolar case) representing the quiescent state of a biological neuron.

artificial reality *See* virtual reality.

artificial repelling force form of virtual force used to resist movement of a teleoperated manipulator or avatar into selected parts of remote space.

ART network *See* adaptive resonance theory network.

ASCII *See* American standard code for information interchange. *See also* ANSI, ISO, unicode.

ASIC *See* application-specific integrated circuit.

ASM *See* algorithmic state machine.

aspect ratio (**1**) the size invariant ratio of length to width for a rectangular box enclosing a shape, the orientation of the box being chosen to maximize the ratio. This measure is used to characterize object shapes as a preliminary to, or as a quick procedure for, object recognition.

(**2**) In a computer display, the ratio of the length of the x-coordinate range to the y-coordinate range.

assembler (**1**) a computer program that translates an assembly-code text file to an object file suitable for linking.

(**2**) a program for converting assembly language into machine code (object code).

assembly code a symbolic notation that traditionally has a one-to-one relationship with the generated machine code. Translated to machine code by use of an assembler. The output of many compilers is assembly code, which must be processed by an assembler before it can be linked. *See also* assembly language.

assembly language a low-level programming language that represents machine code in a symbolic, easier-to-read form of a processor's. *See also* assembler.

assertion (**1**) a Boolean expression for stating the behavior of the program or, if hardware implemented, of a circuit.

(**2**) a logical expression specifying a program state that must exist or a set of conditions that program variables must satisfy at a particular point during program execution. In formal theory, used to specify the weakest preconditions and the postconditions applicable to a block of code.

(**3**) any business rule that is a constraint on the property values of an object. Assertions may be one of three kinds: a pre-condition is any assertion that must hold before the execution of an associated operation(s); a post-condition is any assertion that must hold after the execution of the associated operation(s); and a (class) invariant is any assertion that must hold both before and after the execution of all operations. In logic programming, an assertion can be a logic expression specifying a program feature or a set of features that a program/system has to present. *See also* invariant, proof of correctness.

assessment the process by which the behavior defined by a formal specification is examined and validated against the informal requirements. The term is also used for identifying the process of formal evaluation by means of measures. *See* product assessment, process assessment.

assessment tool a tool that allows the assessment of software products or processes. These can be used for taking under control the design, the development, and the maintenance of a system. Assessment tools may evaluate quality factors and system indexes calculated, directly on the system source, or selected metrics according to predefined assessment procedures.

assignment in an imperative language, a means to change the value in a named or computable location in memory.

assignment problem the problem of finding a matching maximum (or minimum) weight in an edge-weighted graph.

assignment scope the amount of work for which one person will normally be responsible. For instance, the maintenance assignment scope of a certain COBOL system is 15.000 lines per year. Meaning that if the system is 45.000 lines of code we need three people.

assimilation of borrow subtraction without the inclusion of an incoming borrow produces a partial difference and a partial borrow. Assimilation is the combining of a partial borrow and a partial difference.

assimilation of carry addition without the inclusion of an incoming carry produces a partial sum and a partial carry. Assimilation is the combining of a partial sum and a partial carry.

associate mode an operating mode of content addressable memories, in which a stored data item is retrieved that contains a field that matches a given key.

association a relationship between two entities, expressed between concepts but denoting connections between instances. It is a structural relationship that establishes potential interactions. In database, entity relationship modeling, and the Unified Modeling Language the default is that associations are bi-directional. In object modeling based on message passing paradigm, including the OPEN Modeling Language, the default is unidirectional. In object modeling, use of bi-directional relationships has been shown to destroy encapsulation and information hiding which are, together, one of the main tenets of object technology.

association rules a class of rules used in data mining. An *association rule* states an association or relation between sets of data. For example, "When prospectors buy picks, they also buy shovels 14% of the time."

associative (**1**) a mathematical property stating how sequences of identical operations, or operations of identical operator precedence, are to be interpreted.

(**2**) in a language, a specification of how the associative rule is to be applied to computations in that language. *See also* commutative, left-associative, right associative.

(**3**) a rule which is used in compilers to determine evaluation order; while traditionally evaluation order follows the rules of mathematics, including the associative rule, in fact often all that is semantically required is that the application of operations to the computed values follow the associative rule without stating the exact evaluation order.

associative array a data structure, in the abstract an array but in practice which can have any implementation, where the desired value is selected not by a position (numerical subscript) but by specifying a key value by which the element may be retrieved.

associative memory a memory which is interrogated by contents rather than by address or data location. When presented with a word, or part of a word, it returns to the location where that data may be found and possibly the entire word.

Associative memories are most often found in translation buffers or page translation tables of the hardware to support virtual memory. Given the user-space address of a page, they return the physical address of that page in main memory.

Other important applications are in the recognition of destination addresses in computer networks. In 8802.3 networks, a given node may have to recognize its own unique physical address, one or more broadcast addresses, and possibly some other addresses derived from the user network address. A small associative memory can test for all of these addresses in parallel.

An important aspect of associative memories is their very fast operation, with all entries tested in parallel to complete the interrogation within one system clock.

In neural networks, this type of memory is not stored on any individual neuron but is a property of the whole network, which is accomplished by inputting to the network part of the memory.

Also called content addressable memory (CAM). *See also* mask register and response store.

associative processor a parallel processor consisting of a number of processing elements, memory modules, and input/output devices under a single control unit. The capability of the processing elements is usually limited to the bit-serial operations.

associativity (**1**) as a mathematical property, an operator $\oplus$ is associative over the set X if for all $r, s, t \in X$,

$$r \oplus (s \oplus t) = (r \oplus s) \oplus t \, .$$

(**2**) in a cache, the number of lines in a set. An n-way set associative cache has n lines in each set. (Note: the term block is also used for line.)

asymptomatic equipartition property a set of all strings of a given length can be partitioned into a set of "bad states" of low probability; and a set of "good states" such that every string in the latter set has approximately the same probability.

asymptotical complexity the limit for the factors that tend to infinite of computational complexity.

asymptotically self-similar refers to sets for which the self-similarity (at any point) does not become apparent after a single magnification and rotation, but when the rescaling procedure is repeated over and over, the resulting objects converge to a set that is self-similar at a point. The Mandelbrot set and the Julia sets are examples of sets that are asymptotically self-similar at a point.

asymptotically tight when the asymptotic complexity of an algorithm exactly matches the theoretically proven asymptotic complexity of the corresponding problem. Informally, when an algorithm solves a problem at the theoretical minimum. *See also* Θ, asymptotic upper bound, asymptotic lower bound.

asymptotic lower bound an asymptotic bound, as a function of the size of the input, on the best (fastest, least amount of space used, etc.) an algorithm can possibly achieve to solve a problem. That is, no algorithm can use fewer resources than the bound. *See also* Ω, asymptotic upper bound, asymptotically tight.

asymptotic upper bound an asymptotic bound, as a function of the size of the input, on the worst (slowest, most amount of space used, etc.) an algorithm will take to solve a problem. That is, no input will cause the algorithm to use more resources than the bound. *See also* big-O notation, asymptotic lower bound, asymptotically tight.

asynchronous pertaining to two or more processes whose interaction does not depend upon the occurrence of a specific global event.

asynchronous bus a bus in which the timing of bus transactions is achieved with two basic "handshaking" signals, a request signal from the source to the destination and an acknowledge signal from the destination to the source. The transaction begins with the request to the destination. The acknowledge signal is generated when the destination is ready to accept the transaction. Avoids the necessity to know system delays in advance and allows different timing for different transactions. *See also* synchronous bus.

asynchronous communication a type of communication without the presence, at the same time, of the sender and the receiver(s). It is the opposite of synchronous communication. It is usually implemented with a buffer for the receiver process. *See* synchronous communication.

asynchronous event an event that occurs at unpredictable points in the flow-of-control and is usually caused by external sources such as a clock signal.

asynchronous message passing message passing protocol where the sending process allows messages to be buffered and the sending process may continue after the send is initiated; the receiving process will block if the message queue is empty.

asynchronous system a (computer, circuit, device) system in which events are not executed in a regular time relationships; that is, they are timing independent. Each event or operation is performed upon receipt of a signal generated by the completion of a previous event or operation, or upon availability of the system resources required by the event or operation.

asynchronous transfer mode (ATM) method of multiplexing messages onto a channel in which channel time is divided into small, fixed-length slots or cells. In ATM systems the binding of messages to slots is done dynamically, allowing dynamic bandwidth allocation. ATM is asynchronous in the sense that the recurrence of cells containing information from an individual user is not necessarily periodic.

asynchronous updating in artificial intelligence, process by which one unit at a time is selected from within a neural network to have its output updated. Updating an output at any time is achieved by determining the value of the unit's activation function at that time.

Atanasoff, John Vincent (1903–) Born: Hamilton, New York.

Atanasoff is best known for his invention, along with Clifford Berry, of the first digital computer, known as the ABC (Atanasoff-Berry Computer). Unlike the many World War II computer pioneers, Atanasoff's interest in the topic dated to his Ph.D. thesis research at the University of Wisconsin. After graduation Atanasoff taught physics and mathematics at Iowa State College and continued to work on the problem of solving lengthy calculation by electronic means. Legend has it that Atanasoff worked out the basic structure for his new machine while having a drink at an Illinois road house. Clifford Berry, an electrical engineer joined Atanasoff to help with the construction of the device based on Atanasoff's ideas. John Mauchly, another computer pioneer often visited and consulted with Atanasoff. These discussions resulted in a later lawsuit which established Atanasoff as the first person to build an electronic digital computer.

atatic scope a scope which is in effect defined for the entire execution of a program, independent of any lexical or dynamic execution context. Names declared in the static scope (for example, in the outermost block of the program, or at the module level, or in other ways determined by the definition of the language) are potentially accessible by every piece of the program, although various languages may have other rules that limit or permit such accessibility.

AT bus bus typically used in personal computer IBM AT for connecting adapters and additional memory boards. It is also called 16 bit ISA BUS since it presents a data BUS at 16 bit. It presents an additional connector with respect to the classical ISA BUS (at 8 bit) of IBM PCs based on Intel 8088. *See also* EISA.

ATM *See* asynchronous transfer mode.

ATM cell loss probability *See* loss probability.

atmosphere effect atmospheric effects arise because light is affected by the properties of the medium through which it passes. The main effects are attenuation, where distant objects get lower contrast (*see* depth cueing) and blurring, such as might occur with dust, fog, or haze, which scatter the light.

at-most-once property a property of an expression or an assignment statement asserting that it contains at most one reference to a variable changed by another process.

atom *See* atomic formula.

atomic pertaining to an operation which is indivisible into smaller units.

atomic attribute an attribute that cannot be decomposed further into subcomponents.

atomic component a (software) component which cannot be decomposed in smaller parts.

atomic formula in database systems, a formula $p(t_1, \ldots, t_n)$ consisting of a predicate symbol p together with n terms $t_1, \ldots, t_n$; example: father(adam, abel) or smaller (g(1, Y), 17).

atomic instruction an instruction that consists of discrete operations that are all executed as a single and indivisible unit, without interruption by other system events. *See also* test-and-set instruction, atomic transaction.

atomicity the property of transaction processing whereby either all the operations of a transaction are executed or none of them are (all or nothing).

atomic transaction in database systems, the property of indivisibility. An atomic transaction either terminates in its entirety and commits or not at all. *See also* atomic instruction.

atomic type a data type defined as an atomic component. A data type that cannot be decomposed in smaller components.

atomic value a value that cannot be decomposed further.

attachment a syntactic relation between two parts of a sentence where one modifies the meaning of the other.

attention allocation deployment of attentional resources to process and respond to stimuli; resource allocation.

attentional resources finite set of cognitive programs and processes used by humans to process information.

attentive perceptual system perceptual system devoted to processing environment data at the focus of attention; for example, foveal vision.

attenuation (1) the simulation of the atmospheric attenuation from the object to the viewer which affects both the illumination strength and color. The attenuated illumination is computed by $I(\lambda)' = sI(\lambda) + (1 - s)I_{dc}(\lambda)$, where s is a scale factor ranging from 0 to 1, $I(\lambda)$ is the illumination, and $I_{dc}(\lambda)$ is the depth-cue color.

(2) for a light source, a factor in the illumination equation used to simulate surface illumination depending on how far the surface is from the light source. It is defined by: $f_{att} = \min(\frac{1}{c_1 + c_2 d_l + c_3 d_l^2}, 1)$ where d_l is the distance between the source light and the surface, and c_1, c_2, and c_3 are user defined constants associated with the light source.

attitude control subsystem part of a control system functioning to control the pitch of a system.

attractiveness the capability of the software product to be attractive to the user. This refers to the attributes of the software intended to make the software more attractive to the user, such as the use of color and the nature of the graphical design.

attractor the point to which an iterated function tends towards if it does not escape and is not indifferent.

attractor sensitivity the threshold to which a function $f(z)$ is iterated at the point z_0. If the square of its modulus at any point is less than the attractor sensitivity, then the point attracts.

attribute (1) a property associated with an entity. In object-oriented paradigms, an attribute is a component used for defining the class structure. These are called local attributes and are hidden from the outer objects, according to the data hiding mechanism. The attribute has a relationship of is-part-of with the class. In the classes, attributes are inherited from superclasses (inherited attributes) or locally defined. The attributes are the internal components of objects. In most cases, the behavior of a class depends on their behavior.

(2) in a relational algebra, a named column or domain of a relation type.

(3) in assessment, an observable property of an entity; a measure may be defined for an attribute. In this view, attributes can be internal or external.

attribute grammar a formal specification of a language which includes a specification of a set of inherited attributes which propagate downward in a tree and a set of synthesized attributes which are computed upwards in the tree, and a set of equations that describe the values. The equations compute synthesized attributes but may involve the inheritance of attributes whose values are defined by similar equations. To evaluate an attribute grammar involves a convergence algorithm that guarantees a fixed-point at which all the attribute equations are simultaneously satisfied.

audio coding the process of compressing an audio signal for storage on a digital computer or transmission over a digital communication channel.

audio feedback presentation of information by sound.

audit an external review of a software project to assess its compliance with specifications and contractual issues. On its results, the project should be continued or canceled. *See* assessment.

auditory channel sensory channel devoted to processing information afforded by pressure waves traveling through a medium; for example, human hearing.

auditory display audio feedback.

augmentation rule a type of relation in database systems. Given sets of attributes A, B, and C, then if $A \rightarrow B, \models AC \rightarrow BC$.

augmented reality the idea that an observer's experience of an environment can be augmented with computer generated information. Usually this refers to a system in which computer graphics are overlaid onto a live video picture or projected onto a transparent screen as in a head-up display.

augmented reality display display that supplements remote real world scenes by projecting computer graphic images onto video images of them; for example, a display of task edges generated from design drawings, added to a television image.

augmented reality system synthetic environment featuring a combination of virtual and real-world environments; computer assisted teleoperation.

augmenting path a path with alternating free and matched edges which begins and ends with free vertices. Used to augment (improve or increase) a matching or flow. *See also* alternating path.

authentication the process of confirming user identity.

authorization permission. A user is authorized to perform certain actions if the user has sufficient privileges.

authorization graph a graph representing the passing of authorization from one user to another.

autoassociative backpropagation network a multilayer perceptron network which is trained by presenting the same data at both the input and output to effect a self-mapping. Such networks may be used for dimensional reduction by constraining a middle, hidden layer to have fewer neurons than the input and output layers.

autoincrementing (**1**) an addressing mode in which the value in a register is incremented by one word when used as an address.

(**2**) in high-level languages: operation

$$i++ \Rightarrow i = i + 1$$

where i is arbitrary variable, register, or memory location.

(**3**) in machine code, after evaluating the operand address contained in the register, the processor increments the contents of the register by 1, 2, 4, 8, or 16 for a byte, word, longword, quadword, or octaword, respectively.

automata theory a mathematical discipline which studies abstract computational processes. An automaton is an abstract computational device. *See* finite state automaton, pushdown automaton, nondeterministic push-down automaton, and Turing machine.

automated deduction deduction techniques that may be mechanized. Also called automated reasoning.

automated estimation tool a tool that allows the evaluation system features, directly on the system source and/or documentation, by the means of predefined assessment procedures. The evaluation process does not require human interaction.

automated reasoning *See* automated deduction.

automated reverse architectural design any software engineering approach where the architectural design of a system and its components can be recovered automatically using certain tools. Some approaches are able to subsume modules into an automatically derived subsystem structure.

automated verification system a software tool for verifying the system correctness automatically even partially.

automatic (**1**) property pertaining to a process or a device which functions without intervention by a human operator under specified conditions.

(**2**) specification of extent which specifies that the object exists only during the execution of a syntactically specified segment of the program, usually a function or a block.

automatic allocation allocation of memory space to hold one or more objects whose lifetimes match the lifetime of the activation of a module, such as a subroutine. Automatic allocations are usually made upon entry to a subroutine.

automatically guided vehicle robotic vehicle, usually used to move loads from point to point without human intervention.

automatic focusing on an optical disk, the process in which the distance from the objective focal plane of the disk is continuously monitored and fed back to the disk control system in order to keep the disk constantly in focus.

automatic lock a lock acquired by a transaction as a result of access to data, and not explicitly acquired.

automatic optimization automatic generation of an efficient query execution strategy.

automatic programming of tools (APT) a robot programming language.

automatic repeat request (ARQ) an error control scheme for channels with feedback. The transmitted data is encoded for error detection and a detected error results in a retransmission request.

automatic tracking on an optical disk, the process in which the position of the disk head relative to the disk surface is constantly monitored and fed back to the disk control system in order to keep the read/write beam constantly on track.

automatic variable a variable that is "automatically" allocated on the stack when a context is entered and deallocated when the context is executed. Synonym for stack local variable. Also commonly referred to as a local variable, although in general a name with a local scope does not necessarily imply local extent.

automation generic term for machines designed to operate without human input except to reprogram or for maintenance; often, but not always, computer controlled; autonomous robots are a form of *automation,* but not all *automation* involves robotics: robots are reprogrammable to perform a variety of tasks (e.g., the same robot may paint or weld) but automated machines may be capable of only one task (e.g., a lathe).

Automation paradigms include:

a) a continuous flow production process which integrates various mechanisms to produce an item with relatively few or no worker operations, usually through electronic control;

b) self-regulating machines (feedback) that can perform highly precise operations in sequence; and

c) electronic computing machines.

In common use, however, the term is often used in reference to any type of advanced mechanization or as a synonym for technological progress; more specifically, it is usually associated with cybernetics.

automaton (**1**) an abstract computational model which consists of a set of states, a set of input symbols, possibly a store, and a set of mapping functions, called production rules, that map state, input symbol, and (if present) the symbol currently accessible in the store to a new state, and if applicable, a change in the state of the store. Plural automata. *See also* pushdown automaton, finite state automaton, nondeterministic push-down automaton, and Turing machine.

(**2**) a machine that follows a sequence of instructions.

(**3**) any automated device (robots, mechanical and electromechanical chess automata). *See* cellular automaton, state automaton, finite state machine.

autonomic nervous system that portion of the nervous system responsible for controlling functions that usually operate outside of conscious human control, e.g., breathing, heart rate, and digestion.

autonomous agent a computer-controlled virtual actor, capable of interacting with other agents in the virtual environment, and with the virtual environment.

autonomous function during teleoperation, a task, sub-task, or other function completed under computer control.

autonomous robot computer-controlled system programmed to carry out some task without human intervention; for example, a welding robot; distinguishable from other computer-controlled systems by its flexibility, that is, by a capability to carry out more than one task, with reprogramming; for example, a welding robot could also be a painting robot.

autonomy the freedom to be different or behave differently than other nodes within the system.

autoregressive a pth order autoregressive process is a discrete random process that is generated by passing white noise through an all-pole digital filter having p poles. Alternatively, $x[n]$ is a pth order AR process if

$$x[n] = \sum_{i=n-p}^{n-1} \alpha[i]x[i] + q[n] .$$

Autoregressive processes are often used to model signals since they exhibit several useful properties. *See also* moving average.

autostereoscopic display stereoscopic viewing system that does not require viewers to use or wear head-mounted displays or viewing aids, such as shutters or polarized lenses.

availability the probability that a system will be operative at any given instant during a specified time period. The measure of the availability combines reliability and time to restore service. The relevant time domain must therefore be real time, and reliability must be transformed to that domain if expressed relative to some other domain such as execution time. Note that many software failures are intermittent and recovery may be automatic, without significant interruption of service. Therefore, time to recover is not the same as time to repair, since repair may be done off-line. Availability is the capability of the software product to be in a state to perform a required function at a given point in time, under stated conditions of use. Externally, availability can be assessed by the proportion of total time during which the software product is in an up state. Availability is therefore a combination of maturity (which governs the frequency of failure), fault tolerance and recoverability (which governs the length of down time following each failure).

avatar a representation of a user present within a virtual environment; a virtual actor representing a user.

average-case cost the sum of costs of an algorithm over all possible inputs divided by the number of possible inputs. *See also* worst-case cost, amortized cost.

average daily peak hour traffic the mean traffic observed in the peak hour of several days.

averaging the sum of N samples, images, or functions, followed by division of the result by N. Has the effect of reducing noise levels. *See also* blurring, image smoothing, mean filter, noise smoothing, smoothing.

AVL–tree a binary search tree such that the subtrees of each node have heights that differ by at most one.

awk a string-pattern-matching language named after its creators, Alfred Aho, Peter Weinberger, and Brian Kernighan.

axiomatic semantics the meaning of a program as a property or specification in logic.

axis a vector of motion; 2 axes describe a plane; 3 axes describe a volume; machines that can be positioned around 3 axes and oriented around each of them can position and orient their endpoints anywhere in their workspace; also, a joint or degree of freedom of a robotic manipulator.

Specifically, in robotic manipulators, two can be described:

prismatic: a joint whose movement is along the axis of the connecting links, i.e., in and out.

rotational: a joint whose movement causes a link to rotate around an axis perpendicular to the link itself, and usually at one end of the link.

B

b*-tree a B-tree in which nodes are kept 2/3 full by redistributing keys to fill two child nodes, then splitting them into three nodes.

b+Tree a variation of the b-tree structure. A b+tree has two different types of nodes, leaf level nodes and internal nodes. These have different structures and insertion and deletion algorithms are modified accordingly. Each internal node contains search values and tree pointers. Each leaf node contains search values and data pointers. Each leaf node usually also contains a single tree pointer to facilitate range query.

Babbage, Charles (1792–1871) Born: Totnes, England

Babbage is best known for his ideas on mechanical computation. Babbage is said to have been disgusted with the very inaccurate logarithm tables of his day, as well as appalled by the amount of time and people it took to compute them. Babbage attempted to solve the problem by building mechanical computing engines. The government funded Difference Engine was beyond the technology of the craftsman who attempted to build it. Undeterred Babbage followed this failure with the larger and more complex Analytical Engine (also unfinished). The ideas behind the Analytical Engine formed the basis for Howard Aiken's 1944 Mark I computer. Babbage's assistant, Ada Augusta, the Countess of Lovelace and the poet Lord Byron's daughter, is honored as the first programmer for her work and because her meticulous notes preserved the descriptions of Babbage's machines.

Bachman diagram a diagrammatic representation of set types.

back door an unofficial (and generally unwanted) entry point to a service or system.

back end the part of a compiler that depends only on the target language and is independent of the source language. The back end receives the intermediate code produced by the front end and translates it into target language.

back-end usability testing usability testing of a product after design and implementation are substantially complete. This can be done either with the intent of making a decision on whether to release the product or as a starting point for requirements gathering for the next release of the product.

backfacing polygon a polygon whose surface normal points away from the camera position, which can be tested easily by the dot product of the polygon surface normal **n** and the ray **v** from the viewer to the polygon. A polygon is backfacing if $\mathbf{n} \cdot \mathbf{v} < 0$. For closed objects there is no need to draw backfacing polygons as they are always occluded by non-backfacing polygons.

background (**1**) the environment in which a process with low-priority and/or no user interface is under execution.

(**2**) in the Unix operating system, a means of executing a program such that the shell invokes a child process to carry out a user's command and then accepts the user's next command without waiting for its child process to complete. *Compare with* foreground.

background color the intensity level of pixels that are not intersected by any of the displayed surfaces.

background processing the execution of processes in background.

background subtraction for images, the removal of stationary parts of a scene by subtracting two images taken at different times. For 1D functions, the subtraction of a constant or slowly varying component of the function to better reveal rapid changes.

backing memory the largest and slowest level of a hierarchical or virtual memory, usually a disk. It is used to store bulky programs or data (or parts thereof) which are not needed immediately and need not be placed in the faster but

0-8493-2691-5/01/$0.00+$.50

more expensive main memory or RAM. Migration of data between RAM and backing memory is under combined hardware and software control, loading data to RAM when it is needed and returning it to the backing store when it has been unused for a while.

backing storage *See* backing memory.

backlash the amount of free movement in a power transmission system, i.e., movement that is a result of free play in the actuator, transmission system, or links and not driven by the actuator.

backplane *See* backplane bus.

backplane bus a special data bus which is especially designed for easy access by users and allows the connection of user devices to the computer. It is usually a row of sockets, each presenting all the signals of the bus, and each with appropriate guides so that printed circuit cards can be inserted. A backplane differs from a motherboard in that a backplane normally contains no significant logic circuitry and a motherboard contains a significant amount of circuitry; for example, the processor and the main memory.

backprojection an operator associated with the Radon transform

$$g(s, \theta) = \int_{-\infty}^{+\infty} \int_{-\infty}^{+\infty} f(x, y) \delta(x \cos\theta + y \sin\theta - s)\, dx\, dy .$$

The backprojection operator is defined as

$$b(x, y) = \int_0^{\pi} g(x \cos\theta + y \sin\theta, \theta)\, d\theta .$$

$b(x, y)$ is called the backprojection of $g(s, \theta)$. $b(x, y)$ is the sum of all rays that pass through the point (x, y).

backpropagation (**1**) the way in which error terms are propagated in a multilayer neural network. In a single layer feed-forward network, the weights are changed if there are differences between the computed outputs and the training patterns. For multiple layer networks, there are no training patterns for the outputs of intermediate ("hidden") layer neurons. Hence, the errors between the outputs and the training patterns are propagated to the nodes of the intermediate neurons. The amount of error that is propagated is proportional to the strength of the connection.

(**2**) the process of calculating weights by which the perceptron neural network is "trained" to produce good responses to a set of input patterns. In light of this, the perceptron network is sometimes called a "backprop" network.

backpropagation algorithm a supervised learning algorithm that uses a form of steepest descent to assign changes to the weights in a feedforward network so as to reduce the network error for a particular input or set of inputs. Calculation of the modifications to be made to the weights in the output layer allows calculation of the required modifications in the preceding layer, and modifications to any further preceding layers are made a layer at a time proceeding backwards toward the input layer; hence, the name of the algorithm.

backside bus a term for a separate bus from the processor to the second level cache (as opposed to the frontside bus connecting to the main memory).

back-to-back testing a testing phase in which two or more programs are used at the same time by sending to them the same inputs and comparing their outputs in order to see related differences. *See* mutation testing.

backtracking a component process of many search techniques whereby recovery from unfruitful paths is sought by backing up to a juncture where new paths can be explored. Many languages oriented to artificial-intelligence-style problem solving and searching implement *backtracking* mechanisms.

backup (**1**) a copy, duplicate, or version kept for the purpose of saving the contents of mass storage on a different medium.

(**2**) the act or process of creating a backup.

Backus-Naur form (BNF) a notation developed by John Backus and Peter Naur to represent the formal syntactic grammar of Algol-60, and adapted for many other languages since. Also known by its acronym BNF. A production of the form $N \rightarrow x$, for N a member of the set of nonterminal symbols Vn and x is a string of zero or more symbols of the vocabulary of the grammar $V = V_n \cup V_t$. In BNF, nonterminal symbols are represented by a name contained in <> symbols, for example, <number>. Terminals are represented by their own appearance, and the $\rightarrow$ symbol is represented by the sequence ::=. Two or more rules with the same left-hand side may be combined by using the alternation symbol "|", for example < number >::=< digit >|< number >< digit >.

Many formal grammars are based on variants of BNF, following slightly different rules for representing terminal symbols, nonterminal symbols, and the production arrow. Sometimes called Backus-normal form. *See also* context-free grammar, grammar.

backward chaining a style of reasoning in deductive databases, which traverses the arrows of rules in a backward direction, like SLD–resolution: if the head of a rule should be inferred, then all elements of the body of the rule are tried to be inferred.

backward error recovery a software dynamic redundancy technique that uses previously saved correct state information as the starting point after a failure; examples are checkpointing and journaling. Also called rollback.

backwards compatibility a property of hardware or software revisions in which previous protocols, formats, layouts, etc. are irrevocably discarded in favor of "new and improved" protocols, formats, and layouts, leaving the previous ones not merely deprecated but actively defeated. (Too often, the old and new versions cannot definitively be distinguished, such that lingering instances of the previous ones yield crashes or other infelicitous effects, as opposed to a simple "version mismatch" message.) A backwards compatible change, on the other hand, allows old versions to coexist without crashes or error messages, but too many major changes incorporating elaborate backwards compatibility processing can lead to intractably complex software.

backwards ray tracing technique used to render a scene on a view plane by tracing imaginary "eye rays" from the viewer's eye to the surface of the objects in a scene, to determine the objects' visibility. A grid on the view plane is used to cast eye rays from the center of projection (the viewer's eye). It is convenient for the grid to correspond to the pixels of the display screen. For every pixel on the view plane, an eye ray is cast from the center of projection through the center of the pixel and into the scene. The pixel's color is determined by the eye ray's point of first intersection with an object in the scene.

The basic *backwards ray tracing* algorithm can be extended to render shadows in a scene. This extension involves firing an additional ray from the first point of intersection to each of the scene's light sources. If the ray intersects with an object on its path to the light source, then the point of first intersection is in shadow for that light source. The combination of the effect of each ray to the light source determines the first intersection point's color.

balanced code a binary line code which ensures an equal number of logic ones and logic zeros in the encoded bit sequence. Also called a DC-free code because the continuous component of the power spectral density of a balanced encoded sequence falls to zero at zero frequency.

balanced delimiters a pair of symbols which act as delimiters and which must occur in properly balanced pairs. Examples are parentheses (), square brackets [], and angle brackets <>.

balanced merge sort a k-way merge sort in which the number of input and output data streams is the same. *See also* merge sort.

bandpass filter a filter which allows only signals between given set points to propagate, preventing propagation of signals below the lower bound and above the upper bound.

bandwidth (1) generally, the range of frequencies present in a signal; mathematically, the frequency at which output signal amplitude is reduced to 50% of input signal amplitude; colloquially, the amount of information that may be processed or transmitted by a system.

(**2**) in performance analysis, bandwidth represents the maximum rate at which a given device can perform work. This device may be a processor, storage unit, etc., which must handle transactions or other requests at some given rate, or a communications channel, which must transmit data at some given speed.

bang-bang control control scheme which is inherently digital; that is, goes from one state to another with no intermediate stages.

bang metric a complex synthetic metric giving an indication of the "size" of the software. Related to function points.

banker's algorithm a deadlock avoidance scheme in which resource allocations that could potentially lead to deadlock are not made. The algorithm is analogous, in some ways, to the management of lines of credit belonging to customers in a small-town bank. Hence, the moniker.

barrel shifter an implementation of a shifter, which contains $\log_2$ (max number of bits shifted) stages, where each stage shifts the input by a different power of two number of positions. It can be implemented as a combinational array with compact layout that can shift the data by more than one bit using only one gate. For instance, for a 4-bit word, it can execute instructions shl, shl2, shl3, and shl4. This shifter lends itself well to being pipelined.

barrier a synchronization structure that requires all processes to reach it before any of them can proceed any further.

base address (**1**) an address to which an index or displacement is added to locate the desired information. The base address may be the start of an array or data structure, the start of a data buffer, the start of page in memory, etc.

(**2**) as a simpler alternative to a full virtual memory, the code space or data space of a program can be assumed to start at a convenient starting address (usually 0) and relocate in its entirety into a continuous range of physical memory addresses. Translation of the addresses is performed by adding the contents of an appropriate base address register to the user address.

base class in object-oriented programming, a class that has classes derived from it by means of inheritance. The base class contains the subset common to all the derived classes.

base code the quantity of source code of an existing program. Normally used in reference to updates of an original version of software.

based pointer *See* relative pointer.

base index *See* first index.

baseline a major version of a system selected for release to customers and/or for the purpose of measuring some attribute, e.g., reliability. A formally approved version of a configuration item, regardless of media, formally designated and fixed at a specific time during the configuration item's life cycle. Versions of a product can be distinguished both by a baseline identifier and by lower level identifiers. The baseline is altered only when large modifications are made, e.g., functional enhancement. Small changes, e.g., for corrective maintenance, do not alter the baseline. For the purpose of reliability growth monitoring, it is necessary to regard the baseline as unaltered by corrective maintenance actions. The decision to declare a version as a baseline is part of configuration management and is, to some extent, arbitrary.

baseline management in configuration management, the set of actions defining documents and changes to define a baseline at a time instant of the configuration item life-cycle.

base register the register that contains the component of a calculated address that exists in a register before the calculation is performed (the register value in "register+immediate" addressing mode, for example).

base register addressing addressing using the base register. Base register is the same as base address register, i.e., a general purpose register that the programmer chooses to contain a base address.

basic *See* BASIC.

BASIC an acronym for Beginner's All-purpose Symbolic Instruction Code, a very simple language developed at Dartmouth University in the 1960s. *See also* Visual Basic. Also referred to as Basic.

basic block any computation not containing a control transfer. In imperative languages, a sequence of statements not containing conditional computations. A basic block is a fundamental entity used in flow analysis, used for computing code motions and other code optimizations.

basic input-output system (BIOS) part of a low-level operating system that directly controls input and output devices.

basin of attraction the set of all points which when iterated by a function f, attract to the same point.

basis function one of a set of functions used in the transformation or representation of some function of interest. A linear transformation T of continuous functions is of the form

$$y(s) = T\{x(t)\} = \int_{-\infty}^{+\infty} x(t) b(s,t)\, dt$$

where $b(s,t)$ is a basis function. For discrete sequences T would be of the form

$$y[k] = T\{x[n]\} = \sum_{n=-\infty}^{+\infty} x[n] b[k,n] \,.$$

The function to be transformed is projected onto the basis function corresponding to the specified value of the index variable s or k. $y(s)$ is the inner product of $x(t)$ and the basis function $b(s,t)$. For the Laplace transform $b(s,t) = e^{-st}$, and for the Fourier transform $b(\omega,t) = e^{-j\omega t}$. For the discrete-time Fourier transform $b[k,n] = e^{\frac{j2\pi kn}{N}}$, and for the Z-transform $b[z,n] = z^{-n}$.

basis spline a spline curve or surface that can be formulated as a weighted sum of polynomial basis functions. Commonly known as a b-spline.

batch (1) several requests arriving or being served at the same time.

(2) one measurement interval during the execution of a long simulation run. Confidence intervals for a model statistic may be obtained from its sample mean and variance, derived from the output of several independent replications of the simulation experiment, in combination with the Student's t distribution. These samples can be obtained by measuring the statistic during each of several runs of the simulator, or by partitioning a single long run into several batches and separately measuring the statistic over each batch. If these batches are long enough, then batch results can be considered mutually independent.

batch arrival process the process governing the arrival of batches (1). A batch arrival process is usually described by a point process determining the time instances of arrivals and by a discrete random variable describing the number of simultaneous arrivals.

batch file a file containing a series of commands for an interpreter, such as the DOS command interpreter.

batch model model for a system that can be described as a batch system (1). *See also* closed model.

batch processing the processing of data or the accomplishment of jobs, accumulated in advance, in such a manner that the user cannot further influence its processing while it is in progress.

batch system (1) a system allowing requests to arrive or to start being served at the same time.

(2) in computing, a mode in which low-priority jobs are queued for processing, so that only one job in a priority class is active at a time.

bathtub curve a graph that shows that the frequency of malfunctions in hardware components dramatically increases both very early and very late in the life of the component.

baud the signaling rate, or rate of state transitions, on a communications medium. One baud corresponds to one transition per second. It is often confused with the data transmission rate, measured in bits per second.

Numerically, it is the reciprocal of the length (in seconds) of the shortest element in a signaling code. For very low speed modems (up to 1200 bit/s) the baud rate and bit rate are usually identical. For example, at 9600 baud, each bit has a duration of 1/9600 seconds, or about 0.104 milliseconds.

Modems operating over analog telephone circuits are bandwidth limited to about 2,500 baud; for higher user data speeds each transition must establish one or more decodable states according to amplitude or phase changes. Thus, if there are 16 possible states, each can encode 4 bits of user data and the bit rate is 4 times the baud rate.

At high speeds, the reverse is true, with run-length controlled codes needed to ensure reliable reception and clock recovery. For example, FDDI uses a 4B/5B coding in which a "nibble" of 4 data bits is encoded into 5 bits for transmission. A user data rate of 100 Mbit/s corresponds to transmission at 125 Mbaud.

baud rate *See* baud.

Bayes envelope function given the prior distribution of a parameter Θ and a decision function ϕ, the *Bayes envelope function* $\rho(F_\Theta)$ is defined as

$$\rho\left(F_\Theta\right) = \min \phi r\left(F_\Theta, \phi\right) ,$$

where $r(F_\Theta, \phi)$ is the *Bayes risk function* evaluated with the prior distribution of the parameter Θ and decision rule ϕ.

Bayesian classifier a function of a realization of an observed random vector **X** and returns a classification w. The set of possible classes is finite. A Bayesian classifier requires the conditional distribution function of **X** given w and the prior probabilities of each class. A Bayesian classifier returns the w_i such that $P(w_i|\mathbf{X})$ is maximized. By Bayes' rule

$$P(w_i|\mathbf{X}) = \frac{P(\mathbf{X}|w_i)P(w_i)}{P(\mathbf{X})} .$$

Since $P(\mathbf{X})$ is the same for all classes, it can be ignored and the w_i that maximizes $P(\mathbf{X}|w_i)P(w_i)$ is returned as the classification.

Bayesian estimator an estimator of a given parameter Θ, where it is assumed that Θ has a known distribution function and a related random variable X that is called the observation. X and Θ are related by a conditional distribution function of X given Θ. With $P(X|\Theta)$ and $P(\Theta)$ known, an estimate of Θ is made based on an observation of X. $P(\Theta)$ is known as the *a priori* distribution of Θ.

Bayesian mean square estimator for a random variable X and an observation Y, the random variable

$$\hat{X} = E[X|Y] ,$$

where the joint density function $f_{XY}(x, y)$ is known. *See also* mean-square estimation, linear least squares estimator.

Bayesian reconstruction an algorithm in which an image u is to be reconstructed from a noise-corrupted and blurred version v.

$$v = f(Hu) + \eta .$$

An *a priori* distribution $p(u|v)$ of the original image is assumed to be known. The equation

$$\hat{u} = \mu_u + R_u H^T D R_\eta^{-1}\left[v - f\left(H\hat{u}\right)\right] ,$$

where R_u is the covariance of the image u, R_η is the covariance of the noise η, and D is the diagonal matrix of partial derivatives of f evaluated at $\hat{u}$. An initial point is chosen and a gradient descent algorithm is used to find the closest $\hat{u}$ that minimizes the error. Simulated annealing is often used to avoid local minima.

Bayes risk function with respect to a prior distribution of a parameter Θ and a decision rule ϕ, the expected value of the loss function with respect to the prior distribution of the parameter and the observation X.

$$r\left(F_\Theta, \phi\right) = \int_\Theta \int_X L[\theta, \phi(x)] \; f_{X|\Theta}(x|\theta) f_{|\Theta}(\theta)\, dx\, d\theta .$$

The loss function is the penalty incurred for estimating the parameter Θ incorrectly. The decision rule $\phi(x)$ is the estimated value of the parameter based on the measured observation x.

Bayes' rule a rule that relates the conditional probability of an event A given B and the conditional probability of the event B given A:

$$P(A|B) = \frac{P(B|A)P(A)}{P(B)} .$$

BCD *See* binary-coded decimal.

BCNF *See* Boyce-Codd Normal Form.

BCPL Basic Computer Programming Language. Designed by Martin Richards in the late 1960s. The language C is a lineal descendant of BCPL.

beacon signature line or expression that suggests the presence of a programming plan.

beam tracing a method of rendering similar to ray tracing but using an arbitrarily shaped projection, commonly a polygonal cone, rather than a single ray. It is an improvement on ray tracing since it reduces the CPU overhead and reduces aliasing artifacts by taking advantage of known spatial coherence in the beam.

before image a copy of a database object before it is updated.

behavior the set of actions and responses that characterize an object or system. It can be regarded as the set of possible states and reasons for the transitions of the systems with the associated actions that are performed by systems. It can be typically modeled by means of operational approaches, such as state diagrams, Petri nets, etc. In object-oriented programming, the behavior is specified by means of class methods.

behavioral animation computer animation technique depending on modeling an object's actions rather than individually specifying and creating each frame of each motion.

behavioral aspect the aspects of an entity, class, or object related to the evolution of its state along the time.

behavioral feature a specific feature that characterizes the behavior of a software component.

behavioral modeling the representation of the mode of behavior (or a state) of an application and the events that cause transitions from its states.

behavioral repertoire the set of actions which a person or machine is capable of executing at any given time.

behavior model portion of a virtual model describing the actions that an object may take and their implications for the virtual image.

behavior sharing a mechanism of the object-oriented programming. When a class inherits from a superclass, it also inherits the superclass behavior via the superclass methods. This means that the class shares the behavior with its superclass(es). All the instances of a class share the same generic behavior of the class. The difference resides in the internal status of the object in terms of the status of its attributes.

behavior transducer device for sensing human actions and translating those actions into digitized inputs to a computer system.

Belady's anomaly the observation that in FIFO page replacement rule, increasing the number of pages in memory may not reduce the number of page faults.

Bell-LaPadula model a widely used formal model of mandatory access control. It requires that the simple security property and the *-property be applied to all subjects and objects.

benchmark a controlled experiment that measures the execution time, utilization, or throughput for a system using an artificial workload chosen to represent the actual operational use of the system. *Benchmarks* are used to obtain an estimate of performance, to compare similar systems, or to parameterize a model of the system that will be used to determine whether it will operate effectively as a component of a larger system.

benign failure a failure whose severity is slight enough to be outweighed by the advantages to be gained by normal use of the system.

The point is that the benefits gained from using the system (when it is operable) outweigh the severity associated with such a *benign failure* of the system.

Bernoulli distribution a random variable X with alphabet {0, 1} and parameter α such that its probability mass function is

$$p(x) = (1-\alpha)^x \alpha^{1-x} .$$

Berry, Clifford Edward Berry is best known as the co-developer, along with John Vincent Atanasoff, of the first functioning electronic digital computer. Berry was recommended to Atanasoff by the Dean of Engineering at Iowa State College as a most promising student who understood the electronics well enough to help Atanasoff implement his ideas for a computing machine. Unfortunately, Berry's contributions as a computing pioneer were not honored until after his death.

best first search **(1)** a heuristic search technique that finds the most promising node to explore next by maintaining and exploring an ordered open node list.

(2) a search strategy for speech recognition in which the theories are prioritized by score, and the best scoring theory is incrementally advanced and returned to the stack. An estimated future score is included to normalize theories. The search is admissible if the estimated future score is an upper-bound estimate, in which case it can be guaranteed that the overall best-scoring theory will arrive at the end first. Also called A* search.

best-fit memory allocation a memory allocator for variable-size segments must search a table of available free spaces to find memory space for a segment. In "best-fit" allocation, the free spaces are linked in increasing size and the search stops at the smallest space of sufficient size.

best-practice a term to describe a project or a test case that highlights/demonstrates the typical aspects of the practical use of a technology. *Best-practice* projects may include the custom installation of complex systems, or the adoption of a software product in a slightly different content with respect to its classical application.

beta test the process of testing a pre-release (potentially unreliable) version of a piece of software or product by making it available to selected users. This term was used in the early 1960s by IBM, and has become a "de-facto" standard in industrial environment. Software systems often go through two stages of testing: Alpha (in-house) and Beta (out-house). Beta releases are generally made available to a small number of trusted customers.

beta version the software version produced for beta testing. Typically, the beta testing is performed by a set of selected end-users.

Bézier curve a spline curve that (in the usual case of a cubic Bézier curve) is represented by four control points defining a cubic polynomial.

BHCA *See* busy hour call attempts.

bias the systematic (as opposed to random) error of an estimator. *See* rounding bias.

biased exponent an exponent to which a known constant (the bias) has been added. The biasis usually chosen so that the smallest (negative) exponent corresponds to zero.

BiCapitalization The use of capital letters within a name to visually delimit names that are composite words, for example, "the InsertString function".

biconnected graph a graph that has no articulation/cut vertices.

bicubic surface a type of parametric two-variable polynomial surface patch where the polynomials are cubic in both parameters.

bidirectional bus a bus that may carry information in either direction but not in both simultaneously.

bidirectional search a search algorithm that replaces a single search graph, which is likely to grow exponentially, with two smaller graphs: one starting from the initial state and one starting from the goal state.

bifurcation diagram a diagram of an iterated function against the value of a swept constant. In many cases this generates a fractal that tends to have two zones of activity.

big-bang testing an integration testing where both software and hardware components are put together to form the whole system.

big endian a storage scheme in which the most significant unit of data or an address is stored at the lowest memory address. For example, in a 32-bit or four-byte word in memory, the most significant byte would be assigned address i, and the subsequent bytes would be assigned the addresses $i+1$, $i+2$, and $i+3$. Thus, the least significant byte would have the highest address of $i+3$ in a computer implementing the big endian address assignment. "Big-endian" computers include IBM 360, MIPS R2000, Motorola M68000, SPARC, and their successors. *See* little endian, big endian/little endian problem.

big endian/little endian problem the problem caused by the passing code from an architecture based on big-endian ordering of bytes to one adopting the little-endian approach and vice versa. *See also* big endian, little endian.

big-O notation a theoretical measure of the execution of an algorithm, usually the time or memory needed, given the problem size n, which is usually the number of items. Informally saying some equation $f(n) = O(g(n))$ means it is less than some constant multiple of $g(n)$. More formally it means there is some c and k, such that $0 \leq f(n) \leq cg(n)$ for all $n \geq k$. The values of c and k must be fixed for the function f and must not depend on n. *See also* $\Omega(n)$, $\Theta(n)$, little-o notation, asymptotic upper bound, np-complete.

bilateral teleoperator teleoperated manipulator capable of force reflection.

bilateral Z-transform a Z-transform of the form

$$\mathcal{Z}\{x\} = \sum_{n=-\infty}^{+\infty} x[n]z^{-n} .$$

bilevel image coding *See* binary image coding.

bilinear filtering an averaging technique applied to the color values of adjacent pixels so that textures look smooth rather than blocky. It aims to make the texture look more realistic.

bilinear interpolation interpolation of a value in 2D space from four surrounding values by fitting a hyperbolic paraboloid. The value at (x, y), denoted $f(x, y)$, is interpolated using $f(x, y) = ax + by + cxy + d$, where a, b, c, and d are obtained by substituting the four surrounding locations and values into the same formula and solving the system of four simultaneous equations so formed.

binary (**1**) having two states, such as true/false, on/off, or 0/1.

(**2**) arithmetically expressed in radix 2. Fundamental representation of information for computers.

(**3**) in common usage, a synonym for executable file. *See also* octal, hexadecimal.

binary association a formal/mathematical relationship between two sets of entities.

binary code a code, usually for error control, in which the fundamental information symbols which the codewords consist of are two-valued or binary and these symbols are usually denoted by either '1' or '0'. Mathematical operations for such codes are defined over the finite or Galois field consisting of two elements denoted by GF(2). The mathematical operations for such a Galois field are addition and multiplication. For addition over GF(2) one finds that:

$$1+1=0,\ 1+0=1 \text{ and } 0+0=0 .$$

For multiplication over GF(2) one finds that

$$1\cdot 1=1,\ 1\cdot 0=0 \text{ and } 0\cdot 0=0 .$$

See also block coding, convolutional coding, error control coding.

binary-coded decimal (BCD) a notation in which a value is represented as sequences of 4-bit values, where each 4-bit value represents a single decimal digit. Within each 4-bit value, the codes representing the binary values 10 through 16 are illegal. Thus: 0 = 0000, 1 = 0001, 2 = 0010, 3 = 0011, 4 = 0100, 5 = 0101, 6 = 0110, 7 = 0111, 8 = 1000, 9 = 1001.

A BCD value may encode its sign in the low-order 4 bits (often by using one of the illegal values in this position to represent positive or negative sign), or in the high-order 4 bits, depending on the nature of the machine which implements the code.

Many programming languages designed to support commercial applications such as payroll, inventory, and similar tasks provide support for BCD data as a separate data type.

binary constant a constant that has the value 0 or 1.

binary-decimal decimal representation in which each decimal digit is represented by a four-bit binary equivalent; for example, decimal 35 may be represented as 00110101, in which 0011 represents 3 and 0101 represents 5, instead of as 100011 (in conventional binary). Many binary-coded decimal (BCD) representations exist, all of which are convenient for input/output but are not as efficient, in terms of storage, as conventional binary.

binary erase channel a channel where an error detecting circuit is used and the erroneous data is rejected as erasure asking for retransmission. The inputs are binary and the outputs are ternary, i.e., 0, 1, and erasure. Used for ARQ (automatic request for retransmission) type data communication.

binary executable a program file whose instructions are in machine code form.

binary function a function with two arguments. *See also* constant function, unary function, N-ary function.

binary heap a complete binary tree where every node has a key more extreme (greater or less) than or equal to the key of its parent. *See also* heapify, heap, binomial heap.

binary hypothesis testing a special two-hypothesis case of the M-ary hypothesis testing problem. The problem is to assess the relative likelihoods of two hypotheses H_1, H_2, normally given prior statistics $P(H_1)$, $P(H_2)$, and given observations $\mathbf{y}$ whose dependence $p(\mathbf{y}|H_1)$, $p(\mathbf{y}|H_2)$ on the hypotheses is known. The receiver operating characteristic is an effective means to visualize the possible decision rules. *See also* m-ary hypothesis testing, conditional statistic, prior statistic, posterior statistic.

binary image an image whose pixels can have only two values, 0 or 1 (i.e., "off" or "on"). The set of pixels having value 1 ("on") is called the figure or foreground, while the set of pixels having value 0 ("off") is called the background. Examples include black/white photographs and facsimile images.

binary image coding compression of two-level (black/white) images, typically documents. Bilevel coding is usually lossless and exploits spatial homogeneity by runlength, relative address, quadtree, or chain coding. Also called bilevel image coding.

binary large object (BLOB) an unstructured complex object, such as a graphical image or a very long text string, which is required by a database application.

binary lock a lock that has only two states, either locked or unlocked.

binary notation *See* binary.

binary operator any mathematical operator that requires two data elements with which to perform the operation. Addition and Logical-AND are examples of binary operators; in contrast, negative signs and Logical-NOT are examples of unary operators.

binary relation a relation that is between exactly two items at a time, such as "greater than" (>), "not equal to" ($\neq$), "proper subset of" ($\subset$), or "is connected to" (has an edge to) for vertices of a graph. *See also* binary function.

binary search a divide-and-conquer technique in which the range of the search is halved each time. It begins with an interval covering the whole search range. If the search value is less than the item in the middle of the interval, narrow the interval to the lower half. Otherwise narrow it to the upper half. Repeatedly check until the value is found or the interval is empty.

binary search tree a binary tree that is lexicographically arranged so that, for every node in the tree, the nodes to its left are smaller and those to its right are larger.

binary semaphore a semaphore that has only two values, locked and unlocked. *See also* counting semaphore, mutex, semaphore.

binary space partition tree a method for representing a polyhedron that explicitly uses the planes that bound the polyhedron. The technique represents the object as a binary tree, with planes at each non-leaf node. The planes bound a face of the polyhedron, and divide space into two subregions, which in turn can be further bounded by the two children at a node. Leaf nodes (at the edge of the tree) are either completely object or completely free space. A similar idea can be used in 2D for representing polygons. This representation is useful for hidden surface removal and point classification (determining whether a point is inside or outside the object).

binary tree (**1**) formally, a directed graph characterized by nodes which have exactly two arcs leading to other nodes, or no outgoing arcs; a designated node, the root node, which has no incoming arcs; and the property that there is at most one arc leading into any node.

(**2**) recursively defined as a set of nodes $(n_1, \ldots n_k)$ one of which is designated the root and the remaining $k - 1$ nodes form at most two sub-trees.

(**3**) informally (and incorrectly), a tree characterized by nodes which usually have two arcs leading to other nodes. This designation is incorrect usage, but is often used in reference to parse trees even when the language has unary operators or n-ary (for $n > 2$) operators.

binary tree predictive coding predictive image coding scheme in which pixels are ordered in a pyramid of increasingly dense meshes. The sparsest mesh consists of subsamples of the original image on a widely spaced square lattice; succeeding meshes consist of the pixels at the centers of the squares (or diamonds) formed by all preceding meshes. Each mesh has twice the number of pixels as its predecessor. Pixel values are predicted by non-linear adaptive interpolation from surrounding points in preceding meshes. The prediction errors, or differences, are quantized, ordered into a binary tree to provide efficient coding of zeros, and then entropy coded.

binary variable a variable that can only take on the values 0 or 1.

binaural having to do with hearing using both ears.

bind to associate an absolute address, virtual address or device identifier with a symbolic address or label in a computer program.

binding (**1**) mechanism for associating the call to a specific procedure/method. Binding can be static or dynamic. It is called static when it is performed at the compilation/linking time, and it is called dynamic when it is performed at run-time. Dynamic binding is typically used by interpreted languages and by object oriented languages. In some languages the presence of dynamic binding is stated by using specific keywords, such as "virtual" in C++. Hierarchies and sub-hierarchies of abstract classes are the main place in which the dynamic binding is used. *See also* overriding, overloading.

(**2**) the time at which a value is bound to an entity that represents it. For example, the time at which an expression is evaluated. Usually applied to expressions that are used to pass values as parameters to subroutines. Depending on the nature of the language, the binding time can be early, for example, at the time the call is performed, or late, that is, whenever the value is actually required. In languages such as LISP and Algol (Algol's call-by-name binding), the binding time is deferred to the time the value is needed, and consequently other factors may change the actual value computed, each time it is computed. *See* FUNARG.

bingo sort a variant of selection sort which orders items by repeatedly looking through remaining items to find the greatest value and moving all items with that value to their final location. This is more efficient if there are many duplicate values.

binocular having to do with sight using both eyes; in synthetic environments, usually implies presentation of somewhat disparate images to each eye, supporting stereoscopic vision.

binocular imaging the formation of two images of a scene from two different positions so that binocular vision can be performed, in a similar manner to the way humans deploy two eyes.

binocular vision the use of two images of a scene, taken (often simultaneously) from two different positions, to estimate depth of various *point features,* once correspondences between pairs of image features have been established.

binomial coefficients the coefficients of the polynomial resulting from the expansion of $(a+b)^n$. These coefficients are equal to

$$\binom{n}{k} = \frac{n!}{k!(n-k)!},$$

where n is the order of the polynomial and k is the index of the coefficient. The kth coefficient is multiplied by the term $a^k b^{n-k}$.

binomial distribution the binomial distribution is the distribution of a random variable Y that is the sum of n random variables that are *Bernoulli distributed.*

$$Y = X_1 + X_2 + \cdots + X_n .$$

The probability mass function of such a Y is

$$p_Y(k) = \binom{n}{k} p^k (1-p)^{n-k},$$

where p is the parameter of the Bernoulli distribution of any X_i.

binomial heap a priority queue made of a forest of binomial trees with the heap property numbered $k = 0, 1, 2, \ldots, n$, each containing either 0 or 2^k nodes. Each tree is formed by linking two of its predecessors, by joining one at the root of the other. The operations of insert a value, decrease a value, delete a value, and merge or join (meld) two queues take $O(\log n)$ time. The find minimum operation is a constant $\Theta(1)$. *See also* Fibonnaci heap.

bin-packing one of a family of NP-hard problems. This problem is interesting to compiler writers because the register allocation problem is equivalent to a bin-packing problem.

bintree a regular decomposition k-d tree for region data.

biocular presentation of the same image to both eyes. *See* binocular.

biocybernetic controller a device taking electrical signals produced by the nervous system or muscles as direct inputs to a computer or synthetic environment.

biomechanics science of physical and control phenomena associated with movement in biological systems, particularly humans; study of how biological systems move and control their movements.

biorthogonal wavelet a generalization of orthogonal wavelet bases, where two dual basis functions span two sets of scaling spaces, V_j and $\hat{V}_j$, and two sets of wavelet spaces, W_j and $\hat{W}_j$, with each scaling space orthogonal to the dual wavelet space, i.e., $V_j \perp \hat{W}_j$ and $\hat{V}_j \perp W_j$.

BIOS *See* basic input-output system.

bipartite graph a graph in which the vertex set can be partitioned into two sets X and Y, such that each edge connects a node in X with a node in Y.

bipartite matching (1) a perfect matching between vertices of a bipartite graph; that is, a subgraph which pairs every vertex with exactly one other vertex.

(2) the problem of finding such a matching.

bisector for two elements e_i and e_j, the locus of points equidistant from e_i and e_j. That is $\{p|d(p, e_i) = d(p(e_j\}$, where d is some distance metric.

bisimulation an alternative to contextual equivalence, replacing the quantification over all contexts with the more tractable requirements that equivalent expressions should be able to match each other's behavior step-by-step.

bisimulation equivalence an equivalence relation, defined in the context of process algebras, which is a finer equivalence relation than trace equivalence and distinguishes states based on branching properties.

BIST *See* built-in self-test.

bi-stable pertaining to a device with two stable states, e.g., bi-stable multivibrator; circuit that has two possible output states and that will remain in its current state without requiring external inputs; a flip-flop.

bit (1) the fundamental unit of information representation in a computer, short for "binary digit" and with two values usually represented by '0' and '1'. Bits are usually aggregated into "bytes" (7 or 8 bits) or "words" (12 to 60 bits).

A single bit within a word may represent the coefficient of a power of 2 (in numbers), a logical TRUE/FALSE quantity (masks and Boolean quantities), or part of a character or other compound quantity. In practice, these uses are often confused and interchanged.

(2) in Information Theory, the unit of information. If an event E occurs with a probability $P(E)$, it conveys information of $log_2\ (1/P(E))$ binary units or bits. When a bit (binary digit) has equiprobable 0 and 1 values, it conveys exactly 1.0 bit (binary unit) of information; the average information is usually less than this.

bit allocation the allocation of bits to symbols with the aim of achieving some compression of the data. Not all symbols occur with the same frequency. Bit allocation attempts to represent frequently occurring symbols with fewer bits and assign more bits to symbols that rarely appear, subject to a constraint on the total number of bits available. In this way, the average string requires fewer bits. The chosen assignment of bits is usually the one that minimizes the corresponding average coding distortion of the source over all possible bit assignments that satisfy the given constraint. Typically sub-sources with larger variances or energy are allocated more bits, corresponding to their greater importance.

bitBLT *See* bit-oriented block transfer.

BitBlt/RasterOp an abbreviation of bit block transfer. This is an efficient technique for copying rectangular arrays of pixels which exploits the fact that computer memory is organized into multi-bit words.

bite error rate (BER) the probability of a single transmitted bit being incorrectly determined upon reception.

bit error probability the probability that a single bit being transmitted will be received incorrectly.

bit error rate (BER) an estimate for the bit error probability. It can be measured as the ratio of the number of bits received incorrectly vs. the total number of bits received.

bit line used in, for example, RAM memory devices (dynamic and static) to connect all memory cell outputs of one column together using a shared signal line. In static RAM, the "bit" line together with its complemented signal "-bit" feeds a "sense amplifier" (differential in this case) at the bottom of the column serving as a driver to the output stage. The actual cell driving the bit line (and -bit) is controlled via an access transistor in each cell. This transistor is turned on/off by a "word" line, a signal run across the cells in each row.

bit-line capacitance the equivalent capacitance experienced in each "bit line" in a RAM or ROM device. *See also* bit line.

bitmap strictly a one-bit-per-pixel representation for a defined area of a display.

bitmapped image a digital image composed of pixels. Bitmapped images are resolution-dependent; i.e., if the image is stretched, the resolution changes. Also called a raster image. *See also* image, pixel, vector image.

bit normalization the process of shifting a binary pattern to the left until the most significant bit is a 1.

bit-oriented block transfer (bitBLT) a type of processing used mainly for video information characterized by minimal operations performed on large data blocks; a processor designed for such operations. BitBLT operations include transfers, masking, exclusive-OR, and similar logical functions.

bit parallel method to transmit or process information in which several bits are transmitted in parallel: e.g., a bit parallel adder with 4-bit data has 8 input ports for them (plus an initial carry bit); an 8-bit parallel port includes true 8-bit bi-directional datalines.

bit period the time between successive bits in data transmission or data recording. At the transmitter (or recorder) the timing is established by a clock. At the receiver (or reader) an equivalent clock must be recovered from the bit stream.

bit per second (bps) measure of transfer rate of a modem or a bus or any digital communication support. bps and baud are not equivalent because bps is a low-level measure and media; thus, it includes the number of bits sent for the low-level protocol, while baud is typically referred to as a higher level of transmission. *See also* baud and baud rate.

bit plane the binary $N \times N$ image formed by selecting the same bit position of the pixels when the pixels of an $N \times N$ image are represented using k bits.

bit plane encoding lossless binary encoding of the bit planes is termed bit plane encoding. The image is decomposed into a set of k, $N \times N$ bit planes from the least significant bit to $k - 1$ most significant bits and then encoded for image compression.

bit rate a measure of signaling speed; the number of bits transmitted per second. Bit rate and baud are related but not identical. Bit rate is equal to baud times the number of bits used to represent a line state. For example, if there are 16 line states, each line state encodes four bits, and the bit rate is thus four times the baud. *See* baud.

bit serial processing of one bit per clock cycle. If word length is W, then one sample or word is processed in W clock cycles. In contrast, all W bits of a word are processed in the same clock cycle in a bit-parallel system.

For example: a bit serial adder with 4-bit data has one input signal for each bit of data, one bit for carry-in, and two 4-bit shift registers for data.

bit-slice processor a processor organization that performs separate computations (via multiple processing units) separately upon subsections of an incoming channel.

bits per pixel the number of bits used to describe the color or intensity of a pixel. For example, using 8 bits to store a value from the RGB color model would permit 3 bits to be used for both red and green values and 2 bits for the blue value. Blue gets a smaller range because the human eye contains less blue cones and thus is less sensitive to blue variations. True color images have 24 bits per pixel, or 8 bits for each of the red, green, and blue pixels. Typical grayscale images have 8 bits per pixel, giving 256 different gray levels. Compressed image sizes are often represented in bits per pixel, i.e., the total number of bits used to represent the compressed image divided by the total number of pixels.

bitwise an operation, typically a logical operation such as and, or, complement, or exclusive

OR, whose result is determined as a sequence of bits resulting from the combination of the corresponding bits from each of the input operands (if a binary operation) or the input operand (if a unary operation). *See* bitwise AND, bitwise OR.

bitwise AND an operation that combines two values in the computer using the AND operation, applying the operation individually to each of the bits of the values. See also logical AND. Depending on the programming language, the bitwise AND operation may or may not have different semantics from the logical AND operation.

bitwise OR an operation that combines two values in the computer using the OR operation, applying the operation individually to each of the bits of the values. See also logical OR, exclusive OR. Depending on the programming language, the bitwise OR operation may or may not have different semantics from the logical OR operation.

b_k **tree** a binomial tree of order (height) k.

blackboard a public data structure used as a communications area between users or processes. *See* blackboard system.

blackboard architecture an expert systems design in which several independent knowledge bases each examine a common working memory, called a "blackboard".

blackboard system a system where several independent modules interact and coordinate with each other by accessing and posting results on the blackboard.

black box reuse a style of reuse based on object composition without knowing the internal implementation of the reused component (the box). Composed objects reveal no internal details to each other and are thus analogous to 'black-boxes'.

black-box testing testing accomplished on the external features of the entity (subprogram, object). It tests whether an object satisfies the contracts of the interface without regard to the style or detail of the internal implementation.

blend surface a surface added to two or more other surfaces to provide a continuous join between them.

blind deconvolution the recovery of a signal $x[n]$ from $y[n]$ – the convolution of the signal with an unknown system $h[n]$:

$$y[n] = h[n] * x[n] .$$

Occasionally some knowledge of $h[n]$ is available (e.g., that it is a high-pass or low-pass filter). Frequently, detailed knowledge is available about the structure of x. *See also* convolution.

blind search a characterization of all search techniques that are heuristically uninformed. Included among these would normally be state-space search, means-ends analysis, generate and test, depth-first search, and breadth-first search.

blind write a write of an object that has not been previously read by the transaction.

blink in computer display systems, a technique in which a pixel is alternatively turned on and off.

BLISS a language now largely obsolete, but historically important because it was the first language that was an imperative language designed to implement "systems software" (such as operating systems) that was designed without a goto statement. (Earlier languages such as LISP did not possess the goto but were intended for other domains). Designed at Carnegie Mellon University in 1969 by William Wulf and others.

blitter a special-purpose chip or hardware system used for fast implementations of bit-mapped graphics. Blitters are used to copy sections of video memory from one place to another. During the copy operation several source areas may be used and logical operations may be performed on them. One application of blitters is the provision of fast animated graphics, known as sprites.

BLOB *See* binary large object.

block (1) generally, a grouping of data read or written in one contiguous operation.

(2) in architecture, a group of sequential locations held as one unit in a cache and selected as whole. Also called a line. *See also* memory block.

(3) in programming languages, a syntactic construct in most languages which is used to group statements to indicate a unit of execution. Depending on the language, a block may allow or forbid variable declarations within a block, or limit the nature of blocks in which such declarations may appear. In most languages, a block may be permitted to contain no statements at all (an empty block). Languages may limit control transfers into or out of a block of statements.

block anchor *See* anchor record.

block cipher an encryption system in which a successive number of fundamental plain text information symbols, usually termed a block of plain text information, are encrypted according to the encryption key. All information blocks are encrypted in the same manner according to the transformation determined by the encryption key. This implies that two identical blocks of plain text information will always result in the same cipher text when a particular block cipher is employed for encryption. *See also* encryption, stream cipher.

block code a mapping of k input binary symbols into n output symbols.

block coding (1) an error control coding technique in which a number of information symbols, blocks, are protected against transmission errors by adding additional redundant symbols. The additional symbols are usually calculated according to a mathematical transformation based on the so-called generator polynomial of the code. A block code is typically characterized by the parameters (n, k), where k is the number of information symbols per data word, and n is the final number of symbols in the code word after the addition of parity symbols or redundant symbols. The rate of a block code is given by k/n.

Typically, the lower the rate of a code the greater the number of errors detectable and correctable by the code. Block codes in which the block of information symbols and parity symbols are readily discernable are known as systematic block codes. The receiver uses the parity symbols to determine whether any of the symbols were received in error and either attempts to correct errors or requests a retransmission of the information. *See also* binary code, convolutional coding, error control coding.

(2) refers to (channel) coding schemes in which the input stream of information symbols is split into non-overlapping blocks which are then mapped into blocks of encoded symbols (codewords). The mapping only depends on the current message block. *Compare with* trellis coding.

block diagram a system diagram in which the components are represented with geometrical elements. Different meanings are assigned to different geometrical figures. These in turn are connected to each other by means of line speechifying their relationships. Different meanings are assigned to different modalities of drawing lines. Lines specifying the relationships can be oriented or not. Synonymous with configuration diagram, systems resource chart. *See also* box diagram, bubble chart, flow chart, structure chart.

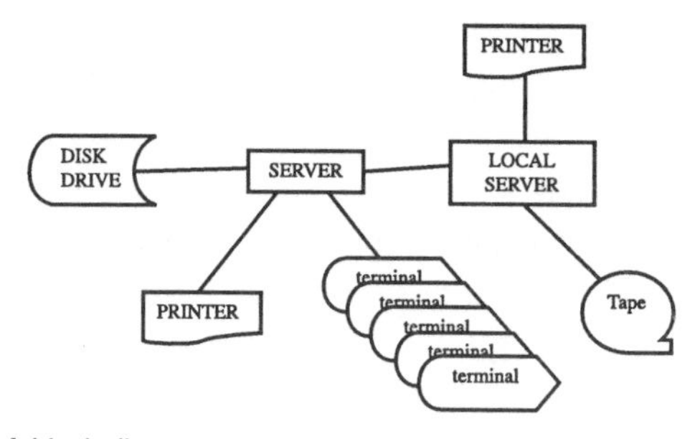

A block diagram.

block driver a device driver that transmits information in blocks.

blocked a state of a task or process in which it is waiting for an occurrence of a specific event to proceed.

blocked transaction a transaction that cannot proceed because it requires a resource currently locked by another transaction.

blocking (**1**) the temporary unavailability of a resource to a request. The term usually refers to an unsuccessful connection set-up or to the unavailability of a path through a switching matrix.

(**2**) a closed queueing network with blocking is one with finite queues.

blocking artifact the visibility in an image of rectangular subimages or blocks after certain types of image processing. Also called blocking effect distortion.

blocking call a call that requires the recipient to be present for the data to be passed.

blocking factor the number of records that fit per block. Given an unspanned organization, the blocking factor is $= \lfloor$ (Block Size/Record Size) $\rfloor$. For a spanned organization, the average number of records per block is taken as the blocking factor.

blocking flow a flow function in which any directed path from s to t contains a saturated edge.

blocking probability the fraction of requests arriving at a system that cannot be served or queued. Blocking can occur in a waiting/loss system if there is no more space left in the queue or in a loss system if all servers are busy. The *blocking probability* is an important grade of service measure.

block I/O the conversion of application reads and writes of records with arbitrary numbers of bytes into reads and writes that can be done based on the block size and alignment required by the underlying hardware.

block matching the process of finding the closest match between a block of samples in a signal and a block of equal size in another signal (or a different part of the same signal) over a certain search range. Closeness is measured by correlation or an error metric such as mean square error. Used in data compression, motion estimation, vector quantization, and template matching schemes.

block multiplexer channel an I/O channel that can be assigned to more than one data transfer at a time. It always transfers information in blocks, with the channel released for competing transfers at the end of a block. *See also* byte multiplexer channel, selector channel.

block structure a syntactic mechanism in which a name has a scope determined by, and limited to, the syntactic context in which it appears. This may or may not be related to the lifetime, or extent, of the value.

block transfer the transmission of a significantly larger quantity of data than the minimum size an interconnect is capable of transmitting, without sending the data as a number of small independent transmissions (the goal being to reduce arbitration and address overhead).

block transform a transform that divides the image into several blocks and treats each block as an independent image. The transform is then applied to each block independently. This occurs in the JPEG standard image compression algorithm, where an image is divided into 8×8 blocks and the DCT is applied independently to each block. Usually the blocks do not overlap each other: that is, they have no signal samples in common.

block truncation coding (BTC) technique whereby an image is segmented into $n \times n$ non-overlapping blocks of pixels, and a two-level quantizer is designed for each block. Encoding is essentially a local binarization process consisting of an $n \times n$ bit map indicating the reconstruction level associated with each pixel. Decoding is a simple process of associating the reconstructed value at each pixel as per the bit map.

blossom an odd length cycle which appears during a matching algorithm on general graphs.

blurring the broadening of image features, relative to those that would be seen in an ideal image, so that features partly merge into one another, thereby reducing resolution. The effect also applies to 1D and other types of signal.

BNF *See* Backus-Naur form.

board the physical structure that houses multiple chips and connects them with traces (busses).

body *See* function body, loop body.

Boltzmann machine an undirected network of discrete valued random variables, where an energy function is associated with each of the links, and for which a probability distribution is defined by the Boltzmann distribution.

Booch diagram a representation of object-oriented design using the Booch method.

Booch method an object-oriented design methodology created by Grady Booch.

Boolean an operator or an expression of George Boole's algebra (1847). A Boolean variable or signal can assume only two values: TRUE or FALSE. This concept has been ported in the field of electronic circuits by Claude Shannon (1938). He had the idea to use Boole's algebra for coding the status of a circuit: TRUE/FALSE as HIGH/LOW as CLOSE/OPEN, etc.

Boolean algebra the fundamental algebra at the basis of all computer operations. A *Boolean algebra* is a ring over a two-valued domain such as 0, 1 or false, true. *See also* Boolean, Boolean OR operation, Boolean AND operation, Boolean complement, Boolean expression, Boolean function, Boolean logic, Boolean operator, Boolean value, Boolean variable.

Boolean AND operation a logical operation on binary variables and constants that produces a one only if both operands are one. It is usually denoted as $\cdot$.

Boolean complement a logical operation on a binary variable or constant that produces a one if the operand is a zero and vice versa. It is usually denoted by a bar over the operand.

Boolean expression an expression consisting solely of Boolean variables and values and Boolean operations, such as and, or, not, implies, etc. Boolean expressions are used for describing the behavior of digital equipment or stating properties/conditions in programs.

Boolean function a function whose range is $\{0, 1\}$. It can be understood to evaluate the truth or falsity of each element of its domain. *See also* binary function, predicate, Boolean expression.

Boolean logic the set of rules for logical operations on binary numbers.

Boolean operator the classical Boolean operators are AND, OR, NOT. There are potentially 16 unique Boolean operators between two binary values, although very few languages implement these operators, other than Exclusive OR. All these operators such as XOR, NAND, NOR, etc. can be easily obtained based on the fundamental ones. In hardware these are implemented with gates. *See,* for example, AND gate.

Boolean OR operation a logical operation on binary variables and constants that produces a one if one or both operands are one. It is usually denoted as $+$.

Boolean value a value which can be set to one of two values, designated true and false. In many implementations of languages, either value may be represented by a set of values in the underlying machine representation.

Boolean variable a variable which holds the value true or false.

BOOM viewer acronym for "binocular omni-oriented monitor"; a system featuring a stereoscopic display mounted on an articulated arm, allowing the display to be positioned to suit the user.

boot to load an operating system.

bootable disk a disk that contains the boot portion of the operating system on it; also called a system disk.

boot area an area on a disk that contains a special code that allows the operating system to start.

Booth's algorithm an algorithm originally developed for the multiplication of signed numbers but now used for high-speed multiplication. Using a large radix, the algorithm reduces the number of multiplicand-multiples that are added and so reduces the number of addition cycles. For radix-r multiplication, the reduction of the number of multiples is by an on-the-fly conversion of the representation of the multiplier from one based on the digit-set 0, 1 to one based on the digit-set $-r, -(r-1), \ldots, 0, \ldots, r-1, r$ for radix-r multiplication and is known as recoding.

booting the process of actually starting the computer's operating system.

boot record (**1**) the structure at the beginning of a hard disk that specifies information needed for the startup and initialization of a computer and its operating system.

(**2**) record kept and displayed by the booting program.

bootstrap (**1**) program, typically stored in a read-only-memory (ROM), which is used to load other programs, most commonly an operating system. From the phrase "to lift oneself by one's bootstraps".

(**2**) to initialize a computer system into a known beginning state by loading the operating system from a disk or other storage to a computer's working memory. This is done by a firmware boot program. Also called boot for short.

(**3**) the process of creating a compiler by writing a small compiler (often in assembly code) that compiles a subset of the language in which the compiler is written. This is used to compile a more powerful compiler written in the subset, which can then compile the compiler written in its own language (although the number of steps required varies from compiler to compiler). A technique more useful in the days before portable compilers.

(**4**) the process of compiling a compiler destined to run on one architecture by using a cross-compiler to compile it on an existing platform, producing executable code for the target system. If the compiler is written in its own language, it can then be compiled on the target system by the compiler created in the first step.

bootstrap code special code stored in the boot area of a disk that allows the operating system to start.

border a string v is a border of a string u if v is both a prefix and a suffix of u. String v is said to be the border of u if it is the longest proper border of u.

borrow-save subtractor an adder in which carries are not assimilated but are saved and assimilated later when they can no longer be saved. Typically used in high-speed dividers, in which saving the carries until the last addition/subtraction cycle enhances performance. Borrow-save subtractors are not as widely used as carry-save adders, which can perform similar actions.

bottleneck the device in a system with the largest utilization. As the number of customers in the system is increased, the bottleneck device will be the one limiting overall system performance.

bottom node the designated node in a lattice that has the property that any path through the lattice starting at the top node (or root node) will reach the bottom node.

bottom-up (**1**) design process by which larger chunks are constructed from smaller ones. An example would be creating a software application by combining together a number of classes from a pre-existing library of classes.

(**2**) a parsing algorithm that constructs the parse tree from the leaves to the root. Bottom-up algorithms include LR and LALR parsers, such as those produced by yacc.

bottom-up analysis an analysis of a context-free grammar which uses rightmost derivation. *See also* LR(k). *Contrast with* top-down analysis.

bottom-up reasoning *See* forward chaining.

boundary a curve that separates two sets of points.

boundary condition a condition representing an extreme point. For example, a computation that fails when presented with zero values to act upon suffers a boundary condition error. Boundary conditions usually introduce special-case code into otherwise straightforward computations, and the lack of the special-case code causes the computation to fail. Boundary conditions may represent discrete conditions (exact values, such as 0, the empty list, the largest possible positive integer, an empty string) or a class of values (for example, computing the tangent of angles very close to 90°). *See also* fencepost error.

boundary representation a paradigm for representing graphical data in terms of the boundaries of the objects involved, e.g., representing a cube as a collection of bounding faces, or a polygon by its edges.

boundary scan a technique for applying scan design concepts to control/observe values of the signal pins of IC components by providing a dedicated boundary-scan register cell for each signal I/O pin.

boundary scan interface a serial clocked interface used to shift in test pattern or test instruction and to shift out test responses in the test mode. *Boundary scan interface* comprises shift-in, shift-out, clock, reset, and test select mode signals.

boundary scan path a technique which uses a standard serial test interface to assure easy access to chip or board test facilities such as test registers (in an external or internal scan path) or local BIST. In particular, it assures complete controllability and observability of all chip pins via shift in and shift out operations.

boundary scan test a technique for applying scan design concepts to control/observe values of signal pins of IC components by providing a dedicated boundary-scan register cell for each signal I/O pin.

boundary value analysis a black box testing method that designs test cases that exercise data boundaries.

bounded distance decoding decoding of an imperfect t-error correcting forward error correction block code in which the corrected error patterns are limited to those with t or fewer errors, even though it would be possible to correct some patterns with more than t errors.

bounding box the smallest regularly shaped box that encloses an object, usually rectangular in shape. Bounding boxes are used to accelerate tests such as visibility or ray-object intersection by providing a pre-test which can eliminate many cases.

bounding volume *See* bounding box.

bounds fault an error that halts the mapper when it detects that the offset requested into an object exceeds the object's size.

box diagram a diagram describing the control flow among components. It consists of rectangles internally subdivided by lines with specific semantics to represent selection conditions, sequential steps, repetitions, etc. Synonymous with program structure diagram, Nassi-Shneiderman chart, Chapin chart. *See also* block diagram, bubble chart, flow chart.

Boyce-Codd Normal Form (BCNF) a relation is in Boyce-Codd normal form (BCNF) if all determinants are candidate keys.

BPR *See* business process reengineering.

branch address the address of the instruction to be executed after a branch instruction if the conditions of the branch are satisfied. Also called a branch target address.

branch and bound algorithm a potentially optimal search technique that keeps track of all partial paths contending for further consideration, always extending the shortest path one level.

branch coverage the process of selecting and executing a series of test cases to ensure that each branch in the code is tested at least once.

branching (1) branching in code execution results when two or more different instruction sequences (branches) may be executed as the result of a data-dependent decision. The degree of branching in code may affect performance optimization by limiting the usefulness of prefetching instructions. *See* instruction prefetch, branch prediction.

(2) a rooted spanning tree in a directed graph, such as the root that has a path in the tree to each vertex.

branching construct a construct in a program which selects one of several potential execution sequences depending on a computation performed. Typical examples from common languages include the if-then-else, case, switch, and guarded condition.

branch instruction an instruction used to modify the instruction execution sequence of the CPU. The transfer of control to another sequence of instructions may be unconditional or conditional based on the result of a previous instruction. In the latter case, if the condition is not satisfied, the transfer of control will be to the next instruction in sequence. In some computer architectures, a "branch" instruction is distinguished from a "jump" instruction in that a "branch" instruction is an instruction using a small number of bits, less than required to represent a complete machine address, to encode a relative offset, and thus can cause a control transfer only a restricted distance from its occurrence. By contrast, a "jump" instruction uses a full machine address and can transfer control to any location. This problem gives rise to the code optimization feature in a compiler which is known as "jump/branch resolution".

branch/jump resolution *See* jump/branch resolution.

branch penalty the delay in a pipeline after a branch instruction when instructions in the pipeline must be cleared from the pipeline and other instructions fetched. Occurs because instructions are fetched into the pipeline one after the other and before the outcome of branch instructions is known.

branch prediction a method of improving processor performance by prefetching instructions that follow the branch point in the code. The branch (*See* branching) believed to be most likely (usually the one most recently executed) will be prefetched. This method can avoid unnecessary pipeline stalls at branches, but requires the capability to undo instructions that have entered the pipeline.

branch target buffer (BTB) a buffer that is used to hold the history of previous branch paths taken during the execution of individual branch instructions. The BTB is used to improve prediction of the correct branch path whenever a branch instruction is encountered. The Pentium BTB is organized as an associative cache memory, with the address of the branch instruction as a tag; it stores the most recent destination address plus a two-bit history field representing the recent history of the instruction.

breadth-first search a control strategy that examines all of the rules or objects on the same level of hierarchy before examining any of the rules or objects on the next lower level.

breakpoint (1) an instruction address at which a debugger is instructed to suspend the execution of a program.

(2) an instruction set by the debugger, overlaying an existing instruction, which causes a control transfer to the debugger.

(3) a critical point in a program, at which execution can be conditionally stopped to allow examination if the program variables contain the correct values and/or other manipulation of data. Breakpoint techniques are often used in modern debuggers which provide nice user interfaces to deal with them. *See* breakpoint instruction.

breakpoint instruction a debugging instruction provided through hardware support in most microprocessors. When a program hits a breakpoint, specified actions occur that save the state of the program, and then switch to another program that allows the user to examine the stored state. The user can suspend the execution of a program, examine the registers, stack, and memory, and then resume the program's execution, which is very helpful in a program's debugging.

Brent-Kung adder a type of carry-lookahead adder; the organization is based on computing carries that are parallel prefixes.

Bresenham's algorithm a technique developed in the framework of raster graphics for generating lines and circles. These algorithms use only integer arithmetic, avoid rounding, and perform an iterative computation of the primitive points by approximating the distance to the nearest pixel center along either the x or y axis. These characteristics make for efficient algorithms.

bridge an edge of a connected graph whose removal would make the graph unconnected.

brightness the perceived intensity of a radiating object.

British museum algorithm a general approach to find a solution by checking all possibilities one by one, beginning with the smallest. The term refers to a conceptual, not a practical, technique where the number of possibilities is enormous. *See also* breadth-first search, brute force search, exhaustive search.

broadcast to send a message to all recipients.

Brownian motion a stochastic process with independent and stationary increments. The derivative of such a process is a white noise process. A Brownian motion process X_t is the solution to a stochastic differential equation of the form

$$\frac{dX}{dt} = b(t, X_t) + \sigma(t, X_t) \cdot W_t \ ,$$

where W_t is a white noise process.

browse to read uncommitted.

brute force search *See* exhaustive search.

BSD acronym for Berkeley Software Distribution, a proprietary version of Unix.

b-spline (**1**) the shortest cubic spline consisting of different three-degree polynomials on four intervals; it can be obtained by convolving four box functions.

(**2**) a multi-segment spline curve representation based on local polynomials having continuity of curve orientation and curvature at the points (knots) where different segments join. Cubic b-splines are popular, but linear, quadratic, quartic, etc. splines are also used. The b in b-spline stands for basis because the b-spline segments are formed from the weighted sum of four local basis functions. The local shape of the spline segment is controlled by four control points; in the case of b-splines, these control points do not lie on the curve itself (i.e., b-splines are not interpolating). One important advantage of b-splines is that the movement of a control point affects only four segments of the curve. B-spline surfaces can be defined from b-spline curves lying in both directions on the surface. Here, 16 non-interpolated control points are needed per patch, but each patch then has tangent and curvature continuity where it joins a neighboring patch. The b-spline is defined over a uniform parameter domain, and is evaluated as a simple polynomial function. More complex forms, such as NURBS, relax these assumptions.

BTB *See* branch target buffer.

BTC *See* block truncation coding.

b-tree a balanced search tree in which every node has between $t-1$ and $2t-1$ children. Insertion and deletion algorithms are modified to ensure the tree remains balanced. For example, on insertion into a full block (i.e., a collision), the block is split and the middle value is propagated up a level to the parent node. All values less than the middle value are placed in the left child of the parent; other values are placed in the right child. This is a good structure if much of the tree is on disk, since the tree height, and hence the number of disk accesses, can be kept small, say one or two. *See also* b*-tree.

bubble chart a data flow diagram in which the elements are represented as circles (bubbles). Relationships among bubbles are represented with arrows and typically represent the flowing of data/information from one bubble to the next.

bucket one or several consecutive disk pages corresponding to one value of a hashing function.

bucket array implementation of a dictionary by means of an array indexed by the keys of the dictionary elements. *See* dictionary.

bucketing method a data organization method that decomposes the space from which spatial data is drawn into regions called buckets. Some conditions for the choice of region boundaries include the number of objects that they contain or their spatial layout (e.g., minimizing overlap or coverage).

bucket sort *See* radix sort.

buddy system a memory allocation strategy which recursively divides allocatable blocks of memory into pairs of adjacent equal-sized blocks called "buddies".

buffer (**1**) a temporary data storage area in memory that compensates for the different speeds at which different elements are transferred within a system. Buffers are used when data transfer rates and/or data processing rates between sender and receiver vary, for instance, a printer buffer, which is necessary because the computer sends data to the printer faster than the data can be physically printed. *See* buffered input/output.

(**2**) a unit for temporary storage of requests. Requests to be stored can be data (in a computer), call or connection requests (in a communication system), etc. A buffer is characterized by its size (in units of number of requests or sum of request sizes) and by its queueing discipline (e.g., FCFS, LCFS, etc.).

buffered input/output input/output which transfers data through a "buffer" or temporary storage area. The main purpose of the buffer is to reduce time dependencies of the data and to decouple input/output from the program execution. Data may be prepared or consumed at an irregular rate, whereas the transfer to or from disk is at a much higher rate, or in a burst.

A buffer is used in "blocked files", where the record size as seen by the user does not match the physical record size of the device.

buffering (**1**) the process of moving data into or out of buffers or to use buffers to deal with input/output from devices. *See* buffer, buffered input/output.

(**2**) assigning buffer space for the duration of the execution of a computer program.

bug simplistically, an error in a program. However, many behavioral aspects of a program may be labeled "bugs" when they are simply behavior the programmer does not wish to have; for example, the failure of columns of numbers to line up in some desired manner. A "bug" can represent anything from an error in the fundamental requirements to an incorrect coding of the algorithm. Often considered an unacceptable term from a software engineering perspective (the correct term is "fault"), the term is attributed to Grace Hopper's discovery of a short-circuit created by a moth in the ENIAC computer.

building block schemata with low order, short defining length and above-average fitness values. Due to the high fitness values of these schemata and their capability to avoid the disruptive effects of crossover and mutation, the number of their representatives in the population will quickly increase. More precisely, the schema theorem states that this number will increase at an exponential rate. *See also* schema theorem.

building blocks hypothesis the hypothesis that the genetic algorithm searches by first detecting biases toward higher fitness in some low-order schemas (those with a small number of defined bits) and converging on this part of the search space. Over time, it then detects biases in higher-order schemas by combining information from low-order schemas via crossover and eventually converges on a small region of the search space that has high fitness. The building blocks hypothesis states that this process is the source of the genetic algorithm's power as a search and optimization method.

built-in predicates for computations in rule bodies often special built-in predicates are needed in practical applications, e.g., the arithmetic predicates $+$, $-$, $*$, $\div$, or the comparison predicates $>$, $<$, $\geq$, $\leq$, $=$, $\neq$.

built-in self-test (BIST) special hardware embedded into a device (VLSI chip or a board) used to perform self testing. On-line BIST assures testing concurrently with normal operation (e.g., accomplished with coding or duplication techniques). Off-line BIST suspends normal operation and it is carried out using built-in test pattern generator and test response analyzer (e.g., signature analyzer).

bulk transfer a class of network traffic that involves the reliable transmission of large volumes of data without a strict delay requirement.

bump map a pattern of surface normal displacements, simulating the undulations of a bumpy surface that is to be mapped onto a geometric surface during rendering.

bump mapping a technique used to increase the realism of a surface by changing how light reflects from that surface. Usually, the surface normal at a given point on a surface is used in the calculation of the brightness of the surface at that point. In bump mapping, the true surface normal $\mathbf{n}$ is perturbed a small amount $\delta\mathbf{n}$ as a function of position on the surface. The perturbation can be regular, so as to give a regular textured shape to the surface, or it can be random, so as to increase the natural appearance of the surface. Part of what gives this technique its appeal is that the original surface maintains its original (usually smooth) shape, and the bump-mapping distortion is specified by a compact function of shape. This is usually much simpler and more compact than specifying the surface texture by explicitly representing the textured surface.

burden rate the cost that must be paid on top of a salary for an employee. As software cost estimator this must be taken into account.

burn-in test the process of letting a computer or other electronic device run for some extended period of time after it has been assembled to flush out major problems quickly.

burstiness a measure for the variability of a process, e.g., a rate process. There are several different mathematical definitions for burstiness, the most common being the ratio of peak rate to mean rate of a process.

burstiness factor used in traffic description, the ratio of the peak bit rate to the average bit rate.

burst refresh in DRAM, carrying out all required refresh actions in one continuous sequence — a *burst. See also* distributed refresh.

burst transfer the sending of multiple related transmissions across an interconnect, with only one initialization sequence that takes place at the beginning of the burst.

bursty arrival process an arrival process with a greater variance than what a Poisson process with the same mean rate would have.

bursty traffic traffic resulting from a bursty arrival process or generated with a high variance of interarrival times or packet sizes.

bus (**1**) a data path connecting the different subsystems or modules within a computer system. A computer system will usually have more than one bus; each bus will be customized to fit the data transfer needs between the modules that it connects.

(**2**) a conducting system or supply point, usually of large capacity. May be composed of one or more conductors, which may be wires, cables, or metal bars (busbars).

bus acquisition the point at which a bus arbiter grants bus access to a specific requestor.

bus arbiter the unit responsible for choosing which subsystem will be given control of the bus when two or more requests for control of the bus occur simultaneously. Some bus architectures, such as Ethernet, do not require a bus arbiter.

bus arbitration function performed by a bus controller to resolve simultaneous competing requests for control of the bus and select one candidate to become bus master. Arbitration is usually by a combination of hard-wired requester priority and the physical position along the bus. Refer to *bus priority*.

bus bandwidth the data transfer rate in bits per second or bytes per second. It is approximately equal to the width of the data bus, multiplied by the transfer rate in bus data words per second. Thus, a 32-bit data bus, transferring 25 million words per second (40 ns clock) has a bandwidth of 800 Mb/s.

The useful bandwidth may be lowered by the time to first acquire the bus and possibly transfer addresses and control information. The bus bandwidth is the transfer rate which it is guaranteed that no user will exceed.

bus broadcast a bus-write operation which is intended to be recognized by more than one attached device.

bus controller the logic which coordinates the operation of a bus. A device connected to the bus will issue a *bus request* when it wishes to use the bus. The controller will arbitrate among the current requests and grant one requester access. The bus controller also monitors possible errors, such as use of an improper address, a device not releasing the bus, and control errors.

bus cycle the sequence of steps involved in a single bus operation. A complete bus cycle may require that several commands and acknowledgments are sent between the subsystems in addition to the actual data that is sent. For example,

1. The would-be bus master requests access to the bus.

2. The bus controller grants the requester access to the bus as bus master.

3. The bus master issues a read command with the read address.

4. The *bus slave* responds with data.

5. The master acknowledges receipt of the data.

6. The bus master releases the bus.

The first two steps may be overlapped with the preceding data transfer. *See* bus controller, bus master.

bus driver the circuits that transmit a signal across a bus.

bus error a problem (external to the CPU) that causes an inability to access one or more words of memory. *See also* address error.

bus grant an output signal from a processor indicating that the processor has relinquished control of the bus to a DMA device.

bus hierarchy a network of buses linked together (usually multiple smaller buses connected to one or more levels of larger buses), used to increase the number of elements that may be connected to a high-performance bus structure.

Bush, Vannevar (1890–1974) Born: Everett, Massachusetts.

Bush is best know as the developer of early electro-mechanical analog computers. His "differential analyzer", as it was called, arose from his position as a professor of power engineering at the Massachusetts Institute of Technology. Transmission problems involved the solution of first and second order differential equations. These equations required long and laborious calculations. His interest in mechanical computation arose from this problem. Bush's machines were used by the military during WW II to calculate trajectory tables for artillery. Vannevar Bush was also responsible for inventing the antecedent of our modern electric meter. He was also scientific advisor to President Roosevelt on the Manhattan Project.

bus idle the condition that exists when the bus is not in use.

business process reengineering (BPR) a business-focused, integrated, and radical approach to redesigning a business system.

business risk one of a set of potential business problems or occurrences that may cause the project to fail.

business-to-business the common modality to specify the business transaction between two companies. It specifies the transactions in the corporate market. More recently, the term is used in the field of electronic commerce for stating that the WWW is defined for supporting the business transactions between companies instead of the end-users. Presently the business-to-business market is growing very fast.

bus interface unit in modern CPU implementations, the module within the CPU directly responsible for interactions between the CPU and the memory bus.

bus line one of the wires or conductors which constitute a bus. A *bus line* may be used for data, address, control, or timing.

bus locking the action of retaining control of a bus after an operation which would normally release the bus at completion. In the manipulation of memory locks, a memory read must be followed by a write to the same location with a guarantee of no intervening operation. The bus must be locked from the initial read until after the update write to give an indivisible read/write to memory.

bus master a bus device whose request is granted by the bus controller and thereby gains control of the bus for one or more cycles or transfers. The bus master may always reside with one subsystem, or may be transferred between subsystems, depending on the architecture of the bus control logic. *See also* bus controller, bus cycle.

bus owner the entity that has exclusive access to a bus at a given time.

bus parking a priority scheme which allows a bus master to gain control of the bus without arbitration.

bus phase a term applying especially to synchronous buses, controlled by a central clock, with alternating "address" and "data" transfers. A single transfer operation requires the two phases to transfer first the address and then the associated data. Bus arbitration may be overlapped with preceding operations.

bus priority rules for deciding the precedence of devices in having bus requests honored.

Devices issue requests on one of several bus request lines, each with a different bus priority. A high priority request then "wins" over a simultaneous request at a lower priority.

The request grant signal then "daisy chains" through successive devices along the bus or is sent directly to devices in appropriate order. The requesting device closest to the bus controller then accepts the grant and blocks its propagation along the bus.

Buses may have handle interrupts and direct memory accesses with separate priority systems.

bus protocol **(1)** a set of rules which two parties use to communicate.

(2) the set of rules which define precisely the bus signals that have to be asserted by the master and slave devices in each phase of a bus operation.

bus request an input signal to a processor that requests access to the bus; a hold signal. Competing *bus requests* are resolved by the bus controller. *See* bus controller.

bus slave a device which responds to a request issued by the bus master. *See* bus master.

bus snooping the action of monitoring all traffic on a bus, irrespective of the address. *Bus snooping* is required where there are several caches with the same or overlapping address ranges. Each cache must then "snoop" on the bus to check for writes to addresses which it holds; conflicting addresses may be updated or may be purged from the cache. Bus snooping is also useful as a diagnostic tool.

bus state triggering a data acquisition mode initiated when a specific digital code is selected.

bus tenure the time for which a device has control of the bus, so locking out other requesters. In most buses the bus priority applies only when a device completes its tenure; even a low priority device should keep its tenure as short as possible to avoid interference with higher priority devices. *See* bus priority.

bus transaction the complete sequence of actions in gaining control of a bus, performing some action, and finally releasing the bus. *See* bus cycle.

bus watching *See* bus snooping.

bus width the number of data lines in a given bus interconnect.

busy pertaining to a system or component that is in use.

busy hour call attempts (BHCA) the average number of call attempts expected per hour on a telephone switch during the busiest time of the day, busy hour, peak hour: Sixty consecutive minutes in a day in which the mean of a traffic measure has its maximum. The traffic measure under consideration can be, e.g., traffic volume or call rate. Different traffic measures often have different peak hours.

busy waiting a processor state in which it is reading a lock and, finding it busy, it repeats the read until the lock is available, with attempt to divert to another task. The name derives from the fact that the program is kept busy with this waiting and is not accomplishing anything else while it waits. The entire "busy loop" may be only two or three instructions. *Busy waiting* is generally deplored because of the waste of processing facilities.

byte in most computers, the unit of memory addressing and the smallest quantity directly manipulated by instructions. The term "byte" is of doubtful origin, but was used in some early computers to denote any field within a word (e.g., DEC PDP-10). Since its use on the IBM "Stretch" computer (IBM 7030) and especially the IBM System/360 in the early 1960s, a byte is now generally understood to be 8 bits, although 7 bits is also a possibility. *See also* word, halfword, longword, quadword.

byte code a sequence of machine instructions for a virtual machine, encoded as a sequence of bytes. Traditionally, a byte code has been executed by an interpreter running on the actual machine, although nothing precludes building actual hardware that executes the byte codes directly. Typically used to create a platform-independent representation of object code. *Contrast with* threaded code.

byte multiplexer channel an I/O channel that can be assigned to more than one data transfer at a time and can be released for another device following each byte transfer. (In this regard, it resembles a typical computer bus.) Byte multiplexing is particularly suited to lower speed devices with minimal device buffering. (IBM terminology.) *See also* selector channel, multiplexer channel.

byte serial a method of data transmission where bits are transmitted in parallel as bytes and the bytes are transmitted serially. For example, the Centronics-style printer interface is byte-serial.

Byzantine agreement *See* Byzantine general's problem.

Byzantine general's problem the problem of reaching a consensus among distributed units if some of them give misleading answers. The original problem concerns generals plotting a coup. Some generals lie about whether they will support a particular plan and what other generals told them. What percentage of liars can a decision-making algorithm tolerate and still correctly determine a consensus?

C

C a language developed at AT&T Bell Laboratories in the 1970s, a derivative of BCPL.

C&P *See* capture and playback.

C++ a language developed at AT&T Bell Laboratories in the 1980s, an object-oriented variant of the C language. A language based loosely on the principles of SIMULA-67.

cable drive in robotic manipulators, transmission of motive forces from actuators to links via wire cables.

cache (**1**) an intermediate memory store having storage capacity and access times somewhere in between the general register set and main memory. A cache is usually invisible to the programmer, and its effectiveness comes from being able to exploit program locality to anticipate memory-access patterns and to hold closer to the CPU; most accesses to main memory can be satisfied by the cache, thus making main memory appear to be faster than it actually is.

There may be more than one cache; if so, they are traditionally numbered by their nearness to the processor. The closest, and usually smallest but fastest, cache is called the "L1 cache", the next one out is called the "L2 cache", and so on. A compiler may be able to take advantage of the presence of a cache to optimize the performance of code.

See code cache, data cache, direct mapped cache, fully associative cache, set-associative cache, cache line, cache line size, and unified cache.

(**2**) a data structure which holds a (typically small) number of values, used to expedite access to larger data structures. When a cache is filled and a new element must be placed in it, a cache replacement algorithm is used to determine which element to discard.

cache aliasing a situation where two or more entries (typically from different virtual addresses) in a cache correspond to the same address(es) in main memory. Considered undesirable, as it may lead to a lack of consistency (coherence) when data is written back to main memory.

cache associativity the number of cache entries that can be searched concurrently.

cache block the number of bytes transferred as one piece when moving data between levels in the cache hierarchy or between main memory and cache. The term *line* is sometimes used instead of block. Typical block size is 16-128 bytes and typical cache size is 1-256 KB. The block size is chosen so as to optimize the relationship of the "cache miss ratio", the cache size, and the block transfer time.

cache coherence the problem of keeping consistent the values of multiple copies of a single variable, residing either in main memory and cache in a uni-processor, or in different caches in a multi-processor computer. In a uni-processor, the problem may arise if the I/O system reads and writes data into the main memory, causing the main memory and cache data to be inconsistent, or if there is aliasing. Old (stale) data could be output if the CPU has written a newer value in the cache, and this has not been transported to the memory. Also, if the I/O system has input a new value to main memory, new data would reside in main memory, but not in the cache.

cache coherence protocol a mechanism to maintain data coherence among multiple caches so that every data access will always return the latest version of that datum in the system.

cache consistency in a system with multiple processors, each with its own local cache, the caches are consistent if each cache always contains exact copies of remote data. The problem of keeping local cache copies up to date with changes in remote data is known as the cache consistency problem.

cache hit when the data referenced by the processor is already in the cache.

0-8493-2691-5/01/$0.00+$.50

cache hit ratio the percentage of time in which a requested instruction or data is actually in the cache.

cache line (**1**) the fundamental quantum of caching. From the notion of a hardware cache, in which some number of bits, bytes, or other storage units are transferred to the next outer cache as a single unit of information. This unit of information is a cache line. Each level of cache in a multilevel system may have its own cache line size.

(**2**) a block of data associated with a cache tag.

cache line size the number of bits, bytes, or other storage units that can be stored in a single cache line. Knowledge of the cache line size and the cache replacement algorithm can allow a compiler to optimize the code to work with, rather than against, the cache. Speedups by factors of 20 or more have been seen for FORTRAN programs accessing array data.

cache memory *See* cache.

cache miss a reference by the processor to a memory location currently not housed in the cache.

cache replacement when a "cache miss" occurs, the block containing the accessed location must be loaded into the cache. If this is full, an "old" block must be expelled from the cache and replaced by the "new" block. The "cache replacement algorithm" decides which block should be replaced. An example of this is the Least Recently Used (LRU) algorithm, which replaces the block that has gone the longest time without being referenced.

cache replacement algorithm a means of determining which element of a cache to replace when the cache is full and a new element must be placed in the cache. Most caches are commonly managed by using a Least Recently Used algorithm, but other techniques including hash associative and set associative can be used.

cache set *See* set-associative cache.

cache synonym *See* cache aliasing.

cache tag a bit field associated with each block in the cache. It is used to determine where (and if) a referenced block resides in the cache. The tags are typically housed in a separate (and even faster) memory (the "tag directory"), which is searched in for each memory reference. In this search, the high order bits of the memory address are associatively compared with the tags to determine the block location. The number of bits used in the tag depends on the cache block "mapping function" used: "Direct-mapped", "Fully associative" or the "Block-set-associative" mapped cache.

cactus stack a representation of a stack in which multiple threads of control share the stack-defined namespace hierarchy; so named because a diagram of the stack relationships resembles a saguaro cactus.

CAD *See* computer-aided design.

calculus of communicating system (CCS) Robin Milner's algebraic theory to formalize the notion of concurrent computation. This process algebra is used to describe process behavior, mainly used in the study of parallelism. It has operators for the parallel composition of processes, for synchronization, to hide events, etc. Programs written in CCS can be compared using the notion of observational equivalence that checks if two processes have the same external behavior. *See* process algebra, CSP.

calendar time synonymous with real-time.

calibration process of adjusting system parameters to optimize accuracy or to ensure conformity to a standard.

call *See* function call.

call-by-address *See* call-by-reference.

call-by-name a mechanism defined by the Algol-60 specification, stating that the appearance of a formal parameter of a function in a function body would evaluate the actual parameter expression each time such a use occurred.

This led to interesting problems in side effects. *See also* call-by-reference, call-by-value.

call-by-name reduction *See* normal-order reduction.

call-by-reference a mechanism for parameter passing in many languages which states that the value of the actual parameter is not passed to the called function, but rather a reference (memory address) to the value is passed instead. Use of the actual parameter in the function body uses or modifies the value thus referenced. Also called call-by-address. *See also* call-by-name, call-by-value, call-by-value-result.

call-by-result a mechanism for parameter passing in many languages which states that the actual parameter must reference a value that can be modified. In the body of the function, the appearance of the formal parameter can only be as the target of an assignment (or it can be passed as a call-by-result parameter to another function). The formal parameter represents a local variable. Upon return from the function, the contents of the local variable representing the formal parameter are copied to the location specified by the actual parameter. *See also* call-by-value, call-by-value-result.

call-by-value a mechanism for parameter passing in many languages that states that the expression defining the value of an actual parameter is computed once and the resulting value is passed to the function being called. In the body of the function, the appearance of the formal parameter represents the copy of the value that has been made. Some languages do not permit the value represented by the formal parameter to be modified in the body of the function.

call-by-value-result a mechanism for parameter passing in many languages that states that the actual parameter to a function must reference a value that can be modified. A copy of this value is made and is passed into the function that is called. Upon return from the function, the value of the formal parameter is copied back to the location from which it was originally copied. While this appears to be similar to call-by-reference, its behavior with respect to side effects is quite different, particularly when aliasing occurs, and it can also support alternate value representations between the actual parameter type and formal parameter type. *See also* call-by-name, call-by-reference, call-by-value.

callee a function which has been called.

caller the code which calls a function.

call graph (1) a human-readable representation of which functions call which other functions. Useful for analyzing complex programs, to reverse-engineer, or otherwise understand their behavior.

(2) a representation of which functions call which other functions used within compilers to determine feasible global optimizations.

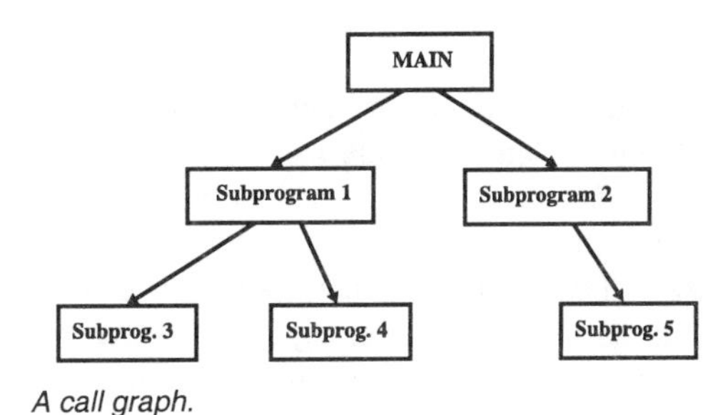

A call graph.

CAM acronym for content-addressable memory or computer-aided manufacturing. *See* associative memory, computer-aided manufacturing.

CAMAC acronym for computer automated monitor and control — an internationally accepted set of standards for electronic instrumentation, which specifies mechanical, electrical, and functional characteristics of the instrument modules.

camera (1) a device for acquiring and storing images, by converting light to electrical signals or by allowing light to react with chemicals on a film. Cameras may operate in optical, infrared, or other wavelength bands.

(2) virtual viewpoint in world space with position and view direction to provide a view of a

scene in the same way as a photographer would position a camera.

camera calibration a process in which certain camera parameters, or equivalently some quantities which are required for determination of the perspective projection on an image plane of a point in the 3D world, are calculated by using the known correspondence between some points in the 3D world and their images in the image plane.

camera model a mathematical model by which the perspective projection on an image plane of a point in the 3D world can be determined.

canceled project a software project that is canceled and never completed. This is a fairly common issue among large projects. About 25% of large projects are cancelled.

candela derived from candle and denoted by the symbol "cd", it is the basic SI unit of luminous intensity. It is defined as the radiation intensity, in a perpendicular direction, of a surface of 1/600000 square meter of a black body at the temperature of freezing platinum under a pressure of 101325 newtons per square meter.

candidate consistency testing the stage of two-dimensional matching where it is checked whether a candidate occurrence of the pattern is checked against the "witness" table.

candidate key any key in a relation that has more than one key.

canned transaction a standard query or update that is run on a regular basis.

Canny operator an edge detector devised by John Canny as the optimal solution to a variational problem with three constraints. The general solution obtained numerically can be approximated in practical contexts by the first derivative of a Gaussian. Canny operator usually refers to the extension to two dimensions of this approximation: i.e., to use a set of oriented operators whose orthogonal cross-sections are a Gaussian and the derivative of a Gaussian. Its advantage is its capability for allowing edges and their orientations to be detected to sub-pixel accuracy. It uses a convolution with a Gaussian to reduce noise and a derivative to enhance edges in the resulting smoothed image. The two are combined into one step — a convolution with the derivative of a Gaussian. A hysteresis thresholding stage is included to allow closed contours to remain closed.

canonical if E and F form a decomposition of image W, meaning $E \cup F = W$, then the pair (E, F) is said to be canonical.

canonical complexity classes the classes defined by logarithmic, polynomial, and exponential bounds on time and space, for deterministic and nondeterministic machines. These are the most central to the field, and classify most of the important computational problems.

canonical form given a set of queries Q, a subset of this set of queries, C, is said to be the canonical form of Q if every query in Q is equivalent to a query in the subset C.

Cantor middle third argument a recursive mathematical procedure involving the removal of the middle third of a line segment.

Cantor set the result of applying the Cantor middle third argument.

canvas a two-dimensional region of graphics information. The canvas may be displayed on screen or be recorded in off-screen display memory.

capability (**1**) an object that contains both a pointer to another object and a set of access permissions that specify the modes of access permitted to the associated object from a process that holds the capability.

(**2**) a unique, global name for an access right to an object.

capability maturity model (CMM) a five-layer model against which an industry can evaluate its organizational maturity with respect to software development. The levels are 1. Initial, 2. Repeatable, 3. Defined, 4. Managed, and 5. Optimizing.

capacity (1) in performance evaluation, the capacity of a system is the maximum rate at which it can perform useful work under the given set of constraints. These constraints could be, for example, the blocking probability or the end-to-end delay of packets in a system.

(2) in communications, the capacity of a channel is the maximum bit rate at which information can be transmitted reliably.

(3) in graph theory, the maximum amount of flow that is allowed to be sent through an edge or a vertex.

capacity miss a category of cache misses denoting the case where the cache is not large enough to hold all blocks needed during execution of a program. *See also* conflict miss, cold start miss.

capture and playback (C&P) a tool for system testing. It allows the capture of all the actions performed by (i) the user on the machine interface (for instance, via mouse, keyboard, etc.) and by (ii) the computer towards the user (visualization and changes on the screen, disk access, etc.). The actions and the response of the program are typically registered on disk by using a specific language. Once the description is saved, it can be reapplied on the system simulating the human's actions (the playback). This allows the comparison of the expected results against the obtained one, and, thus, the performing of the regression testing at low cost. C&P solutions can be intrusive and non-intrusive. The first are applications running on a separate computer that acquire the actions by means of wiring connections on the cables of mouse, video, keyboard, and disk. The second are run directly on the same computer as resident tasks, thus reducing the estimation precision and partially influencing the system behavior, since the CPU is shared with the application under test.

capture error human error caused by habituation to a behavior sequence that is no longer valid; for example, a soccer player accustomed to the sequence "see the ball, run to the ball, kick the ball" would commit an error if recreating that sequence while playing baseball.

capture register internal register which, triggered by a specified internal or external signal, stores or "captures" the contents of an internal timer or counter.

card computer component constructed of plastic laminates that holds dozens or even hundreds of chips and other electronic parts such as resistors, capacitors, transistors, and the like. Also called board.

cardinality ratio specifies the number of relationship instances in which an entity can participate.

carrier the socket into which a chip is inserted. Sometimes called a chip pack.

carry overflow signal which occurs when the sum of the operands at the inputs of the adder equals the base. A binary adder, adding $1 + 1$ will produce a sum of 0 and carry of 1.

carry bit *See* carry.

carry-bypass adder *See* carry-skip adder.

carry chain in an adder, the logic used to pass along carries.

carry-completion adder an asynchronous adder in which the propagation of carries is monitored and the addition is considered complete as soon as there are no more carries to be propagated. This is nominally faster than a corresponding synchronous adder, in which time must be always allowed for the worst-case carry-propagation.

carry flag a single bit of state in the computer that records the state of a previous arithmetic operation.

carry look-ahead adder an adder in which carries are determined prior to the addition proper. An adder stage therefore need not wait for a carry from a lower stage, and the addition process is thus speeded up.

carry-prediction adder *See* carry look-ahead adder.

carry-propagate adder (CPA) an adder in which carries are assimilated as they are produced, in contrast with a carry-save adder in which carries are saved until the very end of the operation at hand.

carry-ripple adder *See* ripple-carry adder.

carry-save adder (CSA) an adder in which carries are not assimilated but are saved and assimilated later when they can no longer be saved. Typically used in high-speed multipliers and dividers, in which saving the carries until the last addition/subtraction cycle enhances performance.

carry-select adder an adder in which two summands are concurrently produced, one under the assumption that the carry-in to the adder (or adder stage) is 0 and one under the assumption that the carry-in is 1. One of the summands is then selected when the carry-in to the adder (or adder stage) is known.

carry-skip adder an adder in which the bit-stages are grouped into blocks, and carries are propagated between blocks, in such a way that carries may skip blocks (that would produce a carry) instead of rippling through. The skipping is faster than the rippling and so speeds up the adder.

CART (**1**) acronym for classification and regression trees. CART is a method of splitting the independent variables into small groups and fitting a constant function to the small data sets. In categorical trees, the constant function is one that takes on a finite small set of values (e.g., Y or N, low or medium or high). In regression trees, the mean value of the response is fit to small-connected data sets.

(**2**) a decision tree technique used for classification of a dataset. Provides a set of rules that you can apply to a new (unclassified) dataset to predict which records will have a given outcome. Segments a dataset by creating two-way splits.

Cartesian coordinates a common system of representing a point in two or more dimensions using an ordered set corresponding to its projection on a spanning orthogonal base set. Commonly encountered Cartesian coordinate systems are the XY plane 2D coordinate system (row, column), 2D image coordinate system, and XYZ 3D scene coordinate system. 3D coordinates in graphics are usually specified with x and y being aligned with x and y on the screen, +x is to the right, +y is upwards, and +z goes into the space "behind" the screen. This is a left-handed coordinate system with the property that most z-values are thereby positive. This is why z-buffers are called z-buffers when they are actually depth-value buffers.

Cartesian product given sets A and B, if $a \in$ A and $b \in$ B, then the ordered pair (a, b) is an element of the Cartesian product of A and B, i.e., $(a, b) \in A \times B$.

Cartesian space mathematical model of space using a 3-axis reference frame; using the Cartesian system, the position of any object may be described as points along the three axes and its orientation by rotations around those axes.

cartridge tape unit sequential storage device used for backups and long-term storage of bulk data.

cascaded error an error that is an artifact of a previous error. In the case of syntax errors, the consequence of the parser not being placed in a state that would allow parsing to resume correctly.

cascading rollback situation in which the rollback of one transaction causes rollback of one or more other transactions (the cascading rollbacks). Cascading rollback can only occur with optimistic concurrency control.

CASE *See* computer-aided software engineering.

case (**1**) typographical convention wherein symbols have alternate representations used by convention in various contexts. *See also* case folding, case-insensitive, case-preserving, case-sensitive.

(**2**) a control construct by which the sequence of operations is determined by the computation of a value, typically of an enumerable type such as integer, character, and the like, and said value is used to select the next computation performed. *See also* switch.

(**3**) the actual statement selected according to definition (2). *See also* otherwise, default.

case based reasoning (CBR) reasoning concerning the nature of a situation based upon an empirical inquiry that investigates a contemporary phenomenon within its real-life context especially when the boundaries between the phenomenon and context are not clearly evident. It is a technique for problem solving that uses previous examples for solving the current problem.

case folding an operation which converts the alphabetical characters in a name to a canonical representation, for example, all upper-case or all lower-case.

case-insensitive a property that indicates that upper-case and lower-case letters are not distinguished as distinct characters in the definition of a name. Many languages ignore upper/lower-case differences in recognizing names in the language. Other languages are case sensitive or case preserving.

case-preserving a property that indicates that although upper-case and lower-case letters are not distinguished as distinct characters in the definition of a name, the name is not transformed internally to a canonical representation (all upper-case or all lower-case), so its appearance in various output listings (such as cross-reference listings) preserves the case distinction.

case-sensitive a property that indicates that upper-case and lower-case letters are treated as distinct characters in the definition of a name. Many languages will distinguish names that differ in case as distinct names. Other languages are case-insensitive or case-preserving.

case study a typical case for verifying the functionalities of a system or the capabilities of an approach.

CASE tool *See* computer-aided software engineering tool.

cast *See* type cast.

casting down the hierarchy an operation by which an object of a general class is declared to be an object of a more restricted subclass. In some languages this can be done even if the object is not an instance of the subclass. Such languages are not type-safe under this casting operation.

casting up the hierarchy an operation by which an object of a specific class is declared to be an object of a more general class.

CAT *See* computer-assisted teleoperation.

cataleptic failure *See* catastrophic failure.

catalog a named collection of schemas. The name of a catalog is used to qualify the names of the schemas in that catalog.

catastrophic code a convolutional code in which a finite number of code symbol errors can cause an unlimited number of decoded symbol errors.

catastrophic encoder a convolutional encoder with at least one loop in the state-transition diagram with zero accumulated code symbol weight, at least one nonzero information symbol, and not visiting the zero state. After decoding, a finite number of (channel) errors can result in an infinite number of errors (catastrophic error propagation).

catastrophic error propagation when the state diagram contains a zero distance path from some nonzero state back to the same state, the transmission of a 1 causes an infinite number of errors. *See* catastrophic encoder.

catastrophic failure a sudden failure that results in a complete inability to perform all required functions of an item. The use of the word "catastrophic" is unfortunate since it confuses failure mode with severity. Software is particularly prone to sudden complete failure, due to the lack of continuity of cause and effect in digital systems: since they operate according to logical rules, not physical laws, a very small change in input or internal state can produce a vast change in output. *See also* wild failure. Synonymous with cataleptic failure.

catch an operation in languages that support a structured nonlocal control transfer in a nested control hierarchy. The control transfer is initiated by a throw operation and intercepted by a catch. Ada and C++ are instances of such languages.

catenation symbols strung together to form a larger sequence, such as the characters in a word and the digits in a number.

cathode ray tube (CRT) a vacuum tube using cathode rays to generate a picture on a fluorescent screen. These cathode rays are in fact the electron beam deflected and modulated, which impinges on a phosphor screen to generate a picture according to a repetitive pattern refreshed at a frequency usually between 25 and 72 Hz.

Cauchy distribution the density function for a Cauchy distributed random variable X is

$$f_X(x) = \frac{1}{1 + x^2} .$$

Note that the moments for this random variable do not exist, and that the cumulative distribution function is not defined. *See also* probability density function, moment.

cause-effect graphing a black-box testing method.

caustic the effect given when light is transmitted through a specular surface and then strikes a diffuse surface. If the specular surface is of high curvature, the light will tend to be focused. When this effect is taken into account, rendered scenes involving liquids or glass are much more photorealistic. Caustics can also arise when light is reflected from a specular surface. The classic example is the caustic on the surface of a liquid. Refraction may make it happen too. The caustic shape is the envelope of the reflected rays.

cautious waiting a currency control algorithm that uses a mixture of waiting and aborting transactions.

CAVE *See* computer augmented virtual environment.

CBR *See* case based reasoning.

CCD *See* charge-coupled device.

CCDF *See* complementary cumulative distribution function.

CCD memory *See* charge-coupled-device memory.

CCITT *See* Comité Consulatif International Télégraphique.

CCS *See* calculus of communicating system.

CD *See* compact disk.

CDF *See* cumulative distribution function.

CD-I *See* compact disk interactive.

CDMA *See* code division multiple access.

CDR *See* critical design review.

CD-ROM a read-only compact disk. *See* compact disk.

CDV *See* cell delay variation.

cell decomposition an approach to path planning where the obstacles are modeled as polygons and the free space is decomposed into cells such that a straight line path can be generated between any two points in a cell.

cell delay variation (CDV) the variation of the end-to-end delay occurring in a packet switching system. ITU [I.371] and the ATM

Forum [TMS4.0] give definitions both for a one-point and a two-point ATM cell delay variation, measured at both ends of a connection (two-point) or at one end (one-point) using a reference clock. Also called delay variation.

cell probe model model of computation where the cost of a computation is measured by the total number of memory accesses to a random access memory with $\lceil \log n \rceil$ bits cell size. All other computations are not accounted for and are considered to be free.

cell switching means of switching data among the ports (inputs and outputs) of a switch such that the data is transferred in units of a fixed size.

cellular automaton a computational paradigm for an efficient description of SIMD massively parallel systems. Cellular automata are designed from many discrete cells, usually assembled in one- or two-dimensional regular arrays, each of which is a standard finite state machine. Each cell may change its state only at fixed, regular intervals, and only in accordance with fixed rules that depend on the cells' own values and the values of neighbors within a certain proximity (usually two for one-dimensional, and four for two-dimensional cellular automata). Cellular automata are a base of cellular computers; fine grain systems that are usually data-driven and used to implement neural networks, systolic arrays, and SIMD architectures.

center of attention that portion of the environment to which the preponderance of human information processing resources is devoted.

center of interest a point in space toward which the virtual camera is always aimed.

center of projection the point within a projector from which all the light rays appear to diverge. When a 3-dimensional object is captured as an image by a 2-dimensional camera, lines may be drawn from each point on the object onto the image, and the point behind the image plane where these lines intersect is the center of projection; the size of an image is inversely proportional to the distance from the *center of projection* to the object.

center of viewpoint *See* center of projection.

center weighted median a type of weighted median filter that is found by repeating only the value at the window center.

central absolute moment for random variable x, the pth central absolute moment is given by $E[|x - E[x]|^p$. *See also* central moment.

centralized arbitration a bus arbitration scheme in which a central bus arbiter (typically housed in the CPU) accepts requests for and gives grants to any connected device wishing to transmit data on the bus. The connected devices typically have different priorities for bus access, so if more than one device wants bus access simultaneously, the one with the highest priority will get it first. This prioritization is handled by the bus arbiter.

centralized mechanism a mechanism in which decisions are made on the local system, based on data located on that system.

centralized memory in a multiprocessor system, memory that is centralized and equally accessible to all processors, as opposed to memory that is distributed among the processors.

central limit theorem in probability, the theorem that the density function of some function of n independent random variables tends towards a normal curve as n tends to infinity, as long as the variances of the variables are bounded: $0 < \sigma \leq v_i \leq \gamma < \infty$. Here σ and γ are positive constants, and v_i is the variance of the ith random variable. *See also* Gaussian distribution.

central moment for random variable X the nth central moment is given by

$$E\left[(X - m)^2\right] = \int_{-\infty}^{\infty} (x - m)^2 f_X(x)dx$$

where $f_X(x)$ is the probability density function of X. *See also* central absolute moment.

central processing unit (CPU) a part of a computer which performs the actual data processing operations and controls the whole computing system. It is subdivided into two major parts:

1. The arithmetic logic unit (ALU), which performs all arithmetic, logic, and other processing operations;

2. The control unit (CU), which sequences the order of execution of instructions, fetches the instructions from memory, decodes the instructions, and issues control signals to all other parts of the computing system. These control signals activate the operations performed by the system.

central server model a model for a multiprogrammed computer system in which a fixed number k of jobs is assumed to move from the central server to one of a fixed number $n - 1$ of peripheral devices. Each job in the central server may move to one of the peripheral devices or stay in the central server. The latter event models the real-life case of one job leaving the system and another arriving. Jobs in the peripheral devices will return to the central server with probability 1.

centroidal profile a method for characterizing and analyzing the shape of an object having a well-defined boundary. The centroid of the shape is first determined. Then a polar (r, θ) plot of the boundary is computed relative to this origin: this plot is the centroidal profile, and has the advantage of permitting template matching for a 2D shape to be performed relatively efficiently as a 1D process.

centroid defuzzification a defuzzification scheme that builds the weighted sum of the peak values of fuzzy subsets with respect to the firing degree of each fuzzy subset. Also called height defuzzification.

Centronics a type of cable connector typically used for connecting printers.

certificate for any graph property $\mathcal{P}$, and graph G, a certificate for G is a graph G' such that G has property $\mathcal{P}$ if and only if G' has the property.

certification the formal processes of obtaining any certification paper according to a specific normative or law. A specific enabled institute or center typically issues a certificate.

CGA *See* color graphics adapter.

CGI *See* common gateway interface.

chain (**1**) to link objects together, as in a linked list.

(**2**) a way of resolving a hash collision by creating a linked list rooted at the location in the hash table where a hash collision occurred. The chain represents an equivalence class of symbols under the hash function.

(**3**) a linearly ordered subset of a partially ordered set.

See also Markov chain.

chain code a method for coding thin contours or lines, for example in a bilevel picture, which encodes the direction of movement from one point to the next. For eight-connected contours, a three-bit code may be used at each point to indicate which of its eight neighbors is the succeeding point.

chaining a collision resolution policy whereby, upon collision, the hashed record is stored in temporary extra storage location (an overflow location). This location is then referenced from the original location. Retrieval of a record is facilitated by hashing followed by a linear search of overflow locations.

chamfer distance a digital distance based on a chamfer mask, which gives the distance between a pixel and those in its neighborhood; then the chamfer distance between two non-neighboring pixels (resp., voxels) is the smallest weighted length of a digital path joining them. The word "chamfer" comes from the fact that with such a distance a circle is in fact a polygon. The n-dimensional Manhattan and chessboard distances are chamfer distances; the Euclidean distance is not. In the 2D plane, the best chamfer distances are given by the $(3, 4)$ and $(5, 7, 11)$ Chamfer masks: in the $(3, 4)$ mask, a pixel is at distance 3 from its horizontal/vertical neighbors and at distance 4 from its diagonal neighbors, while in the $(5, 7, 11)$ mask, it is at distance 5 from its horizontal/vertical neighbors, at distance 7 from its diagonal neighbors, and at distance 11 from its neighbors distant by 1 and 2, respectively, along the two axes. *See* chessboard distance, Euclidean distance, Manhattan distance.

changeability the capability of the software product to enable a specified modification to be implemented. Implementation includes coding, designing, and documenting changes. If the software has to be modified by the end user, changeability may affect operability. Attributes of software that bear on the effort needed for modification, fault removal, or environmental change.

change control an umbrella process that enables a project team to accept, evaluate, and act on changes in a systematic manner.

change dump a selective dump of those memory locations that have changed since some specified event.

change in scale replacing one sampling period by another for some measurement process. Used to detect self-similarity in stochastic processes.

change report a report providing details on the nature of work required to make a change.

change request a document reporting details on the type of change that is requested.

channel (1) a data structure that may be implemented in software or hardware over which processes send messages.

(2) the medium along which data travel between the transmitter and receiver in a communication system. This could be a wire, co-axial cable, free space, etc. *See also* I/O channel.

(3) a rectangular region for routing wires, with terminals lying on two opposite edges, called the "top" and "bottom". The other two edges contain no terminals, but wires may cross these edges for nets that enter the channel from other channels. The routing area of a layout is decomposed into several channels.

channel architecture a computer system architecture in which I/O operations are handled by one or more separate processors known as channel subsystems. Each channel subsystem is itself made up of subchannels, in which control unit modules control individual I/O devices. Developed by IBM, and used primarily in mainframe systems, the channel architecture is capable of a very high volume of I/O operations.

channel capacity a fundamental limit on the rate at which information can be reliably communicated through the channel. Also referred to as "Shannon capacity", after Claude Shannon, who first formulated the concept of channel capacity as part of the noisy channel coding theorem.

For an ideal band-limited channel with additive white Gaussian noise, and an input average power constraint, the channel capacity is $C = 0.5 \log(1 + S/N)$ bits/Hz, where S/N is the received signal-to-noise ratio.

channel coding the process of introducing controlled redundancy into an information sequence mainly to achieve reliable transmission over a noisy channel. Channel coding can be divided into the areas of block coding and trellis coding. Also called error control coding. *See* block coding, trellis coding, and convolutional coding.

channel command word an "instruction" to an I/O channel. The commands consist of parameters (e.g., "operation", "data address", "count") giving the channel processor information on the type of I/O operation requested (e.g., "read" or "write"), where the data is to be read or written, and the number of bytes involved in the data transfer.

In the IBM mainframe architecture there are six different types of channel control words: READ, READ BACKWARD, WRITE, CONTROL, SENSE, and JUMP.

channel control word *See* channel command word.

channel density a measure of physical channel capacity. Orient a channel so that the top and bottom are horizontal edges. Then the density at any vertical line cutting the channel is the number of nets that have terminals both to the left and right of the vertical line. Nets with a terminal on the vertical line contribute to the density unless all of the terminals of the net lie on the vertical line. The channel density is the maximum density of any vertical cut of the channel.

channel I/O an approach to I/O processing in which I/O operations are processed independently from the CPU by a channel system. *See also* channel architecture.

channel matched vector quantization *See* channel optimized vector quantization.

channel optimized vector quantization (COVQ) a combined source-channel code for block-based source coding (vector quantization) and block channel coding. A channel optimized vector quantizer can be designed using a modified version (taking channel induced distortion into account) of the generalized Lloyd algorithm). Also referred to as channel matched VQ. *See also* noisy channel vector quantization.

channel robust vector quantization a vector quantizer that has been made robust against channel errors. *See also* noisy channel VQ.

channel routing the problem of determining the routes, i.e., paths and layers, for wires in a routing channel.

chaos a state of disorder.

chaotic systems systems that when they are in equilibrium are in unstable equilibrium.

Chapin chart *See* box diagram.

character (1) letter, number, or symbol as used on a computer keyboard.

(2) data type that represents an alphanumeric character as a group of bits, usually as an eight-bit byte.

character driver a device driver of a device that transmits data in characters.

characteristic a property by which an item can be assessed. *See* software quality characteristic and feature. *See also* biased exponent.

characteristic function a transformed probability density function,

$$\Phi_{\mathbf{x}}(\omega) = E\left[\exp\left(j\omega^T \mathbf{x}\right)\right]$$

useful in the analytic computation of higher order moments and convolutions of probability densities.

characteristic polynomial a polynomial associated with a square matrix. If A is a square matrix, then its characteristic polynomial $c(A)$ is given by:

$$c(A) = \det(xI - A)$$

where det( ) is the determinant and I is the identity matrix.

character recognition *See* optical character recognition.

character set (1) a specification of the representation of characters in a computer, for example, ANSI X3.4 or ISO-8859.

(2) a specification of the characters of a programming language, independent of their representation.

character string (1) a series of continuous bytes in memory, where each byte represents one character.

(2) data structure corresponding to ordered sequence of characters.

charge-coupled device (CCD) a solid-state device used to record images. A CCD is a digital device which counts the photons that strike it by making use of the photoelectric effect. In a typical CCD array, a large number of such devices is collected into a 2D grid. Each device corresponds to a single pixel, and the number of electrons in the device is linearly related to the brightness or intensity value at that point in the CCD.

charge-coupled-device memory large-capacity shift registers making use of charge-coupled-devices (CCD), i.e., MOS devices in which data bits are stored dynamically as charge between a gate and the substrate. This forms a multi-gate MOS transistor with the source and drain terminals "stretched" apart, and a number of gate terminals in between. The first gate terminal (closest to the source) inserts bits (charge) into the register, and the following gates are controlled with overlapping clocks allowing the charge to move along the array. At the far end, the bit under the final gate terminal is detected as a change in current.

checkerboarding *See* fragmentation.

checking mechanism for controlling the system or component correctness. The checking is typically a control process in which obtained information is verified correct against the known information. The lists of actions to be performed or objects to be taken are typically called checklists. *See* type checking.

checkpoint (**1**) time in the history of execution at which a consistent version of the system's state is saved so that if a later event causes potential difficulties, the system can be restarted from the state that had been saved at the checkpoint. Checkpoints are important for the reliability of a distributed system, since timing problems or message loss can create a need to "backup" to a previous state that has to be consistent in order for the overall system to operate functionally.

(**2**) a point in a transaction, other than the beginning of the transaction, which is saved and can be rolled back to.

(**3**) an internal database mechanism to record that all pending data has been written to data files.

checkpointing a form of backward error recovery in which some subset of the system states (data, program, etc.) is saved at specific points (checkpoints) during the process execution, to be used for recovery if a fault is detected.

checksum a value used to determine if a block of data has changed. The checksum is formed by adding all of the data values in the block together, and then finding the 2s complement of the sum. The checksum value is added to the end of the data block. When the data block is examined (possibly after being received over a serial line), the sum of the data values and checksum should be zero.

checksum character in data communication and storage devices, an extra character added at the end of the data so that the total number of ones in a block is even. The *checksum character* is used to detect errors within the data block.

chessboard distance the distance between discrete points arising from the L^∞ norm. Given two discrete points $x = (x_1, \ldots, x_n)$, $y = (y_1, \ldots, y_n)$ on an n-dimensional integer lattice, the *chessboard distance* between x and y is $\max\{|x_1 - y_1|, \ldots, |x_n - y_n|\}$. So called because it equals the number of moves made by a king when going from one position to another in the game of chess. *See* norm.

chief programmer The lead programmer of the chief programmer team model.

chief programmer team a popular team structure advocated by Brooks. The lead technical person who also supervises the other team members is called the chief programmer.

child term used as an adjective to describe a created entity such as a process or directory.

child directory a directory that descends from another directory; that is, a directory that is the successor of another directory. *Compare with* parent directory.

child node in a directed acyclic graph, a node which has one or more arcs leading into it. The nodes from which these arcs originate are the parent nodes.

child process a process that has been created by another process.

child record type the type of a record acting as a child in a hierarchical database.

Chinese postman problem the problem of finding a minimum length tour that traverses each edge at least once.

Chinese remainder theorem a widely used result in number theory. An integer n can be solved uniquely mod LCM($A(i)$), given modulii ($n mod A(i)$), $A(i) > 0$ for $i = 1..k$, $k > 0$. In other words, given the remainders an integer gets when it is divided by an arbitrary set of divisors, you can uniquely determine the integer's remainder when it is divided by the least common multiple of those divisors.

chip a small piece of semiconductor material upon which miniaturized electronic circuits can be built.

chip carrier a low-profile rectangular component package, usually square, whose semiconductor chip cavity or mounting area is a large fraction of the package size and whose external connections are usually on all four sides of the package.

chip chart this term is often used for the "gray scale" chart used in the process of aligning television camera systems. The gray scale provides logarithmic reflectance relationships.

chip select a control signal input to, e.g., a memory chip, used to make this particular chip "active" in reading or writing the data bus. Read or write is determined by another control input signal: the "R/W-signal". Typically, some of the high order bits from the CPUs address bus are decoded to form the chip select signals.

chip-to-chip optical interconnect optical interconnect in which the source and the detector are connected to electronic elements in two separate chips.

chi square automatic interaction detection (CSAID) a decision tree technique used for classification of a dataset that provides a set of rules to be applied to a new (unclassified) dataset to predict which records will have a given outcome. It segments a dataset by using chi square tests to create multi-way splits.

chi-squared distribution a probability distribution with n degrees of freedom and probability density function

$$f(x) = \frac{x^{\frac{n}{2}-1} e^{-\frac{x}{2}}}{2^{\frac{n}{2}} \Gamma\left(\frac{n}{2}\right)}$$

for $x > 0$ ($f(x)$ is zero elsewhere) and n a positive integer.

choke point a single point through which all traffic must pass.

Chomsky hierarchy a classification of phrase structure grammar-based languages based on the work of Noam Chomsky (1956). A Type 3 language can be specified by a regular expression and recognized by a Finite State Machine. A Type 2 language can be expressed by a context-free grammar and recognized by a push-down automaton. A Type 1 language can be expressed by a context-sensitive grammar and recognized by a nondeterministic push-down automaton. A Type 0 language can be recognized only by a Turing machine. Each level of the hierarchy n contains languages defined by level $n + 1$ as special cases. A language defined by a Type 0 grammar is also called a recursively enumerable language. A language of type n contains all of the languages of type $n + 1$ as a proper subset.

Chomsky normal form a property of a context-free grammar. A context-free grammar, G, is said to be in Chomsky normal form if every rule has the form

$$A \rightarrow BC$$

or

$$A \rightarrow a$$

where A, B, and C are nonterminals and a is a terminal.

chopping in computer arithmetic, a rounding procedure that consists of simply dropping all those digits beyond the ones to be retained. Also known as truncation or round-towards-zero. One of the four rounding methods in IEEE-754.

choreography in computer animation, the timing and sequencing of activity and representation.

chroma (**1**) characterization of how much a color differs from both the pure color and the gray of the same intensity. Also called saturation.

(**2**) the color component of a composite video signal.

chroma keying *See* color keying.

chromatic index the minimum number of colors with which the edges of a graph can be colored.

chromaticity coordinates coordinates based on the Commission Internationale de l'Éclairage (CIE) color scheme, which uses three standard (but physically unrealizable) primary colors called X, Y, and Z. (These are different from red, green, and blue, and are chosen to represent human color matching performance.) Any visible color c can be expressed as a weighted sum of these primary colors: $c = w_x X + w_y Y + w_z Z$. The weights (w_x, w_y, w_z) are called the tristimulus values and are a way of objectively encoding all visible colors. (Actually, each set of weights represents an infinite set of colors which are indistinguishable.) Normalizing the colors by:

$$x = \frac{w_x}{w_x + w_y + w_z}$$
$$y = \frac{w_y}{w_x + w_y + w_z}$$
$$z = \frac{w_z}{w_x + w_y + w_z}$$

generates the chromaticity coordinates (x, y), which are independent of the brightness of the color. Note that $z = 1 - x - y$, so we can recover z, but we have lost the absolute brightness of the color.

chromatic number the minimum number of colors needed to color the vertices of a graph.

chromatic visual field the subset of the visual field within which color vision is possible.

chrominance information describing hue, or the color components orthogonal to the brightness. YUV and YIQ are chrominance/luminance color models.

chromosome an elongated structure in the cell nucleus which physically carries the genes of an organism. In genetic algorithms, the term also refers to a string of symbols in the population which denotes a potential solution to the underlying optimization problem.

chunking a design state in which developers divide the content of their on-line support systems into discrete, easily identifiable units.

CIE diagram the projection of the plane $(X + Y + Z) = 1$ onto the XY plane, where X, Y, Z are the respective tristimulus values as defined by the CIE (*See* tristimulus value and Commision Internationale de l'Eclairage). The CIE diagram shows all of the visible chromaticity values and maps all colors with the same chromaticity but different values (luminances) onto the same points. *See* chromaticity coordinates.

circle detection the location of circles in an image by a computer. Often accomplished with the Hough transform.

circuit a network of input, output, and logic gates, contrasted with a Turing machine in that its hardware is static and fixed.

circuit complexity the study of the size, depth, and other attributes of circuits that decide specified languages or compute specified functions.

circuit relay a relay and filtering program that operates at the transport layer (level four) of the network protocol stack.

circuit switched in communications, a switching mode that reserves some resources for the given connection along the entire length of a specific path. Hence all data in one connection are transmitted over the same path. If the circuit is not blocked during the call setup phase, then subsequent data will not experience any resource contention (i.e., queueing delay or loss). A circuit may be, e.g., a wire in space division multiplexing, a time slot or a set of time slots in time division multiplexing, or a frequency band in frequency division multiplexing.

circuit value problem given an encoding $\overline{\alpha}$ of a Boolean circuit α, inputs $x_1, \ldots, x_n$, and a designated output y, the problem of deciding if output y of α is true on input $x_1, \ldots, x_n$.

circulant matrix a square $N \times N$ matrix $\mathbf{M} = \{m_{i,j}\}$ such that $m_{i,j} = m_{(i+n)\mathrm{mod}N,(j+n)\mathrm{mod}N}$; that is, that each row of M equals the previous row rotated one element to the right. All circulant matrices are diagonalized by the discrete Fourier transform.

circular convolution *See* periodic convolution.

circularity measure the size invariant ratio of area divided by perimeter squared for small shapes, and much used as a preliminary discriminant or measure of shape; so-called because it is a maximum for circular objects.

circular list a list with a designated cell, the list head, and the operations of successor (for a singly linked list) and predecessor (for a doubly linked list) always return a list cell value. A predicate can be used to detect if this returned cell is the designated list head.

circular register buffer a set of general purpose CPU registers organized to provide a large number of registers, which may be accessed a few at a time. The group of registers accessible at any particular time may be readily changed by incrementing or decrementing a pointer, with wraparound occurring from the highest numbered registers to the lowest numbered registers, hence the name circular register buffer. There is overlap between the groups of adjoining registers that are accessible when switching occurs. The overlapping registers can be used for passing arguments during subroutine calls and returns. The circular register buffer is a feature of the SPARC CPU architecture. In the SPARC CPU there are 256 registers, available 32 at a time, with an overlap of eight registers above and eight registers below the current group.

circular self-test path a BIST technique based on pseudorandom testing assured by arranging flip-flops of a circuit (during test) in a circular register in which each flip-flop output is ex-ored with some circuit signal and feeds the input of the subsequent flip-flop. This register simultaneously provides test pattern generation and test result compaction.

circular shift *See* rotate.

circular vection illusion of self-motion produced when placed in a visual field that rotates around one; the observer comes to feel that he is rotating and the surround is stable.

CISC *See* complex instruction set computer.

CISC processor *See* complex instruction set computer.

CISC technology a complex (or complete) instruction set computer (CISC) that has an instruction set with a complete set of addressing modes, including indexed, indirect, and autoincrement addressing instructions, which makes it easier for a programmer writing assembly language directly. Fewer instructions will generally be required to accomplish the same task than with RISC technology, but a longer clock period may be required to accommodate the more complicated instructions, and pipelining instructions may be more difficult.

city-block distance a distance measure between two real valued vectors $(x_1, x_2, \ldots, x_n)$ and $(y_1, y_2, \ldots, y_n)$ defined as

$$D_{\text{city block}} = \sum_{i=1}^{n} |x_i - y_i| \ .$$

City-block distance is a special case of Minkowski distance when $\lambda = 1$. *See* Minkowski distance.

class generally, a definition that captures the concept underpinning a set of instances. These instances have the same abstraction and therefore have the same or similar features. A class consists of an interface (the specification = the external view) and an implementation (the internal view). The internal view is defined as a collection of class attributes with their corresponding types, together with the methods that operate on them. The interface defines the set of services that can be required from the class and is the only way for modifying the internal attributes.

In the context of object-oriented languages, a class represents an abstract data type and its associated operations. Values in such languages represent instances of classes. Generally, new classes can be specified by the programmer, often as being subclasses, also called derived classes, of one or more existing classes, inheriting some or all of the characteristics of the class from which they are derived.

class behavior the sequence of operations performed by one class with respect to the rest of collaborating objects in the system. The *class behavior* is defined by means of class methods.

class definition in object oriented languages, the class definition contains the description of class interface in terms of attributes and methods. The class definition also contains the access level for each class member, like public, protected, or private.

class diagram a common name for a diagram type in object modeling. The diagram shows the static architectural view of the object oriented system, which typically shows classes or types and their interrelationships but may also sometimes show instances.

class feature properties and operations of a class are collectively called features (in UML and Eiffel). In other languages this concept is assigned to class members.

class global member in object-oriented languages, a value which is shared among all instances of objects. In C++, it is called a static member. *Contrast with* instance variable.

class hierarchy a configuration of classes in which all the relationships are inheritance. Ideally these inheritance relationships should be specialization (subtyping) and any uses of implementation inheritance should be flagged accordingly. A class hierarchy may comprehend more class trees. Also called class tree.

classic life-cycle a linear, sequential approach to process modeling (also called waterfall life-cycle).

classification (**1**) the process of finding classification rules.

(**2**) an area of data mining that attempts to predict the category of categorical data by building a model based on some predictor variables.

classification rule any one of a class of rules used in data mining. A *classification rule* partitions the given data into disjoint sets.

classifier (**1**) a method of assigning an object to one of a number of predetermined classes.

(**2**) given a set of patterns of different classes, a system capable of determining the membership of each pattern.

classifier system a rule-based machine learning system where a specific task is achieved through the interactive action of a set of classifiers. Each classifier is capable of accepting a set of specific messages which form its condition part, and will perform a specific action by generating a new message when its condition is satisfied. In addition to the classifiers, the system consists of a set of feature detectors which monitor the environment, a global message list which allows individual classifiers to observe and post messages, and a set of effectors for performing an action in response to the environment. Each message on the list is in the form of a binary string, and the condition and action of each classifier is specified as a symbol string of equal length over the alphabet {0, 1, #}, where the "don't care" symbol allows the classifier to accept more than one message type. To perform a specific task, the feature detectors obtain a set of measurements from the environment and represent these as messages on the global list. Each classifier then matches the messages with its own conditions, and on satisfaction generates new messages which are again posted on the list. External actions are performed by the system when messages on the global list directly activate the effectors. *See also* learning classifier system.

class implementation description of the hidden features which are typically methods (describing the functional aspects) and attributes (data). It is the code of the class.

class instance *See* object, instance.

class library a library of reusable classes. These can be organized in one or more class hierarchies.

class member in a class definition the set of attributes and methods declared in the class are members of it. In an extensive view of the term, it is rarely used for indicating the set of the instances of a class.

class method *See* method.

class specialization specialization occurs in a class hierarchy when the classes are related in the type-subtype configuration. The subtype is a specialized form of the parent type in that (a) the subtype offers all the features of the parent and probably more and (b) polymorphism is present which supports dynamic type substitutability.

class tree *See* class hierarchy.

class variable any variable which pertains to the class itself and, hence, to all instances of that class.

clausal form *See* clause.

clause in logic, a disjunction of literals (negated or not) where all the variables are implicitly universally quantified. A generic clause is:

$$P_1 \vee P_2 \vee \cdots \vee P_n \vee \neg N_1 \vee \cdots \vee \neg N_m$$

where P_i and N_j are literals. *See* Horn clause.

clean cache block a cache block (or "line") where a copy of the information is stored in memory. A clean block can be overwritten with another block without any need to save its state in memory.

cleanroom development a software development method where defect prevention and removal are performed when a software deliverable is being created, before the document is passed to the next step in the software process.

clear **(1)** to set the value of a storage location to zero (often used in the context of flip-flops or latches).

(2) clearing a bit (register) means writing a zero in a bit (register) location. Opposite of "set".

client synonymous with customer or a computer system or process that requests a service of another computer system or process (e.g., a server). In the client-server model, the client is a process that remotely accesses resources of a server. In the object oriented model, a class is considered a client of another class when it uses the other class features. According to C++ syntax, the clients can be grouped into is-a, has-a (or is-part-of), and use-a (or is-referred-by) relationships.

client-server a common form of distributed system in which software is split between server tasks and client tasks. A client sends requests to a server, according to some protocol, asking for information or action, and the server responds. There may be either one centralized server or several distributed ones. This model allows clients and servers to be placed independently on nodes in a network, possibly on different hardware and operating systems appropriate to their function, e.g., fast server/cheap client.

client-server application a common form of distributed system in which software is split between server tasks and client tasks. A client sends requests to a server, according to some protocol, asking for information or action, and the server responds. There may be either one centralized server or several distributed ones. This model allows clients and servers to be placed independently on nodes in a network, possibly on different hardware and operating systems appropriate to their function, e.g., fast server/cheap client.

client-server network a distributed system in which clients are interconnected with servers. A client sends a request to a server that performs the desired operation and returns a result.

client-server principle the architecture in which clients are able to request services of processes executing on remote machines.

client stub system software residing in a client machine to prepare a remote call.

clipping the selective removal of an object disjoint with the display area or the non-visible parts of an object that does intersect the display area. Parts of an object intersecting the display area may lie outside of the display area or be partially or fully obscured by another intersecting object.

clock (**1**) the oscillator circuit which generates a periodic synchronization signal.

(**2**) a circuit which produces a series of electrical pulses at regular intervals which can be used for timing or synchronization purposes.

clock cycle (**1**) one complete event of a synchronous system's timer, including both the high and low periods.

(**2**) the time required by a processor's control unit to execute a single internal instruction. Several clock cycles may be required to execute a machine instruction.

clock doubling a technique in which the processor operates internally at double the external clock frequency.

clock duty cycle the percentage of time that the electronic signal remains in the true or 1 state.

clock pulse a digital signal that, via its rising edge or falling edge, triggers a digital circuit. Flip flops and counters typically require clock pulses to change state.

clock recovery in synchronous systems, the act of extracting the system clock signal from the received sequence of information symbols. *See also* symbol synchronization.

clock replacement algorithm a page replacement algorithm described as follows: A circular list of page entries corresponding to the pages in the memory is formed. Each entry has a use bit which is set to a 1 when the corresponding page has been referenced. A pointer identifies a page entry. If the use bit of the page entry is set to a 1, the use bit is reset to a 0 and the pointer advances to the next entry. The process is repeated until an entry is found with its use bit already reset, which identifies the page to be replaced. The pointer advances to the next page entry for the next occasion that the algorithm is required. The word "clock" comes from viewing the pointer as an arm of a clock. Also known as a first-in-not-used-first-out replacement algorithm.

clock skew the phenomenon where different parts of the circuit receive the same state of clock signal at different times because it travels in wires with different lengths. This skew of the signals causes a processing element to generate an erroneous output. Distribution of the clock by means of optical fibers, waveguides, a lens, or a hologram eliminates clock skew.

clock speed the rate at which the timing circuit in a synchronous system generates timing events.

cloning the activity by which the duplication of an object is performed (in object oriented programming). The cloned object is totally equal to the source object except for the object identificator. This term is frequently synonymous with templating. Specific methods have to be built for providing the possibility to perform the operation of cloning.

close an operation on a file or device that results in the finalization of access to this unit.

closed convex set a set of vectors C such that if $\mathbf{x}, \mathbf{y} \in C$, then $\lambda \mathbf{x} + (1 - \lambda)\mathbf{y} \in C$ for all $0 \leq \lambda \leq 1$.

closed kinematic chain in vision engineering, a sequence of links which forms a loop.

closed loop control control scheme using feedback to guide and correct actions.

closed-loop planner a planning system that periodically makes observations of the current state of its environment and adjusts its plan in accord with these observations.

closed model *See* closed queueing network.

closed network a closed queueing network.

closed queueing network a queueing network model in which the number of customers in the network is a fixed positive integer. The customers circulate among the devices in the network according to specified routing probabilities.

closing a basic morphological operation. Given a structuring element B, the closing by B is the composition of the dilation by B followed by the erosion by B; it transforms X into $X \bullet B = (X \oplus B) \ominus B$. The closing by B is what one calls an algebraic closing; this means that: (a) it is a morphological filter; (b) it is extensive; in other words it can only increase an object. *See* dilation, erosion, morphological filter, structuring element.

closing/opening filter one of an important class of morphological filters. Let γ and φ be opening and closing operators, respectively. The following operators can be obtained by composing γ and φ (i.e., by applying them in succession): $\gamma\varphi$, $\varphi\gamma$, $\gamma\varphi\gamma$, and $\varphi\gamma\varphi$. These are all morphological filters, and collectively they are called closing/opening filters or opening/closing filters. No further operators can be obtained by composing γ and φ. *See* closing, morphological filter, opening.

closure **(1)** for a set of attributes A, the closure of A, denoted A^+, under a set of functional dependencies, FD, is the set of all attributes functionally determined by A.

(2) for a set of functional dependencies FD, the closure of FD, denoted FD^+, is the set of all functional dependencies which can be inferred from FD.

cloud an informal representation of a multiaccess link. The purpose of representing it as a cloud is that what goes on inside is irrelevant to what is being discussed. When a system is connected to the cloud it can communicate with any other system attached to the cloud.

CLT *See* central limit theorem.

cluster **(1)** a group of data points on a space or a group of communicating computer machines. A cluster of computers on a local network can be installed to provide their service as a unique computer. This is frequently used for building large data storage and WEB servers.

(2) in computer disks, a cluster consists of a fixed number of sectors. Each sector contains several bytes, for example 512.

cluster analysis in pattern recognition, the unsupervised analysis of samples in order to cluster them into classes based on (a) a distance metric and (b) a clustering algorithm. Typical algorithms minimize a cluster criterion (e.g., representation error) by grouping samples hierarchically or by iteratively reassigning samples to clusters. The k-means algorithm is an example of the latter. In the case of 2D measurements, cluster analysis becomes a method of image segmentation.

cluster computing a computing paradigm involving massed heterogeneous processors.

clustering **(1)** any algorithm that finds groups of items that are similar. It divides a data set so that records with similar content are in the same group, and groups are as different as possible from each other. Since the categories are unspecified, this is sometimes referred to as unsupervised learning. Popular clustering algorithms include minimum spanning tree and k-means clustering. *See* hierarchical clustering. *See also* distance measure, similarity measure.

(2) grouping workstations together to provide greater throughput and more fault tolerance. A cluster of inexpensive workstations can be organized to provide the aggregate computing power of a large expensive computer, with the additional robustness against failure due to redundancy. However, the communication costs required to share data and other resources may counteract this advantage. *See* workload characterization.

clustering field a non-key field on which a file is physically ordered.

clustering index an index built upon a clustering field.

clutter the name given to background signals which are currently irrelevant to a detection system; clutter is a form of structured noise.

CMM *See* capability maturity model.

CMOS *See* complementary metal-oxide-semiconductor.

coalesce an operation performed on a heap that locates adjacent unused blocks of storage and combines them into larger blocks of unused storage.

coarse grained transaction a transaction that performs a large number of updates.

coarsening to coarsen a problem instance is to alter it, typically restricting to a less complex feasible region or objective function, so that the resulting problem can be efficiently solved, typically by dynamic programming.

coaxial joint joint at which one or more movement axes intersect; a joint capable of more than one movement, as is the human shoulder.

COBOL acronym for common business-oriented language. Developed in the 1950s to solve "data processing" types of applications using a machine-independent language.

cockpit space occupied by users and containing controls and displays necessary for operating a synthetic environment system.

cocktail party effect phenomenon demonstrating perceptual narrowing, which allows one to attend to a particular signal embedded in similar noise, as when one can listen to a single conversation at a cocktail party while many other conversations take place within hearing at the same time.

COCOMO *See* constructive cost model.

CODASYL acronym for Conference on Data System Languages, one of the standards-setting bodies for the COBOL language.

CODASYL model model defined by CODASYL (Conference of Data Systems Language). A network model.

code (**1**) the act of writing software.

(**2**) a technique for representing information in a form suitable for storage or transmission.

(**3**) a mapping from a set of messages into binary strings.

code acquisition the process of initial code synchronization (delay estimation) between the transmitter and receiver in a spread-spectrum system before the actual data transmission starts. It usually requires the transmission of a known sequence. *See also* code tracking.

code and fix a simple approach for program developing on the basis of which the programmers write the code, test and fix the found errors without following a formalized development life-cycle.

codebook a set of code vectors (or codewords) that represent the centroids of a given pattern probability distribution. *See* vector quantization.

codebook design a fundamental problem in vector quantization (VQ). The main question addressed by *codebook design* is "how should the codebook be structured to allow for efficient searching and good performance". Several methods (tree-structured, product codes, M/RVQ, I/RVQ, G/SVQ, CVQ, FSVQ) for codebook design are employed to reduce computational costs. *See* vector quantization, tree structured VQ.

codebook generation a fundamental problem in vector quantization. Codebooks are typically generated by using a training set of images that are representative of the images to be encoded. The best training image to encode a single image is the image itself. This is called a local codebook. The main question addressed here is "what code vectors should be included in the codebook?". *See* vector quantization.

codebook training the act of designing a codebook for a source coding system. The LBG algorithm is often used to design the codebook for vector quantizers.

codec word formed from encoder and decoder. A device that performs encoding and decoding of communications protocols.

code cache a cache that only holds instructions of a program (not data). Code caches generally do not need a write policy, but *see* self-modifying code. Also called an instruction cache. *See* cache.

code combining an error control code technique in which several independently received estimates of the same codeword are combined with the codeword to form a new codeword of a lower rate code, thus providing more powerful error correcting capabilities. This is used in some retransmission protocols to increase throughput efficiency.

code converter a device for changing codes from one form to another.

code division multiple access (CDMA) a technique for providing multiple access to common channel resources in a communication system. CDMA is based on spread spectrum techniques where all users share all the channel resources. Multiple users are distinguished by assigning unique spreading codes to each user. Traditionally, individual detection is accomplished at the receiver through correlation or matched filtering.

code efficiency the unitless ratio of the average amount of information per source symbol to the code rate, where the amount of information is determined in accordance with Shannon's definition of entropy. It is a fundamental measure of performance of a coding algorithm.

code fork from the Macintosh, a synonym for code segment.

code generation (**1**) the process by which the code in a given language is automatically generated by a tool. Typically, the code produced consists of a sort of skeleton of the application. This has to be filled in to complete the whole application. In the literature, there exists several CASE tools for code generation that allow the description of the system by using a higher level model such as: state machines, Petri nets, etc. Code generation is also performed by simpler applications that allow the generation of the code related to the user interface components, such as that for managing dialog windows, menus, etc.

(**2**) that stage of the compilation process in which the object code is generated. This may be done directly as a transformation of the source code but more commonly is based on the intermediate representation. The generated code may be instructions for the target machine, but can also be byte code for a virtual machine. Code generation may include code optimizations. *See also* code selection, peephole optimization.

code motion an operation performed by an optimizing compiler by which a computation is performed at a point in the control flow other than where the statement actually appears in the code. The goal of code motion is to produce a semantically identical program that performs computations in a more efficient fashion.

code optimization any transformation of a program performed by an optimizing compiler whose goal is to eliminate, as much as possible, unnecessary, redundant, and unused computations, while producing a semantically identical program. While traditionally code optimizations are performed internally in a compiler, many code optimization tools are "source-to-source" transformations, reading in a program and producing a semantically identical program that has a lower cost metric along some performance axis (smaller, faster, or both, for example). *See also* peephole optimization.

code rate in forward error control and line codes, the ratio of the source word length to the codeword length, which is the average number of coded symbols used to represent each source symbol.

code reuse the act or discipline of reusing code from one system in another. Code may be reused in many ways. Chunks of code, which may be subroutines or objects, can be reused by incorporating them into other programs either by combination (possibly aggregation) or, in object oriented environments, by inheritance. Reuse is facilitated if the interfaces of the code chunks are clearly specified and well tested and can thus be regarded as "trusted components".

code segment that portion of an executable file that contains executable code. Typically, multiple instances of the program running on the same machine will share a single instance of the *code segment.* Depending on the operating environment, a program may be allowed to have more than one code segment. A *code segment* is typically assumed to be read-only. Also known as a code fork (Macintosh) or text segment (Unix).

code selection that part of the code generation process in which the machine code necessary to implement a computation is represented in the intermediate representation. *See also* machine simulation.

code tracking the process of continuously keeping the code sequences in the receiver and transmitter in a spread-spectrum system synchronized during data transmission. *See also* code acquisition.

code transformation that part of the code generation process that transforms one or more fully or partially specified machine instructions into a new instruction or sequence of instructions. A form of rewrite rule.

code V a widely employed computer code for design of optical systems by Optical Research Associates.

code view any representations of the source code that cover the same information as the code (or parts of it) but in a manner that accelerates the comprehension process. Examples are program slices, call-graphs, data-flow, definition-use graphs, or control dependencies.

codeword the channel symbol assigned by an encoder to a source symbol. Typically the *codeword* is a quantized scalar or vector.

coding (**1**) the process of programming, generating code in a specific language. Note that *coding* in the strict sense is only the task of writing code, but this word is also used for other activities such as design.

(**2**) the process of translating data from a representation form into a different one by using a set of rules or tables. *See* ASCII, EBCDIC, binary.

coding at primary rates for videoconferencing *See* image coding for videoconferencing.

coding of line drawings use of a representation scheme for line drawings. Line drawings are typically coded using chain codes where the vector joining two successive pixels is assigned a codeword. Higher efficiency is obtained by differential chain coding in which each pixel is represented by the difference between two successive absolute codes.

coding redundancy *See* redundancy.

coding system a mathematical or computational method of digitally representing a signal; also, a set of rules by which information may be represented by symbols or sets of symbols, as in written language.

coefficient an old term for significand.

coercion the implicit conversion of a value of one data type to another. Generally, a *coercion* is an implicit operation, often confused with a type cast or transfer function. In some languages, the rules of coercion are fixed by the compiler; in others, they may be redefined or extended by the programmer. *See also* type cast, transfer function.

co-evolution an evolutionary process in which the fitness landscape structure associated with one population may depend on particular evolutionary steps taken in another population.

co-existence the capability of the software product to co-exist with other independent software in a common environment sharing common resources.

cognition the process of acquiring information and selecting and controlling responses to it; decision making or thinking, the application of intelligence to shape behavior; the process of cogitating.

cognitive engineering designing systems to match human cognitive processes.

cognitive function that part of the human system that is responsible for information processing decision making.

cognitive lockup a symptom of cognitive overload in which the cognitive function is unable to adequately perform as required by the situation, and in extreme cases ceases to perform at all.

cognitive map introduced by R. Axelrod to study decision making processes, consists of points, or nodes, and directed links between the nodes. The nodes correspond to concepts. *See also* fuzzy cognitive map.

cognitive metric metric that takes into account the understandability, comprehensibility of a system, module, class. Often these metrics evaluate indexes that can be related to the effort to understand subsystem/class behavior and functionalities and are useful for evaluating reusability, testability, verifiability, and quality of systems and/or classes.

cognitive model mental model; also, a model of the cognitive function.

cognitive overload the condition occurring when cognitive workload exceeds the cognitive processing capacity available.

cognitive science science concerned with how humans acquire and use information; a branch of psychology.

cognitive workload the level of mental effort required during work.

cohesion a measure of how well the different parts of the class "hang together" and create a class which fully describes the concept it is being used to model; i.e., there are no extraneous features and no missing features. *See* temporal cohesion, logical cohesion, procedural cohesion.

coinduction a proof technique associated with bisimulation. To prove that two expressions are equivalent, we assume that they are equivalent and show that no contradiction results.

cold boot restarting the operating system by turning off and then turning on your computer.

cold start (**1**) a complete reloading of the system with no reassumption. All executed processes are lost.

(**2**) the starting of a computer system from a power-off condition.

cold start miss in a cache, a cold start miss occurs when a computer program is referencing a memory block for the first time, so the block has to be brought into the cache from main memory. Also called first reference misses or compulsory misses. When the cache is empty, all new memory block references are cold misses. *See also* capacity miss, conflict miss.

collaboration diagram a diagram that shows how a number of objects work together, by collaboration, in order to fulfill a requested service. A collaboration is a small sized group of objects which interact with each other by client-server in order to realize a specific, single purpose. It is a subcontracting relationship between two classes in which the client object delegates some of its own responsibility to the server class.

collateral execution execution of statements where the order of execution is not necessarily the order in which the statements are written. *See also* execution sequence.

collating sequence the rules of comparison of textual data. Traditionally, the collating sequence has been defined by the representation of characters used on a particular machine. The need for portability of code and for national language support has generally resulted in more rigorous specification of collating sequence, independent of the underlying character representation. *See* locale.

collection (1) a grouping of homogeneous individuals, possibly in space or time. *See* containment.

(2) a style of heap in which allocations of a similar type are allocated from a common pool of similar objects. Often used because it is not only more efficient, but reduces storage fragmentation.

collective recursion a special form of tail recursion, where the results are produced during the recursive calls and nothing is returned. The recursion may be optimized away by executing the call in the current stack frame, rather than creating a new stack frame, or by deallocating the entire recursion stack at once rather than a little at each return. *See also* tail recursion.

collision when, on insertion of a record via hashing, the address returned from the hash function is already occupied. *See* hash collision.

collision avoidance in a robotic system, functions designed to prevent unintended, unexpected, or inappropriate contact with objects in the robot's environment.

collision detection (1) in a robotic system, functions designed to detect imminent contact with objects in the robot's environment.

(2) in virtual reality, used to monitor the relative locations of solid objects. If the virtual environment manager detects that the proximity of two or more objects is sufficiently close, a collision event occurs. As a result of this event, the object's movement can be controlled so their surfaces do not intersect. In an environment which models a natural system, the kinetic energy of a moving object is (partially) transferred to the object it collides with, making the second object move.

collision resolution policy a technique to handle problems caused by collisions; the algorithm used to find another space to store the record currently being hashed.

color visual sensation associated with the wavelength or frequency content of an optical signal.

color blooming phenomenon where the excess charge at a photo receptor can spread to neighboring receptors and change their values in proportion to the overload. For RGB cameras, this effect can modify not only the luminance but also the chrominance of pixels. *See* color clipping, chrominance, luminance.

color burst burst of eight to ten cycles of the 3.579545 MHz (3.58 MHz) chrominance subcarrier frequency that occurs during the horizontal blanking of the NTSC composite video signal. The color burst signal synchronizes the television receiver's color demodulator circuits.

color clipping phenomenon where the intensity of the light on a photoreceptor exceeds some threshold, the receptor becomes saturated, and its response is no longer linear but limited to some bound. For RGB cameras, this effect can modify not only the luminance but also the chrominance of pixels. *See* color blooming, chrominance, luminance.

color discrimination the ability to distinguish among different wavelengths of energy in the visible spectrum.

color graphics adapter (CGA) a video adapter proposed by IBM in 1981. It is capable of emulating MDA. In graphic mode, it allows the image to reach 640×200 (wide per high) pixels with 2 colors or 320×200 with 4 colors.

color image image represented with 16, 24, or more bits/sample.

color image coding compression of color images is usually done by transforming RGB color space into a YC_1C_2 space, where Y represents luminance and C_1 and C_2 are color difference signals. The C_1 and C_2 signals are then subsampled, but coded with the same algorithm as the Y signal. Standard algorithms do not attempt to exploit correlations between the three signals.

color keying using the pixel color of one image to designate that pixel data from another image should replace the first pixel's color. The first image might be a binary image, which would select regions of interest from the second image. Another use is in blue-screening, where an actor works against a blue background. In the output image, the blue pixels get replaced by another image. For example, a weather map can be placed behind the weather presenter who is actually standing in front of a blue screen.

color model a method of specifying a color (position) in color space, often using a coordinate system. Examples include RGB and the Munsell color system.

color representation a method of defining a signal or an image pixel value to be associated with a color index.

color saturation a color with the dominant wavelength located at the periphery of Maxwell's chromaticity diagram. A fully saturated color is pure because it has not been contaminated by any other color or influence.

color signal the portion of a modulated signal that determines the colors of the intended output display.

color space a mathematical space defining a range and encoding of colors. *See* RGB color model, LUV space, HSV, HSL, YIQ color space, YUV color palette, and XYZ color space.

column in relational database systems, a multiset of values that may vary over time. All values of the same column are of the same data type and are values in the same table. A value of a column is the smallest unit of data that can be selected from a table or updated in a table.

column-access strobe *See* two-dimensional memory organization.

column decoder logic used in a direct-access memory (ROM or RAM) to select one of a number of rows from a given column address. *See also* two-dimensional memory organization.

column distance the minimum Hamming distance between sequences of a specified length encoded with the same convolutional code which differ in the first encoding interval.

COM acronym for Common Object Model, or Component Object Model (both definitions are used for this acronym, which describes the same technology). An object methodology developed by Microsoft. A specification for the naming, invocation, and manipulation of objects and the values they contain. *See also* OLE.

combinational circuit a logic circuit whose output is a function only of its inputs. Apart from propagation delays, the output always represents a logical combination of its present inputs. To be contrasted with a sequential circuit, where the output changes to represent the input data only upon a clock or synchronizing signal.

combinational lock interconnections of memory-free digital elements.

combinational logic a digital logic, in which external output signals of a device are totally dependent on the external input signals applied to the circuit.

combinator a lambda-expression with no free variables.

comfort a state of physical and mental well being.

Comité Consulatif International Télégraphique (CCITT) French translation for *International Consultative Committee for Telegraphy and Telephony*. This institution, based in Geneva, Switzerland, issues recommendations concerning all fields related to telecommunications. *See also* IRE.

command (**1**) directives in natural language or symbolic notations entered by users to select computer programs or functions.

(**2**) instructions from the central processor unit (CPU) to controllers and other devices for execution.

(**3**) a CPU command, or a single instruction, ADD, LOAD, etc.

(**4**) an order to an operating system.

commanding human task during human-robot interaction: control via manipulating symbols to trigger behavioral repertoires.

command interpreter a program which accepts user typein and executes actions based on that typein. Command interpreters often implement interpreters for full-fledged programming languages. Also called shell.

command language a language that defines the functions performed by an operating system and is used to write programs for a command interpreter. Synonym: shell language.

command language interpreter a computer program designed to receive and interpret commands.

command language script a program whose instructions are sequences of Unix commands and shell programming constructs; ASCII text files.

command line in a shell, a command typed by a user.

command processor a program that displays the system prompt and processes user commands. It is stored in the command.com file in DOS.

command program any Unix user command that is a separate, stand-alone program; also called external commands.

command prompt a character or text string that indicates the shell is ready and waiting for the user's command.

comment a construct in a program which is ignored by the compiler, and is meant for use as an annotation mechanism by the programmer. They should normally take up from 20 to 50% of source code.

commentary line *See* comment.

comment out to add the appropriate elements to a section of program to mark the code such that it appears to be a comment. Used to temporarily remove a segment of code from the program without actually removing it from the source file.

commercial off-the-shelf (COTS) software components that are purchased from third parts instead of being developed. COTS can be components to be integrated in a system of full functional operating environment. Operating systems are the most widely found form of COTS.

Commision International d'Eclairage (CIE) International standards body for lighting and color measurement. Known in English as the International Commission on Illumination.

commit an operation which updates the state of a computation to be the state computed since the start of a transaction. In general, the state after a commit cannot be undone, unless it is part of a nested transaction. *See also* rollback.

committed transaction a transaction that has been successfully committed.

COMMON from the FORTRAN COMMON declaration, a pool of static storage whose extent is the lifetime of the program, but whose scope is shared explicitly among many functions. A related concept in FORTRAN is named common.

common gateway interface (CGI) a methodology for passing information from a Web-based server to an application it invokes. External programs are known as gateways because they provide an interface between an external source of information and the server. Some languages now support CGI interfaces as part of the language definition.

common memory a portion of main memory that can be used simultaneously by different program units.

common mode failure a failure of apparently independent components or communication links due to an initiating event that affects them all.

common object request broker architecture (CORBA) an Object Management Group specification which provides the standard interface definition between OMG-compliant objects. This approach is typically used for building distributed systems with a common object oriented model. It implements all the mechanisms for communicating among objects and virtualizing the application.

common process framework a process model that encompasses a limited set of problem-solving activities populated by tasks, milestones, software quality assurance points, and deliverables.

communicating sequential processes (CSP) process algebra based on synchronous message passing and selective communications designed by C.A.R. Hoare in 1978. *See* algebra, CCS.

communication channel a means for sending signals to or receiving signals from a remote location.

communication delay latency in transmitted signals; the difference between the time of sending and the time of reception.

communications bit error rate *See* bit error rate, bit error probability.

communications instantaneous the assumption that the time needed to communicate with another process is zero or negligible, i.e., if a message is sent at time t, at the same instant t the message is received by the destination process. This assumption has a sense in discrete time systems if the communication time is less then the time unit.

communications subsystem the hardware and software subsystem that transmits and receives signals between the local and remote environments.

communication theory *See* information theory.

commutative (**1**) rule in mathematics that states that if a result can be computed by an operator applied to a set of operands, it can be applied to the same set of operands in any order, for example, "$A + B = B + A$" is a statement of a commutative operator "+".

(**2**) in compilers, a specification that an operation may be applied to operands following the mathematical rule of commutativity.

(**3**) in compilers, a specification of the evaluation order stating that operands to a commutative operator may be evaluated in any order, although in principle the evaluation order often has little to do with the application of the operation itself.

commutativity an operator $\oplus$ is commutative if $\mathrm{R} \oplus \mathrm{S} = \mathrm{S} \oplus \mathrm{R}$.

compact disk (CD) a plastic substrate embossed with a pattern of pits that encode audio signals in digital format. The disk is coated with a metallic layer (to enhance its reflectivity) and read in a drive (CD player) that employs a focused laser beam and monitors fluctuations of the reflected intensity in order to detect the pits.

compact disk interactive (CD-I) a specification that describes methods for providing audio, video, graphics, text, and machine-executable code on a CD-ROM.

compacting the process of defragmenting a disk. *See* fragmentation.

compaction *See* storage compaction.

compander a point operation that logarithmically compresses a sample into fewer bits before transmission. The inverse logarithmic function is used to expand the code to its original number of bits before converting it into an analog signal. Typically used in telecommunications systems to minimize bandwidth without degrading low-amplitude signals.

comparative analysis a particular form of "what if" question, that is, how a physical system changes in response to the perturbation of one of its parameters.

compare instruction an instruction used to compare two values. The processor flags are updated as a result. For example, the instruction CMP AL,7 compares the contents of register AL with 7. The zero flag is set if AL equals 7. An internal subtraction is used to perform the comparison.

comparison-based algorithm a sorting method that uses comparisons, and nothing else about sorting keys, to rearrange the input into ascending or descending order.

compass gradients a set of eight images that when used in windowed convolution provides an edge filter.

competition in evolutionary computation, the process of comparing the fitness values of individuals in a population to allow selection based on the comparison results. In evolutionary strategies and evolutionary programming, competition is carried out explicitly by ranking the associated fitness values or tournament scores of individuals in descending order. In genetic algorithms, the competition step is implicitly performed in proportional selection by directly representing the fitness values as selection probabilities.

competitive analysis a performance analysis in which an on-line algorithm is evaluated by comparing its performance to the best that could have been achieved if all the inputs had been known in advance.

competitive learning a process in which a network can exhibit self-organization by having hidden layer nodes 'compete' to represent certain features of the input training patterns.

competitive ratio the worst-case ratio between the cost incurred by an on-line algorithm and the optimal cost.

compiler a translator that converts a source program in some specified language to a more directly executable form. Historically, a "compiler" translated a program to machine code while the term interpreter was used to designate translation to a higher-level representation that was executed by "interpreting" the resulting output. Modern usage recognizes the translation process as separate from the execution process, and refers to the translation component of a system as a "compiler". *See* linker, assembler, interpreter, cross-assembler, cross-compiler.

compiler-compiler a term usually designating a parser generator.

compile time a designation used to identify information known at the time a compilation is performed, which allows decisions (such as code generation) to be bound, or partially bound, based on said information. *See also* link time, run-time.

complement (**1**) to swap 1s for 0s and 0s for 1s in a binary number.

(**2**) opposite form of a number system.

complementary arithmetic a method of performing integer arithmetic within a computer, in which negative numbers are represented in such a way that the arithmetic may be performed without regard to the sign of each number.

complementary cumulative distribution function (CCDF) a function describing the probability $p(x)$ of achieving all outcomes in an experiment greater than x.

complementary metal oxide semiconductor (CMOS) (**1**) refers to the process that combines n-channel and p-channel transistors on the same piece of silicon (complementary). The transistors are traditionally made of layers of metal, oxide, and semiconductor materials, though the metal layer is often replaced by polysilicon. There are a number of variations such as HCMOS, high-speed CMOS which scales down the elements compared to the standard MOS process and thus increases the speed and reduces the power consumption for each transistor in the CPU.

(2) a CMOS memory device used in computers to store information that must be available at startup. The information is maintained in the device by a small battery.

complement of a fuzzy set the members outside of a fuzzy set but within the universe of discourse. Represented by the symbol ¬.

Let A be a fuzzy set in the universe of discourse X with membership function $\mu_A(x), x \in X$. The membership function of the complement of A, for all $x \in X$, is

$$\mu_{\neg A}(x) = 1 - \mu_A(x) .$$

See also complement, fuzzy set, membership function.

complement operator the logical NOT operation. In a crisp (non-fuzzy) system, the complement of a set A is the set of the elements which are not members of A. The fuzzy complement represents the degree to which an element is not a member of the fuzzy set.

complete deductive system a deductive system that can produce all true statements of a theory.

complete failure a failure that results in the inability of an item to perform all required functions.

completeness the capability of a system of addressing a behavior for all the possible aspects of its specification.

complete schedule a schedule S of n transactions $T_1, T_2, \ldots, T_N$ where S contains exactly the operations of $T_1, T_2, \ldots, T_N$ and the order of any pair of operations in T_i is the same as that in S. The order of conflicting operations must also be defined.

complete statistic a sufficient statistic T where every real-valued function of T is zero with probability one whenever the mathematical expectation of that function of T is zero for all values of the parameter. In other words, if W is real-valued function, then T is complete if

$$E_\theta W(T) = 0 \forall \theta \in \Theta$$
$$\Rightarrow P_\theta[W(T) = 0] = 1 \forall \theta \in \Theta .$$

complete theory a logical theory in which all true statements have formal proofs within the theory.

complex conjugate if $z = a+bi$ is a complex number then its complex conjugate, denoted $\overline{z}$, is $\overline{z} = a - bi$.

complex instruction set computer (CISC) a processor with a large quantity of instructions, some of which may be quite complicated, as well as a large quantity of different addressing modes, instruction and data formats, and other attributes. The designation was put forth to distinguish CPUs such as those in the Motorola M68000 family and the Intel Pentium from another approach to CPU design that emphasized a simplified instruction set with fewer but possibly faster executing instructions, called RISC processors. One CISC processor, the Digital VAX has over 300 instructions, 16 addressing modes and its instruction formats may take up 1 to 51 bytes.

A CISC processor usually has a relatively complicated control unit. Most CISC processors are microprogrammed. *See also* microprogramming and RISC processor.

One of the benefits of a CISC is that the code tends to be very compact. When memory was an expensive commodity, this was a substantial benefit. Today, speed of execution rather than compactness of code is the dominant force.

See also microprogramming and reduced instruction set computer.

complexity (1) a measure of how complicated a chunk (typically of code or design) is. It represents how complex it is to understand (although this also involves cognitive features of the person doing the understanding) and/or how complex it is to execute the code (for instance, the computational complexity). The complexity evaluation can be performed by considering the computational complexity of the functional part of the system, i.e., the dominant instructions in the most iterative parts of the system. The complexity may be also a measure of the amount of memory used or the time spent in execution of an algorithm.

(**2**) an attribute of an environment or task that drives the decision to apply an autonomous robot vs. a teleoperated robot. Complexity may be spatial, referring to the number of objects in the target environment and to the structure inherent in object location. Complexity may also be temporal, referring to the degree to which the environment is dynamic and to the predictability of changes that occur. Generally, highly complex environments (cluttered and unpredictable) require teleoperated robots and autonomous robots function best in low-complexity environments (sparsely populated and highly predictable).

(**3**) *See* computational complexity.

complexity analysis the analysis of the complexity of a system. This is performed on the basis of the estimated system complexity. The system complexity can be evaluated in several different manners. The complexity is typically estimated by using specific tools on the basis of complexity metrics. A different kind of complexity can also be used in the analysis: the computational complexity. It gives the order of complexity of the algorithm under analysis.

complexity class any of a set of computational problems with the same bounds (Θ) on time and space, for deterministic and nondeterministic machines. *See also* canonical complexity classes, complexity class p-space, Turing machine.

complexity class p space the class of languages that can be accepted by a Turing machine in polynomial space.

complex number a number that has both real and imaginary parts. For example, in the complex number $3 + 4i$, 3 is the real part and 4 is the imaginary part.

complex object an object which cannot be represented by atomic values. One of the motivating reasons for the development of object-oriented database systems was the need to represent complex objects. Complex objects fall into one of two categories. Structured Complex Objects are composed of other objects and hence their structure can be determined by the OODMS. Unstructured Complex Objects are data types that require a large and unpredictable amount of storage capacity, such as a BLOB.

complex plane a map where complex numbers are plotted. It is similar to the Cartesian plane except that the y-axis is labeled as "iy".

complex variable placeholders for complex variables. Usually denoted by some variant of the letter z.

compliance (**1**) the capability of the software product to adhere to standards, conventions or regulations in laws and similar prescriptions. It is a sub-feature of functionality.

(**2**) the degree to which a robotic manipulator can adapt to forces acting on it by means of elasticity or backlash in the mechanical system.

compliance test an operation that compares relevant attributes of a product with applicable standard requirements for determining the achievement of the level of quality required.

compliant motion motion of the manipulator (robot) when it is in contact with its "environment", such as writing on a chalkboard or assembling parts.

complimentary force feedback in telemanipulation, presentation of force feedback that is the sum of force information from the remote world, after application of a low-pass filter, and predicted forces (computer generated) that have passed through a high-pass filter; a synergistic combination of real force information and virtual force information.

component physical part of a subsystem that can be used to compose larger systems. A reusable encapsulation which is typically larger scale than an individual class. Components can be bought and sold and are therefore real not conceptual. Any self-contained item or equipment that is part of a larger item and contributes to the function performed by the larger item. In object-oriented programming, the components are (groupings of) objects.

component diagram a diagram describing the relationships among the components of a systems. In object-oriented programming it is frequently called object diagram and includes relationships of is-part-of and those by means of method calls.

component failure the failure of a component of a larger item. This leads to a component fault and an internal error, which may propagate and lead to a system failure unless error recovery is performed. *See* local fault.

component fault a fault in a component of a larger item. This may lead to an internal error that may propagate and lead to system failure unless error recovery is performed. *See* local fault.

component services provide the underlying component software infrastructure for managing compound documents, i.e., cross-platform portability, and interoperability.

component software an object-based model that facilitates interactions between independent programs. This approach aims to simplify the design and implementation of applications, and simplify human–computer interaction. Component software addresses the general problem of designing systems from application elements that were constructed independently by different vendors using different languages, tools, and computing platforms. The goal is to have end-users and developers enjoying the same level of plug-and-play application interoperability that is available to hardware manufacturers.

component testing the testing of single or groups of components. Synonymous with module testing. *See also* integration test, unit test.

composite attribute an attribute that may be divided into smaller constituents.

composite class in object-oriented programming, a class defined in terms of a set of objects that are instances of other classes. The class components are class attributes.

composite key a key that comprises a set of attributes.

composite object an object made up of one or more (typically two or more) other objects. In this sense, simple values such as integer, floating point, or character may be interpreted as objects. The objects that form a composite object may be selected by a computation (such as a numeric value in an array) or by a name (such as a field name), depending on the rules of the language. *See also* record, variant record, union.

composite state a state hiding a more detailed set of substates.

composite type a data type which is made up of one or more (typically two or more) other types. *See also* record, variant record, union.

compositing the process of combining multiple images into a single image. Usually this is performed in films to make a computer graphics generated character appear on a previously filmed background. The term is also used in traditional photographic manipulation to refer to the process by which cell animation is recorded onto film under a rostrum camera. In film, the "mechanical" process is usually called matte photography (*See* color keying), and the process, when used in film sequences, is ambiguously called traveling matte.

composition a class that contains other objects. In system design, the process of composition is performed when components are defined for composition of already available components, for instance, those taken from libraries or other already available components of the system under development. *See* bottom-up.

compositional modeling a methodology for organizing domain theories so that models for specific systems and tasks can be automatically formulated and reasoned about.

compositional rule of inference generalization of the notion of function. Let X and Y be two universes of discourse, let A be a fuzzy set of X, and let R be a fuzzy relation in $X \times Y$. The *compositional rule of inference* associates a fuzzy set B in Y to A in three steps:

1. cylindrical extension of A in $X \times Y$;
2. intersection of the cylindrical extension with R;
3. B is the projection of the resulting fuzzy set on Y.

If we choose *intersection* as *triangular norm* and *union* as *triangular co-norm*, then we have the so-called *max-min composition* $B = A \circ R$, i.d.

$$\mu_B(y) = \bigvee_x \left[\mu_A(x) \bigwedge \mu_R(x, y)\right] .$$

If we choose *algebric product* for *triangular norm* and *union* as *triangular co-norm*, then we have the so-called *max-product composition* $B = A \tilde{\circ} R$, i.d.

$$\mu_B(y) = \bigvee_x \left[\mu_A(x)\mu_R(x, y)\right] .$$

The compositional rule of inference is the principal rationale behind *approximate reasoning*.

See also approximate reasoning, cylindrical extension of a fuzzy set, fuzzy relation, intersection of fuzzy sets, projection of a fuzzy set.

compound document an object-based model that facilitates automatic transparent updates. A *compound document* is a container for sharing heterogeneous data, which includes mechanisms that manage containment, association with an application, presentation of data/applications, user interaction with data/applications, provision of interfaces for data exchange, and more notably linking and embedding. Data can be incorporated into a document by a pointer (link) to the data contained elsewhere in the document, or in another document. Linking techniques reduce the storage requirements for managing compound documents when these are shared.

comprehensibility the capability of a software to include a set of functionalities. Synonymous with understandability.

compression (1) in information theory, the compact encoding (with a smaller number of bits) I_c of a digital image or signal I obtained by removing redundant or non-significant information, thus saving storage space or transmission time. Compression is termed lossless if the transformation of I into I_c is reversible; otherwise it is termed lossy.

(2) in signal processing, at given bias levels and frequency, the ratio between the small signal power gain ($p_{\text{out}SS}/p_{\text{incident}SS}$) under small signal conditions and the large signal power gain ($p_{\text{out}LS}/p_{\text{incident}LS}$) at a given input power, expressed in dB. As the input amplitude of a signal is increased, the output signal will eventually cutoff and/or clip due to saturation, resulting in compression. If the large signal is insufficiently large to cause cutoff and/or clipping, then the compression will be at or near 0 dB.

$$G_{CR} = 10 \log_{10} (p_{\text{out}SS}/p_{\text{incident}SS}) - 10 \log_{10} (p_{\text{out}LS}/p_{\text{incident}LS}) .$$

compression ratio the ratio of the number of bits used to represent a signal before compression to that used after compression.

compulsory miss *See* cold start miss.

computable a function that can be calculated by an algorithm. The class of computable functions is exactly the class of primitive recursive functions.

computational complexity an algebraic expression describing the relationships among the factors that mainly influence the computational cost of an algorithm or process. There exists several forms and definitions for expressing the theoretical limits of the computational complexity. *See* asymptotical complexity.

computational cost the cost in terms of time (elapsed CPU time) for executing an algorithm or a process.

computational load percentage of time or resources that a program or process takes to execute on a computer.

computation-bound pertaining to a process that performs little input/output but needs significant execution time. Also called compute-bound.

computation error the sum of representation error and arithmetic error.

computation tree logic (CTL) a propositional, branching-time temporal logic for which formulas can be checked in linear time.

compute bound a process that performs little input or output but needs significant execution time.

compute-bound *See* computation-bound.

computed goto a construct, notably in the FORTRAN language, which is a limited form of the multiselect control statement (the case statement) in which the only option is to perform a control transfer to a specific program point. In Algol-60 this was called a switch statement.

computed tomography (CT) *See* tomography.

computer (**1**) an electronic, electromechanical, or purely mechanical device that accepts input, performs some computational operations on the input, and produces some output.

(**2**) functional unit that can perform substantial computations, including numerous arithmetic operations, or logic operations, without human intervention during a run.

(**3**) general or special-purpose programmable system that is able to execute programs automatically. It has one or more associated processing units, memory, and peripheral equipment for input and output. Uses internal memory for storing programs and/or data.

computer-aided design (CAD) field of engineering concerned with producing new algorithms/programs which aid the designer in the complex tasks associated with designing and building a complex system. In the context of graphics, CAD refers to the use of computer based models of objects for visualization or testing as an aid in the design process.

computer-aided engineering (CAE) software tools for use by engineers.

computer-aided manufacturing (CAM) manufacturing of components and products when based heavily on automation and computer tools. *See also* computer-integrated manufacturing.

computer-aided software engineering (CASE) a computer application automating the development of graphical and documentation of application design. *See* computer-aided software engineering tool.

computer-aided software engineering tool (CASE tool) an automated software engineering development tool to assist software engineers during the software life-cycle. It can support the analysis, design, coding, testing, assessment, maintenance, management, and documenting phases of a software project. A CASE tool may provide support in only selected functional areas or in a wide variety of functional areas. CASE tools may be used in several modes: as stand alone tools; in this case, only compatibility with environmental elements should be addressed; in small groups which communicate directly with one another; it may be supposed that integration is predefined, perhaps proprietarily. Adopting the CASE approach to building and maintaining systems involves software tools and training for the developers who will use them.

computer architecture an image of a computing system as seen by a sophisticated computer user and programmer. The above concept of a programmer refers to a person capable of programming in machine language, including the capability of writing a compiler. The architecture includes all registers accessible by any instruction (including the privileged instructions), the complete instruction set, all instruction and data formats, addressing modes, and other details that are necessary in order to write any program. This definition stems from the IBM program of generating the 360 system in the early 1960s. *Contrast with* computer organization. *See also* Flynn's taxonomy.

computer-assisted teleoperation (CAT) form of teleoperation in which augmented reality displays and autonomous functions help the user complete tasks; a hybrid of teleoperation, virtual reality, and autonomous functions.

computer augmented virtual environment (CAVE) a multimedia, immersive display system for virtual environments where the viewer stands inside a room upon whose walls are projected images. The images may be in stereo requiring stereo shutter glasses to be worn.

computer generated hologram a hologram where the required complex amplitude and phase functions are generated by computer and written onto an optical medium.

computer graphics the general term for the art and science of representing images for computer display.

computer hardware description language (CHDL) a programming language used to describe hardware layouts. Examples include VHDL and Verilog; current work in CHDL includes mainly languages for verification, and extensions of existing languages for system description and analog design. CHDL conferences are organized every year.

computer-integrated manufacturing (CIM) manufacturing approach that makes substantial use of computers to control manufacturing processes across several manufacturing cells. *See also* computer-aided manufacturing.

computer model a *computer model* of a device consists of a mathematical/logical model of the behavior of the device represented in the form of a computer program. A good computer model reproduces all the behaviors of the physical device in question, and can be confidently used to simulate the device in a variety of circumstances.

computer organization describes the details of the internal circuitry of the computer with sufficient detail to completely specify the operation of the computer hardware. *Contrast with* computer architecture.

computer program abstract a short description of the capabilities of a computer program. It can be useful for promoting the program or for its selection from a repository of available programs.

computer relay a protective relay which digitizes the current and/or voltage signals and uses a microprocessor to condition the digitized signal and implement the operating logic.

computer simulation a set of computer programs which allows one to model the important aspects of the behavior of the specific system under study. Simulation can aid the design process by, for example, allowing one to determine appropriate system design parameters or aid the analysis process by, for example, allowing one to estimate the end-to-end performance of the system under study.

computer subsystem that part of a robotic system that accepts digitized inputs, performs calculations supporting sensing, motion, and decision-making algorithms, and outputs digitized signals to actuators and displays.

computer supported collaborative work *See* computer supported cooperative work.

computer supported cooperative work (CSCW) comprises the (hardware and software) systems that support groups of people working toward a common goal.

computer vision *See* robot vision.

computer word (**1**) datum consisting of the number of bits that forms the fundamental registers, etc.

(**2**) sequence of bits or characters that is stored, addressed, transmitted, and operated as a unit within a given computer. *Computer words* are 1 to 8 bytes long, but can be longer for special applications.

(**3**) data path of a computer (the size of virtual addresses).

concatenate literally, "linked together". To join two sequences to produce a single sequence. Most commonly applied to joining string values to form a single string, but can also apply to lists as well. Not generally used for unordered sets of values, since it implies an ordering is maintained. Many languages provide string concatenation as a primitive operation in the language, while others provide the facility via subroutine library functions.

concatenated code (**1**) a code that is constructed by a cascade of two or more codes, usually over different field sizes.

(**2**) the combination of two or more forward error control codes that achieve a level of performance with less complexity than a single coding stage would require. Serially concatenated coding systems commonly use two levels of codes, with the inner code being a convolutional code and the outer code being a Reed-Solomon code. Parallel concatenated codes improve performance through parallel encoding and iterative serial decoding techniques. *See also* turbo code.

concave a polygon with the property that some points within its area can be joined by a line segment that passes outside the polygon. *Compare with* convex polygon.

concept assignment the process of matching human-oriented concepts to their implementation-oriented counterparts. This type of conceptual pattern matching enables the maintainer to search the underlying code base for program fragments that implement a concept from the application, which is advantageous since change requests are usually couched in end-user terminology, not in that of the implementation. Concept assignment is pattern-matching at the end-user application semantic level. It is a process of recognizing concepts within the source code and building up an "understanding" of the program by relating the recognized concepts to portions of the program.

concept assignment problem The problem of discovering individual human-oriented concepts and assigning them to their implementation-oriented counterparts for a given problem. People understand a program because they are able to relate the structures of the program and its environment to their human-oriented conceptual knowledge about the world.

concept formation the process of the incremental unsupervised acquisition of categories and their intentional descriptions.

The representative concept formation systems include EPAM, CYRUS, UNIMEM, COBWEB, and SGNN. *See* self-generating neural network.

concept phase (**1**) the time period when the users' needs are reviewed and evaluated on documents during the life-cycle.

(**2**) a part of the start-up phase of a project.

concept recognition a knowledge-based technique that automates the recognition of functional patterns in the code.

conceptual abstraction a representation of the domain model (problem, program, and application) knowledge in the form of informal and semi-formal information. Semi-formal, human-oriented, and domain-specific abstractions play a critical role both in reverse and forward engineering, and therefore also in reengineering. Such conceptual abstractions are fundamental to the reengineering process whether it is a totally manual or partially automated process.

conceptual model in system/software system engineering, a requirements model of the system/software system to be developed, its internal components, and the behavior of both the system and its environment. The model is typically given by using a formal language. *See* conceptual abstraction.

conceptual schema describes the structure of the database at a conceptual level; i.e., entities, relationships, and constraints are specified with no reference to lower level storage details.

concrete class a class that is able to be instantiated.

concrete syntax *Contrast with* abstract syntax.

concrete system a system that implements the behavior of one or more abstract systems. A typical *concrete system* will consist of hardware and executable code, and is what users actually use. Each *concrete system* operates in a particular application context.

concurrency (**1**) a mode of operation in which two or more program units execute simultaneously within a short time interval.

(**2**) the notion of having multiple independent tasks available (tasks in this definition means any work to be done, not a formal computational entity).

(**3**) the extent to which transactions execute in parallel and access the same data.

concurrency control the activity of coordinating transactions that execute in parallel and access the same data.

concurrent pertaining to the mode of operation in which two or more program units execute simultaneously within a short time interval. In computer science, this may mean that the activities are executed in parallel with the appearance of simultaneity if they are executed on the same computer of microprocessor or simultaneously if they are executed on distinct computers of microprocessors.

concurrent flow a multi-commodity flow in which the same fraction of the demand of each commodity is satisfied.

concurrent model virtual modeling technique in which the computations required for different parts of the world are run separately, either as separate processes or on separate processors, but in synchronization; this has the advantage of streamlining computations but the disadvantage of requiring coordination of the multiple model parts.

concurrent processing having one logical machine (which may be a multiprocessor) execute two or more independent tasks simultaneously.

concurrent read and concurrent write (CRCW) shared memory model in which concurrent reads and writes are allowed.

concurrent read and exclusive write (CREW) shared memory model in which concurrent reads but only exclusive writes are allowed.

condition a logical expression on system variables expressing a particular interesting situation.

conditional a statement or expression whose value depends upon some controlling expression. Traditionally if-then-else but, depending upon who is arguing the definition, can also include all branching constructs such as switch, case, guarded condition, cond, and other language-specific constructs.

conditional branch a machine instruction which will transfer control to a designated instruction based on some condition asserted in the machine's state. A compiler writer needs to understand the details of such instructions to generate optimal code. *See also* speculative evaluation.

conditional coding an approach to the solution of the problem of large codewords and lookup tables in block coding. In this scheme one assumes that the receiver already knows the components $b_1, b_2, \ldots b_{N-1}$ of N-tuple b. Current component b_N can now be coded using this information. The assumption that there is statistical dependence between pixels is made.

conditional compilation a mechanism that includes or excludes source statements depending upon externally supplied conditions. These may include the definition or lack of definition of a name, the value of an expression evaluable at compilation time, or options specified to the compilation process, among other mechanisms. The mechanism may be defined as part of the language, or may use an entirely different notation.

conditional instruction an instruction that performs its function only if a certain condition is met. For example, the hypothetical assembly language instruction JNZ TOP only jumps to TOP if the zero flag is clear (the 'not zero' condition).

conditional jamming a version of jamming in which the force-1 operation is carried out only if all the bits to be discarded are 0s.

conditionally addressed ROM read-only memory in which not every address can be used to access a valid word. Usually implemented from a PLA.

conditional statistic a statistic premised on the occurrence of some event. The probability of event E_1 given that E_2 has occurred is denoted by $p(E_1|E_2)$. *See also* Bayes' rule.

conditional-sum adder an adder that consists of several levels organized into blocks such that the block-size starts at two and doubles at each level until there is only a single block. At each level and in each block two summands are generated, one assuming that the carry into the block is 0, and the other assuming that it is 1. At the next level the correct carry into each block is known, the correct summand is selected, and every pair of adjacent blocks forms a new block.

condition code (1) internal flag used in the construction of CPUs.

(2) an implementation of a form of machine state, typically reflecting the results of computations that are not representable in the computational result, such as overflow, underflow, carry, or the results of comparison operations such as greater, less than, or equal. The program may reference these flags to determine whether to branch or not. The precise model of how condition codes reflect the base machine computations is critical to how a compiler generates code.

condition code register register that contains the bits that are the condition codes for the CPU arithmetic or compare instructions.

conditions for self-similarity there are a number of conditions for self-similarity. One of them, referring to the *Hurst parameter,* is as follows: A stochastic process $X(t)$ is statistically self-similar with parameter $H(0.5 \leq H \leq 1)$ if for any real $a > 0$, the process $a^{-H}X(at)$ has the same statistical properties as $X(t)$.

condition synchronization a synchronization technique that involves delaying a process until the state satisfies some predefined Boolean condition.

condition variable (1) a variable set as the result of some arithmetic or logical comparison.

(2) a structure that may appear within a monitor, global to all procedures within the monitor, that can have its value manipulated by wait and signal operations.

cone tracing an alternative to ray tracing in which cones are projected from the camera center through each pixel, where the intersection of the cone and the scene model is used to determine the pixel's color.

configuration (1) generally, the arrangement of a computer system or component as defined by the number, nature, and interconnections of its constituent parts.

(2) in software engineering, the collection of programs, documents, and data that must be controlled when changes are to be made.

(3) operation in which a set of parameters is imposed for defining the operating conditions. The configuration of a personal computer regarding low level features is frequently called set-up. At that level the memory, the sequence of boot, the disk features, etc. are defined. The configuration of a computer also involves that of its operating system. For example, per MS-DOS see CONFIG.SYS and AUTOEXEC.BAT. The configuration of applicative software depends on the software under configuration itself.

(4) for a Turing machine, synonymous with instantaneous description.

configuration audit an activity performed by a software quality audit group with the intent of ensuring that the change control process is working.

configuration control the control of changes to programs, documents, or data.

configuration diagram *See* block diagram.

configuration item hardware or software, or an aggregate of both, which is designated by the project configuration manager (or contracting agency) for configuration management.

configuration management a discipline applying technical and administrative controls throughout the life-cycle to:

1. identify and document physical and functional characteristics of configuration items;

2. limit changes to characteristics of those configuration items;

3. record and report change processing and implementation of the system.

While much effort has focused on managing large suites of source code which can exist in multiple, potentially inconsistent versions, configuration management often extends to documentation (specification, implementation, and end user documentation), test suites, test results, and other related materials. Discussions of languages for specifying configuration management consumed much time during the 1980s.

configuration programming an approach that advocates the use of a separate configuration language to specify the coarse-grain structure of programs. Configuration programming is particularly attractive for concurrent, parallel, and distributed systems that have inherently complex program structures.

configuration status reporting (CSR) an activity that helps software developers to understand what changes have been made and why.

conflict when two or more transactions require incompatible access to a resource, resulting in one or more transactions being blocked.

conflict equivalence when one schedule can be transformed into another schedule by swapping non-conflicting operations.

conflicting operations given two transactions each issuing operations against some database item, if at least one of the requests is a write request, we say operations are conflicting.

conflict miss a cache miss category used to denote the case where, if the cache is direct-mapped or block-set-associative, too many blocks mapped to a set leading to that block can be expelled from the cache, even if the cache is not full, and later retrieved again. These are also called "collision misses". *See also* capacity miss.

conflict serializable when a schedule is conflict equivalent to a serial schedule.

conformance *See* non-conformance, non-conformity.

confusion matrix shows the counts of the actual vs. predicted class values. It shows not only how well the model predicts, but also presents the details needed to see exactly where things may have gone wrong.

congestion an undesirable network situation. When the load on a system approaches capacity, its performance starts to degrade because of interference. Depending on the design of the system, buffers may start to fill up, increasing waiting time, data may be lost as buffers overflow or busy links may reject traffic. Once a network has become congested due to a burst of heavy traffic, it may remain congested for a considerable time, even after traffic subsides.

congestion control as distinguished from flow control, *congestion control* refers to methods for protecting the entire network from becoming congested. Methods for congestion control include traffic shaping to smooth out bursty traffic, flow control to reduce traffic at the point it enters the network from the host, choke packets sent to hosts from downstream nodes, and methods for discarding packets.

conjunction (**1**) the Boolean AND function. *See also* disjunction, implication.

(**2**) less commonly, a bitwise AND.

connect/disconnect bus *See* split transaction.

connected graph an undirected graph which has a path between every pair of vertices.

See also biconnected graph, strongly connected graph, forest, bridge.

connection the establishment of communication channel, in networking, or in concurrent programming. The connection can be based on an M:N relationship among the M senders and the N receivers. The communication mechanisms can be based on very different protocols and supports: Ethernet, modem, radio, satellite, etc. In Internet, to have the connection means the reaching of a location.

connection hijacking the injection of packets into a legitimate connection that has already been set up and authenticated.

connectivity specifies which sets are considered to be connected. Generally it is based on an adjacency relation between pixels (or voxels), so that a set X is connected if and only if for any $p, q \in X$ there is a sequence $p_0, \ldots, p_n$ $(n \geq 0)$ such that $p = p_0, q = p_n$, and for each $k < n$, p_k is adjacent to p_{k+1}. *See* pixel adjacency, voxel adjacency.

consciousness attributes ascribed to *consciousness* usually include self-awareness, a sense of past and future, free will, and most outward signs of intelligent behavior.

conservation law a rule stating that the weighted sum of mean waiting times of requests of different classes in a work conserving system that is not overloaded is independent of the order of service. If ρ_i is the offered load in class i and W_i is a random variable for the waiting time of class i requests, the weighted sum of mean waiting times is

$$\sum_{i=1}^{N} \rho_i \mathrm{E}[W_i] .$$

conservative two phase locking database strategy that requires a transaction to lock all the items it accesses before the transaction begins.

consistency a software specification or a document is consistent when its parts are not in contradiction. It can be regarded as the degree of uniformity.

consistency of interests situation in which there are several decision units with consistent goals. *Compare with* disagreement of interests.

consistency preservation the process of ensuring a database is in a consistent state.

consistency principle a principle from possibility theory relating to the consistency between probability and possibility which states that the possibility of an event is always at least as great as its probability.

consistent backup a backup containing only committed transactions; i.e., updates by active transactions are not included.

consistent estimator an estimator whose value converges to the true parameter value as the sample size tends to infinity. If the convergence holds with probability 1, then the estimator is called strongly consistent or consistent with probability 1.

consistent goals objectives of several decision units in charge of a controlled partitioned system which, when followed, would lead jointly to overall optimal decisions (actions) of these units; independent decision makers contributing to common objectives, with consistent goals, form a team.

const **(1)** abbreviation for constant.

(2) in the C/C++ language, a qualifier to a declaration that states that the value thus declared will not change during the scope of the name. *See also* symbolic constant.

constant a value which cannot change in the course of executing some section of code. Traditionally, a numeric or string value specified in the program (manifest constant), or a name by which such a numeric or string value may be referenced (symbolic constant). In languages supporting nested name scope, a name may have a value that is computed once (traditionally when the name is declared) and which remains unchanged during the scope of the name, but whose value may be different each time the value is computed when the name is declared.

constant angular velocity term normally associated with disk storage units where the disk platters rotate at a constant rotational speed. Because of this, and to have the same amount of data in each track, sectors on the inner tracks are more densely recorded than the outer tracks.

constant bit density on a disk, recording pattern in which the number of bits per unit distance is the same over all tracks.

constant folding a technique by which a compiler recognizes that certain computations may be performed at compilation time and need not be deferred until execution time. While traditionally this has applied to doing simple arithmetic computations such as replacing "2*3" with "6", it may also apply to control structures, conditional computations, and the computation of access paths. For example, if the termination condition for a loop is a compile-time constant and implies the loop can never execute, the entire body of code of the loop might be eliminated.

constant function a function with no arguments or one which always gives the same value. *See also* unary function, binary function, N-ary function.

constant linear velocity term used in conjunction with optical disks where the platter rotates at different speeds depending on the relative position of the referenced track. This allows more data to be stored on the outer tracks than on the inner tracks. Because it takes time to vary the speed of rotation, the method is best suited for sequential rather than random access. *See also* constant angular velocity.

constraint a logical expression, often an equality or inequality relation, used to describe the correct values of one or more system variables, i.e., temperature < 70 IMPLY pressure < 230 is a *constraint* on temperature and pressure of a system. A correct system satisfies all its constraints.

constraint equation an equation which is a Boolean expression that must be true. Generally the goal is to manipulate the values that make up the *constraint equation* to accomplish this. A single constraint equation is usually not hard to solve, but a system of constraint equations is what is usually required. For example, some of the equations may place limits on the values such that the solution set to the system is limited. Many languages have evolved which are based on constraint equations, including languages for ensuring database consistency. Historically, within the domain of problems that can have continuous solutions, linear programming is a technique for solving sets of continuous constraint equations while minimizing some metric (such as cost or weight or other parameter to be optimized).

constraint length in convolutional codes, an indication of the number of source words that affect the value of each coded word. Two typical forms are:

1. A code with constraint length K, in which the value of each coded word is affected by the present source word and up to $K - 1$ previous source words.

2. The number of shifts over which a single message bit can influence the output of a convolutional encoder.

constraint propagation artificial intelligence technique in which a hypothesis generates constraints that reduce the search space over the rest of the data. If no eventual contradiction is derived, then a "match" is achieved.

constraint solving (CS) a technique for solving combinatorial problems such as planning, scheduling, and configuration. A problem is expressed primarily in terms of the variables involved and the constraints on those variables.

constructive algorithm learning algorithm that commences with a small network and adds neural units as learning proceeds until the problem of interest is satisfactorily accommodated.

constructive cost model (COCOMO) a method for evaluating the cost of a software package proposed by Dr. Barry Boehm. There are a number of different types: The Basic COCOMO Model estimates the effort required to develop software in three modes of development (organic mode, semidetached mode, or embedded mode). The Intermediate COCOMO Model is an extension of the Basic COCOMO Model. The Intermediate Model uses an Effort Adjustment Factor (EAF) and slightly different coefficients for the effort equation than the Basic Model. The Intermediate Model also allows the system to be divided and estimated in components. The Detailed COCOMO Model differs from the Intermediate COCOMO model in that it uses effort multipliers for each phase of the project.

constructive solid geometry (CSG) a paradigm for representing 3D shapes in terms of mathematically based compositions of geometric primitives. Any volumetric primitives can be used provided the primitive can satisfy an "inside-outside" test which uniquely partitions points in the space near it. Typically, Boolean set theoretic composition operators (e.g., intersection, union, difference) are used. Affine transformations may be applied to alter the shape of the primitives. For example, the exterior of an igloo may be represented as the union of a sphere and a cylinder, intersected with a cube.

constructor in object oriented languages, a member function that initializes the memory space and the values of the object attributes in order to set the initial object state. *See also* destructor.

consultant a contract programmer or external software specialist.

consumer a program unit that uses data produced by other program units.

contact head *See* disk head.

container abstract data type storing a collection of objects (elements).

containment a grouping of heterogeneous individuals, possibly in space or time. *See also* collection.

content-addressable memory (CAM) *See* associative memory.

contention additional latency incurred as the result of multiple requestors needing access to a shared resource, which can only be used one at a time.

contention protocol class of a multiple access protocol where the user's transmissions are allowed to conflict when accessing the communication channel. The conflict is then resolved through the use of a static or dynamic conflict resolution protocol. Static resolution means that the conflict resolution is based on some preassigned priority. A static resolution can be probabilistic if the statistics of the probabilities are fixed. A common example is the p-persistent ALOHA protocol. The dynamic resolution allows for the changing of the parameters of the conflict resolution algorithm to reflect the traffic state of the system. A common example is the Ethernet protocol.

context the privilege, protection, and address-translation environment of instruction execution.

context-free grammar a phrase structure grammar that is a Type 2 grammar in the Chomsky hierarchy. Characterized by productions of the form $N \rightarrow x$, where N is a member of the nonterminal symbols of the grammar and x is a sequence of zero or more symbols from the grammar's vocabulary, V; that is, x can be defined as a member of V^*, where $*$ is the Kleene Star. *See also* pushdown automaton.

context of use the users, goals, tasks, equipment (hardware, software, and materials), and the physical and social environments in which a product is used.

context-sensitive grammar a phrase structure grammar which is a Type 1 grammar in the Chomsky hierarchy. Characterized by productions of the form $\mu_1 N \mu_2 \rightarrow \mu_1 A \mu_2$, where μ_1 and μ_2 are sequences of zero or more members of the grammar's vocabulary, N is a member of the set of nonterminal symbols, and A is a nonempty sequence of symbols of the grammar's vocabulary. *See also* nondeterministic pushdown automaton.

context switch the housekeeping operations performed when the currently executing program is removed from the processor and a new program is loaded. *Context switching* implies the saving of the current status of the microprocessor and of the operating system for loading the new status. *See* overhead, scheduler.

context units a set of memory units added to a feedforward network that receives information when an input is presented to the network and passes this information to the hidden layer when the next input is presented to the network.

contiguous pertaining to the allocation of data or program elements which are stored in a memory area without address gaps.

contiguous file a file that is stored on a disk without any unused space in between file storage units.

continuous Hopfield network a Hopfield network with the same structure as the discrete version, the one difference being the replacement of the linear threshold units by neurons with sigmoidal characteristics. Any initial setting of the neuron outputs leads to a motion in the network's state space towards an attractor which, so long as the weights in the network are symmetric, is a fixed point. This allows the network to be employed for the solution of combinatorial optimization problems (its main application) by arranging the network's weights so that an optimal solution lies at a fixed point of the network's dynamics. *Compare with* discrete Hopfield network.

continuous improvement the process of tuning the software development process in order to achieve better results in the future versions. The improvement is based on the assessment of the systems development and in performing corresponding actions for correcting problems and improving the general process behavior.

continuous metrication the process of continuous monitoring of software characteristics by using software metrics to evaluate system status.

continuous motion illusion perception of flickering, sequential images as uninterrupted movement, as in motion pictures or television.

continuous signal a continuous function of one or more independent variables such as time, that typically contains information about the behavior or nature of some phenomenon.

continuous simulation a model involving differential equations.

continuous simulation language a simulation language characterized by having data types and operations suited to the task of simulating a system using continuous mathematics, such as fluid flow, stress analysis, and the like.

continuous speech recognition the process of recognizing speech pronounced naturally with no pauses between different words.

continuous time a model for time that associates with the time instant a value in $\mathcal{R}$ (the set of the real numbers). Since real numbers cannot be represented in finite machines, the values of the time instants are generally mapped to $\mathcal{Q}$ (the set of rational numbers).

continuous tone image coding a process that converts a digitized continuous tone image to a binary bit stream which has fewer bits than the original image for the purpose of efficient storage and transmission. *See* still image coding.

contour (1) an image curve, often used to represent the set of points where a given function has a given constant value. A familiar example is a contour line on a topographic map. Here the contour denotes where the land has a given elevation. Another type of map contour might denote the boundary between increasing and decreasing population density. The equivalent concept in 3D is the level surface or isosurface.

(2) the edge that separates an object from other objects and the background. It must consist of one or several closed curves, one for the outer contour, and the others (if any) for the inner contours surrounding any holes. *See* contour filling, contour following, edge.

contour filling an object contour is generally built with an edge detector, but such a contour can be open because some of its pieces, not recognized by the edge detector, may be missing. In order to close the contour, missing pieces can be added by an operator filling small holes in a contour. *See* contour, edge detection.

contour following an operator which, starting from a contour point, follows the closed curve made by that contour. *See* contour.

contour model a model of static and dynamic scope which uses nested rectangles to illustrate static scope and directed arrows to indicate dynamic scope.

contract a *contract* spells out the obligations and benefits afforded to both the client and supplier in an interaction.

contract-driven lifecycle a lifecycle model, originally used in SOMA and OPEN, in which the focus is on a set of activities which are objectified. The model describes a process that is configurable. Developers using the process move from one activity to another in such a way that their path is governed by the satisfaction of contracts associated with the activity objects.

contract software a software developed by an external organization under contract.

contrast the range of colors in an image. Increasing the *contrast* of a color palette makes different colors easier to distinguish, while reducing the *contrast* makes them appear washed out.

contrast enhancement alteration of the contrast in an image to yield more details or more information. *See also* contrast, histogram stretching, histogram equalization.

contrast sensitivity the responsiveness of the human visual system to low contrast patterns. In psychophysics, the threshold contrast is the minimum contrast needed to distinguish a pattern (such as a spatial sinusoid) from a uniform field of the same mean luminance, and the contrast sensitivity is the inverse of the threshold contrast. *See also* human visual system.

contrast sensitivity function function expressing the relationship between human temporal contrast sensitivity and brightness; temporal contrast sensitivity shifts toward higher frequencies with increasing brightness, affecting flicker fusion frequency.

contravariant a type that varies in the inverse direction from one of its parts with respect to subtyping. The main example is the contravariance of function types in their domain. For example, assume $A <: B$ and vary X from A to B in $X \rightarrow C$; we obtain $A \rightarrow C :> B \rightarrow C$. Thus, $X \rightarrow C$ varies in the inverse direction of X.

control (**1**) the study and practice of controlling or making a system behave in a specific manner.

(**2**) intervention, by means of appropriate manipulated inputs, into the controlled process in the course of its operation; some form of observation of the actual controlled process behavior is usually being used by the controller.

control algorithm the algorithm used to control a system.

control bus contains processor signals used to interface with all external circuitry, such as memory and I/O read/write signals, interrupt, and bus arbitration signals.

control cable special cable that connects a device to its controller, and passes control signals between them.

control channel used to transmit network control information. No user information is sent on this channel. *Compare with* traffic channel.

control character (**1**) generally, a character, such as `<CTRL-d>`, that is generated by simultaneously pressing the control key and some other key.

(**2**) with respect to the terminal driver, a character that is processed as an instruction to the driver itself.

control-display gain for continuous controls, the relative relationship between a control movement and the corresponding reaction on an associated display. Also called control-display ratio.

control-display ratio *See* control-display gain.

control flow sequence of operations performed during the execution of a program. The *control flow* can be traced for storing the control flow evolution. *See also* data flow.

control flow diagram *See* control flow.

control flow trace *See* execution trace.

control instruction machine instruction that controls the actions of a processor such as setting flags to enable specific modes of operation. Generally *control instructions* do not perform computations. Sometimes control instructions include instructions that can effect sequential execution of a program, such as branch instructions.

control law basic principle used for a control algorithm.

controlled redundancy the duplication of data in a database in a controlled manner.

controller (**1**) device that is used to transfer information between a peripheral and the CPU.

(**2**) a device used to make inputs during manual control.

control line in a bus, a line used in a computer bus to administrate bus transfers. Examples are bus request (a device wants to transmit on the bus) or bus grant (the bus arbiter gives a device transmit access on the bus).

controlling human task during human-robot interaction: continuous manual control.

control loop a graphical or mathematical representation of a feedback system.

control memory a semiconductor memory (typically RAM or ROM) used to hold the control data in a micro programmed CPU. This data is used to control the operation of the data path (e.g., the ALU, the data path busses, and the registers) in the CPU. If the control memory uses RAM, the CPU is said to be micro programmable, which means that the CPUs instruction set can be altered by the user and the CPU can thus "emulate" the instruction set of another computer. Same as control store and micromemory.

control order the qualitative nature of system response to a user input. In zero order or position control, the system moves the controlled element to the same relative position as the input device; in first order or rate control, the system moves the controlled element at a velocity proportional to the position of the input device; in second order or acceleration control, the system changes the acceleration of the controlled element at a rate proportional to the position of the input device; etc. Mathematically, control systems can be modeled using gain (K) and the Laplace operator (s). *Control order* is a discrete set of steps K (zero order or position control), K/s (first order or rate control), K/s^2 (second order or acceleration control), etc.

Humans can cope with *control order* up to third order, but with increasing workload and likelihood of instability as control order increases. By far the majority of human-controlled systems are zero or first order systems. For example, a mouse controlling a computer cursor operates in position control and aircraft controls for pitch, yaw, and roll operate in rate control.

control point one of a set of points which controls the shape of a curve by its position. The curves may go through some (*see* Bézier curve, an interpolating spline) or all (e.g., the Catmull-Rom interpolating splines) of the control points. Positioning is often interactive and the points are combined by blending functions to generate the shape desired. *See also* b-spline and Bézier curve. Note the distinction between knots and control points: in an interpolating spline, knots and control points are at the same positions in space. In a quadratic or higher order approximating spline, they are in different places: the knots lie on the curve and control points lie near the knots, but not on the curve.

control policy *See* control rule.

control rule decision mechanism (sequence of such mechanisms) used — within the considered control layer — to specify on-line the values of the control inputs; for example, the values of the manipulated inputs in the case of the direct control layer or the set-point values in the case of the optimizing control layer. Also known as control policy.

control scene initial entity, a given, world, which is then partitioned into the controlled process and its (process) environment; control scene is the initial world which is of interest to a control engineer or a system analyst.

control statement in an imperative language, a statement that modifies the control flow of the execution without necessarily computing a value.

control store *See* control memory.

control structure essentially the same as the control system. This term is used when one wants to indicate that the controller is composed of several decision units, suitably interlinked. Decision units of a control structure usually differ in their tasks, scope of authority, and access to information. Depending on the context one speaks of a control structure or of a decision structure. Decentralized control, multilayer control, and hierarchical control are examples of *control structures*.

control subsystem the hardware and software that make up a controller.

control system (**1**) the entity comprising the controlled process and the controller. *Control system* is influenced by the environment of the process both through the free inputs to the process itself and through any current information concerning the behavior of these free inputs which is made available to the controller.

(**2**) an arrangement of interconnecting elements which interact and operate automatically to maintain a specific system condition or regulate a controlled variable in a prescribed manner.

control variable historically, in a counted iteration, the variable that was used to maintain the count. In a general iterator, the variable that maintains the current state of the iterator.

convergence eye movements conducted to cause the left and right lines of gaze to intersect on an object, generally autonomic.

conversational pertaining to a mode of operation in which the interaction resembles a human dialogue.

conversational system a computer system that is able to carry on a spoken dialogue with a user in order to solve some problem. Usually, there is a database of information that the user is attempting to access, and it may involve explicit goals such as making a reservation.

conversion (**1**) the migration of software from one platform to another.

(**2**) the translation of software from one language to another.

(**3**) transformation of data format, e.g., from binary to decimal format.

convex fuzzy set (**1**) a fuzzy set that has a convex type of membership function.

(**2**) a fuzzy set in which all α-level sets are convex. *See also* α-level set.

convex hull for a given set of points, the smallest convex set that contains all the points.

convex polygon a polygon with the property that any line segment joining two points belonging to the polygon area is completely inside the polygon.

convolution the mathematical operation needed to determine the response of a system from its stimulus signal and its weighting function. The convolution operation is denoted by the symbol '$*$'. The convolution of two continuous time signals $f_1(t)$ and $f_2(t)$ is defined by

$$f_1(t) * f_2(t) = \int_{-\infty}^{\infty} f_1(\tau) f_2(t - \tau)\, d\tau$$
$$= \int_0^t f_1(\tau) f_2(t - \tau)\, d\tau$$
$$\text{if } f_1(t), f_2(t) = 0, t < 0 .$$

The integral on the right-hand side of the above equation is called the convolution integral, and exists for all $t \geq 0$ if $f_1(t)$ and $f_2(t)$ are absolutely integrable for all $t > 0$. f_1 is the weighting function that characterizes the system dynamics in the time domain. It is equivalent to the response of the system when subjected to an input with the shape of a Dirac delta impulse function. Laplace transformation of the weighting function yields the transfer function model for the system.

The convolution of two discrete time signals $f_1[k]$ and $f_2[k]$ is defined by

$$f_1[k] * f_2[k] = \sum_{i=-\infty}^{\infty} f_1[i] f_2[k - i]$$
$$= \sum_{i=0}^{k} f_1[i] f_2[k - i]$$
$$\text{if } f_1[k], f_2[k] = 0, k < 0 .$$

The summation on the right-hand side of the above equation is called the convolution sum. Convolution is useful in computing the system output of LTIL systems. *See also* windowed convolution.

convolutional code (**1**) a code generated by passing the information sequence to be transmitted through a linear finite-state shift register and the coder memory couples the currently processed data with a few earlier data blocks. Thus, the coder output depends on the earlier data blocks that have been processed by the coder.

(**2**) a channel code based on the trellis coding principle but with the encoder function (mapping) determined by a linear function (over a finite-alphabet). The name convolutional code is motivated by the fact that the output sequence is a (finite-alphabet) convolution between the input sequence and the impulse response of the encoder.

convolutional coding a continuous error control coding technique in which consecutive information bits are combined algebraically to form new bit sequences to be transmitted. The coder is typically implemented with shift register elements. With each successive group of bits entering the shift register, a new, larger set of bits is calculated for transmission based on current and previous bits. If for every k information bits shifted into the shift register, a sequence of n bits is calculated, the code rate is k/n. The length of the shift register used for storing information bits is known as the constraint length of the code. Typically, the longer the constraint length, the higher the code protection for a given code rate. *See also* block coding, error control coding.

convolution integral *See* convolution.

co-occurrence matrix an array of numbers that relates the measured statistical dependency of pixel pairs. *Co-occurrence matrices* are used in image processing to identify the textural features of an image.

Cook reduction a reduction computed by a deterministic polynomial time oracle Turing machine.

Cook's theorem the theorem that the language SAT of satisfiable Boolean formulas is NP-complete.

Cook-Torrance shading *See* Torrance-Cook shading.

Coons patch a form of parametric bicubic spline representation for surface patches. It allows explicit control of patch boundary position and tangent plane continuity. It is an example of a lofted surface.

Cooper-Harper Scale pencil and paper instrument for measuring subjective mental workload.

coordinated rotation digital computer (CORDIC) algorithm for calculating trigonometric functions using only additions and shift operations.

coordinate frame a set of orthogonal axes that is used to define a point in space.

coordinate system a minimal set of mutually orthogonal vectors which span a given space. All points or vertices in the space may then be represented using a linear combination of these spanning vectors.

coordinate transformations a matrix that transforms points or vectors from one coordinate frame to another.

coordinating unit *See* coordinator unit.

coordination working together. The integration of multiple parts to provide synergy to optimize task performance.

coordinator unit control (decision) agent in a hierarchical control structure, being in charge of decisions (control instruments) influencing operation of the local decision units; *coordinator unit* performs either iterative or periodic coordination of the local decisions and is often regarded as the supremal unit of the hierarchical control structure. Also called coordinating unit.

co-prime polynomials polynomials which have no common factors. For example, polynomials $(s^2 + 9s + 20)$ and $(s^2 + 7s + 6)$ are co-prime, while (s^2+5s+6) and $(s^2+9s+14)$ are not since they have a common $(s+2)$ factor.

coprocessor a processor that is connected to a main processor and operates concurrently with the main processor, although under the control of the main processor. *Coprocessors* are usually special-purpose processing units, such as floating point, array, DSP, or graphics data processors.

copy back *See* copy-back.

copy-back in cache systems, an operation that is the same as write-back — a write operation to the cache that is not accompanied with a write operation to main memory. In *copy-back* the data is written only to the block in the cache. This block is written to main memory only when it is replaced by another block.

copy semantics semantics which require that operations which appear to modify a value act as if they are working on a copy of the value. *Contrast with* reference semantics. *Copy semantics* can be implemented more efficiently on a system that uses reference semantics in most situations but uses lazy evaluation to determine when to actually do the copy to preserve copy semantics.

CORBA *See* common object request broker architecture.

CORDIC a set of algorithms that use additive and multiplicative normalization to evaluate elementary functions. The algorithms may also be seen as realizing rotations in a geometric space. First used in the Coordinate Rotational Digital Computer. *See* coordinated rotation digital computer.

core obsolete. A term used in the U. S. to refer to the main memory of the machine. The operating image of a process (sometimes used to refer to the part residing in physical memory) often written to disk if the program crashes (dumping core). Since magnetized ferrite rings (cores) were once used in main memory to store a single bit each, the name remained and now core memory means the same as main memory, although currently, main memory is chip-based. *See also* magnetic core memory.

core dump a file that contains the memory image of a process at the moment that it encountered an irrecoverable error; used in conjunction with a debugger such as **sdb** to determine the cause of the fatal error.

core memory *See* magnetic core memory.

Coriolis forces forces created during rotational movements and which may contribute to simulator sickness when head movements are made during body rotation.

corner detection the detection of corners, often with a view to locating objects from their corners, by a process of inference, in digital images.

coronal projection a projection image formed on a plane parallel to the chest and perpendicular to the transverse and sagittal planes.

coroutine (1) a routine that begins execution at the point it was last suspended, and that is not required to return control to the calling program unit. The use of multiple coroutines was historically called cooperative multitasking.

(2) a form of control characterized by the characteristic that a function's execution context (stack frame) has an extent (lifetime) which is not limited by its execution.

correction layer control layer of a multilayer controller, usually situated above the direct control layer and below the optimizing control layer, required to make such modifications of the decisions supplied by the optimizing layer — before these decisions are passed to the direct layer — that some specified objectives are met; for example, a *correction layer* of the industrial process controller may be responsible for such adjustment of a particular set point value that an important constraint is satisfied by the controlled process variables — in the case when the optimizing layer is using an inaccurate model of this controlled process.

corrective maintenance the maintenance carried out after fault recognition and intended to put an item into a state in which it can perform a required function. *See* perfective maintenance, adaptive maintenance.

correctness (1) a measure of a system or components about its freedom from faults.

(2) a measure of the conformity of a system or document with respect to early specified requirements.

correctness proof any mathematical technique for verifying specifications and estimating the specification correctness.

correlation the product of one function and shifted versions of another function; the maximum of the correlation can be used to find the best relative shift of two functions.

correlation coefficient a measure of the ability to predict one random variable x as the linear function of another y. The correlation coefficient

$$\rho = \frac{E[xy] - E[x]E[y]}{\sigma_x \sigma_y}$$

satisfies $-1 \leq \rho \leq 1$, where $|\rho| = 1$ implies a deterministic linear relationship between x and y, and $\rho = 0$ implies lack of correlation. *See also* correlation.

correspondence motion detection motion analysis using the time disparity between successive monocular images. Also called matching. *Contrast with* spatio-temporal motion detection. *See also* gradient edge detector.

co-set *See* owner-coupled set.

cost analyst an analysis for cost estimation of the software to be built.

cost avoidance the reduction of expenses such as defect removal costs as a result of taking some explicit action.

cost based optimization optimization of queries based on some estimate of the cost of execution. The cost of a set of different, yet logically equivalent, evaluation strategies is estimated. The cheapest is used for future executions of the query.

cost/benefit analysis the analysis of benefits and costs related to the implementation of a product.

cost estimation describes a suite of techniques that takes early artifacts of the software development process and, from these, calculates a first estimate of overall cost. In the early stages of a software project, some estimate of the total cost, overall effort required, and, hence, personnel requirement (and other resources) is needed. Cost estimation COCOMO and Function Points are two cost estimation models used in a traditional development.

cost formula any formula applied in cost-based optimization to return an estimate of the cost associated with the execution of a query.

cost function a nonnegative scalar function which represents the cost incurred by an inaccurate value of the estimate. Also called penalty function.

cost of failure a measure of the severity of the consequences of failure. Depending on the type of system, different scales may be used, e.g., duration of down time, consequential cash loss, number of lives lost, etc. Cost of failure to user must be distinguished from cost of maintenance to vendor.

cost of measurement the cost (e.g., the person month spent or the total cost) of the project part related to system assessment performed by measuring software characteristics with metrics.

cotree the complement of a tree in a network.

COTS *See* commercial off-the-shelf.

counted iteration an iteration which steps a numeric value through a sequence of values in a specified range. A counted iteration will typically allow the programmer to specify the initial value, final value, and step size. Many languages permit the step size to be negative. The FORTRAN DO-loop and the for-loop in other languages are examples of language constructs to support counted iteration.

counter **(1)** a variable or hardware register that contains a value which is always incremented or decremented by a fixed amount, and always in the same direction (usually incremented by one, but not always).

(2) a simple Moore finite state machine that counts input clock pulses. It can be wired or enabled to count up and/or down, and in various codes.

counting rules the rules according to which source lines of code, LOC, are counted. Important to give such a definition since depending on the definition the outcome can vary by about 500%. Examples are: the counting of LOC without considering comments, the counting of LOC considering both instruction and comments, the counting of LOC excluding blank lines, the counting of LOC considering only the lines containing statements and not delimitators (begin, end, {, }, etc.), etc.

counting semaphore a semaphore that takes more than two values. Counting semaphores can be constructed from two binary semaphores and an integer variable. *See also* binary semaphore.

coupling a measure of the degree to which chunks are connected together. In a traditional environment, the chunks are subroutines, and in an object-oriented environment, they are classes. Highly coupled systems are generally agreed to be hard to understand and maintain.

covariant a type that varies in the same direction as one of its parts with respect to subtyping. For example, assume $A <: B$ and vary X from A to B in $D \rightarrow X$; we obtain $D \rightarrow A <: D \rightarrow B$. Thus, $D \rightarrow X$ varies in the same direction as X.

covering given a finite collection of subsets of a finite ground set, the process of finding an optimal subcollection whose union covers the ground set. *See* packing.

cover set of functional dependencies a set of functional dependencies A is said to cover another set of functional dependencies B, if the closure of A is equal to the closure or B, i.e., $A^+ = B^+$.

CPA *See* carry-propagate adder.

CPU *See* central processing unit.

CPU time the time that is required to complete a sequence of instructions. It is equal to the (cycle time) × (number of instructions) × (cycles per instruction). The types of instructions can be floating point, fixed point, cache, or branch.

CPU utilization a measure of the percentage of non-idle processing.

CPU wait the idle time spent waiting for a particular device external to the processing platform's processor to complete a transaction request. This is that part of application latency incurred due to some process external to the processor. This latency could be the result of waiting on retrieval of information from a storage device, a communications interface, or waiting on a user's response.

Cramer-Rao bound a lower bound on the estimation error covariance for unbiased estimators. In particular, the estimation error covariance $\Lambda_e(\mathbf{x})$, which is a function of the unknown quantity $\mathbf{x}$ to be estimated, must satisfy $\Lambda_e(\mathbf{x}) \geq I(\mathbf{x})$, where I is the Fisher information matrix. If an estimator achieves the *Cramer-Rao bound* with equality for all $\mathbf{x}$, it is efficient; if an efficient estimator exists, it is the maximum likelihood estimator. *See also* bias, maximum likelihood estimation.

crash (**1**) the sudden and complete failure of a computer system or component leading to temporary or permanent loss of part of the data.

(**2**) the sudden, complete failure of a database management system.

(**3**) the event that occurs when a read/write head touches the surface of a platter. A *crash* usually causes irreparable damage to the disk.

CRC *See* cyclic redundancy code.

CRC character a type of error detection code commonly used on disk and tape storage devices. Data stored on a device using CRC has an additional character added to the end of the data that makes it possible to detect and correct some types of errors that occur when reading the data back.

CRC-code code that employs cyclic redundancy checking. *See* cyclic redundancy code.

CRCW *See* concurrent read and concurrent write.

creator in object-oriented programming, the constructor for producing instances from a class.

credit assignment problem the problem of determining the contribution of each individual stage in a series of decisions toward the final realization of a successful outcome. In learning classifier systems, this problem, which takes the specific form of determining which classifiers are important in contributing toward the performance of an effective action by the system, is resolved by the bucket brigade algorithm. *See also* learning classifier system.

creeping user requirements the phenomenon that during development customers and developers add new functionality to the software. A growth rate of 1% per month is not uncommon.

crest factor the ratio of the peak value of a signal to its rms value.

CREW *See* concurrent read and exclusive write.

crisp set in fuzzy logic and approximate reasoning, this term applies to classical (nonfuzzy Boolean) sets which have distinct and sharply defined membership boundaries. *See also* fuzzy set.

criterion a required level of quality measure against which attributes of an object (e.g., a product) or of a process (e.g., the design cycle) are judged to evaluate the level of quality achieved.

critical design review (CDR) a review for verifying the design of one or more configuration components. It is mainly devoted to evaluate the compatibility among entities, to evaluate the risks.

critical failure a failure that is assessed as likely to result in injury to persons, significant material damage, or other unacceptable consequences. Precisely which failure modes are critical will obviously depend on the type of system.

critical incident any event which triggers an evaluative response in an observer; sets of *critical incidents* may be recorded as a measure of the quality of human or telerobot performance.

critical item in configuration management, a configuration item or its part that requires specific care for its critical aspects.

criticality classification of the consequences, or likely consequences, of a failure/fault mode, or classification of the importance of a component for the required service of an item.

critically sampled sampled at the Nyquist frequency.

critical path (**1**) in a precedence graph, the longest path from the source to the sink. In the context of a parallel job, the length of a path is the sum of the execution times for the units of computations that appear on the path.

(**2**) with respect to PERT charts, the set of activities that must be completed in sequence and on time to have the entire project completed on time.

critical path problem the problem of finding the longest path from any source to any sinks in a directed acyclic graph which has weights, or numeric values, on vertices.

critical piece first a system development model where the most critical aspects are analyzed and developed first with respect to the others. *See* bottom-up, top-down.

critical section a section of code which may not be executed concurrently or reentrantly by more than one thread of control while operating on specific data items. Many languages provide critical section constructs as part of the core syntactic structure. *See also* semaphore, mutex, monitor.

critical software a software for which the safety is strongly relevant and its failure could produce damages for the users. *See* real-time system, critical task, critical system.

critical success factor (CSF) a factor that must succeed for the business to prosper. If the CSF is not met, the organization ceases trading.

critical system a system which possesses a critical (or safety-critical) mode of failure which could have impact on safety or on economic aspects. For example, critical on-board avionics systems are defined as those that, if they fail, will prevent the continued safe flight and landing of the aircraft (e.g., those responsible for pitch control).

critical task a task whose failure may lead to great damage. *See* real-time system.

cross-assembler a computer program that translates assembly language into machine code for a target machine different from the one on which the cross-assembler runs.

crossbar switch a structure that allows N units to communicate directly with each other, point to point. Which pairs are connected depends on how the switch is configured at that point in time. Crossbars are usually implemented for small (8 or less) numbers of nodes, but not always.

cross-compiler a compiler which executes on one machine architecture that creates programs that will be executed on a different architecture.

cross-correlation a measure of the correlation or similarity of two signals. For random processes $x(t)$ and $y(t)$, the cross-correlation is given by: $R_{xy}(t_1, t_2) = Ex(t_1)y(t_2)$.

crossed disparity term applied to the retinal disparity of the images of objects closer to the observer than the fixation point.

crossing number the even number obtained by adding the number of changes in binary value on going once around a particular pixel location; *crossing number* is useful for finding skeletons of shapes and for helping with their analysis: it is a measure of the number of "spokes" of an object emanating from a specified location.

cross-modality display display of information normally sensed by one modality to another, e.g., the visual display of force information during teleoperation.

crossover a specific recombination operator in genetic algorithm where segments of chromosomes are exchanged between two individuals in the population to form two new individuals. The simplest form of this operator, known as one-point crossover, selects a random position along the chromosome and exchanges the symbol sub-strings beyond this point between two parents. For example, if the two chromosomes 10011000 and 01110001 are chosen for crossover at the position between the 4th and 5th bit, the resulting new individuals become 10010001 and 01111000. This operation is generalized in multi-point crossover where more than one random location is chosen along the chromosome and sub-strings are exchanged alternately between these points. In uniform crossover, the sub-strings are reduced to single symbols and the decision to exchange each symbol is based on the comparison between a sampled random variable and a threshold. *See also* genetic algorithm, recombination.

crossover model mathematical model of information transmission through humans.

crossover point a *crossover point* of a fuzzy set A is the point in the universe of discourse whose membership in A is 0.5.

crosspoint a point at which two overlapping neighboring fuzzy sets have the same membership grade.

crosspoint level the membership grade of a fuzzy set at a crosspoint.

cross-reference (**1**) a tool that gathers and makes available to other tools and possibly the user information about names in an application. Typically, this information identifies where both names are defined and where they are used. More sophisticated cross-referencers accurately associate references to a name with the proper definition.

(**2**) a list of names referencing the names to one or more source files, which indicates those source lines that define a name and those that reference it. It is uncommon to have such a system for modern languages, but very useful in legacy languages such as FORTRAN, COBOL, and assembly code.

cross validation a method for estimating the true error rate of a theory learned from a set of instances. The data are divided into N equal-sized groups and, for each group in turn, a theory is learned from the remaining groups and tested on the hold-out group. The estimated true error rate is the total number of test misclassifications divided by the number of instances.

CRT *See* cathode ray tube.

cryptography the art and science of secret writing.

CS *See* constraint solving.

CSA *See* carry-save adder.

CSAID *See* chi square automatic interaction detection.

CSCW *See* computer supported cooperative work.

cse acronym (by convention, usually all in lower case), Common SubExpression, a computation which is recognized as being identical to another computation performed in the same compilation. A *cse* is usually detected during the flow analysis.

CSF *See* critical success factor.

CSG *See* constructive solid geometry.

CSP *See* communicating sequential processes.

CSR *See* configuration status reporting.

CT computed tomography. *See* tomography.

ctce acronym (by convention, usually all in lower case), Compile Time Constant Expression, a computation whose value can be determined by having the compiler perform the evaluation.

CTL *See* computation tree logic.

cuberille a representation of 3D space consisting of a regular array of cubes, often referred to as voxels.

cubic voxel *See* voxel.

cue a term from psychology describing a source of perceptual information, as in depth cues like retinal disparity and texture gradients.

culling a process to remove whole polygons that cannot be seen from the viewpoint and do not therefore need to be considered by the hidden-surface removal algorithm.

cumulative defect removal efficiency sum of all the defects found by all available methods (review, testing, etc) before delivery to the customer.

cumulative distribution function (CDF) a function describing the probability $p(x)$ of achieving all outcomes in an experiment less than or equal to x. For a random variable $\mathbf{x}$, the probability that $\mathbf{x}$ is less than or equal to some value x, denoted $F_{\mathbf{x}}(x)$. $F_{\mathbf{x}}(-\infty)$ is zero and increases monotonically to $F_{\mathbf{x}}(+\infty) = 1$. For a continuous probability density function $p(x)$, the CDF is

$$F_{\mathbf{x}}(x) = \int_{-\infty}^{x} p(t)\, dt \ .$$

The CDF is used in image processing to carry out histogram equalization. *See also* histogram equalization, probability density function.

current the rate of flow of electrons. Measured in amperes.

cursor (1) the symbol on a computer screen that indicates the location on the screen that subsequent input will affect.

(2) a movable, visible mark used to indicate a position of interest on a display surface.

(3) an iterator used to iterate over the result collection without consuming much of the application cache when the result of a query is too large to fit in the application cache.

cursor stability a frequently implemented, transaction isolation level, not defined in the SQL92 standard.

curvature a geometric property that describes the degree that a surface or a curve is bent. The curvature of a curve is the magnitude of the rate of change of the unit tangent vector with respect to arc length. The curvature of a surface is given in terms of a metric tensor which embodies two principal (planar) curvatures, κ_1 and κ_2. The curvature of a surface is sometimes characterized by the so-called Gaussian curvature $\kappa = \kappa_1\kappa_2$.

curvature function a function which gives curvature values at different locations of a curve or surface.

custom benchmark a benchmark created for the purpose of evaluating different database platforms for a specific workload to be implemented at a site.

customer recipient of a product provided by the supplier. In a contractual situation, the "customer" is called the "purchaser". The "customer" may be, for example, the ultimate consumer, user, beneficiary, or purchaser. The "customer" can be either external or internal to the organization. The customer is not necessarily identical to the user. The person or organization who purchases an item from a vendor.

customer satisfaction survey periodic formal survey of how well customers receive a certain software product.

customer support answering questions and providing support for solving problems for customers who purchased a product.

cut an annotation used in PROLOG programs to bypass certain nondeterministic computations.

cutting plane a valid inequality for an integer polyhedron that separates the polyhedron from a given point outside it.

cutting theorem states that for any set $\mathcal{H}$ of n hyperplanes in $\Re^k$, and any parameter r, $1 \leq r \leq n$, there always exists a $(1/r)$-cutting of size $O(r^k)$. In two dimensions, a $(1/r)$-cutting of size s is a partition of the plane into s disjoint triangles, some of which are unbounded, such that no triangle in the partition intersects more than n/r lines in $\mathcal{H}$. In $\Re^k$, triangles are replaced by simplices. Such a cutting can be computed in $O(nr^{k-1})$ time.

cut vertex *See* articulation vertex.

cybersurgery application of synthetic environments technology to surgery, whether for training, planning, or actual operations inside a human or animal.

cycle *See* clock cycle.

cycle search a storage allocation technique in which each search for a suitable block begins with the block following the one last allocated.

cycles per instruction (CPI) a performance measurement used to judge the efficiency of a particular design.

cycle stealing an arrangement in which a DMA controller or I/O channel, in order to use the I/O bus, causes the CPU to temporarily suspend its use of the bus. The CPU is said to hesitate. *See also* direct memory access.

cycle time time required to complete one clock cycle (usually measured in nanoseconds).

cyclic access in devices such as magnetic and optical disks (and older bubble memories) that store data on rotating media, the property that any individual piece of data can be accessed once during each cycle of the medium.

cyclic code a linear block code where every cyclic shift of a codeword is another codeword in the same code. This property is an outcome of the significant algebraic structure that underlies these codes.

cyclic convolution *See* periodic convolution.

cyclic redundancy check *See* cyclic redundancy code.

cyclic redundancy code (CRC) a coding scheme for detecting errors in messages. When using the CRC technique, the modulo-2 sum of the message bits is calculated after grouping them in a special way and then appended to the transmitted message and used by the receiver to determine if the message was received correctly. The number of bits used in the CRC-sum is typically 8 or 16, depending on the length of the message and the desired error-detection capability. CRCs are generated using a shift register with feedback.

cyclomatic complexity a technique invented in 1976 by Thomas McCabe in which graph theory is used to construct a measure of the complexity of a chunk of code as a prerequisite to constructing testing strategies. *See* McCabe's metric.

cylinder a stack of tracks where these tracks are at a constant radial position (the same track number) on a disk or disk pack.

cylindrical extension of a fuzzy set let A be a fuzzy set in a Cartesian product space X^i, and X^n be another Cartesian product space including X^i. Then the cylindrical extension of A in X^n is a fuzzy set in X^n, denoted as $cext(A; X^n)$, with membership function defined equal to membership in X^i. *See also* fuzzy set, membership function, projection of a fuzzy set.

D

D-adjacent an entry reachable for a $(d-1)$-extremal entry through a unit vertical, horizontal, or diagonal-mismatch step.

daemon a process that operates on behalf of the operating system rather than on behalf of any particular user.

DAG *See* directed acyclic graph.

DAG shortest paths technique to find the single-source shortest-path problem in a weighted directed acyclic graph. The procedure is as follows:

1. Do topological sort on the vertices by edge so vertices with no incoming edges are first and vertices with only incoming edges are last.
2. Assign an infinite distance to every vertex ($\text{dist}(v) = \infty$) and a zero distance to the source.
3. For each vertex v in sorted order, for each outgoing edge $e(v, u)$, if $\text{dist}(v) + \text{weight}(e) < \text{dist}(u)$, set $\text{dist}(u) = \text{dist}(v) + \text{weight}(e)$ and the predecessor of u to v. *See also* Dijkstra's algorithm.

daily traffic profile a function giving the dependence of a traffic measure as a function of the time of day. The granularity is usually 1 hour or 15 minutes. The traffic measure can be traffic volume or call rate.

daisy chain (1) a type of connection when devices are connected in series.

(2) a hardware configuration where a signal passes through several devices. A signal will be passed through a device if that device is not requesting service, or not passed through if the device is requesting service.

(3) an interrupt-prioritizing scheme in which the interrupt acknowledge signal from the CPU is connected in series through all devices. A shared interrupt-request line connects all devices to the CPU with a single common line. When one (or several) device activates its request line, the CPU will (after some delay) respond with an acknowledge to the first device. If this device did request an interrupt, it will be serviced by the CPU. However, if the device did not request an interrupt, the acknowledge is just passed through to the next device in the daisy chain. This process is repeated until the acknowledge signal has passed through all the connected devices on this chain. The scheme implements prioritized service of interrupts by the way the devices are electrically connected in the *daisy chain:* the closer a device is to the CPU, the higher its priority.

A more general case exists where several daisy chains are used to form priority groups, where each chain has a unique priority. The CPU will service interrupts starting with the daisy chain having the highest priority. In this scheme, any device may be connected to more than one priority group (chain), using the interrupt priority level appropriate for the particular service needed at this moment.

danger the probability of a given hazard leading to an accident within a given time interval under given conditions. It is assumed that no accident has occurred at the start of the interval.

dangling pointer in a language supporting dynamic value allocation but which does not support garbage collection, a reference to an object which has been discarded or destroyed. The general result of a dangling pointer is that it now references a section of memory whose contents are meaningless as far as the semantics of the object referenced, but which may nonetheless be meaningful to some other class now defined by the storage (including the "free space list" of unallocated storage).

DASD *See* direct access storage device.

data any information, represented in binary, which a computer receives, processes, or outputs.

data access fault a fault, signaled in the processor, related to an abnormal condition detected during data operand fetch or store.

0-8493-2691-5/01/$0.00+$.50

data acquisition (**1**) method used for capturing information and converting it to a form suitable for computer use.

(**2**) process of measuring real-world quantities and bringing the information into a computer for storage or processing.

data administration responsible person or group for all the data necessary to manage an enterprise.

database one or more large structured sets of persistent data. The simplest *database* might be a single file containing many records, each containing the same set of fields, where each field has a certain fixed width. Usually a database is associated with software to update and query the data, and therefore it is considered as one of the components of a database management system.

In the hypertext language, it means a collection of nodes managed and stored in one place and all accessible via the same server. On the World Wide Web this is called a web site. Regarding logic programming, a database is the set of all the facts and rules comprising a logic-programming program.

database administrator (DBA) the person responsible for administration of a database and the DBMS.

database computer a special hardware and software configuration aimed primarily at handling large databases and answering complex queries.

database delete a database command resulting in the deletion of a table entry (usually a table row, or a particular piece of data within a table row) as a result of a delete SQL command.

database insert a database command resulting in the insertion of a table entry (usually a table row, or a particular piece of data within a table row) as a result of an insert SQL command.

database management system (DBMS) system software that abstracts file operations and provides an interface that facilitates the creation and maintenance of a database.

A *database management system* is a complex set of software programs that controls the creation, organization, storage, and retrieval of data (fields, records, and files) in a database. It also controls the security and integrity of the database. The DBMS accepts requests for data from the application program and instructs the operating system to transfer the appropriate data. When a DBMS is used, information systems can be changed much more easily as the organization's information requirements change. New categories of data can be added to the database without disruption to the existing system.

database query a database command usually resulting in the retrieval of a particular data item from a table. The retrieval occurs typically as a result of a query SQL command.

database schema a description of the data to be stored in the database. In relational databases, for example, the *database schema* comprises a collection of relational schemas.

database state the data currently stored in the database.

database update a database command usually resulting in replacing an already existing data item within a table row with another. This results from an update SQL command.

data bottleneck a computer calculation in which the speed of calculation is limited by the rate at which data is presented to the processor rather than by the intrinsic speed of the processor itself. Ultra high speed parallel processors are very frequently limited in this way.

data buffer *See* buffer.

data bus a bus that transports data values between parts of a computer. *See* bus.

data cable special cable that connects a device to its controller and passes data between them.

data cache a small, fast memory that holds data operands (not instructions) that may be reused by the processor. Typical data cache sizes currently range from 8 kilobytes to 8 megabytes. *See* cache.

data centered program understanding instead of focusing on the control structure of a program (such as call graphs, control-flow graphs, and paths), *data centered program understanding* focuses on data and data-relationships.

data communications equipment (DCE) a device (such as a modem) that establishes, maintains, and terminates a session on a network.

data complexity a measure of the complexity of the data structures. A simple version can be the estimation of the needed memory space of the data structure. More complete versions also have to consider the relationships among data structures in the systems. *See* complexity.

data compression theorem Claude Shannon's theorem presenting a bound to the optimally achievable compression in (lossless) source coding. *See also* Shannon's source coding theorem.

data declaration the instruction for static instantiation of a variable of a given type.

data definition the set of instructions for describing the structure of a new type of data. Once a data type is defined, the definition can be used to declare instances of that type.

data definition language (DDL) used by the database designers to define both the conceptual schema and the internal schema for the database.

data dependence a relation between two statements or operations where one operation must precede the other because the first produces or uses data that the second uses or overwrites.

data dependence graph a graphical representation of the data dependence relations in a program, procedure, or body of code.

data dependency the normal situation in which the data that an instruction uses or produces depends upon the data used or produced by other instructions such that the instructions must be executed in a specific order to obtain the desired results.

data detection in communications, a method to extract the transmitted bits from the received signal.

data dictionary a set of data descriptions that can be shared by several applications. The dictionary of names used in the data flow analysis. Names include: names for processes, name of variables, names for signals, etc.

data element counting element in function point measure.

data encapsulation the inclusion in a unique structure of a set of data instances. *See* data hiding, encapsulation.

data flow the sequence of data transmission, use, and transformation that is performed during the program execution.

data flow analysis a process to discover the dependencies between different data items manipulated by a program. The order of execution in a data driven language is determined solely by the data dependencies.

data flow architecture (**1**) a computer architecture that operates by having source operands trigger the issue and execution of each operation, without relying on the traditional, sequential von Neumann style of fetching and issuing instructions.

(**2**) an MIMD architecture where control flow is determined by the availability of data. *See also* token, activity packet.

data flow computer a form of computer in which instructions are executed when the operands that the instructions require become available rather than being selected in sequence by a program counter as in a traditional von Neumann computer. More than one processor is present to execute the instructions simultaneously when possible.

data flow diagram (DFD) a graphical notation used to describe how data flows between processes in a system. An important tool for most structured analysis techniques.

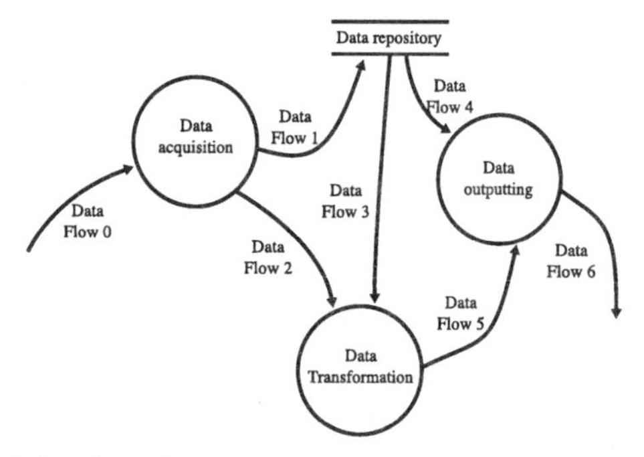

A data flow diagram

data fork the name the Macintosh uses to describe a data segment.

data fragmentation in distributed databases, refers to the subdivision of the database into smaller sections, termed fragments, that may be stored at different sites.

data fusion analysis of data from multiple sources — a process for which neural networks are particularly suited.

data hazard in pipelined processors, a dependency between instructions that coexist in the pipeline.

data hiding modularization produces chunks in code (subroutines, functions, objects). These chunks have interfaces (to the rest of the system). When *data hiding* is practiced, all data are hidden inside these chunks and are not visible from outside; i.e., they are not visible in the interface and there is no use of any global visibility concept (as typically used in procedural languages such as COBOL, FORTRAN, and C). *See* data encapsulation.

data integration a means of connecting software tools by using a common database. All intermediate results and tool output are stored in the database for use by other tools. The result is a very tight integration scheme but one that is expensive to implement.

data link layer the layer that gets information from one machine to a neighbor machine (a machine on the same link).

datalog an adjective describing a (definite) rule r where there are no function symbols in r and all variable symbols in the head of r occur in the body of r as well. This is important when we reason with r by forward chaining, since there remain no free head variables when the body is unified with ground atomic formulas.

data logger a special purpose processor which gathers and stores information for later transfer to another machine for further processing.

data manipulation language (DML) a language provided by a database management system to allow users to insert, delete, retrieve, and update the data in a database. DMLs fall into two categories. High-level or non-procedural DMLs are used on their own to specify database operations. Low-level or procedural DMLs are embedded in a general-purpose programming language. High-level DMLs are generally used to retrieve a set of data, whereas low-level DMLs are generally used to retrieve a single record at a time.

data marshalling the process of packing and sending data to a server over a network connection by a client.

data mining (**1**) a class of analytical applications that search for hidden patterns in a database. *Data mining* is the process of sifting through large amounts of data to produce data content relationships. Data mining tools use a variety of techniques including case-based reasoning, data visualization, fuzzy query and analysis, and neural networks.

(**2**) an information extraction activity whose goal is to discover hidden facts contained in databases. Using a combination of machine learning, statistical analysis, modeling techniques, and database technology, *data mining* finds patterns and subtle relationships in data and infers rules that allow the prediction of future results.

data modeling the phase of analysis where the entities that the system has to manage are mapped to data structures. Attributes and relations between entities are defined in this phase. A typical data modeling language is entity relationships diagrams.

data object an input or output that is user visible.

data-oriented methodology an application development methodology that considers data the focus of activities because it is more stable than processes.

data page a disk page in an access method which contains data records.

data path the internal bus via which the processor ships data, for example from the functional units to the register file and vice versa.

data pipeline a mechanism for feeding a stream of data to a processing unit. Data is pipelined so that the unit processing the data does not have to wait for the individual data elements.

data preprocessing the processing of data before it is employed in network training. The usual aim is to reduce the dimensionality of the data by feature extraction.

data processing inequality information theoretic inequality, a consequence of which is that no amount of signal processing on a signal can increase the amount of information obtained from that signal. Formally stated, for a Markov chain $X \to Y \to Z$,

$$I(X;Z) \leq I(X;Y) .$$

The condition for equality is that $I(X;Y|Z) = 0$, i.e., $X \to Z \to Y$ is a Markov chain.

data re-engineering tools that perform all the re-engineering functions associated with source code (reverse engineering, forward engineering, translation, re-documentation, restructuring/normalization, and re-targeting), but act upon data.

data register a CPU register that may be used as an accumulator or a buffer register or as an index register in some processors. In processors of the Motorola M68000 family, data registers are separate from address registers in the CPU.

data segment the portion of a binary executable file that contains data. It may contain any type of data including ASCII strings, binary integers, binary floating point numbers, or pointers. Referred to in the Macintosh as a data fork. Some executable file formats permit more than one data segment, allowing some data segments to be read-only and shared, some to be read/write and shared, and some to be read/write with a copy made for each new instance of the program that runs.

data stripe storage methodology where data is spread over several disks in a disk array. This is done in order to increase the throughput in disk accesses. However, latency is not necessarily improved. *See also* disk array.

data structure a particular way of organizing a group of data, usually optimized for efficient storage, fast search, fast retrieval, and/or fast modification.

data structure diagram a diagram for describing data and its relationships. *Compare with* data flow diagram. *See also* entity relationship diagram.

data tablet a device consisting of a surface, usually flat, and incorporating means for selecting a specific location on the surface of the device and transmitting the coordinates of this location to a computer or other data processing unit that can use this information for moving a cursor on the screen of the display unit.

data terminal equipment (DTE) a device, such as a subscriber's computer, Exchange workstation, or Exchange central system, that controls data flow to and from a computing system. It serves as a data source, data sink, or both, and provides for the data communication control function according to protocols. Each DTE has an address which is a 12-digit number uniquely identifying the subscriber's connection to the network. The term DTE is usually used when referring to the RS-232C serial communications standard, which defines the two end devices of the communications channel: the DTE and the DCE (Data Communications Equipment). The DCE is usually a modem and the DTE is a UART chip of the computer.

data type the type that characterizes data with the definition of its structure and the allowed operations. Examples of data type are integer, character, and floating point.

data visualization the set of techniques used to turn a set of data into visual insight. It aims to give the data a meaningful representation by exploiting the powerful discerning capabilities of the human eye. The data is displayed as 2D or 3D images using techniques such as colorization, 3D imaging, animation, and spatial annotation to create an instant understanding from multi-variable data.

data warehouse a database typically used in decision support. The data stored is usually derived and collated from a set of online databases.

Daubechies wavelets a class of compactly supported orthogonal and biorthogonal wavelets, first proposed by Ingrid Daubechies, that can be obtained by imposing sufficient conditions on wavelet filters.

daughter board a computer board that provides auxiliary functions, but is not the main board (motherboard) of a computer system (and is usually attached to the motherboard).

DBA *See* database administrator.

DBMS *See* database management system.

DCE *See* data communications equipment, distributed computing environment.

DCT *See* discrete cosine transform.

DDL *See* data definition language.

DDL Compiler one of the components of a database management system. The DDL Compiler is used to process DDL statements to identify and store the schema description.

deadline a time constraint that asserts that a task has to be completed within a time instant; generally this instant is specified relatively to the occurrence of an event; i.e., after the reception of a failure event the recovery procedure has to be completed within 3 seconds. A *deadline* can be classified as a soft deadline: when the deadline is not satisfied the system value decreases but does not go to zero.

deadline monotonic scheduling a scheduling technique that assigns priorities to tasks according to the principle: the shorter the task deadline the higher its priority.

deadlock a condition where two or more resources block one another and cause the system to appear to cease functioning. This condition can occur in database systems where file access can be the cause of the deadlock, or in real-time multitasking where access to critical regions can be the cause. A common example is a program communicating to a server, which may find itself waiting for output from the server before sending anything more to it, while the server is similarly waiting for more input from the controlling program before outputting anything.

Deadlock can be dealt with through the adherence to one of several strategies that avoid the situation (*see* deadlock avoidance) or through the elimination of one of the four causes of deadlock (mutual exclusion, hold-and-wait, no preemption, and circular wait). Deadlock can also be dealt with by ignoring it (in systems of low criticality) or by aborting one or more of the deadlocked processes when the deadlock is detected. (*See* deadlock detection.)

Also known as the deadly embrace or death spiral. *See also* livelock.

deadlock avoidance a dynamic technique which relies on examining each new resource request for a deadlock, and on denying requests that could lead to a deadlock. The banker's algorithm is one example.

deadlock detection a technique that relies on granting resource requests always when they are available, with an operating system checking periodically for deadlocks. Deadlocks can be detected through the use of watchdog timers and process monitors.

deadlock detector an algorithm to detect the presence of a deadlock.

deadlock prevention a technique assuring that one of the necessary conditions for deadlock is not met.

deadly embrace *See* deadlock.

deallocation generally, the act of freeing dynamically allocated storage explicitly. *Contrast with* garbage collection.

deassert to return an enabling signal to its inactive state.

death process a birth and death process with birth rate zero. Often used to describe the instationary behavior of a system after switching off the arrival process, e.g., the number of users still logged on to a computer after logins have been disabled, or to describe the waiting process of a selected request in a queueing system.

death spiral *See* deadlock.

debug to detect, locate, and correct faults in a computer program. *See also* fault.

debugee when a debugger is being used to debug a program, the program being debugged, and hence the program on which the debugger is operating, is called the debugee.

debugger a computer program used to debug. Debuggers are designed to help in debugging another program by allowing the programmer to step through the program, examine data, and check conditions. There are two basic types of debuggers: machine-level and source-level. Machine-level debuggers display the actual machine instructions (disassembled into Assembly language) and allow the programmer to look at registers and memory locations. Source-level debuggers let the programmer look at the original source code (C for example), examine variables and data structures by name, and so on.

debugging **(1)** locating and correcting errors in a circuit or a computer program.

(2) determining the exact nature and location of a program error, and fixing the error.

debug port the facility to switch the processor from run mode into probe mode to access its debug and general registers.

DEC acronym for Digital Equipment Corporation, a major player in the evolution and widespread adoption of mini, and later personal, computers.

decidability a property of sets for which one can determine whether something is a member in a finite number of steps. Decidability is an important concept in computability theory. A set is said to be "decidable" if a program can be written to determine whether an element is in the set and the program will always terminate with an answer YES or NO after a finite number of steps.

decidable *See* undecidable.

decidable problem a decision problem that can be solved by a GOTO program that halts on all inputs in a finite number of steps. For emphasis, the equivalent term *totally decidable problem* is used. The associated language is called recursive.

decimation an operation which removes samples with certain indexes from a discrete-time signal and then re-indexes the remaining samples. Most frequently, *decimation* refers to keeping every nth sample of a signal. Also know as down-sampling.

decision aid a procedural guideline for making proper choices under predetermined conditions, sometimes computerized and added to displays as overlays

decision boundary a boundary in feature space which separates regions with different interpretations or classes; e.g., the boundary separating two adjacent regions characterizing the handwritten characters 'E' and 'F'. In practice, the regions associated with neighboring classes overlap; consequently, most decision boundaries lead to some erroneous classifications, so an error criterion is used to select the "best" boundary. *See also* classifier, Bayesian classifier.

decision level the boundary between ranges in a scalar quantizer. On one side of the decision level, input values are quantized to one representative level; on the other side, input values are quantized to a different representative level.

decision mechanism rules and principles by which the information available to a given decision unit (control unit) is processed and transformed into a decision; typical decision mechanisms within control systems are fixed decision rule and optimization-based mechanisms. Decision mechanisms can assume a hierarchical form; then one may talk about a hierarchical decision structure.

decision problem a computational problem with a yes/no answer. Equivalently, a function whose range consists of two values {0, 1}.

decision support generic term for queries that help planners decide what to do next, e.g., which products to push, which factories require overtime and so on.

decision support system a system whose purpose is to seek to identify and solve problems.

decision tree a tree-shaped structure that represents a set of alternatives in a decision process. These decisions generate rules for the classification of a dataset. A number of domain-specific languages use decision trees as the input language. *See* CART.

decision tree analysis decomposing a problem into alternatives represented by branches where nodes (branch intersections) represent a decision point or chance event having probabilistic outcome. Analysis consists of calculating expected values associated with the chain of events leading to the various outcomes.

declaration a syntactic construct that defines something but (traditionally) performs no computations. For example, a declaration may introduce a new name whose purpose is to hold a value, or may introduce a new type. Note that in some languages, a declaration may actually imply some amount of computation, for example, to allocate space for the value, or, in the case of object-oriented languages, to execute a constructor that performs the necessary computations for the creation of the value. Other examples of declarations are compiler directives, type definitions, and macros. *See* data declaration.

declarative language a non-procedural language in which the user has to specify the program by means of a collection of facts and rules. This approach is mainly focused on describing what the system has to do instead of describing how as in classical programming languages and approaches, which are operational. *See* denotational model, descriptive model, operational model.

decode cycle the period of time during which the processor examines an instruction to determine how the instruction should be processed. This is the part of the fetch-decode-execute cycle for all machine instruction.

decode history table a form of branch history table which is accessed after the instruction is decoded and when the branch address is available so that the table need only store a Boolean bit for each branch instruction to indicate whether this address should be used.

decoder **(1)** a logic circuit with N inputs and 2^N outputs, one and only one of which is asserted to indicate the numerical value of the N input lines read as a binary number.

(2) a device for converting coded information to a form which can be understood by a person or machine.

decoder source the coded signal input to the decoder. In information theory, the decoder source is modeled as a random process.

decoding (1) in a CPU, determining which set of microinstructions corresponds to a given macroinstruction.

(2) the operation of the decoder. The inverse mapping from coded symbols to reconstructed data samples. *Decoding* is the inverse of encoding, insofar as this is possible.

(3) the process of producing a single output signal from each input of a group of input signals.

decompiler a program that converts compiled code back to a high-level representation. *See also* disassembler. How successful a decompiler is depends upon the nature of the source language and the object code, and the availability of symbolic information to aid in debugging. For example, languages that compile to a byte code and have a rich runtime symbol table will be easier to decompile than optimized machine code.

decomposable searching problem a searching problem with query Q is decomposable if there exists an efficiently computable associative and commutative binary operator @ satisfying the condition: $Q(x, A \cup B) = @(Q(x, A), Q(x, B))$. In other words, one can decompose the searched domain into subsets, find answers to the query from these subsets, and combine these answers to form the solution to the original problem.

decomposition the process of creating a program in terms of its components by starting from a high level description and defining components and their relationships. The process starts from the highest level to reach the definition of the smallest system components and their relationships, passing through several intermediate structural abstractions. For example, the service time in a queueing system can sometimes be expressed as a sum of waiting time and service time.

decorrelation the act of removing or reducing correlation between random variables. For random vectors, a discrete linear transform is often used to reduce the correlation between the vector components.

decoupling the process of making more independent a set of program/system modules.

decrement to reduce the value of a variable or content of a register by a fixed amount, normally one.

decryption a process, implemented in hardware or software, for reconstructing data previously coded by using a cryptography algorithm, that is encrypted data. These algorithms are typically based on keywords or codes.

dedicated overview display a display devoted to presentation of a view of the entire robotic work site, which helps users maintain situation awareness and avoid perceptual narrowing on the end-effector, a common phenomenon during teleoperation.

deduction problem solving approach that infers information that is a logical consequence of the data.

deductive database a database given by a set DB of facts and rules. Also called logic database, definite logic program.

deductive system a set of rules used to reason about facts; new statements are deduced from others using inference rules and axioms.

deep binding the implementation of dynamic name binding in languages such as LISP wherein the most recently defined names in the dynamically nested control hierarchy are added to the head of a list of name-value pairs in a stack-like fashion, and locating the value to which a name is bound involves searching said list until the first occurrence of the name is located, obtaining the associated value. *See also* shallow binding.

deep copy a mirror image of one object created by creating another instance of the same class containing the same information. The new object shares the same information as the source object, but has its own buffer and resources, and thus the destruction of either object will not affect the remaining one.

deep pixel a cluster of information associated with that point within an object. Pixel in this sense is more closely synonymous with "node" or "datum" than it is with the traditional definition of a pixel as a picture element, although the information stored in the deep pixel may be visual.

default (**1**) an action to be taken if some specified condition does not arise.

(**2**) a value which will be used as an actual parameter if an explicit value is not specified. *See* default parameter.

(**3**) in C, a specification of an action to be taken in a switch statement (C's version of a case selection statement). *See also* otherwise.

default negation a way to express negative knowledge in rule bodies: $A B_1 \wedge \cdots \wedge B_n \wedge \natural C_1 \wedge \ldots \natural C_m$. The meaning of a default negative literal $\natural C_i$ is usually derived by the closed world assumption, which says that $\natural C_i$ is true if we cannot infer — by other means — that the atomic formula C_i is true.

default parameter some languages permit the specification of a value to be supplied to a function if the programmer does not code an explicit value. For example, C++ allows such a specification.

defect non-fulfillment of an intended usage requirement, or reasonable expectation, including one concerned with safety or a nonconformance between the input products and the output products of a system development phase.

The main purpose of verification activities (e.g., inspection) is to detect defects so that they can be corrected before subsequent development phases. If not detected and corrected during development, defects may give rise to one or more faults in the delivered system, and hence to failure in operation. Some defects (e.g., inappropriate comments in source code) cannot give rise to faults, but may adversely affect maintainability or other quality characteristics.

defect origin the origin where a defect (that can be found later in the life cycle) was created.

defect potential the potential number of defects that can be encountered during the entire life cycle.

defect prevention the set of technologies to minimize the risk of human error by software staffs.

defect removal the action performed to remove defects in software. This process is a part of the software maintenance process.

defects count the number of defects identified in a system or sub-system in a given period.

defect severity the relative impact of a defect on the software use.

deferred addressing *See* indirect addressing.

deferred update protocol where database updates are written to a log, prior to committing the updates to the database.

definite fact *See* fact.

definite logic program *See* deductive database.

definition (**1**) the point in a program at which a name and its associated properties are established. *Compare with* use.

(**2**) a declaration of a type.

(**3**) a declaration of a function, specifying its parameters (if any) and their types, and its result (if any) and its type. *See* prototype, signature.

(**4**) the point at which a variable is assigned a value, prior to which its value was not defined.

defuzzification the process of transforming a fuzzy set into a crisp set or a real-valued number.

defuzzifier a fuzzy system that produces a crisp (non-fuzzy) output from the results of the fuzzy inference engine. The most used defuzzifiers are:

1. maximum defuzzifier that selects as its output the value of y for which the membership of the output membership function $\mu_B(y)$ is maximum;

2. centroid defuzzifier that determines the center of gravity (centroid), $\overline{y}$ of B, and uses this value as the output of the fuzzy inference system.

See also fuzzy inference engine, fuzzy inference system.

degree the number of edges incident to a vertex in a graph.

degree of a relationship the number of entities involved in a relationship.

degree of automation the relative amount of autonomous, i.e., computer-controlled activity during completion of a mission; level of automation.

degree of freedom for a robotic manipulator, a movement axis or joint; six degrees of freedom are required to position and orient an end-effector or tool anywhere within a robot's work envelope.

degree of membership the degree to which a variable's value belongs in a fuzzy set. The degree of membership varies from 0 (no membership) to 1 (complete membership).

degree of relation the number of attributes in a relation.

degree of visual angle the angle subtended by an object of a given width a given distance away from the viewer.

degrees of isolation four degrees from 0 to 3 representing increasing transaction isolation. Degree 3 is serializable.

delay (**1**) generally, the difference between a system stimulus (such as an input) and the associated event.

(**2**) the difference in time between the arrival and departure of a job from a system/subsystem that occurs due to the processing and/or queueing of the job at the system/sub-system.

(**3**) the time required for a signal to propagate along a wire (propagation delay).

delay center a queueing network is a device that produces pure delay without any queuing. The residence time of a customer at a delay center is exactly equal to the customer's demand at the center.

delay variation *See* cell delay variation.

delegation In components-based modeling *delegation* is the mechanism according to which a component defined in terms of a set of sub-components or referring to other sub-components by specific links delegates the performance of an action to another related component. In object oriented programming, the delegation allows maintenance of low class cohesion.

delete an operator in many languages which frees an instance of an object and returns its storage to the heap (or free storage pool). In object-oriented languages, can invoke the corresponding destructor for objects of the type being deleted. Some languages do not have an explicit delete operator but instead provide a general mechanism for returning storage to the heap. For example, the free function in C, and other languages do not have any mechanism for freeing objects, instead relying upon garbage collection to reclaim unused storage (for example, LISP and Java).

delete authorization authorization to delete data from relations.

deleted code source code deleted from an existing application. Most of the time done during enhancement.

deletion anomaly this type of anomaly may occur in unnormalized relations. The deletion of data may result in the loss of more data than was intended.

deletion marker a field associated with each record. The status of this field indicates whether the record is deleted or not. The records are logically deleted but not physically deleted.

delimiter a symbol, or sequence of symbols, that separates lexical entities in a programming language. The exact set of symbols that constitute delimiters is established by each programming language. Some delimiters are defined to occur in matching pairs, known as balanced delimiters.

deliverable artifact transferred from the developer to the user to the financier. Deliverables may include requirement specifications for signoff, interim analysis and design documents, and final working code. In some cases, the *deliverable* may be also software demonstrators.

delivered product quality the actual quality of the delivered product after testing completion including validation.

delivered system the end product of the system development process, which is sold to a customer and/or provided to users.

delivery productivity the output productivity of a team by using the delivered software as a base line. Suppose that a team reuses a lot of code, then development productivity gives a totally different picture than delivery productivity. This is a rationale for the need of this measure.

delta backup *See* incremental backup.

delta frame the difference between two consecutive images. Often used in video compression algorithms that exploit the temporal coherence of image sequences.

delta-modulation a special case of differential pulse code modulation (DPCM), where the input signal is quantized using one bit per sample and prediction of order one.

demand fetch in a cache memory, the name given to fetching a line from the memory into the cache on a cache miss when it is requested and not before.

demand paging the transfer of a page from auxiliary storage at the moment of need for this page. The first reference to each page will always cause a "page fault" (page not in main memory). After these initial page faults, most of the pages needed by the program are in main memory.

demodularization the process of fusing two or more modules to a unique module. It is typically performed to improve performance.

demonstration a test phase in which the system is tested against actual test cases by end-users at the site of the software builder or in a controlled environment. It is frequently opened to the public and thus is performed for getting consensus and feedback about the system developed and proposed.

DeMorgan's law *See* DeMorgan's theorem.

DeMorgan's theorem a formula for finding the complement of a Boolean expression. It has two forms:

1. $\overline{A \vee B} = \overline{A} \wedge \overline{B}$
2. $\overline{A \wedge B} = \overline{A} \vee \overline{B}$ where A and B are Boolean variables and $\wedge$ represents logical AND and $\vee$ represents logical OR and the overbar represents the logical complement.

demultiplexer a logic circuit with K inputs and I controls which steers the K inputs to one set of 2^I sets of output lines. *Compare with* multiplexer.

demultiplexing the inverse operation of multiplexing which enables the transmission of two or more signals on the same circuit or communication channel.

Denavit-Hartenberg a standard form of the homogeneous transformations.

denial of service an attack whose primary purpose is to prevent legitimate use of the computer or network.

denormalization the reverse of normalization. Relations are merged thereby possibly introducing redundancy and the potential for anomalies. Denormalization is often used to increase the efficiency of certain queries as the number of required joins is reduced.

denormalized number a number (representation) in which the significand is not normalized; in IEEE-754, *denormalized numbers* have significands with magnitudes in the range (0, 1), i.e., representation patterns of the form $0 \cdot xxx \ldots$, and minimal exponents. Denormalized numbers are useful when underflow occurs.

denotational *See* descriptive.

denotational model a formal model based on the mathematical specification of the system. Synonymous with descriptive model.

denotational semantics the meaning of a program as a compositional definition of a mathematical function from the program's input data to its output data.

dense index an index is said to be dense if it contains an index entry for every record in the underlying file.

density the amount of information that can be successfully stored per unit area on a disk. Device drivers software interfaces to special hardware.

density estimation the problem of modeling a probability distribution from a finite set of examples drawn from that distribution.

density function (DF) an alternative name for probability density function (PDF).

departure process the point process described by the time instants when departures from a system occur.

dependability (**1**) a system feature that combines such concepts as reliability, safety, maintainability, performance, and testability. *Dependability* is described in terms of several external attributes of a system, such as safety, security, usability, maintainability, recoverability, availability, and extendibility. Dependability is used only for general descriptions in non-quantitative terms. No single measure can be defined for dependability. The importance of the different attributes, and hence the target values for the corresponding measures, depends on the type of system.

(**2**) the trustworthiness of a system such that the service delivered by it can be justifiably relied upon. The service delivered by a system is its behavior as perceived by its users.

dependency a logical constraint between two operations based on information flowing among their source and/or destination operands; the constraint imposes an ordering on the order of execution of (at least) portions of the operations. For example, if the first operation in a sequential program produces a result that is an operand of the second operation of the program, that second operation cannot be performed until the first operation has been completed, since its operand value will not be available earlier. *See also* functional dependency.

deployment diagram a diagram describing the main processes of a system (distributed or not) and their related relationships.

depoissonization to interpret a solution to a poisson model problem. *See* poissonization.

depth (**1**) the longest chain of sequential dependencies in a computation.

(**2**) in computer vision, the distance to a surface, as perceived subjectively by the observer. Also, the number of bits with which each pixel is represented in a digital image.

depth buffer *See* Z-buffer.

depth complexity a measure of the complexity of an algorithm. It is equivalent to the number of pieces of data written to a framebuffer divided by the total number of pixels in the framebuffer, when a whole frame is rendered.

depth cue source of information used to make distance judgements, including binocular cues like retinal disparity and convergence and monocular cues like interposition, shadows, perspective, texture gradients, and motion parallax.

depth cueing objects closer to the viewer appear brighter and more distinct than distant objects. Thus, more distant objects or distant parts of objects are displayed with less intensity to simulate this phenomenon and enhance perception of depth.

depth estimation the ability to make absolute judgements of distance to and between objects in the visual field.

depth-first search a search strategy for a tree or trellis search where processing is performed depth first, i.e., a particular path is processed through the depth of the tree/trellis as long as it fulfills a certain threshold criterion (e.g., based on the Fano metric). If a path fails the threshold test, a new path is considered. Also known as sequential search.

depth map a map of depth in a scene corresponding to each pair of coordinates in an image of the scene.

depth of field the range of depths over which objects in the field of vision are in acceptable focus.

depth perception the ability to make comparative judgements of distance to and between objects in the human visual field; the construction of a 3-dimensional understanding of objects in the visual field from the 2-dimensional images on the retinas, using binocular cues (e.g., retinal disparity, convergence) and monocular cues (e.g., size, linear perspective, surface texture gradients, interposition).

dereference given a reference to a value, the operation that obtains the value itself. Depending on the language, a dereference operation may be implicit in the semantics of the language or may require that an explicit operator be coded by the programmer.

derivation a tree of judgements obtained by applying the rules of a type system.

derived attribute an attribute that may be derived or calculated from another attribute (or attributes) of the entity. It is not explicitly stored in the database.

derived class in object-oriented programming, a class that is derived from a superclass in terms of one (single inheritance) or more (multiple inheritance) classes. The derived class contains all the features of the superclass, also having new features or redefining existing features. Also known as a subclass. *See* subclass.

derived measure a measure calculated as a function of other measures, or obtained by the analysis of other measures. *See* indirect measurement.

derived type a type derived for specialization from another type. The derived type is a specialization from the conceptual point of view and may be an expansion from the structural point of view. *See* subtype.

descendant in object-oriented programming, a class that is derived from bases class by means of inheritance is called *descendant.* It is another term for subclass.

descriptive an adjective for a specification model that represents the system with a set of descriptions (e.g., properties) written with a formal language. Rarely can the specification be executed. Descriptive specification languages can be algebraic or logic. The verification of specifications is usually based on theorem proving.

descriptive model *See* denotational model.

descriptor an object describing an area of space within memory. A descriptor contains information about the origin and length of the area.

design the phase of software development following analysis, concerned with how the problem is to be solved. During the design, the system architecture is defined identifying the structures, the interfaces of system components, and their detailed relationships.

design diversity the independent design and development of redundant components of a subsystem with the intention of ensuring that design failures of the different redundant components occur independently.

design failure a failure of an item due to the activation of a design fault.

design fault a fault due to the inadequate design of an item. A *design fault* is caused by human error during system development. Design faults in software are usually latent, transient, recurrent, and systematic. Design faults give rise to failure during operation when activated by a certain combination of conditions referred to as the trigger. Since these conditions are encountered at random during operation, design failures can appear to be random events.

design for reuse a process of design with particular attention to defining software components that can be reused. These have to be self-contained minimizing the cohesion among components, general enough, and well defined and documented. In object-oriented programming, the *design for reuse* means also to minimize the number of specialization relationships.

design hierarchy in object oriented programming, this is a class hierarchy. Typically, it also includes is-part-of and is-referred-by relationships.

design model a mathematical model that is used to design a controller. The *design model* may be obtained by simplifying the truth model of the process. The truth model is also called the simulation model. The truth model is usually too complicated for controller design purposes. The controller performance is tested using the truth model.

design pattern a small set of collaborating classes identified as providing a common solution to a problem in a particular context.

design recovery a subset of reverse engineering in which domain knowledge, external information, and deduction or fuzzy reasoning are added to the observations of the subject system to identify meaningful higher level abstractions beyond those obtained directly by examining the system itself. *Design recovery* recreated design abstractions from a combination of code, existing design documentation (if available), personal experience, and general knowledge about problem and application domains. The recovered design abstractions must include conventional software engineering representation such as formal specifications, module breakdowns, data abstractions, data flows, and program description languages. In addition, they must include informal linguistic knowledge about problem domains, application idioms, and the world in general.

design review the process in which the design is presented to project personnel — managers, developers, users, etc. — for discussion and approval.

design scheme abstract knowledge structure that contains information about a class of design solutions, the elements at each level, and their decomposition into subelements.

design specification a document that describes the design. It typically includes formal descriptions in graph, diagrams, or formal languages.

design time time spent for the system design.

design to cost plan to match the capabilities of a software system to the costs.

design-use cycle the course of which the user, to perform a given task, inputs data in one or more dialogue steps and receives feedback for each step with regard to the processing of the data concerned.

design walkthrough a formal technical review of the design.

desk checking private review and debugging by individual programmers of their own deliverables.

deskilling loss of skill through lack of practice; potentially a problem for users of supervisory control systems who spend long periods observing robots operate but who must, at infrequent intervals, intervene to carry out difficult or unexpected tasks or to re-teach or re-program the system.

desk top metaphor human-computer interface design approach that represents computer objects and functions to objects and functions found on a desktop, by analogy. For example, using an iconic representation of a file folder to symbolize a subdirectory on a hard disk.

destination operand destination operand is where the results of an instruction are stored; e.g., the instruction MOV AL,7 uses AL as the destination operand (7 is the source operand).

destructive read reading process in which the information is lost from memory after being read out.

destructor in an object-oriented language, a specification of computations to be performed when an object is destroyed. *See also* constructor.

detailed design a design activity that focuses on the creation of an algorithm.

determinant an attribute (or set of attributes) that functionally determines the values of another set of attributes.

deterministic permitting at most one next move at any step in a computation.

deterministic algorithm an algorithm whose execution is completely determined by its input.

deterministic arrivals when the interarrival time between two customers arriving at a queue is a fixed constant. Denoted as "D" arrivals in queueing theory.

deterministic data transfer the situation in which transfer of data occurs from one node to the next with a constant service time, at fixed time intervals, whenever an input is present. Systems with deterministic data transfer, typical of computer systems when looked at on the time scale of a clock period, are generally less tractable mathematically than "memoryless" models with exponentially distributed service times and interarrival times.

deterministic finite automaton (DFA) *See* deterministic finite state automaton.

deterministic finite state automaton (DFSA) a finite state automaton that has the property that the mapping from a (state, input symbol) pair is a unique next state. *Contrast with* nondeterministic finite state automaton. Generally, the "deterministic" qualifier is dropped and such automata are simply referred to as finite state automata.

deterministic finite state machine *See* deterministic finite state automaton.

deterministic finite tree automaton a deterministic finite state automaton that accepts finite trees rather than just strings. The tree nodes are marked with the letters of the alphabet of the automaton, and the transition function encodes the next states for each branch of the tree. The acceptance condition is modified accordingly. *See also* nondeterministic finite tree automaton, deterministic tree automaton.

deterministic system a system where for each possible state, and each set of inputs, a unique set of outputs and the next state of the system can be determined. *See* event determinism, temporal determinism.

deterministic time function events occur at exact time intervals of $1/\mu$, for a rate of μ events per time unit. In a queueing system with a deterministic time function for both the arrival and service rates (D/D/1), queues do not grow until the arrival rate exceeds the service rate, at which point conceptually they grow to infinity.

deterministic tree automaton a deterministic finite automaton that accepts infinite trees rather than just strings. The tree nodes are marked with the letters of the alphabet of the automaton, and the transition function encodes the next states for each branch of the tree. The expressive power of such automata varies depending on the acceptance conditions of the trees. *See also* deterministic finite tree automaton.

developer an organization that performs development activities (including requirements analysis, design, testing through acceptance) during the software life cycle process.

development the sum of all activities that are necessary to build a software product.

development context the context in which a system is developed. It typically includes system/subsystem structure (GUI, non GUI, embedded, Real-Time, etc.); application field (toy, safety critical, etc.); tools and languages for system development (C, C++, developing tools, etc.); development team (expert, young, mixture, small, large, to be trained, etc.); adoption of libraries; development methodology; assessment tools; etc.

development life cycle *See* life-cycle.

development process the process by means of which the software is developed from its requirement analysis to the final product. It typically includes requirement definition, analysis, design, coding, assessing, risk evaluating, and testing software, etc. The process of development is structured on the basis of a specific life-cycle (spiral, waterfall, rapid prototyping) that states the evolution of the phases of the development process and the rules that have to be followed for the development process.

development productivity measuring what a team builds as the basis for productivity, opposed to delivery productivity.

deviation (**1**) a departure from a specified goal, requirements, target, profile.

(**2**) an approved change to the planned activity.

(**3**) a measure of the dispersion among the elements in a set of data.

device (**1**) a hardware entity that exists outside of the motherboard, and is accessed through device drivers. Devices often relate to I/O (floppy drives, keyboards, etc.).

(**2**) the special file associated with a peripheral, such as a printer or terminal, or a reference to the peripheral itself.

device controller (**1**) a device used to connect a peripheral device to the main computer; sometimes called a peripheral controller.

(**2**) software subroutine used to communicate with an I/O device.

device driver an operating system module (usually in the kernel) that controls access to a particular device. Device drivers perform the basic functions of device operation.

device register register in an I/O device that may be read or written by the processor to determine status, effect control, or transfer data.

dexterity in humans, skill or proficiency in using the hands; in robotic manipulators, the capability for rapid, accurate movements through a wide range of motion; akin to responsiveness in teleoperated manipulators.

dexterous having dexterity.

dexterous workspace region of space within which a robotic manipulator may position and orient its end-effector or a tool attached to the end of the manipulator.

DF *See* density function.

DFA *See* deterministic finite automaton.

DFD *See* displaced frame difference, data flow diagram.

D flip-flop a basic sequential logic circuit, also known as bistable, whose output assumes the value (0 or 1) at its *D* input when the device is clocked. Hence, it can be used as a single bit memory device or a unit delay.

DFSA *See* deterministic finite state automaton. *See also* NDFSA.

DFS forest a rooted forest formed by depth-first search.

DFT *See* discrete Fourier transform.

Dhrystone *See* Dhrystone benchmark.

Dhrystone benchmark synthetic benchmark program consisting of a representative instruction mix used to test the performance of a computer. Dhrystone was developed by R.P. Wecker in 1984 and tests the integer performance of a computer system. The name is a play on the Whetstone benchmark. *See also* Whetstone benchmark.

diagnostic (**1**) a program or process for detecting faults or failures. It can be referred to program, manual, and messages.

(**2**) pertaining to the detection and isolation of faults or failures. For example, a diagnostic message or a diagnostic manual.

diagonalization a proof technique for showing that a given language does not belong to a given complexity class, used in many separation theorems.

diagrammatic reasoning spatial reasoning, with particular emphasis on how people use diagrams.

dialogue modeling the part of a conversational system that is concerned with interacting with the user in an effective way. This includes planning what to say next and keeping track of the state of completion of a task such as form filling. Important considerations are the ability to offer help at certain critical points in the dialogue or to recover gracefully from recognition errors. A good dialogue model can help tremendously to improve the usability of the system.

diambiguating rule a rule stated separately from the rules of a context-free grammar that specifies the correct choice of syntax tree structure when more than one structure is possible.

diameter of a graph the distance between two vertices u and v in a graph is the sum of weights of the edges of the shortest path between them. (For unweighted graph, it is the number of edges of the shortest path.) The diameter of a graph is the maximum among all distances between all possible pairs of vertices.

dichotomic search Search by selecting between two distinct alternatives (dichotomies) at each step. *See also* binary search.

dictionary container storing elements from a sorted universe supporting searches, insertions, and deletions. *See* container.

dictionary lock a lock acquired on a database schema object.

difference given two sets A and B, the difference, i.e., $(A - B)$, is defined to be the set of tuples that occur in A but not in B.

difference engine a mechanical calculator developed by Babbage in 1823.

difference of Gaussian filter a bandpass filter whose point spread function is the difference of two isotropic Gaussians with different variances. The result is a "Mexican hat" shape similar to the Laplacian of a Gaussian. *See* Marr-Hildreth operator. Various physiological sensors, including some filters in early vision, appear to have difference of Gaussian (DOG) point spread functions.

differential coding a coding scheme that codes the differences between samples. *See* predictive coding.

differential equation an equation involving a function and its derivatives.

differentiator a filter that performs a differentiation of the signal. Since convolution and differentiation are both linear operations, they can be performed in either order.

$$\begin{aligned}(f * g)'(x) &= f'(x) * g(x)\\ &= f(x) * g'(x)\end{aligned}$$

Thus, instead of filtering a signal and then differentiating the result, differentiating the filter and applying it to the signal has the same effect. This filter is called a *differentiator.* A low-pass filter is commonly differentiated and used as a differentiator.

diffuse local reflection models separately evaluate a diffuse, a specular, and an ambient component. The diffuse component is the light that is reflected from a point equally in all directions, simulating a matte or plastic-like surface.

diffuse reflection the portion of light that falls on a facet (small piece of the surface) which is radiated diffusely in all directions.

diffuse shading the illumination of an object where light is reflected equally in all directions, with the intensity varying based on surface orientation with respect to the light source. This is also called Lambertian reflection, since it is based on Lamber's law of diffuse reflection.

diffusion model one of the models adopted in parallel genetic algorithms where individual processors are assigned to each individual in the population. Also known as the fine-grain model, this approach restricts the processes of crossover and selection within a local neighborhood of each individual associated with a prespecified topology. In this way information is allowed to propagate slowly or, in other words, to diffuse throughout the population. Important parameters of this model include the structure and size of the neighborhood, and the specific modification of the original crossover and selection operator to accommodate this local interactivity. *See also* migration model, parallel genetic algorithms.

digital image (**1**) sampling of an image into discrete units in both space and intensity (or color) for processing with a digital computer; an array of pixels.

(**2**) a function of two discrete variables.

digital optical computing optical computing that deals with binary number operations, logic gates, and other efforts to eventually build a general-purpose digital optical computer. In digital optical computing, new optical devices are sought to replace elements in an electronic computer. The digital optical computer may be primarily based on already known computer architectures and algorithms.

digital-optical computing that branch of optical computing that involves the development of optical techniques to perform digital computations.

digital reality *See* virtual reality.

digital tree a tree that stores digital information (e.g., strings, keys, etc.). There are several types of digital trees, namely: trees, PATRICIA trees, digital search trees, and suffix trees.

digital video a sequence of digital images representing a sampled video signal.

digital video interactive (DVI) a system for producing digitized, compressed representation of video data.

digitization a process applied to a continuous quantity that samples the quantity first, say, in time or spatial domain, and then quantizes the sampled value. For instance, a continuous-time signal can be first sampled and then quantized to form a digital signal, which has been discretized in both time and magnitude.

digitize the action of converting information from analog to digital form.

digitizer *See* data tablet.

digit-online arithmetic arithmetic in which the operands are fed into the hardware unit one digit at a time, from the most significant digit to the last, and results are also produced one digit at a time, in a similar order. The implementation is usually pipelined.

digit-pipelined arithmetic *See* digit-online arithmetic.

digraph *See* directed graph.

Dijkstra's algorithm an algorithm to find the shortest paths from a single source vertex to all other vertices in a weighted, directed graph. All weights must be nonnegative. *See also* all pairs shortest path.

dilation a fundamental operation in mathematical morphology. Given a structuring element B, the dilation by B is the operator transforming X into the Minkowski sum $X \oplus B$, which is defined as follows:

1. If both X and B are subsets of a space E,
$$X \oplus B = \{x + b \mid x \in X, b \in B\} .$$
2. If X is a gray-level image on a space E and B is a subset of E, for every $p \in E$ we have
$$(X \oplus B)(p) = \sup_{b \in B} X(p - b) .$$
3. If both X and B are gray-level images on a space E, for every $p \in E$ we have
$$(X \oplus B)(p) = \sup_{h \in E} [X(p - h) + B(h)] ,$$
with the convention $\infty - \infty = -\infty$ when $X(p-h), B(h) = \pm\infty$. (In the two items above, $X(q)$ designates the gray level of the point $q \in E$ in the gray-level image X.) *See* erosion, structuring element.

dilation equation the equation

$$\pi(t) = \sum_{n=-\infty}^{\infty} a(n)\pi(2t - n) ,$$

with $\pi(t)$ being the scaling function and $a(n)$ being the coefficients. It states the fact that, in multiresolution analysis, a scaling space is contained in a scaling space with finer scale.

DIM in the original dialect of BASIC, a declaration that was used to declare a name to be an array instead of a scalar value. In later dialects of BASIC, often used as a means of declaring names of any specific type.

dimension the number of subscripts required to access a single item in an array. *See also* slice.

diminished-radix complement the generalization, to higher radices, of one's complement notation. In radix r, the complement of a given representation is obtained by subtracting from $r - 1$ the value of each digit. In the binary system the radix complement is called the 2s complement and the diminished radix complement is called the 1s complement.

DIP *See* dual in-line package.

DIP switch set of micro-switches (on/off or deviators) that are compliant with the DIP for the position of their pin (connections); thus, they can be installed in standard sockets for integrated circuits. These arrays of rocker- or slider-type switches are used to set certain system parameters. They are often used in place of setup programs on older systems.

direct access device in which individual data can be retrieved by bypassing all preceding data. *Compare with* sequential access.

direct-access storage storage in which an item can be accessed without having to first access all other items that precede it in the storage; however, sequential access may be required of a few preceding systems. An example is a disk, in which blocks may be accessed independently, but access to a location within a block is preceded by access to earlier locations in the block. *See also* sequential-access storage.

direct access storage device (DASD) typically describes permanent storage devices of the rotating hard drive variety configured to a processor through a small computer serial interface (SCSI).

direct addressing address of the operand (data upon which the instruction operates) of the instruction is included as part of the instruction.

direct cost any cost directly related to a project or unit: consumables, equipment, travel and subsistence, protection knowledge, and intramuros assistance. External services, third party assistance, and subcontracting in some cases are not considered as direct costs.

direct drive the practice of driving a joint directly from an actuator without any form of gear reduction.

direct-drive motor a motor connected directly to a robot link without any form of transmission reduction.

direct-drive robot a mechanical arm where all or part of the active arm joints are actuated with the direct drive. Due to the fact that many actuators are best suited to relatively high speeds and low torques, a speed reduction system is required. Gears are the most common elements used for reduction. A robot with a gear mechanism is called a geared robot. Gears are located at different joints; therefore, geared robots usually have a transmission system which is needed to transfer the motion from the actuator to the joint.

directed acyclic graph (DAG) a graph represented abstractly by a collection of nodes and directed arcs or directed edges such that there exists no path from any node back to itself. Often used in compilers to represent the program during intermediate states of compilation.

directed arc an arc which has a specific direction. Synonym for directed edge.

directed edge *See* directed arc.

directed graph a graph whose edges are ordered pairs of vertices. That is, each edge goes from one vertex to another, rather than merely connecting the vertices. Also known as digraph. *See also* directed acyclic graph, undirected graph.

direct file *See* hash file.

direct fuzzy control the use of fuzzy control directly in the inner control loop of a feedback control system.

direct graph a graph in which the entities are related with directed lines, arrows.

directional lighting a light source that radiates in such a way that rays from it are nonparallel.

direction cube a technique used for representing spatial directions, often used by recursive direction decomposition algorithms. The cube is placed at the origin and aligned so that the coordinate axes are orthogonal to the faces. Each face of a cube is subdivided into a number of squares. Each square represents a collection of similar directions. Subdividing the squares on a face increases the resolution of the directions.

direct manipulation interface a human–computer interface in which objects on a screen are manipulated using a pointing device, allowing inputs to be simple human actions rather than depending on a coding system.

direct mapped cache a cache where each main memory (MM) block is mapped directly to a specific cache block. Since the cache is much smaller than the MM, several MM blocks map to the same cache block.

If, for example, the cache can hold 128 blocks, MM block k will map onto cache block k modulo 128. Because several MM blocks map onto the same cache block, contention may arise for that position. This is resolved by allowing the new block to overwrite the old one, making the replacement algorithm very trivial in this case.

In its implementation, a high speed random access memory is used in which each cache line and the most significant bits of its main memory address (the tag) are held together in the cache at a location given by the least significant bits of the memory address (the index). After the cache line is selected by its index, the tag is compared with the most significant bits of the required memory address to find if the line is a required line and to access the line.

direct measure a measure of a feature by means of the direct counting of unit elements of the feature. For instance, the measure of the number of components, the measure of the length of a piece of code in terms of lines. *See* indirect measure.

direct measurement *See* direct measure.

direct memory access (DMA) an input/output scheme where access to the computer's memory is afforded to other devices in the system without CPU intervention.

DMA is used in a computer system when transferring blocks of information between I/O devices (e.g., disk memory) to/from the main memory with minimal intervention from the

CPU. A DMA controller is used and can, after initiation by the CPU, take control of the address, control, and data busses. The CPU initiates the DMA controller with parameters such as the start address of the block in main memory, number of bytes to be transferred, and the type of transfer requested (Read or Write). The transfer is then completely handled by the DMA controller, and the CPU is typically notified by an interrupt when the transfer service is completed. While the DMA transfer is in progress, the CPU can continue executing the program doing other things. However, as this may cause access conflicts of the busses between the CPU and the DMA controller, a memory bus controller handles prioritized bus requests from these units. The highest priority is given to the DMA transfer, since this normally involves synchronous data transfer which cannot wait (e.g., a disk or tape drive). Since the CPU normally originates the majority of memory access cycles, the DMA control is considered as "stealing" bus cycles from the CPU. For this reason, this technique is normally referred to as cycle stealing. *Compare with* memory-mapped I/O, programmed I/O.

direct metric *See* direct measure.

direct mode a memory addressing scheme in which the operand is the data contained at the address specified in the address.

directness degree of correspondence between the user's mental goals and the functions and interaction facilities provided by the system.

directory a list of files with reference to their locations.

directory hierarchy all directories that are accessible to a system's users, that is, all directories that descend from a system's root directory.

directory look aside table (DLT) *See* translation look aside buffer.

direct perception false notion that sensation is directly mapped onto areas of the brain; perception is more complicated and requires more processing than this conception implies and is also much more of a seeking, rather than receiving, process.

direct vision unaided viewing, that is, viewing a task scene directly without the use of video, sometimes possible with teleoperated manipulators.

dirty bit a status bit used to indicate if a block (e.g., cache block, page, etc.) at some level of the memory hierarchy has been modified (written) since it was first loaded in. When the block is to be replaced with another block, the dirty bit is first checked to see whether the block has been modified. If it has, the block is written back to the next lower level. Otherwise, the block is not written back.

dirty block a modified block. A system usually tracks whether or not an object has been modified — is dirty — because it needs to save the object's contents before reusing the space held by the object. For example, in the file system, a system-cache buffer is dirty if its contents have been modified. Dirty blocks must be written back to the disk before they are reused.

dirty data data which has been updated and not yet committed. Implicitly, the reading of such data.

dirty page a page in memory which has been altered since last loaded into main memory. *See also* dirty bit.

dirty read a read of uncommitted changes to data.

dirty write an update of a database object, overwriting earlier updates.

disable action that renders a device incapable of performing its function; the opposite of enable.

disagreement of interests situation in which there are several decision units with conflicting goals. *Compare with* consistency of interests.

disambiguation the process of resolving an ambiguity by selecting one (or a subset) of the possible set of interpretations for a part of a text.

disassembler a program which takes an object file and creates the corresponding assembly code that would generate the same object file. The degree to which this is successful depends upon how much symbolic information might be available for the process. The original code need not have been generated by an assembler. Most debuggers have some kind of built-in disassembler, allowing the programmer to view an executable program in terms of human-readable assembly language. *See also* decompiler.

disc *See* magnetic disk. Also spelled "disk".

discrete cosine transform (DCT) an image transform similar to the discrete Fourier transform, but which provides better separation for images with strong pixel-to-pixel correlation.

There is a family of DCTs, of which the DCT-II described above is the one commonly used. These other types of DCTs, specifically the DCT-IV, are sometimes used in calculating fast transforms. The $N = 8$ element DCT is particularly important for image data compression and is central to the JPEG and MPEG standards. As a matrix, the 8- element DCT is

$$\left(\begin{array}{rrrr} 0.3536 & 0.3536 & 0.3536 & 0.3536 \\ 0.4904 & 0.4157 & 0.2778 & 0.0975 \\ 0.4619 & 0.1913 & -0.1913 & -0.4619 \\ 0.4517 & -0.0975 & -0.4904 & -0.2778 \\ 0.3536 & -0.3536 & -0.3536 & 0.3536 \\ 0.2778 & -0.4904 & 0.0975 & 0.4157 \\ 0.1913 & -0.4619 & 0.4619 & -0.1913 \\ 0.0975 & -0.2778 & 0.4157 & -0.4904 \end{array} \right.$$

$$\left. \begin{array}{rrrr} 0.3536 & 0.3536 & 0.3536 & 0.3536 \\ -0.0975 & -0.2778 & -0.4157 & -0.4904 \\ -0.4619 & -0.1913 & 0.1913 & 0.4619 \\ 0.2778 & 0.4904 & 0.0975 & -0.4517 \\ 0.3536 & -0.3536 & -0.3536 & 0.3536 \\ -0.4157 & -0.0975 & 0.4904 & -0.2778 \\ -0.1913 & 0.4619 & -0.4619 & 0.1913 \\ 0.4904 & -0.4157 & 0.2778 & -0.0975 \end{array} \right)$$

discrete-event simulation describes a class of modeling tools that emulate the low-level operation of a system and record various measurements of its behavior along a particular sample path. Such simulations are distinct from continuous system simulations, which largely involve finding transient solutions to sets of differential equations, and can handle more general systems than closed-form analytical modeling approaches.

discrete event simulation language a type of simulation language used to simulate a problem characterized by discrete events, typically events which occur over a designated time domain. *Contrast with* continuous simulation language. The language SIMULA was one of the more notable of the early languages, and GPSS and SIMSCRIPT are widely used systems that support discrete event simulation.

discrete Fourier transform (DFT) an operation that separates an image into its frequency components. *See also* fast Fourier transform.

discrete fuzzy set a fuzzy set that includes only those sample points of a continuous variable.

discrete Hadamard transform *See* Hadamard transform.

discrete Hopfield network a single layer, fully connected network that stores (usually bipolar) patterns by setting its weight values w_{ij} equal to the (i, j) entry in the sum of the outer products of the patterns. The network can be used as an associative memory so long as the number of stored patterns is less than about 14% of the number of neural elements. *Compare with* continuous Hopfield network.

discrete random variable a random variable that has a finite number of values, or a number of values that is countable. An example is the number of times that a coin must be flipped until head comes up.

discrete simulation a model using finite difference equations.

discrete sine transform (DST) a unitary transform mapping N samples $g(n)$ to N coefficients $G(k)$ according to:

$$G(k) = \sqrt{\frac{2}{N+1}} \sum_{n=0}^{N-1} g(n) \sin \frac{nk\pi}{N+1} ,$$

with inverse

$$g(n) = \sqrt{\frac{2}{N+1}} \sum_{k=0}^{N-1} G(k) \sin \frac{nk\pi}{N+1} .$$

As with the discrete cosine transform there is a family of DSTs, the other members of which are rarely used. While the DST is closely related to the DCT, the latter is the form which has attained supremacy for image data compression.

discrete time model for time that associates the time instant with a value in $\mathcal{N}$ or $\mathcal{Z}$ (the set of the natural numbers or integer numbers). Two situations are possible: the time instant is associated with occurrence of events; in this case the real temporal distance from instant i and $i+1$ depends on the rate of occurrences of events. The other case is when time is sampled at regular intervals so the mapping to real time is given multiplying the number by a real constant that represents the time quantum (1ms, 10ms, 1s...).

discrete time Fourier series representation of a periodic sequence x_n with period N by the sum of a series of harmonically related complex exponential sequences:

$$x_n = \frac{1}{N} \sum_{k=0}^{N-1} X_k e^{\frac{j2\pi kn}{N}} .$$

The X_k are the Fourier series coefficients, obtained by

$$X_k = \sum_{n=0}^{N-1} x_n e^{-\frac{j2\pi kn}{N}} .$$

discrete time signal a signal represented by samples at discrete moments of time (usually regularly spaced). The samples may take values from a continuous range, so the term is usually used to differentiate a sampled analog signal from a digital signal which is quantized.

discrete Walsh-Hadamard transform *See* Hadamard transform.

discrete wavelet transform (DWT) a computation procedure that calculates the coefficients of the wavelet series expansion for a given finite discrete signal.

discretionary access control access control method in which individual privileges are granted to users. Discretionary refers to the fact that the users, at their discretion, can specify to the system who can access their files.

discriminant a synonym for variant selector in a record. This term is used in many programming languages, notably Ada.

discriminator *See* discriminant.

disjoint set a set whose members do not overlap, are not duplicated, etc.

disjunction (**1**) from the notation of propositional calculus, the Boolean OR function. *See also* conjunction, implication.

(**2**) less commonly, a bitwise OR.

disjunctive normal form a representation of a Boolean expression that involves a logical sum of products (maximum of minima).

disjunctive rule a rule $r(A_1 \vee \ldots \vee A_k$ $B_1 \wedge \cdots \wedge B_n)$ that denotes a statement in which at least one of the atomic formulas A_i in the head of r should be inferred, if all atomic formulas B_j in the body of r have been inferred already. Both the head and the body of r can be empty. For example, the rule father (X, P) ∨ mother(X, P)parent(X, P) states that a parent P of a person X is either the father or the mother.

disk *See* magnetic disk. Also spelled "disc".

disk actuator a mechanical device that moves the disk arms over the disk surface(s) in order to position the read/write head(s) over the correct disk track.

disk arm a mechanical assembly that positions the head over the correct track for reading or writing a disk device. The arm is not movable on a fixed-head disk but is on a moving-head disk.

disk array a number of disks grouped together, acting together as a single logical disk. By this, multiple I/O requests can be serviced in parallel, or the bandwidth of several disks can be harnessed together to service a single logical I/O request.

disk cache a buffer memory area in main memory used to hold blocks of data from disk storage. The cache can hide much of the rotational and seek latencies in disk accesses because a complete data block (disk sector) is read or written together. *Disk caches* are normally managed by the machine's operating system, unlike a processor cache, which is managed by hardware.

disk controller unit that carries out the actions required for the proper operations of a disk unit.

disk drive assembly consisting of electronics and mechanical components, to control disk and disk-head movement and to exchange data, control, and status signals with an input/output module, as required for the proper reading or writing of data to or from a disk, i.e., the head disk assembly plus all the associated electronics.

diskette a floppy disk is a flexible plastic diskette coated with magnetic material. It is a smaller, simpler, and cheaper form of disk storage media than the hard disk, and also easily removed for transportation.

disk format the (system-dependent) manner in which a track of a disk is partitioned so as to indicate for each sector the identity, the start and end, synchronization information, error-checking information, etc. A disk must be formatted before any initial writing can take place.

disk head read/write head used in a disk drive. Such a head may be fixed-gap, in which the head is positioned at a fixed distance from the disk surface; contact-head, in which the head is always in contact with the surface; or aerodynamic, in which the head rests lightly on the surface when the disk is motionless but floats a small distance above when the tape is rotating. Typically, contact heads are used in floppy disks, and aerodynamic heads are used in Winchester disks.

In earlier systems, "fixed-head disks" having one read/write head per track were used in some disk systems, so the seek time was eliminated. However, since modern disks have hundreds of tracks per surface, placing a head at every track is no longer considered an economical solution.

disk latency time between positioning a read/write head over a track of data and when the beginning of the track of data passes under the head.

disk operating system (DOS) a set of procedures, services, and commands that can be used by the computer user for managing its resources with special attention to disk managing. The most famous DOS for personal computers is the MS-DOS (Microsoft-DOS). It is a mono task and mono user operating system.

disk pack a stack of disk platters which can be removed for off-line storage.

disk platter metal disks covered with a magnetic material for recording information.

disk scheduling algorithms used to reduce the total mechanical delays in disk accesses, as seen by a queue of simultaneous I/O requests, e.g., if a "shortest-seek-time-first" scheduling algorithm is used, seek times can be reduced. That is, among the queue of pending I/O requests, the one next serviced is the one requiring the shortest seek time from the current location of the read/write head. The *disk scheduling* algorithm is run by the computer operating system.

disk sector the smallest unit that can be read or written on a track; adjacent sectors are separated by a gap. A typical track has 10 to 100 sectors.

disk spindle a stack of disk platters.

disk striping the notion of interleaving data across multiple disks at a fine grain so that needed data can be accessed from all the disks simultaneously, thus providing much higher effective bandwidth.

disk track concentric circle over which a read/write head moves; adjacent tracks are separated by a gap. A typical disk has hundreds to thousands of tracks.

disocclusion the uncovering of an object. *See* occlusion.

disorientation a dimension of simulator sickness with symptoms such as dizziness and vertigo.

dispatch to allocate time on a processor to jobs or tasks that are ready for execution.

dispatcher a program within an operating system, the purpose of which is to dispatch.

dispatching discipline in scheduling theory, the order in which different requests are served.

displaced frame difference (DFD) the difference between a given digital image frame and its estimate obtained by using the motion compensation technique. It is useful in image (sequence) data compression and motion estimation.

displacement-feedback display remote force information displayed by moving an object (usually the controller) in the operator's environment a distance proportional to the remote force.

displacement joystick isotonic joysticks; joysticks that move.

displacement map a pattern of surface position displacements to create the undulations of a bumpy surface that is to be mapped onto a geometric surface during rendering.

displacing priority a priority scheme in which a higher-priority request entering a queue displaces an already waiting request of lower priority. The displaced request is lost.

display (1) any device used to communicate information to a human user.

(2) a technique used to implement up-level addressing in languages such as Algol. Consists of making the frame pointers for the enclosing blocks available to the nested procedure while it is executing.

display adapter *See* video card.

display aid for depth hardware or software enhancement to visual displays, used to provide depth cues or to increase the salience of existing ones, with the aim of improving depth perception.

display-control correspondence spatial correspondence referring specifically to the relationship between control inputs and display responses.

display intercorrelation the degree to which different displays, whether different in format or modality, agree or disagree.

dispose an operation in Pascal that frees a dynamically allocated object.

dissolve an animation effect that is a transition between two sequences involving a fade from one directly to the other.

distal attribution an identification of self to include the external world. Because our own bodies have a physical reality and a phenomenological reality which can differ, a tool grasped in one's hand may become phenomenologically part of one's body even though it is not physically part of it.

distance *See* chamfer distance, chessboard distance, Euclidean distance, Hamming distance, Hausdorff distance, inter-feature distance, Manhattan distance.

distance between symbol strings a measure of the difference between two symbol strings. The most frequently used distance measures between two strings include Hamming distance, Edit distance, Levnshtein distance, and maximum posterior probability distance. *See* Hamming distance, edit distance or Levenshtein distance, maximum posterior probability distance.

distance measure a function $d(x, y)$ defined on a metric space, such that

1. $d(x, y) \geq 0$, where = holds iff $x = y$,
2. $d(x, y) = d(y, x)$, and
3. $d(x, y) \leq d(s, z) + d(z, y)$.

See similarity measure.

distance profile for convolutional codes, the minimum Hamming weight of all sequences of a specific length emerging from the zero state. A distance profile for one code is superior to that of another if

1. all values of the distance profiles of the two codes up to a certain depth p (lower than the constraint length) are equal; and
2. the superior distance profile code has higher values of the distance profile for all depths above the given depth p.

distance transform a map of all the pixels in a shape showing the closest distance of each point in the shape from the background; also, an image in which the distance maps of all the shapes in the image are indicated. *See also* distance.

distinguished copy used in concurrency control for distributed databases. For any data item that is replicated, one such copy is chosen as a distinguished copy, in order to access any copy of the item, a transaction must obtain a lock on the distinguished copy.

distinguished element the owner record in any instance of the set type is known as the distinguishing element.

distributed (1) computations performed on more than one computer.

(2) information stored on more than one computer.

(3) in the APL language, the "/" operator applied to another operator and an operand performs the specified operator over each of the member values of the operand.

distributed actuator an actuator positioned at many places, usually near the links or control surfaces they move, rather than clustered in one location. This has the advantage of reducing the need for power transmission from actuator to joint, but has the disadvantage of increasing link inertia.

distributed arbitration a scheme used for bus arbitration where multiple bus masters can access the bus. Arbitration is not done centrally (by a bus arbiter), but instead done in a distributed fashion. A mechanism to detect when more than one master tries to transmit on the bus is included. When this happens, one (or all) stops transmitting and will re-attempt the transmission after a short (e.g., random) time delay.

distributed computing an environment in which multiple computers are networked together and the resources from more than one computer are available to a user. *See* distributed computing environment (DCE).

distributed computing environment (DCE) an industry-standard, comprehensive and integrated set of services that supports the development, use, and maintenance of distributed computing technologies. DCE is independent of the operating systems and network types. It provides interoperability and portability across heterogeneous platforms, and provides security services to protect and control access to data. DCE also provides services that make it easy to find distributed resources; for instance, Directory Service, a DCE component, is a central repository for information about resources in a distributed system. Distributed Time Service (DTS), another DCE component, provides a way to synchronize the times on different hosts in a distributed system. DCE gives a model for organizing widely scattered users, services, and data. It runs on all major computing platforms and is designed to support distributed applications in heterogeneous hardware and software environments. It is particularly important for the World Wide Web and security of distributed objects.

distributed computing system a system whose different parts can run on different processors.

distributed control control system featuring a number of cooperating sub-systems, each responsible for some specific aspect of control, rather than a single omnipotent controller.

distributed database a database in which the data and system components are stored at a set of sites connected via a network.

distributed database management system a database management system that manages a database that is distributed across the nodes of a computer network and makes this distribution transparent to the users.

distributed deadlock deadlock that occurs over more than one site. *See* deadlock.

distributed file system *See* remote file system.

distributed mechanism a mechanism in which decisions are made on the local system, based on data located on that system.

distributed memory denotes a multiprocessor system where main memory is distributed across the processors, as opposed to being equally accessible to all. Each processor has its own local main memory (positioned physically "close"), and access to the memory of other processors takes place through passing of messages over a bus. The term "loosely coupled" could also be used to describe this type of multi-processor architecture to contrast it from shared memory architectures, which could be denoted as "strongly coupled".

distributed memory architecture a multiprocessor architecture in which physical memory is distributed among the processing nodes, as opposed to being in a central location, equidistant from all processors.

distributed operating system an operating system in which components of the kernel reside on or across more than one machine.

distributed processing execution of tasks that run on multiple interconnected computers and cooperate with each other for achieving a common objective.

distributed refresh in a DRAM, carrying out refresh operations one at a time, at regular intervals. Requires that all rows be refreshed in a time less than the time before which any given row needs to be refreshed. *See also* burst refresh.

distributed simulation a network of distinct computers that, when linked together and properly coordinated, create a multi-user SE.

distributed system a configuration of hardware that is physically separated upon which runs a single software application. Connectivity often can be underpinned by an Object Request Broker (ORB) architecture, or by several other mechanisms and solutions.

distributional complexity the expected running time of the best possible deterministic algorithm over the worst possible probability distribution on the inputs.

distribution frequency the observed frequency of a discrete random variable having a certain value. The set of all observed distribution frequencies gives the empirical distribution of a discrete random variable.

distribution function *See* cumulative distribution function.

distribution transparency the degree to which the user is aware of the distributed nature of the database. If the user is unaware of the distribution, then the database is said to have a high degree of *distribution transparency.*

distributivity an operator $\oplus$ is said to be distributive over $\otimes$ if $R \oplus (S \otimes T) = (R \oplus S) \otimes (R \oplus T)$.

disturbance an unplanned, unanticipated, or undesirable system state.

dithering one of many processes for reducing the total number of colors present in an image while retaining visual fidelity. *Dithering* can be done by interleaving pixels of selected colors to locally approximate the desired color. Dithering can be applied to either a color or a grayscale color space and may be necessary due to a limited number of colors available on the display device.

The two main approaches are matrix dithering, where a matrix of black and white dots is associated with each gray level, and error-diffusion dithering, where each gray level pixel in turn is assigned an available color, and the error is spread to its unprocessed neighbors (as in the Floyd-Steinberg) method. *See* halftone.

diversity for a fault tolerance system, the implementation of the same function or functionality by using different techniques and solutions.

divide-and-conquer a paradigm of algorithm design in which an optimization problem is solved by reducing it to subproblems of the same structure.

divide and marriage before conquest a variant of divide and conquer in which subproblems created in the "divide" step are merged before the "conquer" step.

divide-by-zero error occurring when a division per zero is operated. This case is mathematically undefined. In many cases, this problem is detected directly at the level of microprocessor that activates an exception and leaving true a status flag. The exception can be managed for recovering the error and avoiding the interruption of the program execution. Also known as divide-per-zero.

divide-per-zero *See* divide-by-zero.

division (**1**) the arithmetic operation of division. However, in programming languages and integer data types, the result of division involving one negative operand usually involves truncation to zero, rather than truncation to negative infinity.

(**2**) a set theoretic operation. Given two relations A and B, $A \div B$ is defined to be the set of tuples t that occur in A in composition with every tuple of B.

DLT *See* directory look aside table.

DMA *See* direct memory access.

DML *See* data manipulation language.

documentation the human readable information produced during the development process of a system. Often used to better understand what and how the system does or will do. Usually it is printed on paper, but (to save trees) it can be recorded on other supports (CD-ROM) and printed only if necessary. *See* user documentation.

document management discipline concerned with locating, creating, filing, versioning, and tracking documents and the information they contain. Documents may be realized electronically or in hard-copy form.

document set a collection of documents from which relevant information is returned via an information retrieval system. The documents are usually very loosely structured or are unstructured.

DOG filter *See* difference of Gaussian filter.

do-loop (**1**) in the FORTRAN language, a counted iteration. *See* for-loop as the equivalent in other languages.

(**2**) in other languages, a loop characterized by a predicate condition evaluated upon completion of the loop. *Compare with* while-loop. *See also* do|while, do|until.

domain (**1**) the inputs for which a function or relation is defined. For instance, 0 is not in the domain of reciprocal $(1/x)$.

(**2**) the possible values of a variable. *See also* range, total function.

(**3**) for an attribute, the range of values that an attribute may have.

(**4**) a specific phase of the software life cycle in which a developer works. Domains define developers' and users' areas of responsibility and the scope of possible relationships between products. The domain is also a functional area covered by a family of systems or problems.

domain analysis the set of activities aiming at identifying, collecting, organizing, analyzing, and representing the relevant information in a domain, based on the study of existing systems and their development history, knowledge captured from domain experts, underlying theory, and emerging technologies within the domain. *Domain analysis* aims at producing domain models and analyzing commonalties and variants among a family of products. The domain analysis is typically more general than the problem analysis. The first is devoted to analyzing a class of problems and systems, while the latter is focused on finding a specific model for a specific problem.

domain architectural model an architecture applicable to a family of applications belonging to the domain. Sometimes called the "generic architecture".

domain constraint a constraint that states that if any value a_i occurs as an instance of an attribute defined to be of domain A, then $a_i \in$ domain of A.

domain engineering an encompassing process which includes domain analysis and the subsequent methods and tools that address the problem of development through the application of domain analysis products (e.g., domain implementation).

domain model the result of domain analysis. A *domain model* is a definition of domain abstractions (objects, relationships, functions, events, etc.). It consists of a concise and classified representation of the commonalties and variabilities of the problems in the domain and of their solutions. It is a representation of a family. Domain models include domain requirements models (the problem) and domain architecture (the solution). It is assumed that the domain model (problem, program, and application) knowledge can be usefully represented as patterns of informal and semi-formal information, which are called conceptual abstractions.

domain of discourse the set of values to which a first-order variable may be assigned.

domain relational calculus a calculus based on the predicate calculus that uses domain variables that range over the domains of variables.

domain requirements model a model of the requirements for a family of applications belonging to the domain (sometimes called "generic requirements"). Such a model may be a model of user requirements, or a model of software (or system) requirements.

domain-specific language any language which contains specific constructs to support a particular application domain. For example, languages for writing discrete event simulation, payroll systems, or operating systems are all instances of domain-specific languages.

domain theory a collection of general knowledge about some area of human knowledge, including the kinds of entities involved and the types of relationships that can hold between them, and the mechanisms that cause changes (e.g., physical processes, component laws, etc.). Domain theories range from purely qualitative to purely quantitative to mixtures of both.

domain variable a variable that can become instantiated with a value from a given domain. To create a relation of degree n in the relational calculus, n domain variables are used.

dominance a derived class dominates a base class in the sense that in the presence of a name collision between them, the client code resolves the conflict in favor of the derived class.

don't care (1) a function that can be taken either as a minterm or a maxterm at the convenience of the user.

(2) a "wildcard" symbol matching any other symbol of a given alphabet.

don't care nondeterminism the arbitrary choice of one among multiple possible continuations for a computation.

don't know nondeterminism situations in which there are equally valid choices in pursuing a computation.

doomsday rule an algorithm to find the day of the week for any date. It is simple enough to memorize and do mentally.

dormant server a server that can cause requests to be queued even if it is not busy. A dormant server is often used to model initial set-up times before jobs in a queue are being processed.

DOS *See* disk operating system.

dot in Unix and DOS, a file name that refers to the directory that contains it.

dot-dot in Unix and DOS, a file name that refers to the parent of the directory that contains it.

dot-matrix printer a printer that produces readable characters by imprinting a large number of very small dots.

dot pitch the center-to-center distance between adjacent green phosphor dots of the red, green, blue triad in a color display.

dots per inch (DPI) a measure of the density of line-printer plots in dots per inch.

double-buffering a buffering technique in which two identical copies of some data structure are kept. One data structure is accessed by only one process at a time, while they other may be accessed by one or more processes. *Double-buffering* is used to deal with execution rate differences in producer and consumer processes.

A specific example is in terminal-to-computer communication, where a number of remote terminals are connected to a single computer through a "multiplexer". This unit connects *n* low-speed bit-serial transmission lines onto a single high-speed bit-serial line (running n times faster) using STDM (synchronous time division multiplex) of the connection. In the multiplexer, each low-speed line is connected to a "one-character buffer", converting the received low-speed bitstream to a high-speed bitstream using two shift-registers (buffers). The first one receives the low-speed line character bits (8 bits), clocked by the low-speed "receive clock". When a complete character has been received, it is moved to the second shift-register, where it is stored until it can be shifted out on the high-speed line in the appropriate time slot. Meanwhile, the next character is assembled in the first shift-register. For full-duplex operation (simultaneous two-way communication), a similar structure is needed for the opposite (computer-to-terminal) connection.

A second application comes from image processing. In this case, double-buffering serves as a mechanism for duplicating the frame-buffer memory by using a two-buffer system in which the image in one buffer is displayed while the image in the other buffer is computed. The newly created image is then displayed by swapping buffer pointers rather than having to copy memory. Double-buffering allows the CPU to have uninterrupted access to one of the buffers while the video controller has access to the other.

double buffering *See* double-buffering.

double-extended precision *See* double precision.

double-frequency recording *See* magnetic recording code.

double indirect mode a memory addressing scheme similar to indirect mode but with another level of indirection.

double precision in a given computer system, this is twice the number of bits used in the smallest format for the representation of floating-point numbers; in IEEE-754, the double-precision format is 64 bits wide. Double-extended precision uses more bits but no more than twice double precision; in IEEE-754, double-extended precision is between 79 and 128 bits wide.

double-sided assembly a packaging and interconnecting structure with components mounted on both the primary and secondary sides.

double-sided disk a disk in which both sides of a platter are covered with magnetic material and used for storing information.

double word a binary representation of floating point values which typically occupies two machine words, and consequently has more bits of precision than a single precision. Historically, double precision floating points had twice as many bits as single precision floating points, but various machines chose to interpret the bits in various ways. For example, some machines supported more bits of exponent in double precision representation, and most forms of floating point representation more than double the number of bits of mantissa. Double precision floating point operations have usually been slower than single precision operations, hence the data type is reflected directly into most programming languages. *See* single precision, extended precision, IEEE floating point.

doubly-linked list a list in which each list cell has two pointers, one to a predecessor list cell and one to a successor list cell. *Contrast with* singly linked list.

do|until a loop construct which executes the loop body at least once, and continues to do so until a specified predicate becomes "true". *See also* do|while, while|do.

do|while a loop construct which executes the loop body at least once, and continues to do so as long as a specified predicate remains "true". *See also* do|until, while|do.

download to bring data from a remote source to local storage.

down-sampling *See* decimation.

downsizing (**1**) transforming a large system moving the accent from a centralized approach for managing the information to the adoption of personal computers.

(**2**) the term is also used for indicating the process of removing people from the organization in order to reduce the organization size.

down time the time period during which it has been in non-operative conditions. *See* mean time to repair.

DPDA deterministic push-down automaton. *See* push-down automaton.

DPI *See* dots per inch.

DRAM *See* dynamic random access memory.

drive train mechanism or mechanisms for transferring power from actuator to joint.

driving simulator a simulator designed to dynamically emulate the task of driving a vehicle.

drop authorization authorization to drop a relation.

dropping probability *See* loss probability.

drum memory an old form of backing up memory. Similar to magnetic disks in operation, but here the magnetic film is deposited on the surface of a drum instead of a disk.

DST *See* discrete sine transform.

DTE *See* data terminal equipment.

dual for a planar graph, G, a graph with a vertex for each region in G and an edge between vertices for each pair of adjacent regions. The new edge crosses the edge in G which is the boundary between the adjacent regions.

dual force feedback force feedback scheme in the presence of time delay, in which actual (but delayed) forces are presented to the inactive hand and predicted forces are presented to the controlling hand.

dual in-line package (DIP) a standard case for packaging integrated circuits. The package terminates in two straight, parallel rows of pins or lead wires. This standard has been more recently substituted by surface-mount standards. *Compare with* single inline packaging.

dual linear program for the linear program under linear program, the program $\max_y \{b \cdot y : A^T y \leq c \text{ and } y \geq \bar{0}\}$. Every linear program has a corresponding linear program called the dual. For any solution x to the original linear program and any solution y to the dual, we have $c \cdot x \geq (A^T y)^T x = y^T (Ax) \geq y \cdot b$. For optimal x and y, equality holds. For a problem formulated as an integer linear program, feasible solutions to the dual of a relaxation of the program can serve as witnesses.

dual model a specification model that integrates aspects of both descriptive and denotational models. Synonymous with hybrid model. *See* formal model.

dual port memory memory system that has two access paths; one path is usually used by the CPU and the other by I/O devices.

dummy record type *See* linking record type.

dump to copy the contents of memory to the external medium.

duplex channel two-way simultaneous (and independent) data communication, e.g., between a computer and remote terminals.

durability the property of becoming or being persistent, and continuing to be so.

DUT acronym for device under test.

Dutch national flag a computational problem used to benchmark computer performance. The problem is to rearrange elements in an array into three groups, bottom, middle, and top. During execution, the top group grows down from the top of the array, the bottom grows up from the bottom, and the middle is kept just above the bottom. Begin by moving a previously unexamined element to a temporary location. If it belongs in the top, swap it with the element just below the top, then repeat for the element swapped into the temporary and similarly if it belongs in the middle. If it belongs in the bottom, swap it with the middle element just above the bottom, then swap that with the element just above the middle. Complexity is $O(2n)$ moves.

DVI *See* digital video interactive.

DWT *See* discrete wavelet transform.

dyadic decomposition wavelet decomposition in which the dilation is a dyadic number, i.e., 2^j with j being the decomposition level; this corresponds to an octave frequency band division in the frequency domain.

dynamic pertaining to something that is performed during the program execution.

dynamic allocation (1) allocation of memory space that is determined during program execution. Dynamic allocation can be used to designate space on the stack to store objects whose lifetimes match the execution interval of a subroutine; this allocation is performed upon entry to the subroutine.

(2) in database systems, when resources, including locks, are allocated to a transaction as required by the transaction.

dynamically checked language a language where good behavior is enforced during execution.

dynamical systems a sub-field of mathematics that is concerned with the repeated application of an algorithm.

dynamic analysis defect removal method where defects are found during execution of the software. Opposed with static analysis.

dynamic assertion an assertion about the system that depends on elements that can change their truth-value with time.

dynamic binding (1) a means by which the data type of a value associated with a specific name may change based on the most recent assignment to that name.

(2) a means by which the interpretation of a name not defined by a function (either as a formal parameter or a declared name within the function) depends upon the context from which the function was called.

(3) in object-oriented systems, the mechanism of identifying at the execution time a spe-

cific procedure code/method that has to be executed for an object. It is the technical implementation of the polymorphism and overloading mechanisms of classical object-oriented languages. For polymorphism, dynamic binding is typically specified by the programmer by using specific statements — virtual in C++. It is also referred to as late binding.

dynamic checking a collection of run time tests aimed at detecting and preventing forbidden errors.

dynamic equilibrium the final state (displacement or velocity) of a system when all inputs are constant.

dynamic hashing a technique that allows hashing to a dynamic file. This technique uses the binary representation of the hash field to facilitate hashing. A collision occurs when a record is to be placed in an already full disk block. On collision, the block is split into two and its contents redistributed according to the next significant bit. An index (binary tree structure) is created as the file expands.

dynamic inheritance in object-oriented programming, the mechanism by which an attribute is statically defined for all objects instantiated from a class and for all objects instantiated from its subclasses. The change of this value implies a change in the data status of the whole object involved. This may dynamically change their behavior. This feature is typically obtained by using static values in the definition of classes, and is not available in all object-oriented languages.

dynamic linking (**1**) deferring the determination of the association between a symbol used in a program module and the object to which it refers until that object must be accessed (i.e., during program execution).

(**2**) a method of calling functions whose addresses are not known at link time, but which are determined only at execution time.

dynamic memory term used to describe memory that needs to be refreshed periodically due to gradual discharge of the capacitive storage elements.

dynamic memory allocation the run-time assignment of small units of memory to an active program. Used typically to support growing structures such as lists.

dynamic model in object-oriented analysis, the description of the object state, behavior, and relationships with other objects in the system.

dynamic object instantiation the creation of an instance of a variable performed at run-time. The memory for this variable is not reserved at compile time, but should be taken from the heap of free store.

dynamic path reconnect technique used in IBM's high-end computer systems to allow a "subchannel" to change its channel path each time it cycles through a disconnect/reconnect with a given device. This enables it to be assigned to another available path, rather than just wait for the currently allocated path to become free.

dynamic priority system a preemptive priority system where the task priorities can change during program execution. *Contrast with* fixed priority system.

dynamic programming an algorithmic technique in which an optimization problem is solved by caching subproblem solutions (memorization) rather than recomputing them. *See also* greedy algorithm, principle of optimality.

dynamic random access memory (DRAM) a semiconductor memory using one capacitor and one access transistor per cell (bit). The information is stored dynamically on a small charge on the cell capacitance, and can be read or written through the "access transistor" in the cell. Since the charge will slowly leak away (through semiconductor junctions), the cells need to be "refreshed" once every few milliseconds. This is typically done using on-chip circuitry. DRAMs have very high storage density, but are slower than SRAMs (static RAMs). *See also* burst refresh, distributed refresh.

dynamics the study of bodies in motion.

dynamic scheduling (**1**) the scheduling process performed at run-time, without progressive knowledge of timing constraints of the tasks to be scheduled. The scheduler can refuse the execution of a new process if the resources needed are not available or the timing constraint cannot be satisfied. *See* static scheduler.

(**2**) automatic adjustment of the multiprocessing program at run time that reflects the actual number of CPUs presently available. For instance, a DO loop with 100 iterations is automatically scheduled as 2 blocks with 50 iterations on a two-processor system, as 10 blocks with 10 iterations on a ten-processor system and as one block on a single-processor machine. This allows the user to run multiprocessor programs on single-processor computers.

(**3**) creating the execution schedule of instructions at run-time by the hardware, which provides a different schedule than strict program order. When a processor issues instructions to functional units out of program order, the processor can dynamically issue an instruction as soon as all its operands are available and the required execution unit is not busy. Thus, an instruction is not delayed by a stalled previous instruction unless it needs the results of that previous instruction.

dynamic scope a term applied to indicate that the interpretation of a name depends upon the execution context which is in effect at the time the name is evaluated. *See* dynamic binding.

dynamic storage storage which is allocated and freed as needed. This may be stack storage or heap storage.

dynamic type-checking type checking performed at run time devoted to the identification of the type of unknown data at compile time, like the generic types defined in some object-oriented languages.

dynamization transformation a data structuring technique that can transform a static data structure into a dynamic one. In so doing, the performance of the dynamic structure will exhibit certain space-time tradeoffs.

dysfluency a portion of a spoken sentence that is not fluent language. *Dysfluencies* can include false starts, filled pauses, or grammatical constructs due to a changed plan midstream. Dysfluencies are particularly problematic for speech recognition systems.

E

E a lexical convention which is used in many languages to represent scientific notation. When used in a number, it indicates an exponent, for example 1.3E2 represents 1.3' 102. *See also* floating-point.

eager evaluation a technique by which anticipated computations are performed whether or not they may actually be used. In the presence of certain kinds of parallelism, *eager evaluation* can utilize otherwise unused computational resources. Often used in hardware with instruction pipelines where it is known as speculative evaluation. *See also* lazy evaluation.

EAPROM electrically alterable programmable read-only memory. *See* electronically erasable programmable read-only memory.

earcon sound having meaning in the context of a human-computer interface; a play on the word icon.

Earle latch a type of latch in which both storage and computational functions are combined, thus reducing the delay from what it would be if the two were separated. The latch is useful in the implementation of pipelined arithmetic units.

earliest deadline first (EDF) a dynamic scheduling strategy that relies on scheduling the task that currently has the earliest deadline.

Early's algorithm shows that the worst-case upper bound for parsing a Type 2 grammar is n2 in stack space for linear time or n2 in time for bounded stack space, for n productions in the grammar.

early stopping a technique applied to network training that is aimed at assuring good generalization performance. Training on a finite set of data eventually leads to overfitting and this can be avoided by periodically assessing generalization performance on a set of test data. As training proceeds, network performance on both the training set and the test set gradually improves, but eventually performance on the test set begins to deteriorate indicating that training should be stopped. The set of network weights that gives the best performance on the test data should be employed in the trained network.

early vision the set of (mainly perceptual) processes occurring at an early stage of the vision process, typically at the retinal level.

early warning system identifying problems far ahead, like is done in the spiral model for the development life-cycle.

EAROM *See* electrically alterable read only memory.

ease of use term describing the degree to which human–machine interactions are efficient, error-free, and non-fatiguing.

easy of learning *See* learnability.

EBCDIC *See* extended binary-coded-decimal interchange code.

EBNF *See* extended BNF.

ECC *See* error correcting code.

Eckert, John Presper Eckert (1919–1995) is best known as one of the designers of ENIAC (Electronic Numerical Integrator And Calculator), an early computer. Like many computer pioneers during World War II, Eckert was looking for more efficient ways of calculating trajectory tables for artillery and ranging systems for radar. Eckert graduated from the University of Pennsylvania and remained there to work with John Mauchly. Eckert and Mauchly later formed a company and continued to develop and refine their machine. Eckert eventually sold the company to what would become the Sperry Rand Corporation. Here he produced UNIVAC I (Universal Automatic Computer), one of the first commercially successful computers. The chief improvement of this machine over its predecessors was in its use of a stored memory.

0-8493-2691-5/01/$0.00+$.50

ECO *See* engineering change order.

ecological psychology school of psychology that seeks to understand human perceptual and psychomotor phenomena as responses to information present in the environment and as driven by goal-directed needs for interaction with the environment.

e-commerce any activity related to commerce via the Internet. This is an emerging method for establishing and completing business transactions. It includes both the promotion and the money transaction and can be implemented by using several different techniques such as email, WWW, facsimile (fax), etc.

economy of scale a connection level multiplexing effect which indicates that in order to reach the same performance, the amount of network resources needed per customer is less if the resources are shared between more customers. *See also* Erlang B, Erlang C.

EDF *See* earliest deadline first.

edge (1) a connection between two vertices of a graph. Synonym for arc. *See also* directed graph, undirected graph, weighted graph, hyperedge.

(2) a substantial change, over a small distance, in the values of an image's pixels — typically in the gray level values. Edges can be curved or straight and are important because they are often the boundaries between objects in an image.

edge coloring an assignment of colors to the edges of a graph such that no two edges incident to a common vertex receive the same color.

edge crossing two nonincident edges cross in a graph if their geometric representations intersect. The number of crossings in a graph drawing is the number of pairs of edges which cross.

edge detection locating significant changes in an intensity (gray level) image, generally using local differential image properties. *See also* Canny operator, gradient edge detector, gradient edge detector, Marr-Hildreth operator, Sobel operator, straight edge detection.

edge elements the basis functions that are associated with the edges of the discretizing elements (such as triangles, tetrahedrals, etc.) in a numerical method such as finite element method.

edge enhancement a type of image processing operation where edges are enhanced in contrast, such as by passing only the high spatial frequencies in an image.

edge filter an operation that takes in a gray-scale image and yields a binary image whose 1-valued pixels are meant to represent an edge within the original image.

edge merging the process of replacing the edge of a polygon with the adjacent edges of neighboring polygons to prevent cracks appearing during rendering.

edge recombination in genetic algorithm optimization, a special-purpose crossover operator designed to be used with permutation representations for sequencing problems. The edge-recombination operator attempts to preserve adjacencies between neighboring elements in the parent permutations.

edge-triggered flip-flop a flip-flop which changes state on a clock transition from low to high or high to low rather than responding to the level of the clock signal.

edit to modify the contents of a computer file.

edit buffer an area of memory that holds the text for a file being edited.

edit distance the smallest number of insertions/deletions/substitutions required to change one string into another. The edit distance between two strings A and B is defined as

$$D_{\text{edit}} = \min\{a + b + c\} ,$$

where B is obtained from A with a replacements, b insertions, and c deletions. There is an infinite number of combinations $\{a, b, c\}$ to achieve this. One of the ways to find the minimum from these is dynamic programming. Edit distance is also called Levenshtein distance.

edit operation on a string, the operation of deletion, insertion, or substitution performed on a single symbol. On a tree T, the deletion of a node v from T followed by the reassignment of all children of v to the node of which v was formerly a child, or the insertion of a new node along some consecutive arcs departing from a same node of T, or the substitution of the label of one of the nodes of T with another label from Σ. Each edit operation has an associated nonnegative real number representing its cost.

editorial review a development stage, usually late in the process, in which on-line support systems are examined for consistency, style, mechanics, and completeness.

edit script a sequence of viable edit operations on a string.

edit session the sequence of events that occurs after invoking and prior to exiting the editor.

EEPROM *See* electrically erasable programmable read-only memory.

EER model *See* extended entity relationship model.

effective address (**1**) the computed address of a memory operation.

(**2**) the final actual address used in a program. It is usually 32 or 64 bits wide.

The *effective address* is created from the relative address within a segment (that is, relative to the base of a segment), which has had applied all address modifications specified in the instruction word. Depending on the configuration of the memory management unit, the effective address may be different than the real address used by a program, i.e., the address in RAM, ROM, or I/O space where the operation occurs.

(**3**) when a memory location is referenced by a machine instruction, the actual memory address specified by the addressing mode is called the effective address, e.g., in an instruction using an indirect mode of addressing, the effective address is to be found in the register or memory location whose address appears in the instruction.

effective bandwidth the bandwidth by which the link rate of a communication link offering certain delay and loss values has to be increased to accommodate an additional variable bit rate connection in order to still guarantee the same delay and loss values. The effective bandwidth of a connection is a function of its traffic characteristics, the delay and loss requirements, the link capacity and traffic characteristics already present on the link. The effective bandwidth of a connection usually is more than its mean bit rate and it can be less than its peak bit rate.

effectiveness the capability of the software product to enable users to achieve specified goals with accuracy and completeness in a specified context of use.

effective precision *See* accuracy.

efferent filter a filter on information being transmitted from a system.

efficiency (**1**) the relative volume of useful work done per unit time.

(**2**) the capability of the software product to provide appropriate performance, relative to the amount of resources used, under given conditions. Resources may include other software products, the software and hardware system configuration, and materials (e.g., print paper, diskettes) and services of operating, maintaining, or sustaining staff. For a system that is operated by a user, the efficiency is the combination of functionality, reliability, usability, and efficiency. *Efficiency* can be measured externally by quality in use.

efficient estimator an unbiased estimator which achieves the Cramer-Rao bound. *See also* Cramer-Rao bound.

effort the amount of work that is associated with carrying out a software project. Most of the time this is measured in person hours or months, and in some cases also in money.

effort estimation a part of cost estimation. Since

$$\text{cost} = \text{effort} \times \text{payment rates} ,$$

cost estimation techniques are really effort estimation techniques.

effort planning the process of predicting and allocating effort for a software project. During the project definition phase, *effort planning* takes care of the amount of effort (i.e., person month) allocated for each project task and the related schedule. This process is typically supported by Gantt diagrams.

EGA *See* enhanced graphics adaptor.

ego presence sense of personal presence in a remote or virtual world; telepresence.

eight connected *See* pixel adjacency.

eighteen connected *See* voxel adjacency.

eight queens a computational problem used in performance estimation. The problem is to place eight chess queens on an 8×8 board such that no queen can attach another and then efficiently find all possible placements. *See also* N queens.

EISA *See* extended industry standard architecture.

elastic deformation a deformation that, when unloaded, returns to its original configuration.

elasticity a measure of the compliance of a system.

elastic traffic traffic that adapts its transmission rate to the available bandwidth. An example is traffic transmitted via TCP, which uses acknowledgements and flow control to determine the available bandwidth between a sender and a receiver.

elbows-down manipulator robotic manipulator with a reference stance similar to that of the human arm, that is, with the proximal link vertical with its distal joint below the level of the proximal joint, and its distal link horizontal and at about the same level as the end effector.

elbows-up manipulator robotic manipulator with a reference stance with the proximal link horizontal with its distal joint at about the same level as the proximal joint, and its distal link vertical with the end effector below the link.

electrical actuator an actuator that transforms electrical input (voltage or current) to a mechanical output (velocity or force).

electrically alterable read only memory (EAROM) a PROM device which can be erased electronically. More costly than the EPROM (erasable PROM) device, which must be erased using ultraviolet light.

electrically erasable programmable read-only memory (EEPROM) a term used to denote a programmable read-only memory where the cells are electronically both written and erased. Also known as EAROM. *See also* EPROM.

electrically programmable read-only memory (EPROM) programmable read-only memory that is electronically written but which requires ultraviolet light for erasure.

electrogoniometer electronic device used to measure human joint angles.

electromyography the sensing of electrical impulses generated by muscles as they contract and expand.

electronic commerce *See* e-commerce.

electro-pneumatic valve a valve that controls the flow of air in response to an electrical signal.

electro-tactile stimulation electrical stimuli presented to the skin as a means for providing tactile feedback.

element a component of a list, set, or collection. *See* entity.

elementary functions functions such as sine, cosine, exponential, logarithm, and so forth.

elementary type a data type which is defined as part of the language. *Compare with* composite type.

elitism a selection strategy in which the individual with the maximum fitness value is always chosen and incorporated into the new population. In other words, the maximum fitness value in the population can never decrease from one generation to the next.

ellipse detection the detection of ellipses in digital images, often with a view to locating elliptical objects or those containing ellipses; ellipse detection is also important for the location of circular features on real objects following orthographic projection or perspective projection.

ellipsis the designation "...". The use of the designation "..." in certain formal language definitions indicates an indefinite repetition of a preceding construct.

Emacs an advanced, customizable, and extensible text editor that is generally the editor of choice on Unix systems and is available on a wide variety of platforms. Its extensibility has been used to make it language-knowledgeable and to integrate it with other programming tools.

embedded computer (**1**) a computing machine contained in a device whose purpose is not to be a computer. For example, the computers in automobiles and household appliances are embedded computers.

(**2**) a device, consisting of a microprocessor, firmware (often in EPROM), and/or FPGAs/EPLDs, which is dedicated to specific functions, and becomes an inseparable component of a device or system, in contrast to devices that are controlled by stand-alone computers. Embedded computers use embedded software, which integrates an operating system with specific drivers and application software. Their design often requires special software-hardware co-design methods for speed, low power, low cost, high testability, or other special requirements.

embedded computer system *See* embedded computer, embedded system.

embedded query language a query language, such as embedded SQL, that can be used from within a general-purpose programming language environment, such as C, C++, Java, or Pascal. In the context of the embedded query, the general-purpose programming language is called the host language.

embedded software a specialized software system that may be part of a larger system on a machine that can work without human intervention. Typically, *embedded software* is stored in ROM for functioning on a single microprocessor board. It includes all the software programs that are built for covering specific applications; for example, microwaves, VCRs, car controller, aircraft controllers, etc. Some embedded software systems can operate in an environment with a light operating system. Typically the operating system is even missing.

embedded SQL *See* embedded query language.

embedded system *See* embedded software, firmware.

emittance the light emitted by a surface. This may have different intensities and spectral characteristics in different directions.

empirical distribution function *See* histogram.

emulate executing a program compiled to one instruction set on a microprocessor that uses an incompatible instruction set, by translating the incompatible instructions while the program is running.

emulation a model that accepts the same inputs and produces the same outputs as a given system. To imitate one system with another. *Contrast with* simulation.

emulation mode state describing the time during which a microprocessor is performing emulation.

emulator (1) the firmware that simulates a given machine architecture.

(2) a device, computer program, or system that accepts the same inputs and produces the same outputs as a given system. *Compare with* simulator.

enabled state in a dataflow architecture when all necessary tokens have arrived and the input lines are full. Also called the ready state.

encapsulation property of a program that describes the complete integration of data with legal processes relating to the data. Modularization produces chunks that are said to be encapsulated, i.e., they have an identifiable boundary. This boundary may be transparent (and thus the internal features are visible everywhere) or may be opaque (in which case we also have information hiding).

encode network a perceptron network designed to illustrate that the hidden layer nodes play a crucial role in allowing the network to learn about special features in the input patterns. Once it has learned about the "generalized" features of the training patterns, it can respond usefully in new situations.

encoder a displacement sensor that provides a digital output signal proportional to the relative displacement of the degree of freedom.

encoding (1) the act of placing information to be transmitted in a form which can be transmitted over a particular medium and which will be recognizable by the receiver. *See* coding, encoding.

(2) in computing systems, to represent various pieces of information by a defined sequence of the binary digits 0 and 1, called bits. To apply the rules of a code.

encryption the transformation employed to transform information to be transmitted (plaintext) into a format which is unintelligible (ciphertext or a cryptogram). The ciphertext can then be transmitted via a communication channel without revealing the contents of the plaintext. This is achieved by means of an encryption key. A system for performing encryption is also known as a cipher. The information to be encrypted is referred to as plaintext, and the encrypted message resulting from encryption is referred to as ciphertext. The intended receiver of the ciphertext also has the encryption key and by having both the ciphertext and the encryption key available, the original plaintext can be recovered. *See also* encryption key, block cipher, stream cipher, public key cryptography, private key cryptography.

encryption key a codeword used for decryption of ciphertext into plaintext in encryption systems. Ciphertext can then be transmitted via a communication channel without revealing the contents of the plaintext. The plaintext can only be recovered by someone in possession of the encryption key. *See also* encryption, block cipher, stream cipher.

end-around borrow in subtraction, a borrow out of the least significant bit that is then immediately included in the most significant position of another subtraction (usually, in the same subtractor).

end-around carry (1) in addition, a carry out of the most significant bit that is then immediately included in the least significant position of another addition (usually, in the same adder).

(2) technique used in one's complement arithmetic, in which a carryover of the result of an addition or subtraction beyond the leftmost bit during addition (or subtraction) is "wrapped around" and added to the result.

end effector gripper or tool attached to the distal end of a robotic manipulator and used to make contact with the remote world. Position and orientation of the manipulator are referred to the end effector. The frame attached to the end effector is known as the end effector frame. The end effector can be a gripper or it can be attached to the end effector. The orthonormal end effector frame consists of three unit vectors a-approach (it approaches the object), o-orientation (which is normal to the sliding plane between the fingers of the gripper), and n-normal vector to two others so that the frame (n, o, a) is right-handed.

endian an adjective used with a qualifier to indicate how long values are constructed from smaller values that reside in memory. A "big-endian" machine constructs long values by placing the smaller parts from the lower addresses at the left end of the value (the "big" end).

endoscopic surgery surgical procedure featuring very small cameras and cutting tools inserted into the body through small slits and operated by remote control.

endoscopy *See* endoscopic surgery.

end-point control Control scheme using a coordinate frame centered on the end-point of a robotic manipulator or a tool in its grasp.

endpoint detection the process of isolating spoken words, typically used subsequently for word recognition.

end product a product which is provided to a user. Examples of software products are object code, user manuals, and support documentation.

end-to-end delay the time needed by a request to pass from one reference point to another in a network. The end-to-end delay usually consists of four components:

1. propagation delay (determined by the velocity of an electrical or optical wave in the physical medium),
2. transmission delay (the time it takes for a finite sized request to enter the physical medium, determined by the bit rate of the medium),
3. processing delays, and
4. queueing delays.

Also called transfer delay.

engine class in object-oriented programming, one of a class of the systems with a high degree of autonomy that manage several objects in the system. Engine classes typically produce only one instance for the whole system and this is an active object.

engineering change order (ECO) a mini-specification that describes the technical aspects of a change.

engineering psychology branch of psychology devoted to the study of humans as functional systems and system components.

Engset traffic traffic from a finite number of sources which each show negative exponentially distributed holding and interarrival times.

enhanced graphics adaptor (EGA) a video adapter proposed by IBM in 1984. It is capable of emulating CGA and MDA. It allows to reach 43 lines with 80 columns. In graphic mode, it allows to reach 640×350 pixels (wide per high) with 16 colors selected from a pallet of 64.

enhanced small device interface *See* enhanced small disk interface.

enhanced small disk interface (ESDI) mass storage device interface similar to MFM and RLL except that the clock recovery circuits are in the peripheral device rather than in the controller. Originally designed by Maxtor.

enhancement improvement of signal quality without reference to a model of signal degradation. *See* image enhancement.

ensemble the collection of different realizations of a stochastic process. Stochastic processes can be regarded by looking at the behavior of one process as a function of time or by looking at a number of different realizations of a process at the same time or a mix of both.

ensemble processor a parallel processor consisting of a number of processing elements, memory modules, and input/output devices under a single control unit. It has no interconnection network to provide interprocessor or processor-memory communications.

ensemble statistics statistics generated by looking at an ensemble of stochastic processes instead of observing one process for a long time. Ensemble statistics are especially important for dealing with periodic processes where the evolution of one process is a periodic function of time whereas the ensemble is distinguished by different phase relations.

entity an object or thing about which data is to be stored. According to the International Organization for Standardization's open systems interconnection (OSI) terminology for a layer protocol machine, an *entity* within a layer performs the functions of the layer within a single computer system, accessing the layer entity below and providing services to the layer entity above at local service access points. An entity may be an activity, a process, a product, an organization, a system or person, or any combination thereof. In object-oriented programming, an entity is part of the definition of a class (group) of objects. In this instance, an entity might be an attribute of the class (as feathers are an attribute of birds), or it might be a variable or an argument in a routine associated with the class. In database design, an entity is an object of interest about which data can be collected. In a retail database application, customers, products, and suppliers might be entities. An entry can subsume a number of attributes. Product attributes might be color, size, and price; customer attributes might include name, address, and credit rating. Synonymous with item.

entity attribute a specific characteristic/property of a design entity.

entity integrity constraint a constraint that states that no primary key value can have a NULL value.

entity relationship diagram *See* entity-relationship diagram.

entity-relationship diagram (E-R diagram) a diagram describing a set of entities and their logical relationships. The relationships considered can be from 1:1 to N:M. This diagram is frequently used for describing relationships among data structures in databases. More recently, is-a relationships can be specified in some extended versions of the E-R diagrams as well.

Some common components of E-R diagrams include:

1. Strong Entity Type: usually depicted as a rectangle in Entity-Relationship diagrams.
2. Weak Entity Type: usually depicted as a double rectangle in Entity-Relationship diagrams.
3. Relationship Type: usually depicted as a diamond.
4. Identifying Owner Relationship Type: usually depicted as a double diamond.
5. Attribute: Usually represented as an oval connected to the entity type via a straight line.
6. Composite Attribute: usually depicted as an oval joined to an entity type with each of its constituents represented as ovals connected via straight lines.
7. Multi-Valued Attribute: depicted in ER diagrams as a double oval.
8. Derived Attribute: usually depicted as a connected oval drawn with a dashed line in an ER diagram.
9. Primary Key: usually underlined in an ER diagram.
10. Partial Key: usually underlined with a dashed line in an ER diagram.
11. Total Participation: depicted by a double line from entity type to relationship type.
12. Partial Participation: depicted by a single line from entity type to relationship type.
13. Cardinality Ratios: illustrated by placing signifies on line joining entities to relationship types. Ratios may be (1:1), (1:N), (N:M).

entity-relationship model (ERM) a conceptual model for database design. The entity-relationship (ER) model views the world as consisting of entities and relationships between them, where an entity is a thing that can be distinctly identified, e.g., a chamber, a wire, a point, a track, a vertex, etc. A relationship is an association between entities, e.g., a point that belongs to a track is an association between a point and a track.

entity type a collection of entities with common attributes.

entropy a way to measure variability other than the variance statistic. Some decision trees split the data into groups based on minimum entropy.

entropy coding a term generally used as equivalent to lossless source coding. The name comes from the fact that lossless source coding can compress data at a rate arbitrarily close to the entropy of the source.

entry mask bit pattern associated with a subroutine entry point to define which processor registers will be used within the subroutine and, therefore, which should be saved upon entry to the subroutine. Some processor designs perform this state saving during the execution of the instruction that calls a subroutine.

entry point point in a computer program at which the execution starts.

enumerator a construct that generates, in turn, each member of a set of values. The classic for-loop construct is an enumerator over a range of integers. A linked list has a successor function that allows each of the list elements to be enumerated. A set of values does not need to be an ordered set to possess an enumerator. An enumerator over an ordered set guarantees that a transitive order relationship is maintained between two successive elements of the enumeration. An enumerator over an unordered set does not provide any such guarantee.

environment (**1**) in Unix, POSIX, and many other operating systems, a set of named values maintained by the operating system, and visible to all programs via suitable query operations.

(**2**) in the abstract specification of a program or program fragment, the state that exists at the time the program or fragment executes.

(**3**) in system/software engineering, the conditions (part of the real world) under which something is expected to be built or operated.

EP *See* evolutionary programming.

epipolar geometry the constraint that the locations of points in one image must lie along a particular line in order to correspond to the same scene point as a given point in another image.

epistasis interaction between genes at different loci in which the effect of one gene (the hypostatic gene) is inhibited by the presence of another (the epistatic gene). In evolutionary computation, the term can be generalized to include other types of interaction between genes apart from inhibition.

EPROM *See* erasable programmable read-only memory.

epsilon (**ϵ**) in a discussion of formal grammars, a metasymbol used to represent the sequence of no symbols, or the empty string. *See also* lambda.

equalization a method used in communication systems to compensate for the channel distortion introduced during signal transmission.

equijoin a join operation where the join condition is the equality operator.

EQUIVALENCE a syntactic construct in FORTRAN that allows the programmer to specify one or more alternate interpretations of a block of storage. A primitive approach to what is now termed a variant record or union.

equivalence in continuous-valued logic similar to equivalence in conventional logic. Equivalence, in continuous-valued logic between two variables x and y, which are continuous in the open interval $(0, 1)$, can be defined as:

$$e(x, y) = \max\{\min(x, y), \min((1 - x), (1 - y))\} .$$

See equivalence in logic.

equivalence in logic for two Boolean variables x and y is defined as

$$(x \equiv y) = (\bar{x} \wedge \bar{y}) \vee (x \wedge y) .$$

erasable optical disk a magneto-optical disk which can be both read/written and erased. A thermo-magneto process is used for recording and erasing information. The recording process uses the "laser power modulation" or the "magnetic field modulation" technique.

erasable programmable read-only memory (EPROM) a nonvolatile chip memory, it is used in the place of PROM. EPROMs present a glass on the case that allows the user to see the chip. They can be erased by exposing the chip at the ultraviolet light for typically 20 minutes. Once erased they can be reprogrammed. The programming has to be performed by using a special algorithm and a supplementary V_{pp}. Also called UVPROM. *See also* EEPROM.

erase character in Unix and related systems, the terminal driver control character used to delete a character from the current line; the default is #.

E-R diagram *See* entity-relationship diagram.

EREW *See* exclusive reads and exclusive writes.

ergonomics the study of people at work and the application of the knowledge gained to designing safer, more effective, and more pleasant workplaces.

Erlang B the fraction of time m parallel servers in an M/M/m loss system are all busy for a given offered load A. Also known as Erlang's function of the first kind or Erlang B function:

$$B(m, A) = \frac{\frac{A^m}{m!}}{\sum_{k=0}^{m} \frac{A^k}{k!}} .$$

This formula is widely used to dimension telephone trunks with m lines for a given blocking probability B at offered load A. A is the product of mean arrival rate and mean holding time and is measured in the pseudo unit Erlang.

Erlang C the fraction of time m parallel servers in an M/M/m waiting system with infinite queue capacity are all busy. This is the waiting probability for an arbitrary request. In:

$$C(m, A) = \frac{\frac{mA^m}{m!(m-A)}}{\sum_{k=0}^{m-1} \frac{A^k}{k!} + \frac{mA^m}{m!(m-A)}} ,$$

A is the product of mean arrival rate and mean holding time and is measured in the pseudo unit Erlang.

Erlang traffic traffic from an infinite number of sources which each show negative exponentially distributed holding and interarrival times.

ERM *See* entity-relationship model.

erosion an important basic operation in mathematical morphology. Given a structuring element B, the erosion by B is the operator transforming X into the Minkowski difference $X \ominus B$, which is defined as follows:

1. If both X and B are subsets of a space E,

$$X \ominus B = \{z \in E \mid \forall b \in B, z + b \in X\} .$$

2. If X is a gray-level image on a space E and B is a subset of E, for every $p \in E$ we have

$$(X \ominus B)(p) = \inf_{b \in B} X(p + b) .$$

3. If both X and B are gray-level images on a space E, for every $p \in E$ we have

$$(X \ominus B)(p) = \inf_{h \in E} \left[X(p + h) - B(h) \right] ,$$

with the convention $\infty - \infty = +\infty$ when $X(p+h), B(h) = \pm\infty$. (In the two items above, $X(q)$ designates the gray-level of the point $q \in E$ in the gray-level image X.) *See* dilation, structuring element.

erroneous program from the Ada formal definition, a program that is syntactically correct and that passes all semantic analysis, but whose result is unspecified.

erroneous state *See* error, fault.

error (**1**) a discrepancy between a computed, observed, or measured value or condition and the true, specified, or theoretically correct value or condition. *See* anomaly, bug, defect, exception.

(**2**) an incorrect internal state of a system, due to a component failure or an external failure. Unless redundancy or fault tolerant design enables a successful error recovery to be performed, a system failure may result. An error can be caused by a faulty item, e.g., a computing error made by faulty computer equipment.

(**3**) manifestation of a fault at logical level. For example, a physical short or break may result in logical error of stuck-at-0 or stuck-at-1 state of some signal in the considered circuit.

error control coding *See* channel coding.

error correcting code (ECC) code used when communicating data information in and between computer systems to ensure correct data transfer. An error correcting code has enough redundancy (i.e., extra information bits) in it to allow for the reconstruction of the original data, after some of its bits have been subject of error in the transmission. The number of erroneous bits that can be reconstructed by the receiver using this code depends on the Hamming distance between the transmitted codewords. *See also* error detecting code.

error correction capability of a code is bounded by the minimum distance and for an (n, k) block code, it is given by $t = [(d_{\min} - 1)/2]$ where $[x]$ denotes the largest integer contained in x.

error detecting code code used when communicating data information in and between computer systems to ensure correct data transfer. An error detecting code has enough redundancy (i.e., extra information bits) in it to allow for the detection of bits if they have been subject of error in the transmission. The number of erroneous bits that can be detected by the receiver using this code depends on the Hamming distance between the transmitted codewords. *See also* error correcting code.

error detection the process of detecting if one or more errors have occurred during a transmission of information. Channel codes are suitable for this purpose. The family of CRC-codes is an example of channel codes specially designed for error detection. *See* error detecting code.

error detection capability of a code is bounded by the minimum distance and for an (n, k) block code, it is given by $d_{\min} - 1$.

error extension the multiplication of errors which might occur during the decoding of a line coded sequence, or during the decoding of a forward error control coded sequence when the number of symbol errors exceeds the error correction capability of the code.

error-prone module a module in a software system containing a lot of errors. At first glance, one might think that errors are normally distributed in a software system. Empirical research has pointed out that on the average 50% of the errors are located in about 3% of the modules.

error recovery (**1**) the detection, containment, and correction of an erroneous state by a fault tolerant system, to avoid a system failure. For example, handling arithmetic errors by a well-defined value substitution, or allowing the user to specify an action to be taken when specific errors occur. *See also* backward error recovery, forward error recovery, redundancy, fault-tolerance.

(**2**) in a compiler, a method of attempting to achieve a state that will allow program analysis to continue, in particular, an attempt to avoid cascaded errors. Error recovery may be performed at the lexical, syntactic, or semantic level.

error report a report, usually in a fixed format, containing a description of an error (also known as problem report).

error tolerance the capability of a system or entity to continue its normal operating conditions despite the occurrence of errors. *See* fault tolerance.

ES *See* evolutionary strategy.

escape (**1**) a construct which changes the flow of control so as to "escape" from the current execution context. *See also* exit, return, leave.

(**2**) when the result of iterating a function at a point tends toward infinity or minus infinity.

escape character a character that changes the interpretation of one or more succeeding characters, for example to represent certain characters within a string constant.

escape sequence a mechanism used to change the interpretation of input. Although typically used within strings to allow the representation of characters not otherwise possible according to the rules of the language, it can also be used to represent characters outside the base character set of the language (for example, to allow a text editor to use ideographic characters and allow a compiler to have a representation in a smaller character set), or change the input to a different language, for example, to generate an assembly-code insertion in a high-level language program.

escaping a character a technique that masks the special meaning of meta-characters and thus causes the shell to treat it as a literal; sometimes called quoting a character.

ESDI *See* enhanced small disk interface.

ESM *See* extended state machine.

estimated quality software quality that is estimated and predicted for the end software product quality at each stage of development, and which is based on fundamental design quality and intermediate product quality.

estimating model a model for estimation system features. For example, costs, quality, etc. *See* metric, measure, assessment.

Ethernet a widely used Local Area Network (LAN) technology that provides best-effort delivery for variable-length frames. *Ethernet* runs at a variety of speeds (currently 10 Mbps, 100 Mbps, and 1000 Mbps) across many types of physical media (coaxial cable, twisted pair, and optical fiber). Ethernet supports both shared half-duplex collision domains, in which hosts use the Carrier Sense Multiple Access (CSMA) protocol to serialize their transmissions, and collision-free point-to-point full duplex links between each host and a switch port.

Euclidean distance the straight line distance between two points. In a plane with p_1 at (x_1, y_1) and p_2 at (x_2, y_2), it is $\sqrt{(x_1 - x_2)^2 + (y_1 - y_2)^2}$. *See also* rectilinear, Manhattan distance, l_m distance.

Euclidean Steiner tree a tree of minimum Euclidean distance connecting a set of points, called *terminals,* in the plane. This tree may include points other than the terminals, which are called Steiner points. *See also* Steiner tree, rectilinear Steiner tree.

Euclidean traveling salesman problem a problem from the study of algorithms. The problem is to find a path of minimum Euclidean distance between points in a plane which includes each point exactly once and returns to its starting point. *See also* traveling salesman problem, spanning tree.

Euler angles three sets of coordinates that describe the orientation of a rigid body.

Eulerian graph a graph that has a Euler tour.

Euler number a topological invariant of an object having an orientable surface. Assuming that the surface is endowed with the structure of a graph with vertices, edges, and faces (where two neighboring faces have in common either a vertex or an edge with its two end-vertices, their interiors being disjoint): the Euler number is $V - E + F$, where V, E, and F are, respectively, the number of vertices, edges, and faces; this number $V - E + F$ does not depend on the choice of the subdivision into vertices, edges, and faces. For a bounded 2D object in a Euclidean or digital plane, the Euler number is equal to the number of connected components of that object, minus the number of holes in it. For 2D binary digital figures on a bounded grid, the Euler number can easily be computed by counting the number of occurrences of some local configurations of on and off pixels. Also called genus.

Euler tour problem a problem in graph theory that asks for a traversal of the edges that visits each edge exactly once.

Euro problem a problem derived from the introduction of the Euro as a currency unit in countries of the European Union. With this new requirement, all software involved with currency had to be updated. Moreover, in a transitory phase all prices have to be shown in the original currency unit and in Euros. After this period, the currency will have to be shown only in Euros. Thus, the software has to be updated twice. In countries where the currency is represented with integers, there is also the problem of changing the data representation of currency from integer to fixed point decimal.

evaluation systematic examination of the extent to which an entity is capable of fulfilling specified requirements. The requirements may be formally specified, as when a product is developed for a specific user under a contract, or specified by the development organization, as when a product is developed for unspecified users, such as consumer software, or the requirements may be more general, as when a user evaluates products for comparison and selection purposes.

evaluation item an element of the environment to be evaluated and/or measured. An evaluation item can be a software product, a software process, a project, software engineer, effect of product use, etc. The part to be measured and the part to be evaluated are not necessarily the same, i.e., attributes of a deliverable software product can only be measured after it is developed. In this case, attributes of the deliverable product should be estimated by using values obtained by measuring substitute attributes that may be attributes of another target entity.

evaluation level a set of evaluation methods to be applied and evaluation results to be achieved for the software product commensurate with the rigor required by the evaluation objectives.

evaluation module encapsulation of the definition of an evaluation method applied to product or process information in order to measure software characteristics or sub-characteristics by applying metrics, check pass-fail criteria, and deliver an evaluation report and a cost report. A package of evaluation technology for a specific software quality characteristic or sub-characteristic that includes evaluation methods and techniques, inputs to be evaluated, data to be measured and collected, acceptance criteria, and supporting procedures and tools.

evaluation order a specification of the order in which operands of an expression are to be computed. Various languages are more or less restrictive in their formal specification of what an *evaluation order* should be. Evaluation order is often confused with the application of the operators to the computed results, and with the application of the associative and commutative rules of mathematics (which often do not hold for computer representations of values, such as floating point, whose properties are often confused with the properties of real numbers; floating point addition, for example, is not in general commutative or associative, although for some computations it might be). Also applied to the order of evaluation of actual parameters in a function call. Evaluation order may be critical when side effects are considered; for example, the evaluation of one operand has a side effect that affects the computation of another operand. In some cases, the presence of short-circuit evaluation in a language may influence evaluation order, or evaluation order requirements may inhibit the use of short-circuit evaluation.

evaluation technique there exist various types of evaluation techniques for deductive databases, all of which try to achieve an efficient evaluation of queries by set-oriented techniques from relational databases and by avoiding redundancy, e.g., magic sets, counting, some mixed forms of the latter two techniques, wavefront techniques, etc.

event (**1**) any occurrence that results in a change in the sequential flow of program execution. *See* asynchronous, synchronous event.

(**2**) an occurrence signaling a change of state of a system or component.

(**3**) in probability theory, a specific instance taken from some sample space, normally with an associated probability or probability density; also commonly an idealized infinitesimal point in (x, y, z, t) space at which some occurrence is taken to happen. For example, a flash of light at time t_0 at position (x_0, y_0, z_0) — event (x_0, y_0, z_0, t_0) — will lead at a later time t_1 to a wave of light passing a point (x_1, y_1, z_1) — namely event (x_1, y_1, z_1, t_1).

event determinism system characteristic in which the next state and outputs of a system are known for each set of inputs that trigger events.

event driven (**1**) a systems architecture for real-time systems. For example, a program, such as a graphical user interface, with a main loop that just waits for events to occur. Each event has an associated handler that is passed the details of the event, e.g., mouse button 3 pressed at position (553,176). For example, X window system application programs are event-driven.

(**2**) a type of specification used for reactive systems where how the system has to react to external events must be specified.

event-driven simulation a form of discrete-event simulation in which the program structure consists of a main loop that at each iteration removes the first event from an ordered list of pending events and updates the simulation clock and system state accordingly. *See also* time-driven simulation.

event flag a synchronization construct with two values, 0 and 1, and two operations, set and clear.

event relationships the description of relations between events that occurred in a system. They can be of two types: ordering and timing descriptions. *See* ordering among events, timing constraint.

event trace an ordered set of events that represents the history of events received/produced by a system. If E is the set of possible events, the event trace can be represented as an element of E^* (the set of all possible sequences of E elements). For example, if $E = \{\text{start, receive, send}\}$ an event trace on E can be: start receive send receive send receive send

In this way the event trace only records the order of events and not the time distance between each other.

eventually operator a temporal logic operator used to state that there exists a time instant where a formula is true relative to the evaluation instant. Usually there are two operators, one for the future and one for the past, and they are represented with a diamond (white for future and black for past). For example:

$$(\diamond P) \wedge (\diamond Q)$$

This formula is true if there exist two future time instants, one where P is true and one where Q is true (generally the instants are distinct).

evolution the theory, proposed by Charles Darwin and independently by Alfred R. Wallace, which explains the emergence of existing species through the process of natural selection. More specifically, an implicit selection mechanism is implemented by nature which rewards or penalizes particular species in terms of how they adapt to a specified set of environmental conditions: the better adapted species, or those species with a higher fitness, are allowed to survive and reproduce, while the not so well adapted ones perish. The apparent changes in physical traits of particular species in response to the external environment is explained in terms of mutation in the genotype of the species due to random replication errors in reproduction. These changes are manifested externally in the phenotype as modified physical and behavioral traits, the desirability of which is in turn evaluated by their subsequent enhancement or degradation of the survival capability of the species. This selection process is accentuated by competition between the new and original species for the finite resources in the environment.

evolutionary computation a class of computational algorithms that mimic the process of natural evolution to search for a solution of an optimization problem. Random multiple potential solutions to the problem are initially generated and maintained in a population, with a fitness value defined in terms of the original optimization cost function associated with each of the solutions. The objective of the algorithm is to generate new individuals with higher associated fitness values, and thus represents better solutions to the optimization problem, from the original individuals in the population. New individuals are generated by processes of random perturbation and combination, the most common of which include recombination and mutation. The original and new individuals then undergo a process of competition and selection in which their associated fitness values are evaluated and compared with each other. The overall objective of the selection process, which can be either deterministic or stochastic, is to encourage the proliferation of individuals with above-average fitness in the population, while gradually displacing the below-average individuals from the collection. This process is repeated in a designated number of generations to allow convergence of the individuals in the population to a set of optimal solutions for the current optimization problem. Compared with conventional optimization algorithms, the advantages of this approach include the adoption of multiple search points on the error surface, which allows many regions of the search space to be explored simultaneously, and its stochastic nature which allows search points to spontaneously escape from local optima. Specific implementations of evolutionary computational algorithms mainly differ in their choice of representations for the individuals in the population, their preference for recombination, mutation or variations of these operators for generating new individuals, and the choice of specific deterministic or stochastic mechanism for the competition and selection processes. Important representatives of this class of algorithms include genetic algorithms, evolutionary strategies, and evolutionary programming.

evolutionary model a software development life-cycle is classified as an evolutionary model of development when the final product is obtained for successive refinement by adding features to an early prototype. New functionalities and features are committed for each evolution. Synonymous with prototyping.

evolutionary programming (EP) a representative algorithm in the field of evolutionary computation proposed by L. Fogel and further developed by D.B. Fogel. Similar to ES, in EP the individual potential solutions in the population are also represented as real-valued vectors without previous encoding. As a result, the primary emphasis of EP is also on the role of phenotypic changes for fitness improvement. However, the most important difference between EP and ES is that no recombination operations are adopted in EP. In other words, EP relies exclusively on mutation to generate new individuals in the population. Mutation is implemented as the addition of Gaussian random numbers to individual components of the real-valued vectors, with the associated variances specified as a transformation of the fitness function. Also unlike ES, a probabilistic mechanism known as tournament selection is adopted which assigns a non-zero probability for individuals with below-average fitness to be incorporated into the new population in anticipation of non-stationary fitness landscape. In each generation, mutation is applied to the original μ individuals in the population to form μ new individuals, and tournament selection is performed on the combined 2μ individuals to create the new population, which can therefore be considered as a stochastic form of $(\mu + \mu)$ ES without recombination. *See also* evolutionary strategy.

evolutionary strategy (ES) a representative algorithm in the field of evolutionary computation, proposed by I. Rechenberg and H-P. Schwefel, in which individual potential solutions in the population are represented as real-valued vectors. New individuals are primarily generated by directly applying the operations of recombination and mutation to the vectors without the use of any encoding process. In other words, the primary emphasis of ES is on the role of phenotypic modifications to bring about fitness improvement. Between the above two operations, recombination generates new individuals by combining two or more of the original vectors in the population. On the other hand, mutation modifies the individual vector by adding a Gaussian random number to each vector component. To facilitate this process, the vector components, which are also known as object variables, are associated with a set of strategy parameters which specify the covariance matrix of the underlying Gaussian density function. These strategy parameters are in turn modified by a self-adaptation process to allow the adaptation of the mutation operation to the current fitness landscape. Unlike GA and EP, ES adopts a deterministic selection process where the fitness values are ranked in descending order and individuals corresponding to values at the top of the list are incorporated into the new population. More specifically, assume that there are μ individuals in the original population, and λ new individuals are produced in each generation, the two primary selection methods adopted are known as $(\mu + \lambda)$ and (μ, λ) strategies. For (μ, λ) selection, fitness ranking is only performed on the λ new individuals, and the first μ individuals on the list are selected. On the other hand, for $(\mu+\lambda)$ selection, the fitness ranking is performed on all $\mu + \lambda$ individuals and the first μ individuals are selected from the combined list. *See also* evolutionary programming.

evolution planning the process of identifying constraints, formulating an evolution strategy, scheduling and monitoring an evolution project. Formulating an evolution strategy involves deciding whether to keep an application in service with continued maintenance support, to reengineer the application, or to redevelop the application. A hybrid approach may be most appropriate. This decision must be made after performing cost-benefit analysis, and calculating the risks involved for each strategy.

exact coding coding that involves coding methods which reproduce the picture at the receiver without any loss. This method is also called information-lossless or exact coding techniques. Four methods of exact coding are run-length coding, predictive coding, line-to-line predictive differential coding, and block coding. However, several coding schemes use these in a hybrid manner.

exact mean value analysis *See* mean value analysis.

exact string searching the algorithmic problem of finding all occurrences of a given string usually called "the pattern" in another larger "text" string.

exception **(1)** an unusual condition arising during program execution that causes the processor to signal an exception. This signal activates a special exception handler that is designed to handle only this special condition. Division by zero is one exception condition. Some vendors use the term "trap" to denote the same thing.

(2) an event that causes suspension of normal program execution. Types include addressing exception, data exception, operation exception, overflow exception, protection exception, and underflow exception. It can be a fatal error, like divide by zero, or an error that can be handled and managed by special classes devoted to error recovering. The exception can be recovered by means of specific hardware and software mechanisms.

exception handler a special block of system software code that reacts when a specific type of exception occurs. If the exception is for an error that the program can recover from, the program can recover from the error and resume executing after the exception handler has executed. If the programmer does not provide a handler for a given exception, a built-in system exception handler will usually be called, which will result in terminating the process that caused the exception. Finally, the reaction to exception can be halting of the system. As an example, bus error handler is the system software responsible for handling bus error exceptions.

exception reporting a way of reporting to management where only exceptional situations are reported. It is related to early warning systems.

exchangeable disk *See* removable disk.

exclusive lock a lock that prevents concurrent writers and readers to an object.

exclusive OR Boolean binary operator typically used for comparing the status of two variables or signals. Sometimes written "XOR".

The truth table for $\oplus \equiv$ XOR is as follows:

X	Y	$X \oplus Y$
F	F	F
F	T	T
T	F	T
T	T	F .

exclusive reads and exclusive writes (EREW) shared memory model, in which only exclusive reads and exclusive writes are allowed.

executable another term for a binary executable program file.

executable file a representation of a program that is in a form suitable for execution. This may be as a file of compiled and linked machine code modules, as a sequence of byte codes, and in some cases as simple text files such as shell scripts. Sometimes referred to, when the representation suggests it, as a binary executable or binary.

executable lines of code the total number lines of code executed by the computer in the completion of a software function. This differs from the total number of source lines of code (SLOC) as the number of ELOC takes into account the number of executions of software lines of code contained within loops (for example). Hence, the number of ELOC is generally not less than or equal to SLOC, and potentially much greater than SLOC, depending on the size and quantity of loop contained within the software function. ELOC is used typically in the modeling of CPU time, as a more direct relation exists between the ELOC and the number of machine instructions per executed line of code.

executable specification a specification method that produces a model of the system that can be executed by an abstract machine. The behavior of the system can be checked by using: simulation to see if the system responds correctly to the external inputs; model checking to see if a property (generally expressed using a temporal logic) is satisfied by the system; bi-simulation to see if the system has the same behavior of a more abstract model.

execute to perform the execution of an instruction. In most cases, this means to begin sequencing through the microinstruction program corresponding to the macroinstruction to be executed.

execute permission grants or denies the associated class of users the right to execute a program file or to use a directory in a path name.

execution the activity of carrying out instructions of a computer program on a processor.

execution cycle sequence of operations necessary to execute an instruction.

execution profile a measurement of the proportion of total execution time that is spent executing code within each subsystem or module of a software item.

execution semantics those rules of semantic analysis which cannot be applied until the program executes. Typical examples of execution semantics include type checking, null pointer detection, array bounds checking, and arithmetic overflow detection.

execution sequence the order in which the statements or other executable entities of a program are executed. Some languages, particularly domain-specific languages such as systems of constraint equations, spreadsheet languages, event code in discrete event simulation languages, and the like may have no specified execution sequence, while other languages, such as machine code, may have very rigid and carefully specified execution sequences.

execution strategy a strategy for executing a query. This comprises an ordering of the operations with possible parallelism of operations. Often specifies algorithms for execution of individual components.

execution time **(1)** amount of time it takes a computation to complete, from beginning to end. The measure chosen will depend on the type of system. Generally, it will not be equivalent to real time. Possible measures are processor time consumed, number of instructions executed, etc.

(**2**) time during which an instruction or a program is executed. The portion of one machine cycle needed by a CPU's supervisory-control unit to execute an instruction.

execution trace a record reporting information about the execution of a program. It may include details about the execution of each single instruction as well as only some details about specific aspects.

execution unit in modern CPU implementations, the module in which actual instruction execution takes place. There may be a number of execution units of different types within a single CPU, including integer processing units, floating point processing units, load/store units, and branch processing units.

executive a computer program, usually part of an operating system, that controls the execution of other programs and regulates work flow.

exhaustive search a search strategy that systematically examines every possible path through a decision tree or network. For example, for the maximum-likelihood (exhaustive search) detection of a sequence of k bits, all 2^k possible bit sequences are considered and the one with the largest likelihood is selected. Also called brute force search.

existential quantification a first order logic operator used to quantify a variable over a finite or infinite set. It is used to state that a formula is true for one or more values of a variable. It is usually represented with $\exists$. *See* universal quantification.

exit (**1**) a construct in a language which transfers control from the current context to some enclosing context. Sometimes this is the equivalent of a return from a function. More often, it is a specification to terminate a loop under conditions other than those specified by the loop construct, or to leave a block of code by transferring control to the end of the block (a limited form of goto).

(**2**) an operation in a program, whether part of the language specification or provided as a library call, that terminates execution of the program.

exoskeletal master master controller that the user in some way wears around the limbs or body, or which is fastened to the body rather than simply grasped.

expandability *See* extensibility.

expanded memory a term pertaining to older PC-based systems. Expanded memory specification, EMS, was developed for adding memory to PCs (the so called LIM-EMS). PCs were limited in memory to 640 Kbytes even if the 8088/8086 CPU's limit is 1 Mbyte. Thus, in order to overcome this limit, the additional memory was added by using a paging mechanism: up to four windows of 16 Kbytes of memory included into the 640 Kbytes to see up to 8 Mbytes of memory divided into pages of 16 Kbytes. To this end, special memory boards were built. Currently, the MS-DOS is still limited to 640 Kbytes, but the new microprocessors can address even several gigabytes over the first megabyte. Thus, to maintain the compatibility with the previous version and the adoption of the MS-DOS, the presence of expanded memory is simulated by means of specific drivers.

expanding phase *See* growing phase.

expected value of a random variable ensemble average value of a random variable that is given by integrating the random variable after scaling by its probability density function (weighted average) over the entire range.

expert system a computer program that contains a knowledge base and a set of algorithms or rules that infer new facts from knowledge and from incoming data. An expert system is an artificial intelligence application that uses a knowledge base of human expertise to aid in solving problems. The degree of problem solving is based on the quality of the data and rules obtained from the human expert. Expert systems are designed to perform at a human expert level.

In practice, they will perform both well below and well above that of an individual expert.

The expert system derives its answers by running the knowledge base through an inference engine, a software program that interacts with the user and processes the results from the rules and data in the knowledge base. Expert systems are used in applications such as medical diagnosis; equipment repair; investment analysis; financial, estate, and insurance planning; route scheduling for delivery vehicles; contract bidding; counseling for self-service customers; and production control and training.

explanation in artificial intelligence, a tree-structured graph such that each leaf node is either a prior belief or a property of prior nodes. The root node is the training classification assigned to the example.

explicit declaration a mechanism for declaring a name and its properties. In most programming languages, the use of *explicit declaration* is mandatory. In other languages, implicit declaration will declare the name and assign properties according to the rules of the language. Some languages (such as FORTRAN) allow explicit declaration in addition to the implicit declaration normally supported, or require explicit declaration when the programmer wishes the name to have properties different from those that would be assigned by the implicit declaration rules.

explicit lock a lock acquired by issuing a statement to specifically acquire the lock.

explicitly typed language a typed language where types are part of the syntax.

explicit surface a surface representation in which the z coordinate is expressed as a function $z = f(x, y)$ of the x and y coordinates.

explicit value a value associated with the bit string according to the rule defined by the number representation system being used.

exploitation an optimization process where the next search point is chosen such that the greatest reduction in cost is achieved. The new search point is usually chosen within a restricted neighborhood of the current point using a deterministic procedure. In this way, information regarding the local topology of the fitness landscape can be incorporated to determine the search direction for maximum cost reduction. Compared with exploration techniques, this approach usually achieves fast convergence at the expense of possible termination at local minima. As a result, this technique is usually adopted in the final stages of the optimization process to further refine the search after the location of promising regions by a previous exploration step. *See also* exploration.

exploration an optimization process where search points can be located in any regions of the search space. A controlled random process is usually adopted to generate new search points such that there exists a finite probability for each new point to reach any location in the search space and thus prevent it from being trapped in a local minimum. This approach is usually adopted in the initial stages of the search to prevent premature convergence to non-optimal solutions. *See also* exploitation.

exponent **(1)** in a floating point number, represented as a computation $m \times b^e$, the value e. *See also* exponent radix, mantissa.

(2) a shorthand notation for representing repeated multiplication of the same base. 2^4 is exponential notation to multiply two by itself four times : $2^4 = 2 \cdot 2 \cdot 2 \cdot 2 = 16$. 4 is called the exponent, indicating how many times the number 2, called the base, is used as a factor.

(3) in programming languages, the component of a floating-point number that signifies the integer power by which the significand is multiplied in determining the value of the represented number. For example, in the number 1.2E4, the exponent is 4.

exponential **(1)** any function that is the sum of constants times other constants to the power of the argument: $f(x) = \Sigma_{i=0}^{k} c_i b_i^{x p_i}$.

(2) in complexity theory, the measure of computation, $m(n)$ (usually execution time or memory space), is bounded by an exponential function of the problem size, n. More formally $m(n) = O(k^n)$. *See also* logarithmic, linear, polynomial.

exponential back-off the technique used in CSMA-CD (Ethernet) networks as specified by the IEEE standard 802.3 for how long to wait before trying to send again after a collision has occurred. With binary exponential back-off, a process will wait a random number of time slots before trying to send again, where the random number is selected between 0 and 2 number of previous collisions, up to 1024 time slots.

exponential distribution a probability density function having the following exponential behavior:

$$f(x) = \begin{cases} \lambda e^{-\lambda x} & x \geq 0 \\ 0 & x < 0 \end{cases},$$

where $\lambda > 0$. This distribution can describe a number of physical phenomena, such as the time for a radioactive nucleus to decay, or the time for a component to fail. *See also* probability density function, Cauchy distribution, Gaussian distribution.

exponent radix in a floating point number, represented as a computation $m \times b^e$, the value b (typically 2, 8, 10, or 16, depending on the machine architecture).

exponent spill the computation of an exponent that lies outside the permissible range for exponents. It may lead to underflow or overflow.

export to transform a database into a portable format.

exported information which is made visible outside the confines of an entity. For example, a function name may be exported from a compilation unit, a method may be exported from a class. Related to public. *Compare with* private, protected.

exposure time the length of time for which a hazard exists before being detected and corrected or avoided, or neutralized by a change in environmental conditions. The longer the exposure time, the more likely a hazard is to lead to a mishap. It is generally measured by real time, since it depends on conditions in the real world.

expression a computation performed by a program. The exact nature of what constitutes an "expression" (as distinct from a statement or a declaration) is language-specific. In some languages a program consists solely of a single "expression" (for example, LISP). Traditionally, an expression is a computation involving some subset of the syntactic units of the language, usually consisting of the computation of a value based upon the application of, as present in the language, functions, operators, or both, to objects representing values, and producing as a result a new object.

expurgated code a code constructed from another code by deleting one or more codewords from the original code.

extendability *See* extensibility.

extended binary-coded-decimal interchange code (EBCDIC) character code developed by IBM and used in mainframe computers. It is closely related to the Hollerith code for punched cards.

extended BNF (EBNF) an extension to the BNF that adds bracketing metasymbols [...] and {...} to indicate optional and repeated structures, respectively. These can be written as (...)? and (...)* to remain consistent with standard regular expression notation.

extended code a code constructed from another code by adding additional symbols to each codeword. Thus an (n, k) original code becomes an $(n + 1, k)$ code after the adding of one redundant symbol.

extended entity relationship (EER) model a model in which constructs such as inheritance, generalization, specialization, and categories are added to the basic ER model.

extended industry standard architecture (EISA) a bus architecture designed for PCs using an Intel 80386, 80486, or Pentium microprocessor. EISA buses are 32 bits wide and support multiprocessing. The EISA bus was designed by IBM competitors to compete with Micro Channel architecture (MCA). EISA and

MCA are not compatible with each other, the principal difference between EISA and MCA is that EISA is backward compatible with the ISA bus (also called the AT bus), while MCA is not. Therefore, computers with an EISA bus can use new EISA expansion cards as well as old AT expansion cards, while computers with an MCA bus can use only MCA expansion cards.

extended light source a light source with surface area which will cast shadows with both umbra and penumbra and thus is more difficult to model than a point source.

extended memory in PCs, the memory located over the first megabyte of memory. This kind of memory is seen in the PC as continuous. Operating systems such as Windows NT, LINUX, and OS/2 are capable of working in protected mode, and thus at 32 bits. When the MS-DOS is adopted, to allow the exploitation of this memory as EMS, special software drivers have to be used. This overcomes the limit of 640 Kbytes imposed by the real-mode and adopted by MS-DOS.

extended precision a floating point value which has more than 64 bits of representation. In IEEE floating point, an extended precision number has 80 bits of representation.

extended state machine (ESM) an extension of finite state machine with a richer representation of states. Generally these are introduced variables that can be manipulated in transitions, i.e., if event *e* occurs, variable *count* is incremented.

extended storage *See* solid state disk.

extendible hashing a technique that allows hashing to a dynamic file. The binary representation of the hash field is used to hash files to the correct disk block. On collision, the file is doubled in size and contents redistributed according to the next significant bit. A lookup table is created as the file expands.

extensibility the capacity of a system, component, or class to be readily and safely extended in its behavioral specification/operational and/or structural capabilities.

extension a part of the file name, usually following a period, that assigns the file to a particular category.

extension language a language that can be used to extend an existing system, commonly an application program. An extension language may be an existing language or one constructed for that purpose. Examples of extension languages are Tcl and Visual Basic for Applications. *Compare with* language extension.

extension principle a basic identity for extending the domain of nonfuzzy or crisp relations to their fuzzy counterparts.

extent (1) the lifetime of a value or object in a system. May or may not follow the scope of the name defining the value. In systems with dynamic allocation of objects, the extent of an object is its lifetime until it is deallocated or garbage collected.

(2) an unordered collection of references to the instances of a type or a class. In an OODBMS, the extent must be explicitly maintained by the database.

external attribute an attribute of a system which characterizes its interaction with the environment. External attributes are those that most directly concern users. Certain internal attributes may be indicators of external attributes, although not identical to them.

external containment a class is subject to external containment each time the data members are referenced by pointer or reference and, therefore, are stored externally.

external declaration a construct in many programming languages which permits, or requires, the programmer to declare a name and (usually) its associated properties.

external description the specification of the system or of a subpart of it that describes how the part interacts with the other parts of the system and with the environment. Only elements that are visible from the outside such as inputs and outputs can be referred to in the external description. *See* external input, external output, external interface.

external event event occurring outside the CPU and I/O modules of a computer system that results in a CPU interrupt. Examples include power fail interrupts, interval timer interrupts, and operator intervention interrupts.

external failure an undesirable event in the environment which adversely affects the operation of an item.

external fragmentation in segmentation, leaving small unusable areas of main memory that can occur after transferring segments into and out of the memory.

external hashing a form of hashing where the value returned from the hash function address is a disk block.

external index an auxiliary data structure added to a main data structure to improve operations, such as a search on a secondary key. *See also* inverted index, hash table, search tree.

external input an element used to acquire information from the outside or from another part of the system.

external inquiry one of the factors used for calculating function points. It is denoted by a query/response pair.

external interface (**1**) the set of elements of a part of a system that is visible from the outside (from the other parts or from the environment). Typical elements are inputs and outputs.

(**2**) a factor used for calculating function points. It denotes the data or control information that passes the application's boundary.

external interrupt a signal requesting attention that is generated outside of the CPU.

external measure an indirect measure of a product derived from measures of the behavior of the system of which it is a part. An external attribute can only be measured by observing the system in operation. The system includes any associated hardware, software (either custom software or off-the-shelf software), and users. The number of faults found during testing is an external measure of the number of faults in the program because the number of faults are counted during the operation of a computer system. External measures can be used to evaluate quality attributes closer to the ultimate objectives of the design.

external memory secondary memory of a computer.

external metric *See* external measure.

external output an element used to produce information to the outside of the system or to another part of the system.

external quality the extent to which a product satisfies stated and implied needs when used under specified conditions.

external schema a description of a user view, i.e., the part of the database that interests a particular user (or set of users).

external sorting the situation when the file to be sorted is too large to fit in main memory.

external sort merge *See* sort merge.

external specification *See* external description.

external symbol a symbol that is not defined in a program module, but whose value or location may be needed when the program executes. An *external symbol* may be the consequence of implicit declaration (such as in FORTRAN or C), or may have to be specifically stated by the programmer. External symbols are made accessible by linking.

extrapolation one of several methods to estimate the values of a sequence $r(k)$ for lags $|k| > p$ from the given values of $r(k)$ for $|k| \leq p$.

extremal some of the entries of the auxiliary array used to perform string searching. An entry is d-extremal if it is the deepest entry on its diagonal to be given value d.

extreme point a corner point of a polyhedron.

eye tracker device used to sense and record the aiming point of the eye.

eye tracking the use of an eye tracker.

F

F a representation convention in the C language that is used to designate that a literal is to be interpreted as single-precision floating point. May follow an exponent if one is present. *See also* L.

face/facet/patch normal a term from descriptive geometry. A solid object can be constructed from many surface pieces which fit together. Each piece is called a face/facet/patch. Its normal is the direction from the surface of the object that is perpendicular to the piece's surface.

facet a small piece (usually a planar polygon) of a larger surface. *See* faceting, triangulation.

faceting the technique used to construct a surface from multiple facets.

facsimile encoding a bilevel coding method applied to the encoding and transmission of documents. Facsimile systems may include support for grayscale image coding too, which is described under still image coding.

fact a statement that an atomic formula holds. As an example, if "father" is a rule, then father (adam, abel) is a fact.

factor (**1**) in many programming languages, this refers to an expression with an infix operator where the operator represents a multiplication, division, or similar operation such as modulo or integer divide.

(**2**) for a string v, a string u if $u = u'vu''$ for some strings u' and u''.

factorial for an integer n, written $n!$, it is $n \times n - 1 \times \ldots \times 2 \times 1$. By definition $0! = 1$. *See also* gamma function, Stirling's approximation.

factoring (**1**) the process of decomposition. *See* system decomposition.

(**2**) the action of reorganizing procedure on modules in order to arrange them in more appropriate locations to reduce cohesion among modules.

factory (**1**) the plant in which the software is produced.

(**2**) in object-oriented languages, an instance of a class that creates instances of other objects.

fading a method of switching between video sources, or images, using a black image as an intermediate. Fading without this intermediate is called a dissolve.

failed transaction a transaction that does not complete because of hardware, software, or other failure.

failover (**1**) a system property by which, in the event of a failure of one component, the function provided by that component is taken over by another.

(**2**) moving a virtual machine or resource from one element to another when the first element can no longer support it.

fail safe a mechanism in which, following detection of a hazardous state, a mishap can be avoided despite a possible loss of service. The possibility of designing an item to "fail safe" obviously depends on its having a safe mode of failure. A train running along the tracks has a (generally) safe mode of failure (i.e., stop). An airliner flying at 30,000 feet does not!

fail soft the condition of a system that continues to provide main functionalities even in the presence of some failure. For example, when a computer system presents some problems with the color of the screen, it may continue to perform computation and interact with the user with the available pallet of colors.

fail-stop processor a processor which does not perform incorrect computation in the event of a fault. Self-checking logic is often used to approximate fail-stop processing.

failure the inability of a computer system to perform its functional requirements, or the de-

0-8493-2691-5/01/$0.00+$.50

parture of software from its intended behavior as specified in the requirements. *Failure* can also be considered to be the event when either of these occurs, as distinguished from "fault" which is a state. A failure is an event in time. A failure may be due to a physical failure of a hardware component, to activation of a latent design fault, or to an external failure. Following a failure, an item may recover and resume its required service after a break, partially recover and continue to provide some of its required functions (fail degraded), or it may remain down (complete failure) until repaired.

failure cause of a system the combination of a fault and a trigger which activates it in order to initiate the failure mechanism. This applies to both design failures and physical failures, although in the latter case, for a simple item, the trigger may be trivial. The circumstances during design, manufacture, or use which have led to a failure.

failure mechanism of a system the process of error propagation following a component failure which leads to a system failure. The failure of a component results in a local error which means that its users (i.e., other components) must cope with a fault (temporary or permanent) at that level. If unable to effect error recovery, they themselves will fail, generating a higher-level error. This process of "error-fault-failure" will continue through all subsystem levels until a failure occurs at the system interface. This conceptual model of error propagation applies to both design and physical failures.

failure mode the observed set of symptoms by which a failure is recognized. The term is quite obsolete, and fault mode should be used.

failure rate the number of failures of a specified category in a given period, in a given number of computer runs, or in some other given rate of measure. The *failure rate* of a system or component varies during its lifetime, at first decreasing as problems are detected and repaired and finally increasing due to deterioration. Between these two periods the rate usually remains steady.

failure ration *See* failure rate.

failure semantics refers to the likely and/or allowable processor failure behaviors. *Failure semantics* depend, to a great extent, on the error detection mechanisms implemented in the system.

failure severity the seriousness of the effect of a failure. This may be measured on an ordinal scale (e.g., classification as major, minor, or negligible) or on a ratio scale (e.g., cost of recovery, length of down-time). This value may also depend on the frequency of the error.

fairness (**1**) the concept of providing equivalent or near-equivalent access to a shared resource for all requestors.

(**2**) an attribute of a scheduler which is concerned with the divesting of processing resources to competing jobs in a way that does not discriminate against any job based on its service demands. There are several different criteria for fairness, e.g., equal share of resources, equal blocking probability.

(**3**) the degree to which a scheduling or allocation policy is equitable and non-discriminatory in granting requests among processes competing for access to limited system resources such as memory, CPU, or network bandwidth.

falling edge (**1**) the region of a waveform when the wave goes from its high state to its low state.

(**2**) the high-to-low transition in voltage of a time-varying digital signal.

fall time (**1**) in digital electronics, the period of duration of the transition of a digital signal from a stable high voltage level to a stable low voltage level.

(**2**) in optics, the time interval for the falling edge of an optical pulse to transition from 90 to 10% of the pulse amplitude. Alternatively, values of 80 and 20% may be used.

false a value which represents the "false" value in two-valued logic. Programming languages are, in the abstract, not constrained to a particular representation of the notion of "false" and traditional choices have been "the constant 0", "any even value", "the constant −1", or "any negative value". *Compare with* true.

false color the replacement of a color in a colored image by a different color, usually not present in the original image. Used to highlight regions or distinguish pixels of similar colors. *See also* pseudocolor.

false coloring *See* pseudocolor.

false sharing the situation when more than one processor accesses different parts of the same line in their caches but not the same data words within the line. This can cause significant performance degradation because cache coherence protocols consider the line as the smallest unit to be transferred or invalidated.

false start in speech processing, a type of dysfluency that occurs when a word or phrase is abruptly ended prior to being fully uttered, and then verbally is replaced with an alternative form. *See also* dysfluency.

family a set of systems/products sharing some commonalties (equivalent to product line).

family design an architecture applicable to a family of applications belonging to the domain. Sometimes called the "generic architecture".

fan beam reconstruction reconstruction of a computed tomography image from projections created by a point source that emits a fan- or wedge-shaped beam of radiation. *Fan beam reconstruction* enables data to be gathered much more quickly than by using a linear beam to produce parallel projections. *See also* computed tomography, image reconstruction, projection, Radon transform.

Fano's algorithm a sequential decoding algorithm for decoding of trellis codes.

Fano's inequality information theoretic inequality bounding the probability of incorrectly guessing the value of one random variable based on observation of another. If P_e is the probability of incorrectly guessing a random variable $X \in \mathcal{X}$, based upon observation of the random variable Y, then

$$H\left(P_e\right) + P_e \log\left(|\mathcal{X} - 1\right) \geq H(X|Y) .$$

Named after its discoverer, R.M. Fano (1952-). Used in proving the weak converse to the channel coding theorem (Shannon's second theorem).

fan out **(1)** the blocking factor of a multilevel index.

(2) in digital circuits, the number of output stages connected to some circuit component or subsystem.

far pointer in some languages, running on machines that have more than one representation of a machine address, designates those addresses that can access a memory location beyond the area which can be encoded in the shorter form of the address. Although commonly used in the context of Microsoft Windows running on segmented-address Intel processors, the concept predates that usage by at least 20 years. *Compare with* near pointer.

fast Fourier transform (FFT) a fast version of the discrete Fourier transform.

fast Hadamard transform a way of improving compression using the Hadamard matrix via factorization.

father wavelet the scaling function in the coarsest resolution in wavelet analysis.

fathoming the process of pruning a search tree.

fault (1) in software engineering, the state of an item characterized by inability to perform a required function, excluding the inability during preventive maintenance or other planned actions, or due to lack of external resources. A fault is often the result of a failure of the item itself, but may exist without prior failure.

(2) in hardware, a physical defect or imperfection of hardware. Typical circuit faults are short opens in conductor, defects in silicon, etc. *See* disturbance.

fault activation the event in which a latent fault gives rise to a failure in response to a trigger.

fault detection the process of locating distortions or other deviations from the ideal, typically during the process of automated visual inspection, e.g., in products undergoing manufacture.

fault masking a condition in which the occurrence of a fault is masked.

fault mode an observable state of an item, which can give rise to failure under certain operating conditions.

fault prevention any technique or process that attempts to eliminate the possibility of having a failure occur in a hardware device or software routine.

fault recovery coverage factor the probability, given that an error has occurred, that the system recovers from a specified type of failure.

fault report an error report that does not originate from the developers but from the users of a software system. *See also* error report.

fault resilient (1) a synonym for fault tolerance.

(2) as opposed to fault tolerance, the ability to recover from failures and prevent loss of data by using undamaged members of a computing cluster, albeit with a loss of performance compared to a cluster with no failed components.

fault secure a system that is capable of avoiding the production of failure in the presence of a set of failures.

fault tolerance the capability of the software product to maintain a specified level of performance in cases of software faults or of infringement of its specified interface. The specified level of performance may include fail safe capability. Increasing *fault tolerance* usually involves providing redundant components as well as back-up and recovery systems. Methods for error detection and correction must also be part of a fault tolerant system.

fault tree a method of analyzing a system failure as a tree of AND and OR gates with leaf inputs representing single events.

FCFS *See* first-come-first-served.

FCM *See* fuzzy c-means, fuzzy cognitive map.

FDDI *See* fiber distributed data interface.

FDT *See* formal description techniques.

feasibility study an evaluation of possible strategies which might include do nothing, modify current system, build new software system, or even build new manual system. The conclusion of this study should be a clear view about the feasibility of a project/system, in general. Thus, criticisms and other qualitative evaluation are typically included.

feasible solution any element of the feasible region of an optimization problem.

feature (1) in software engineering, an identified property of a software product that can be related to the quality characteristics. Examples of features include path length, modularity, program structure, and comments. Used in object-oriented modeling as a collection of attributes and operations (*See* class feature).

(2) in image processing, a measurable characteristic of an object in an image. Simple examples would be area, perimeter, and convexity. More complex features use vectors; examples include moments, Fourier descriptors, projections, and histogram-based features. Features are frequently used to recognize classes of object, and sets of simple features can be collected into a vector for this purpose. Using both area and perimeter, for instance, one can quickly distinguish between a circle and a triangle. May also refer to a characteristic of a whole image; such a feature could then be used in image database analysis.

feature detection the detection of smaller features within an image with a view to inferring the presence of objects. This type of process is cognate to pattern recognition. Typically, it is used to locate products ready for inspection or to locate faults during inspection. A feature can be detected by finding points having optimal response to a given combination of local operations such as convolutions or morphological operators. *See also* object detection.

feature extraction a method of transforming raw data, which can have very high dimensionality, into a lower dimensional representation which still contains the important features of the data.

feature map a fixed geometrical structure (often two-dimensional) for unsupervised learning which maps the input patterns to different output units in the structure so that similar input patterns always trigger nearby output units topographically. *See* self-organizing system, self-organizing algorithm.

feature measurement the measurement of features, with the aim of recognition or inspection to determine whether products are within acceptable tolerances.

feature orientation measurement of the orientation of features, either as part of the recognition process or as part of an inspection or image measurement process.

feature recognition the process of locating features and determining what types of features they are, either directly or indirectly, through the location of sub-features followed by suitable inference procedures. Typically, inference is carried out by application of Hough transforms or association graphs.

FEC *See* forward error correction.

federated DBMS a distributed database in which every site has an independent centralized DBMS.

feedback (**1**) conditions within a system or at the outputs of a system having an effect on the inputs of the system. Feedback may cause a system to become more stable (e.g., when information about congestion causes input flows to be reduced), or less stable (e.g., when requests to retransmit packets are increased because packets were lost).

(**2**) a technique used in information retrieval where returned documents are graded by the user (usually as relevant or irrelevant, positive and negative feedback, respectively). The judgments are used to try to improve the performance of the retrieval.

feedback hierarchy term referring to the hierarchical and multi-stage character of human feedback. There may be, for example, feedback loops assessing the quality of information processing and response selection, the quality of a goal-directed movement, and the outcome of that movement for system functionality, etc.

feedback linearization a modern approach to robot arm control that formalizes computed torque control mathematically, allowing formal proofs of stability and design of advanced algorithms.

feedback queue a queue in which customers can re-join the arrival stream after receiving service. Customers that complete their service can either leave the system or can be fed back to the queue. How the customer can be fed back to the queue depends upon the policy. For example, in a queue with "Bernoulli feedback", upon receiving service a customer is fed back to the queue with probability p and leaves the system with probability $(1 - p)$.

feed-forward (**1**) a neural network in which the signals flow in one direction only, from the inputs to the outputs.

(**2**) signals transmitted to a system under control by a control system and used to affect the state or behavior of the system.

feedforward network *See* feedforward neural network.

feedforward neural network one of two primary classifications of neural networks (the other is recurrent). In a feedforward neural network, the **x**-vector input to the single functional layer of processing elements (this input typically occurs via a layer of input units) leads to the **y**-vector output in a single feedforward pass. An example of the feedforward neural network is a multi-layer perceptron.

feet-per-second (FPS) a measure of the speed of an animation which refers to the number of feet (30.48 cm) of cinema film displayed in one second. *See also* frames-per-second.

female connector a connector presenting receptacles for the insertion of the corresponding male connector that presents pins.

fencepost error a particular kind of boundary condition error caused by performing a computation one fewer or one more time than the correct computation. For example, executing a loop once too many or once too few times because the limit test was written as "less than or equal" instead of "less than" or vice-versa. Named after the error of "How many fenceposts, 10 feet apart, are required to create a fence 100 feet long?", in which the correct answer is 11.

ferrite core memory *See* magnetic core memory.

fetch-and-add instruction for a multiprocessor, an instruction which reads the content of a shared memory location and then adds a constant specified in the instruction, all in one indivisible operation. Can be used to handle multiprocessor synchronizations.

fetch cycle the period of time during which an instruction is retrieved from memory and sent to the CPU. This is the part of the fetch-decode-execute cycle for all machine instruction.

fetch-execute cycle the sequence of steps that implement each instruction in a computer instruction set. A particular instruction is executed by executing the steps of its specific fetch-execute cycle. The fetch part of the cycle retrieves the instruction to be executed from memory. The execution part of the cycle performs the actual task specified by the instruction. Typically, the steps in a fetch-execute cycle are made up of various combinations of only three operations:

1. the movement of data between various registers in the machine
2. the addition of the contents of two registers or the contents of a register plus a constant with the results stored in a register
3. shift or rotate operations upon the data in a register.

fetching the process of reading instructions from a stored program for execution.

fetch-on-demand *See* fetch policy.

fetch-on-miss *See* fetch policy.

fetch policy policy to determine when a block should be moved from one level of a hierarchical memory into the next level closer to the CPU.

There are two main types of fetch policies: "fetch on miss" or "demand fetch policy" brings in an object when the object is not found in the top-level memory and is required; "prefetch" or "anticipatory fetch policy" brings in an object before it is required, using the principle of locality. With a "fetch on miss" policy, the process requiring the objects must wait frequently when the objects it requires are not in the top-level memory. A "prefetch" policy may minimize the wait time, but it has the possibility of bringing in objects that are never going to be used. It can also replace useful objects in the top-level memory with objects that are not going to be used. *See also* cache and virtual memory. The prefetching may bring data directly into the relevant memory level, or it may bring it into an intermediate buffer.

FFT *See* fast Fourier transform.

fiber distributed data interface (FDDI) an American National Standards Institute standard for a high-performance local area network, running at 100 Mbps over distances of up to 200 km and up to 1000 stations connected. Incorporates token ring processing and supports circuit-switched voice and packetized data. For its physical medium, it uses fiber optic cable, in a dual counter-rotating ring architecture.

fiber optic cable a glass fiber cable that conducts light signals and can be used in token ring local area networks and metropolitan area networks. Fiber optics can provide higher data rates than coaxial cable. Fiber optic cables are also immune to electrical interference.

fiber optic interconnect interconnect that uses an optical fiber to connect a source to a detector. An optical fiber is used for implementing a bus. The merits are large bandwidth and high speed of propagation.

Fibonacci sequence a sequence of numbers that begins with 0, 1, then proceeds by adding the preceding two numbers in a sequence to get the next.

Fibonnaci heap a heap made of a forest of trees. The amortized cost of the operations create, insert a value, decrease a value, find minimum, and merge or join (meld) two heaps is a constant $\Theta(1)$. The delete operation takes $O(\log n)$. *See also* priority queue.

fidelity in synthetic environments and simulation, the degree to which a computer-generated world looks like (physical fidelity) and works like (functional fidelity) the real world it represents.

field (1) in a composite object (for example, a record), a named member of the composite.

(2) in abstract algebra, a set containing two operators, each with an inverse, that map values in the set to values in the set. Real numbers are a field. Floating point numbers are not.

field memory video memory required to store the number of picture elements for one vertical scan (field) of video information of an interlaced scanned system.

The memory storage in bits is computed by multiplying the number of video samples made per horizontal line times the number of horizontal lines per field (vertical scan) times the number of bits per sample. A sample consists of the information necessary to reproduce the color information.

Storage requirements can be minimized by sampling the color video information consisting of the luminance (Y) and the two color difference signals, (R - Y) and (B - Y). The color signal bandwidth is less than the luminance bandwidth that can be used to reduce the field memory storage requirements. Four samples of the luminance (Y) signal are combined with two samples of the (R - Y) signal and two samples of the (B - Y) signal. The preceding video sampling technique is designated as 4:2:2 sampling and reduces the field memory size by one-third. Field memory for NTSC video sampled at 4 times the color subcarrier frequency at 8 bits/pixel would require 3.822 Megabits of RAM when 4:2:2 sampling is used.

field-programmable gate array (FPGA) a programmable logic device which consists of a matrix of programmable cells embedded in a programmable routing mesh. The combined programming of the cell functions and routing network defines the function of the device. Reprogrammability of FPGAs makes them generic hardware and allows them to be reprogrammed to serve many different applications. FPGAs consist of SRAMS, gates, latches, and programmable interconnects.

field rendering in interlaced video, a single image frame is sent as two fields composed of even scanlines and odd scanlines. Field rendering refers to a method of rendering where fields are rendered separately in order to reduce motion artifacts.

FIFO *See* first-in-first-out.

FIFO memory commonly known as a queue. It is a structure where objects are taken out of the structure in the order they were put in. Compare this with a LIFO memory or stack. A FIFO is useful for buffering data in an asynchronous transmission where the sender and receiver are not synchronized; the sender places data objects in the FIFO memory, while the receiver collects the objects from it.

fifth generation language (5GL) a programming language chracterized by the underlying use of natural language processing, database systems, and expert systems.

fifth normal form (5NF) a property used in characterizing databases. A relation R is in fifth normal form with respect to an F (set of functional, multi-valued, and join dependencies) if for every nontrivial join dependency $*(R_1, R_2, \ldots R_N) \in F^+$, R_i is a superkey of R.

file a collection of related records.

file cache a cache used to buffer data being transferred between file storage (e.g., disk) and main memory.

file descriptor an identifier that allows accessing a file from within a computer program.

file format the structure of the computer file in which an image is stored. Often the format consists of a fixed-size header followed by the pixel values written from the top to the bottom row and within a row from the left to the right column. However, it is also common to compress the image. *See also* graphics interchange format, header, image compression, tagged image file format (TIFF).

file handle *See* file descriptor.

file name a sequence of characters that identifies a file, that is, the name portion of a directory entry (link).

file name generation a method of referring to files and directories using representative text patterns instead of literal file names.

file protection a mechanism to prevent unauthorized file use by setting rights to read, write, and execute.

file system a means of storing, organizing, and accessing files in a computer system.

file transfer protocol (ftp) a client-server protocol which allows a user on one computer to transfer files to and from another computer over a TCP/IP network. Also the client program the user executes to transfer files.

fill a technique for coloring areas bounded by line edges. The algorithms that fill interior-defined regions (the largest connected region of pixels whose values are the same as a given starting pixel) are called flood fill algorithms.

filled Mandelbrot set a Mandelbrot set where colors are not used.

filled pause in speech processing a type of dysfluency that occurs when words such as "er" or "umm" are used. *See also* dysfluency.

fill factor of an index, the percentage of space used in the index.

FILO *See* first-in-last-out.

filter (**1**) a computer program used to modify the data submitted to its input and sending it to the output.

(**2**) an optical device that selectively attenuates the intensity of light passing through it according to the light's properties. Common filters attenuate light according to either wavelength or polarization state.

(**3**) an electronic device that modifies some characteristic of an input signal for conditioning or other purposes. *See* high-pass filter, low-pass filter, bandpass filter.

filtering the process of eliminating object, signal, or image components which do not match up to some pre-specified criterion, as in the case of removing specific types of noise from signals. More generally, the application of an operator (typically a linear convolution) to a signal.

final query tree the query tree following the application of the heuristic optimization rules.

final state in an automaton, a final state is a state which designates that the input stream has been recognized. Synonym for accepting state.

final test electrical test performed after assembly to separate "good" devices from "bad".

fine-adjusting movement in a target acquisition movement, high-bandwidth, low-amplitude sub-movement used to make final trajectory corrections.

fine grained transaction a transaction that performs a small number of updates.

finger (1) in robotics, by analogy the most distal part of a gripper, which makes contact with grasped objects and applies grasping forces.

(2) a Unix shell command.

finitary tree a tree with a finite number of children at every node.

finite automaton *See* finite state machine.

finite difference equation a recursive equation that describes a function at time t in terms of the function at previous time samples $t-1$, $t-2$, and so on.

finite-extent sequence the discrete-time signals with finite duration. The finite-extent sequence $\{x(n)\}$ is zero for all values of n outside a finite interval.

finite state automaton (FSA) *See* finite state machine.

finite state machine (FSM) a mathematical model that is characterized by a mapping function from state, input symbol to a new next state. If the mapping is always to a unique next state, the automaton is designated as a deterministic finite state automaton (DFSA). If the mapping is to two or more next states, the automaton is designated as a nondeterministic finite state automaton (NDFSA). If a qualifier specifying determinism is not specified, generally a deterministic FSA is intended. A FSM describes many different concepts in communications such as convolutional coding/decoding, CPM modulation, ISI channels, CDMA transmission, shift-register sequence generation, data transmission, and computer protocols.

More formally, a finite state machine can be uniquely defined by a set of possible states, S, an output function, $y = f(x, S_c)$, and a transition function, $S_n = g(x, S_c)$. An FSM is computationally equivalent to a regular expression or a Type 3 grammar in the Chomsky hierarchy and also to a restricted Turing machine where the head is read-only and shifts only from left to right.

Also known as finite state automata (FSA), finite machine, state machine. *See also* state transition diagram.

finite state VQ (FSVQ) a vector quantizer with memory. FSVQs form a subset of the general class of recursive vector quantization. The next state is determined by the current state S_n together with the previous channel symbol u_n by some mapping function.

$$S_{n+1} = f\left(u_n, S_n\right), n = 0, 1, \ldots$$

this also obeys the minimum distortion property

$$\alpha(\mathbf{x}, s) = \min{}^{-1} d(\mathbf{x}, \beta(u, s))$$

with a finite state $\mathbf{S} = [\alpha_1, \alpha_2, \ldots, \alpha_k$, such that the state S_n can only take on values in S. The states can be called by names in generality.

finite type a data type whose value set is finite; for example, Boolean values, enumeration sets, character values. Note that types such as integer values and floating point values are considered infinite types, even though on any specific machine and specific compiler the available values might be enumerable and hence finite.

fire wall an electronic barrier restricting communications between two parts of a network.

firm deadline a deadline that can be sporadically missed without severe consequences.

firm real-time system a real-time system which can fail to meet one or more deadlines without system failure. *Compare with* soft real-time system, hard real-time system.

firmware the combination of a hardware device and computer instructions or computer data that reside as read-only software on the hardware device. The firmware is typically executed at the computer boot and used by the operating systems for configuring and using computer hardware components.

first-class object an object that can be stored in data structures, passed as arguments, and returned as the result of function calls. In functional languages, functions are first-class objects.

first-come-first-served (FCFS) (1) a queueing discipline in which customers are serviced in the order in which they arrive.

(2) a scheduling strategy that relies on processing program units in the order of arrival, without preemption. *See also* last-in-first-served.

first difference for a sequence $\{x(n)\}$ the sequence obtained by simply subtracting its $(n-1)$th element from its nth element, i.e.,

$$y(n) = x(n) - x(n-1) .$$

first-fit memory allocation a memory allocation algorithm used for variable-size units (e.g., segments). The "hole" selected is the first one that will fit the unit to be loaded. This hole is then broken up into two pieces: one for the process and one for the unused memory, except in the unlikely case of an exact fit, there is no unused memory.

first index the lowest level of a multi-level index.

first-in-first-out (FIFO) (1) a policy that items are processed in order of arrival.

(2) a queuing discipline whereby the entries in a queue are removed in the same order as that in which they joined the queue. *See also* last-in-first-out.

first-in-last-out (FILO) a queuing rule whereby the first entries are removed in the opposite order as that in which they joined the queue. This is typical of Stack structures and equivalent to last-in-first-out (LIFO).

first normal form (1NF) a concept used to characterize databases. A relation is said to be in first normal form (1NF) if it contains no multi-valued attributes.

first-order control form of control in which user inputs affect the position of the controlled object; position control.

first order hold (FOH) for a signal $f(k)$, the sequence of straight lines connecting the sample points of $f(k)$. It interpolates the values between two adjacent samples $f(k)$ and $f(k+1)$ using a linear approximation given by

$$x(t) = x(kT_s) + \frac{t - kT_s}{T_s}\left(x((k+1)T_s) - x(kT_s)\right) .$$

first-order logic logic in which predicates may have arguments and formulas may be quantified.

first order logic (FOL) the extension of propositional logic with universal and existential quantifiers. *See* universal quantification, existential quantification.

first-order type system a language system that does not include quantification over type variables.

Fisher information a quantitative measurement of the ability to estimate a specific set of parameters. The Fisher information $J(\theta)$ is defined by

$$J(\theta) = E_\theta\left(\frac{\partial \ln f_\theta(y)}{\partial\theta}\right)^2 = -E_\theta\left(\frac{\partial^2 \ln f_\theta(y)}{\partial^2\theta}\right)$$

where Y is a N-dimensional vector indexed by a vector of parameters θ. *See also* Cramer-Rao bound.

fitness in evolutionary computation, a quantity that measures the likelihood of an individual to be selected into the next generation. In evolutionary systems and evolutionary programming, the fitness function is usually identical to the objective function associated with the original optimization problem. On the other hand, a scaling operation on the original objective function is required in genetic algorithms to ensure the positivities of the fitness values before applying proportional selection.

Fitts' law for goal-directed movements, an equation describing the relationship of movement amplitude (A) and target size (W) to the time (MT) required to complete the movement ($MT = a + b\log_2[2A/W]$).

Fitts' task a simple tapping task developed for study of Fitts' Law and also sometimes used to characterize telemanipulator performance.

fixed cost a cost that is always in place for a software project. For instance, requirements always need to be developed. The concept of a fixed cost is important because some vendors of CASE tools or methodology claim enormous productivity improvements. Many of the issues have fixed prices, such as writing documentation. The vendors often do not take the fixed costs into account and come up with overrated improvements by using their language or tool.

fixed-gap head *See* disk head.

fixed-grid method the process of space decomposition into rectangular cells by overlaying a grid on the space. If the cells are congruent (i.e., of the same width, height, etc.), then the grid is said to be uniform.

fixed-head disk a disk in which one read/write head unit is placed at every track position. This eliminates the need for positioning the head radially over the correct track, thus eliminating the "seek" delay time. Rarely used today because modern disks consist of hundreds of tracks per disk surface, making it economically infeasible to place a head unit at every track.

fixed length field the size of the field is of a defined length.

fixed-length instruction the machine language instructions for a computer all have the same number of bits.

fixed partitioning a form of static parallel job scheduling in which the partition size allocated to a job is determined at system generation time and can only be changed by restarting the system.

fixed-point number system a number-representation system in which the position of the radix (base) point is fixed for all representations. Usually used to represent integers and significands of floating-point numbers.

fixed-point processor a processor capable of operating on scaled integer and fractional data values.

fixed-point register a digital storage element used to manipulate data in a fixed-point representation system whereby each bit indicates an unscaled or unshifted binary value. Common encoding schemes utilized in fixed-point registers include unsigned binary, sign-magnitude, and binary coded decimal representations.

fixed-point representation (**1**) a number representation in which the radix point is assumed to be located in a fixed position, yielding either an integer or a fraction as the interpretation of the internal machine representation. *Contrast with* floating-point representation.

(**2**) a method of representing numbers as integers with an understood slope and origin. To convert a fixed-point representation to its value, multiply the representation by the slope (called the DELTA in Ada) and add the origin value. *Fixed-point representation* provides fast computations for data items that can be adequately represented by this method.

fixed point unit an execution unit within a CPU processor that is responsible for all integer arithmetic instruction computations.

fixed priority system a preemptive priority system where the task priorities cannot be changed once the system is implemented. *Compare with* dynamic priority system.

fixed-radix number system a number system in which only a single radix is employed.

fixed rate system a software system where interrupts occur only at fixed frequencies.

fixed resolution hierarchy an image processing scheme in which the original and reconstructed image are of the same size. Pixel values are refined as one moves from level to level. This is primarily used for progressive transmission. Tree-structured VQ and transform-based hierarchical coding are two of the fixed resolution hierarchies.

fixup a program is combined with other modules (linking) or placed into memory for execution (loading). *See also* relocatable.

flag (**1**) a bit used to set or reset some condition or state in assembly language or machine language; for instance, the inheritance flag or the interrupt flag. As an example, each maskable interrupt is enabled and disabled by a local mask bit. An interrupt is enabled when its local mask bit is set. When an interrupt's trigger event occurs, the processor sets the interrupt's flag bit.

(**2**) a variable that is set to a prescribed state, often "true" or "false", based on the results of a process or the occurrence of a specified condition. Same as indicator.

flag register (**1**) a register that holds a special type of flag.

(**2**) a CPU register that holds the control and status bits for the processor. Typically, bits in the flag register indicate whether a numeric carry or overflow has occurred, as well as the masking of interrupts and other exception conditions.

flash EEPROM *See* flash memory.

flash memory a family of single-transistor cell EPPROMs. Cell sizes are about half that of a two-transistor EEPROM, an important economic consideration. Bulk erasure of a large portion of the memory array is required.

The mechanism for erasing the memory is easier and faster than that needed for EEPROM. This allowed their adoption for making memory banks on PCMCIA for replacing hard disks into portable computers. Recently it has been used also for storing BIOS on PC main boards. In this way, the upgrading of BIOS can be made by software without opening the computer by non-specifically skilled people.

flat address a continuous address specification by means of a unique number. Operating systems such as Windows NT, OS/2, and Mac OS use such a type of address. MS-DOS adopts the real-mode that adopts a segmented memory.

flat dilation *See* moving maximum.

flat erosion *See* moving minimum.

flat execution profile with respect to an executable program, the amount of time required by and the number of calls to a particular procedure.

flat file (**1**) typically, a file saved in ASCII text format. Such files are normally read using standard input and output class structures available in all common programming languages.

(**2**) a file that does not contain any relational information of references within it. Examples are text files and pictures. For instance, ISAM (Indexed Sequential Access Method) are flat files for database access.

flat panel display a very thin display screen used in portable computers. Nearly all modern flat-panel displays use liquid crystal display technologies. Most LCD screens are backlit to make them easier to read in bright environments.

flat shading shading a polygonal patch with a single color and intensity. The shade chosen is a function of a variety of factors, such as light source position, viewer position, and surface normal, according to the shading model used. A single shade is how the patch would appear if the surface is genuinely planar, rather than just being approximated by polygons, and if several viewing environment conditions hold (distant viewer and light source).

Fletcher-Powell algorithm an iterative algorithm of Fletcher and Powell (1963) for finding a local minimum of the approximation error. In its original form, the Fletcher- Powell algorithm performs an unconstrained minimization.

flexibility the degree of facility with which a system or component can be modified or adapted to work in different conditions with respect to those for which it has been developed.

flexible manufacturing system (FMS) *See* advanced manufacturing system.

flicker fusion the perception by the human visual system of rapidly varying lights (flicker) as being steady (fused). Flicker fusion is why fluorescent lights and scenes in movies and television appear to have constant illumination.

flicker fusion frequency the frequency at which the human visual system resolves a flickering light as a single, continuous light; varies with luminance but is about 60 Hz for typical interior lighting conditions. *See* contrast sensitivity function.

flight simulator a simulator of aircraft operations.

flip-flop any of several related bistable circuits which form the memory elements in clocked, sequential circuits. They are analogous to toggle switches or latching relays.

In an "edge-triggered" flip/flop, information is stored in the device at the transition (positive or negative) of a clock-signal. A flip/flop is typically constructed using two "latches" in series. The first one (the "master") opens and closes by one clock-phase, and the second (the "slave") opens and closes using another (non-overlapping) phase. A flip/flop may be implemented using a static or dynamic logic style. In the latter case, the size is only half of the static version, but the information is lost (due to leakage) after some time, needing "refresh" (or storing new data) for proper operation.

floating-point an approximate representation of a numeric value as an exponent e and mantissa m such that the actual value is $m \times b^e$, where b is the exponent radix. Since the representation is in binary, and of fixed precision, it is not precise. Most programming languages support computations using floating-point numbers. *See also* scaled integer arithmetic, IEEE floating point.

floating-point constant a value that is written in a program to indicate a floating-point value. Most languages have adopted the FORTRAN representation of scientific notation, of a decimal number followed by the letter E followed by a signed exponent; for example, the value 9999.7 can be written as 9999.7, 999.97E1, 99.997E2, or 9.9997E3. Some languages allow a specification of the precision of the representation as part of the constant. For example, C allows a suffix "F" or "f" to mean single-precision floating point (9999.7F) and "L" or "l" to mean extended precision floating point (9999.7L).

floating-point number system the computer-equivalent of standard scientific notation. A number is represented as $s \times r^e$, where e is a signed integer known as the exponent, r is an unsigned integer known as the base or radix, and s is a signed fixed-point number known as the significand; r is usually fixed for a given computer. Nowadays most computers use the IEEE-754 standard for floating-point arithmetic.

floating-point operations per second (FLOP) a measure of processing speed, as in megaFLOPs or gigaFLOPs.

floating-point register a register that holds a value that is interpreted as being in floating-point format.

floating-point representation *See* floating-point.

floating-point unit an execution unit within a CPU processor that is usually responsible for all floating-point arithmetic instruction computations.

flock of birds a sensor based on pulsed magnetic fields that provides the position and orientation of a sensor, or many sensors, with respect to a fixed reference frame. Made by Ascension Technologies.

flood fill *See* fill.

floor plan an approximate layout of a circuit that is made before the layouts of the components composing the circuit have been completely determined.

floor planning designing a floor plan.

floor plan sizing given a floor plan and a set of components, each with a shape function, finding an assignment of specific shapes to the components so that the area of the layout is minimized.

FLOP *See* floating-point operations per second.

floppy disk a flexible plastic disk coated with magnetic material. Enclosed in a cardboard jacket having an opening where the read/write head comes into contact with the diskette. A hole in the center of the floppy allows a spindle mechanism in the disk drive to position and rotate the diskette. A *floppy disk* is a smaller, simpler, and cheaper form of disk storage media, and is also easily removed for transportation.

flow analysis that aspect of the compilation process that constructs a graph representing control flow in a program. Each node of the graph represents a sequence of computations, called a basic block, and each arc of the graph represents a potential control transfer from one basic block to another.

flow chart a traditional graphic representation of an algorithm or a program, in using named functional blocks (rectangles), decision evaluators (diamonds), and I/O symbols (paper, disk) interconnected by directional arrows which indicate the flow of processing. Synonymous with flow diagram. *See also* block diagram, box diagram, bubble chart.

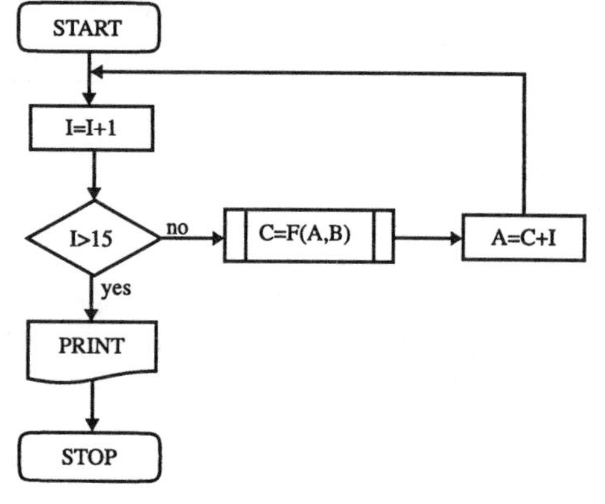

A flow chart.

flow control methods for preventing a sender from flooding a receiver in a point to point link, generally involving feedback from the receiver to the sender about the receiver's ability to receive more packets.

flow dependency *See* true data dependency.

flow diagram *See* flow chart.

flow equivalent service center a single service center in a queueing network that replaces an aggregate subsystem. It behaves exactly like the aggregate in terms of both the average and the distribution of the interdeparture times from the aggregate.

flow graph a representation of a program, most commonly used as an internal representation within a compiler, in which the nodes of the graph represent computations executed without control transfers (basic blocks), and the arcs or edges of the graph represent potential control paths connecting the basic blocks.

flow of control *See* control flow.

Floyd-Warshall algorithm an algorithm to solve the all pairs shortest path problem in a directed graph by multiplying an adjacency-matrix representation of the graph multiple times.

fluff-word *See* stop-word.

fluoroscopy a mounted fluorescent screen on which the internal structure or parts of an optically opaque object may be viewed as shadows formed by the transmission of X-rays through the object.

flush the act of clearing out all actions being processed in a pipeline structure. This may be achieved by aborting all of those actions, or by refusing to issue new actions to the pipeline until those present in the pipeline have left the pipeline because their processing has been completed.

flushed when one of the instructions in the pipeline is a branch instruction and the prefetched instructions are removed because of obsolescence. *See* pipelining.

Classification for Computer Architectures

	Single-Data Stream	Multiple-Data Stream
Single-instruction stream	von Neumann architecture/uniprocessors RISC	Systolic processors Wavefront processors Vector processors
Multiple-instruction stream	Pipelined architectures Very long instruction word processors	Dataflow processors Transputers

flushing-to-zero *See* sudden underflow.

flying head disk a disk storage device which uses a read/write head unit "flying" over (i.e., very closely above) the disk surface on a thin air bearing. Used in "Winchester" disks (a sealed "hard disk"). This is in contrast to floppy disks, where the head unit is actually in physical contact with the disk when reading or writing. *See also* disk head.

Flynn's taxonomy a classification system that organizes computer processor types as either single-instruction stream or multiple-instruction stream and either single-data stream or multiple-data stream. The four resultant types of computer processors are known as SISD, MISD, SIMD, and MIMD (see table). Due to Michael J. Flynn (1966).

focal point point in space at which the eyes are aimed (converged) and focused.

focus of control typically of the applications based on a windowing system. For a window to have the focus means that the corresponding application is capable of receiving input from the keyword. The process is not practical in foreground.

focus of expansion the point in an image from which feature points appear to be diverging when the camera is moving forward or objects are moving towards the camera.

fogging The blending of a color, often light gray, with parts of an image such that the farther objects become increasingly obscured. (*See* atmosphere effect.) In other words, the contrast between the fog color and objects in the image becomes lower the deeper an object appears in the scene. Fogging may be used to provide a back-clipping plane where objects too distant to be seen clearly are removed to speed up the rendering of a scene.

FOH *See* first order hold.

FOL *See* first order logic.

folding a type of hash function where an arithmetic or logical function is applied to the hash field value (or part of it) to calculate the hash address.

forbidden error the occurrence of one of a predetermined class of execution errors; typically the improper application of an operation to a value such as "not (3)".

force-1 rounding *See* jamming.

force control robotic manipulator control strategy based on regulating the interaction force between the manipulator and the environment.

forced error a human error that is the result of system demands exceeding human response capabilities.

forced flow law the average number of times a device in a queuing network is visited per unit time is equal to the product of the average number of visits to the device made by a job and system throughput.

force distribution feedback display of the pattern of forces or torque acting on a remote manipulator.

force feedback display of force or torque acting on the remote part of a teleoperator.

force reflection proportional force feedback display using the controller to push against the user.

force-reflective telemanipulation use of a teleoperated manipulator with force reflection.

force-torque display any device used to present information about forces or torques acting on a remote manipulator or avatar, whether the display modality is visual, haptic, or auditory.

foreground (1) a process that is currently interacting with the user in a shell, as opposed to background processes, which get suspended when they require input from the user.

(2) a term describing a Unix command that is using your terminal for input and output, so that you will not get another Unix prompt until it is finished.

(3) in image processing and displays, the foreground of a window is a single color or a pattern. The foreground text or graphics is displayed against the background of the window.

foreground processing the execution of processes in foreground.

foreign key a set of attributes in a relation R is a foreign key of R if it has the same domain as the primary key attributes of another relation schema R1 and that the value of a foreign key in any tuple of R occurs as a value of a tuple in R1 or else has the value NULL.

forest an acyclic graph.

forest editing problem the problem of transforming one of two given forests into the other by an edit script of minimum cost.

forest of trees in object-oriented languages that allow more than one root class, the set of class hierarchies necessary to implement the system.

fork (1) the action of a process creating another (child) process without ending itself.

(2) linguistic construct supported in some languages that allow for multithreaded execution, to indicate the point at which sections of the program can execute in parallel. *See also* join.

(3) in Unix, a process that can create a new copy of itself (a child) which has its own existence until it either terminates (or is killed) or its parent terminates. The prepare handler is called before the processing of the fork subroutine commences. The parent handler is called after the processing of the fork subroutine completes in the parent process. The child handler is called after the processing of the fork subroutine completes in the child process.

(4) on the Macintosh, a segment of an executable file. A program on the Macintosh usually has three such segments, the code fork, the data fork, and the resource fork.

for-loop (1) a counting loop which increments or decrements a value from a designated starting value to a designated terminal value.

(2) a loop designated by the keyword "for", which iterates over any value set.

Synonymous with do-loop.

formal description techniques (FDT) a formal method for developing telecommunications services and protocols. FDTs range from abstract to implementation-oriented descriptions. All FDTs offer the means for producing unambiguous descriptions of OSI services and protocols in a more precise and comprehensive way than natural language descriptions.

FDTs provide a foundation for analysis and verification of a description. The target of analysis and verification may vary from abstract properties to concrete properties. Natural language descriptions remain an essential adjunct to formal description, enabling an unfamiliar reader to gain rapid insight into the structure and function of services and protocols. Examples of FDTs are LOTOS, Z, SDL, and Estelle.

formal language a language with a defined syntax and semantics and for which the verification of its functionalities can be performed by using a mathematical approach. *Contrast with* natural language. *See* formal method.

formal method a software specification and production method based on a precise mathematical syntax and semantics that comprises:

1. a collection of mathematical notations addressing the specification, design, and development phases of software production,

which can serve as a framework or adjunct for human engineering and design skills and experience

2. a well-founded logical inference system in which formal verification proofs and proofs of other properties can be formulated

3. a methodological framework within which software may be developed from the specification in a formally verifiable manner. Formal methods can be operational, denotational, or dual (hybrid).

See also formal verification, model checking.

formal model a model at the basis of a formal method.

formal parameter variables used for passing data to a software module by calling a procedure or function. *See* parameter.

formal specification a specification given by using a formal language and model.

formal technical review a structured meeting conducted by software engineering with the intent of uncovering errors in some deliverable or other software product.

formal verification establishing properties of hardware or software designs using logic, rather than (just) testing or informal arguments. This involves formal specification of the requirement, formal modeling of the implementation, and precise rules of inference to prove, say, that the implementation satisfies the specification. *See also* model checking, temporal logic, formal method.

format (1) in FORTRAN, the FORMAT statement specifies how to transform external (textual) values to internal (binary) values, and internal values to external values.

(2) generally, to convert an internal representation to an external representation.

FORTH a stack-oriented interpretive language with polymorphic operators.

FORTRAN acronym for FORmula TRANslator. A language first defined by John Backus et al. in 1956, used primarily for scientific computations.

forward-backward procedure an efficient algorithm for computing the probability of an observation sequence from a hidden Markov model.

forward chaining style of reasoning in deductive databases, which infers the head of a rule if the body of the rule has already been inferred. Also called bottom-up reasoning.

forward declaration a syntactic construct that allows an entity to be made known to the compiler before it is actually encountered in the source. For example, if function A calls function B, but function A precedes function B lexicographically in the source, function A is said to have a forward reference to function B. A *forward declaration* of B must precede the definition of function A. In some cases, such as mutual recursion, it is impossible to produce an ordering of declarations and use of functions that would avoid the need for a forward declaration.

forward engineering the traditional process of moving from high-level abstractions and logical implementation-independent designs to the physical implementation of a system.

forward error correction (FEC) an error control system used for simplex channels (only a forward channel) where the channel code is used to determine the most likely transmitted sequence of information symbols. *Compare with* automatic repeat request.

forward error recovery one of several software dynamic redundancy techniques that attempts to continue operation with the current system state even though it may be faulty. Also called roll-forward, forward error recovery is highly application dependent.

forwarding (1) sending a just-computed value to potential consumers directly, without requiring a write followed by a read.

(2) to immediately provide the result of the previous instruction to the current instruction, at the same time that the result is written to the register file. Also called bypass.

forward kinematics identification of Cartesian task coordinates given robot joint configuration.

forward recovery *See* forward error recovery.

forward reference a reference in a program to an entity that has not yet been defined at the time the reference is made. For example, if function A precedes function B, but function A calls function B, the reference to function B is a *forward reference* if function B has not already been defined. Note that some languages do not permit this. Such languages must make provisions for having a forward declaration.

fountain life-cycle a visualization of the life-cycle which exhibits parallel, iterative, and incremental features. It was originally proposed in 1990 as a good description for what is observed to occur in object-oriented developments.

four connected *See* pixel adjacency.

Fourier transform a linear mathematical transform from the domain of time or space functions to the frequency domain. The discrete version of the transform (DFT) can be implemented with a particularly efficient algorithm (fast Fourier transform or FFT). The discrete Fourier transform of a digital image represents the image as a linear combination of complex exponentials.

The Fourier transform of a continuous time period signal $f(t)$ is given by

$$F(\omega) = \int_{-\infty}^{\infty} f(t)e^{-j\omega t}dt, -\infty < \omega < \infty .$$

If the signal $f(t)$ is absolutely integrable and is well behaved, then its Fourier transform exists. For example, the rectangular pulse signal

$$f(t) = \begin{cases} 1, -\frac{T}{2} \leq t < \frac{T}{2} \\ 0, \text{otherwise} \end{cases}$$

has Fourier transform $F(w) = \frac{2}{\omega} \sin \frac{\omega T}{2}$.

The inverse Fourier transform of a signal is given by

$$f(t) = \frac{1}{2\pi} \int_{-\infty}^{\infty} F(\omega)e^{j\omega t}d\omega$$

See also optical Fourier transform, two-dimensional Fourier transform.

fourth generation language (4GL) an application specific language. The term was invented to refer to non-procedural high level languages built around database systems. The first three generations were developed fairly quickly, but it was still frustrating, slow, and error prone to program computers, leading to the first "programming crisis", in which the amount of work that might be assigned to programmers greatly exceeded the amount of programmer time available to do it. Meanwhile, a lot of experience was gathered in certain areas, and it became clear that certain applications could be generalized by adding limited programming languages to them. Thus were born report-generator languages, which were fed a description of the data format and the report to generate and turned that into a COBOL (or other language) program which actually contained the commands to read and process the data and place the results on the page. Some other successful 4th-generation languages are: database query languages, e.g., SQL; Focus, Metafont, PostScript, RPG-II, S, IDL-PV/WAVE, Gauss, Mathematica and data-stream languages such as AVS, APE, Iris Explorer.

fourth normal form (4NF) a concept used in characterizing databases. A relation R is in fourth normal form with respect to a set of functional dependencies FD, if for every nontrivial multi-valued dependency $X \rightarrow Y \in F^+$, X is a superkey for R.

four-way interleaved splitting a resource into four separate units that may be accessed in parallel for the same request (usually in the context of memory banks).

fovea region at the center of the retina where cone cells are most densely concentrated and which is the functional center of human vision.

FPGA *See* field-programmable gate array.

FPS *See* frames-per-second, feet-per-second.

fractal an image with an infinite amount of self-similarity. From the Latin *fractus,* meaning broken or irregular, a fractal is a rough geometric shape that is self-similar over multiple scales, i.e., its parts are approximate copies of larger parts, or ultimately of the whole.

A *fractal* has statistical self-similarity at all resolutions and is generated by an infinitely recursive process. In reality, those fractals generated by finite processes may exhibit no visible change in detail after some stage so they are adequate approximations. So, for computer graphics we can extend the definition to include anything that has a substantial measure of exact or statistical self-similarity. This is illustrated by three stages of the construction of the von Koch snowflake, where each straight edge is repeatedly replaced by a copy of the entire figure. Fractals are good models of many natural phenomena such as coast lines, clouds, plant growth, and lightning as well as artificial items such as commodity prices and local-area network traffic. *See also* fractal coding, self-similarity.

fractal coding the use of fractals to compress images. Regions of the image are represented by fractals, which can then be encoded very compactly. Fractal coding provides lossy compression that is independent of resolution and can have extremely high compression ratios, but it is not yet practical for general images. Some of the methods are covered by patents. *See also* compression, compression ratio, fractal, image compression, image resolution, lossy compression, model-based image coding, spatial resolution, vector image.

fractal dimension fractional dimension of geometric images. Defined as the logarithm of the number of self-similar pieces divided by the logarithm of the magnification needed to obtain them.

fraction in computer arithmetic, old term for a significand whose magnitude is less than unity.

fractional Brownian motion a nonstationary generalization of Brownian motion; it is a zero-mean Gaussian process having the following covariance:

$$E\left[B_H(t)B_H(s)\right] = \frac{\sigma^2}{2}\left(|t|^{2H} + |s|^{2H} - |t-s|^{2H}\right).$$

This process has interesting self-similarity and spectral properties; it is a fractal with dimension $D = 2 - H$. *See also* Brownian motion.

fractional solution typically, a solution to a relaxation of a problem.

fragment and replicate a parallel join algorithm which involves partitioning one of the input relations across the available processors, replicating the second input relation across all processors, executing them locally, and then merging the results.

fragmentation (1) the creation of chunks of unused space throughout memory as a result of dynamic memory management or on a disk as the result of dynamic file creation and deletion.

Internal fragmentation occurs when memory blocks are rounded up to fix block sizes, e.g., allocated in sizes of power of 2 only. For example, if 35 K of data is allocated a 64 K block, the difference (64 − 35 = 29 K) is wasted. External fragmentation occurs between allocated segments, as a result of allocating different sized segments for processes entering and leaving memory. This latter fragmentation is also called checkerboarding. *See* storage fragmentation.

(2) in object-oriented databases, storing related data in separate physical locations for the purposes of improving performance and efficiency.

fragmentation schema definition of a set of fragments that includes all the tuples and attributes in the database such that the database can be reconstructed via the application of union and outer join operators.

frame (1) a single image or component of a sequence of images which, displayed rapidly in succession, give the illusion of a moving picture.

In video, a frame represents a single complete scan of the image; it often consists of two interlaced fields. The standard frame rate for TV standards is 25 frames/s (European standards, e.g., PAL and SECAM) or 30 frames/s (U.S. and Japanese standards, e.g., NTSC).

(2) in paging systems, a memory block whose size equals the size of a page. Frames are allocated space according to aligned boundaries, meaning that the last bits of the address of the first location in the frame will end with n zeros (binary), where n is the exponent in the page size. Allocating frames for pages makes it easy to translate addresses and to choose a frame for an incoming page (since all frames are equivalent).

(3) time interval in a communication system over which the system performs some periodic function. Such functions can be multiple access functions (e.g., TDMA multiple access frame) or speech processing functions (e.g., speech coding frame), interleaving frame, or error control coding frame.

(4) a knowledge representation scheme that associates an object with a collection of features (e.g., facts, rules, defaults, active values, etc.).

(5) a set of four vectors giving position and orientation information.

See also stack frame.

frame buffer a buffer of memory for storing image information.

frame error rate the percentage of frames that contain at least one error. A frame in this context is the smallest data item that can be retransmitted independently. The bit error rate influences the maximum frame size feasible for a desired frame error rate.

frame grabber a device that is attached to an electronic camera and which freezes and stores images digitally, often in gray-scale or color format, typically in one or three 8-bit bytes per pixel, respectively.

frame memory video memory required to store the number of picture elements for one complete frame of electronically scanned video information. The memory storage in bits can be computed by multiplying the number of video samples made per horizontal line, times the number of horizontal lines per field (vertical scan), times the number of bits per sample, times the number of fields per frame. A sample consists of the information necessary to reproduce the color information.

The NTSC television system consists of two interlaced fields per frame. Storage requirements are usually minimized by sampling the color video information consisting of the luminance (Y) and the two color difference signals, $(R - Y)$ and $(B - Y)$. The color signal bandwidth is less than the luminance bandwidth that can be used to reduce the field memory storage requirements. Four samples of the luminance (Y) signal are combined with two samples of the $(R - Y)$ signal and two samples of the $(B - Y)$ signal. The preceding video sampling technique is designated as 4:2:2 sampling and reduces the field memory size by one-third.

The number of memory bits that are required to store one NTSC frame is two times the number of bits required to store one NTSC field. Field memory for NTSC video sampled at four times the color sub-carrier frequency at 8 bits/pixel would require 7.644 Megabits of RAM when 4:2:2 sampling is used.

frame pointer a designated machine register which is used to refer to a stack frame. The register may be either a register dedicated to that purpose by the design of the machine, or a general register designated for that purpose by agreed-upon convention, or a general register designated for that purpose determined solely by the implementor of a compiler. Not all languages, nor all implementations, require a frame pointer.

frame processing image processing applied to frames or sequences of frames.

frame rate number of times per second that a simulation, virtual world, or display updates an important factor in the quality of a system and in the bandwidth and computing power required. The frame rate is limited by both the speed at which image data can be created and the rate at which video images can be presented on a display. For example, the NTSC system redraws at 30 Hz, PAL is 25 Hz, and computer displays are now usually 72 to 75 Hz.

frame size a term used to refer to the dimensions of the array of pixels forming a frame of an animation or, alternatively, the memory requirement and hence indirectly the resolution and dimensions.

frames-per-second (FPS) (**1**) a measure of the speed of an animation in terms of the number of complete, fully rendered images or frames which can be displayed in one second.

(**2**) the same, except FPS refers to the number of feet (30.48 cm) of cinema film displayed in 1 second.

frame store a device that stores images digitally, often in gray-scale or color format, typically in one or three 8-bit bytes per pixel, respectively; a *frame store* often incorporates, or is used in conjunction with, a frame grabber.

frame synchronization a method to obtain rough timing synchronization between transmitted and received frames. The degree of synchronization obtained depends on the level of *frame synchronization* (super-frame, hyper-frame). The level of frame synchronism also determines the actual method and the place where the synchronism takes place.

frame synchronizer (**1**) a time-base corrector effective over one frame time. Video data may be discarded or repeated in one-frame increments; in the absence of an input, the last received frame is repeated, thus producing a freeze-frame display.

(**2**) a device that stores video information (perhaps digitally) to reduce the undesirable visual effect caused by switching nonsynchronous sources.

framework a domain-specific application shell or skeleton software system. It is the dual of a software component, whereas a component is a reusable part that is incorporated into new applications, a *framework* is a reusable whole, for which new components are supplied in order to tailor it to new applications. A framework is a large-scale, implemented system of classes, in contrast with a pattern, which is a small-scale, language-independent design model or an idiom, which is a language-specific coding style.

free distance the minimum Hamming distance between two convolutionally encoded sequences that represent different valid paths through the same code trellis. The *free distance* equals the maximum column distance and the limiting value of the distance profile.

free edge an edge that is not in a matching. *See also* matched edge, free vertex.

free-form *See* free-form surface.

free-form surface a surface that does not have a simple geometric description (e.g., not a plane or quadric surface). It is usually represented using a spline surface or a triangulated surface.

free-page list a linked-list of information records pointing to "holes" (i.e., free page frames) in main memory.

free storage memory, usually in a heap, which has not been allocated, or which has been allocated but has been subsequently de-allocated and is available again. *See also* garbage collection.

free storage pool *See* heap.

free store the dynamic memory allocation space used by new and delete operators of C++. It is the equivalent of the heap memory for procedural languages and models.

free vertex a vertex not on a matched edge in a matching, or one that has not been matched. *See also* matched vertex, free edge.

freeware software that is available for free, usually from a host site on the World Wide Web.

freeze quantifier used in temporal logics to get the time associated with a state. This type of quantification is used for the specification of timing constraints. For example, the following formula

$$\Box teE \rightarrow \Diamond taA \wedge ta < te + 10$$

states that in every instant with time te if E holds, then exists an instant with time ta where A holds and this instant is at least 10 time units distant from E.

frequency the number of times that a periodic function or vibration repeats itself in a specified time or space. It is often measured in cycles per second, cycles per centimeter, or cycles per degree of visual arc.

frequency domain sampling a procedure that is a dual of the time-domain sampling theorem, whereby a time-limited signal can be reconstructed from frequency-domain samples.

frequency-modulation recording *See* magnetic recording code.

Fresnel equation an equation used to determine the attenuation of unpolarized light reflected from a surface, given the refractive index of the surface material and the angle of incidence of the light relative to the surface normal.

friend in C++, a class or function that is not part of the declaration of a class, but is permitted to access its protected members. The friend relationship creates tight coupling among classes. This can be regarded as a violation of the data hiding.

friend class *See* friend.

front-back reversals misperception in which the user sees an object as reversed in depth, common with some wire-frame representations.

front end the part of a compiler that depends only on the source language and is independent of the target language. The front end translates and analyzes the source program.

frontside bus the term used to describe the main bus connecting the processor to the main memory (c.f. backside bus).

frustrum of vision the visible region of 3D space. Projecting rays from the viewer through all pixels in the image plane defines an infinite pyramid-like solid shape within which all visible objects appear. The pyramid is then truncated by a distant plane to eliminate the space which is too far away to render, and by a closer plane which eliminates object too close to render. The space in between is the frustrum of vision.

FSA acronym for finite state automaton. *See* finite state machine.

FSM *See* finite state machine.

ftp *See* file transfer protocol.

full adder a combinational logic circuit that takes two operand bits and a carry bit as input, and produces one summand bit and one carry bit that are the results of adding the three input operands.

full backup a complete backup of an entire database.

full path in Unix and related systems, a pathname that starts with a slash character (/), that is, a pathname that uses the root directory as its reference point and refers to a distinct file regardless of the current directory. Synonymous with absolute path. *Compare with* relative path.

full replication the storing of the whole database at each site.

full-size card card that is twice as long as half-size cards and might require a full-size slot on the motherboard. Also called full cards.

full subtractor a one-bit subtractor; it takes two operand bits and a borrow bit as input, and as output it produces one difference bit and one borrow bit that are the result of subtracting one input operand from the other, with the third as a borrow-in.

fully associative cache a cache that is divided into a number of sets, each set consisting of groups of lines with each line having its own stored tag (the most significant bits of the address). A set is accessed first using the index (the least significant bits of the address). Then all the tags in the set are compared with that of the required line to find the line in the cache and to access the line. *See also* direct mapped cache, associativity.

fully dynamic graph problem problem where the update operations include unrestricted insertions and deletions of edges.

fully polynomial approximation scheme an approximation scheme in which the running time of A_ϵ is bounded by a polynomial in the length of the input and $1/\epsilon$.

FUNARG acronym, function argument. A problem in the LISP language, and other languages using similar dynamic binding, which refers to the problem of determining the values of a function argument whose names should be bound to the values that were defined for the names at the time of the call, rather than the values for the names at the time the argument value is requested at a deeper evaluation context. A special case of determining binding time.

function (**1**) a computation that takes some arguments or inputs and yields an output. Any particular input yields the same output every time. More formally, a mapping from each element in the domain to an element in the range.

(**2**) generally synonymous with subroutine or procedure. In some languages, this term designates a subroutine that can return a value. *See also* relation, total function, Boolean function, constant function, unary function, binary function, trinary function, N-ary function, procedure.

(**3**) the implementation of an algorithm in the program with which the user or the program can perform part or all of a work task. A subprogram function is typically typed, in the sense that it can be used in expressions as a parameterized variable. A function can also be a feature of the system or component. For instance, a system can have a primary function to collect data.

functional abstraction the process, by automation or by hand, of generating a functional specification in abstract terms out of a segment of source code.

functional aspect an aspect of the system that has to do with data transformation. *See* data flow diagram.

functional block an ASIC design approach. ASIC designs using this methodology are more compact than either gate arrays or standard cells because the blocks can perform much more complex functions than do simple logic gates.

functional decomposition (**1**) the division of processes into modules.

(**2**) a technique used during planning, analysis, and design; creates a functional hierarchy for the software.

functional dependency a relationship between attributes in database systems. Given attributes A and B, a functional dependency exists between A and B, if and only if there is only one value of A for any given B. We say B is functionally dependent on A. We say A functionally determines B. A functional dependency between A and B is denoted by $A \rightarrow B$.

functional design a type of design that serves to define behavior on an abstract form that can be understood, communicated, and analyzed as much as possible without binding the implementation. It will be expressed using design languages.

functional design implementation the description of an abstract system in an implementation language such as C++ plus object code. It is the result of mapping a functional design into the implementation language.

functional fidelity degree to which a simulation accurately represents cause and effect relationships found in the target real world within the simulated world; the degree to which a simulation "works right".

functionality a set of attributes that bear on the existence of a set of functions and their specified properties. The functions are those that satisfy stated or implied needs. This set of attributes characterizes what the software does to fulfill needs, whereas the other sets mainly characterize when and how it fulfills needs.

functional model a model viewpoint: that part of the software system that describes the functionality aspects of the system to be delivered.

functional redundancy checking an error detection technique based on checking bus activity generated by a processor (master) with the use of another slave processor which monitors this activity.

functional requirement a system requirement that specifies a function that a system or system component must be capable of performing. Functional requirements define the behavior of the system, that is, the fundamental process or transformation that software and hardware components of the system perform on inputs to produce outputs.

functional unit a module, typically within a processor, that performs a specific limited set of functions. The adder is a *functional unit.*

function analysis (**1**) one of the oldest analysis methods for eliciting a software architecture.

(**2**) the process of identifying what a system does and what human users must do to operate the system; a predecessor to task analysis and an important early stage in system design because it provides information for function allocation (including human vs. machine allocations) and crew size decisions.

function body the computation that comprises the implementation of a function.

function call a syntactic construct by which the execution of a defined function is invoked. When a function call occurs, values may be passed to the function by binding the computation of these values, called the actual parameters, to the means by which these values are referenced, called the formal parameters. Control passes to the function body, and does not resume following the call until control passes back via a return.

function composition the process of applying a function to the result of another function or itself.

function iteration repeated composition of a function.

function-oriented design a design methodology based on a functional viewpoint of a system starting with a high-level view and progressively refining it into a more detailed design.

function-oriented interface a way of structuring the interaction on the basis of the functions of an application. Typically, a function (or an entire application) is selected first, and then the data for that function are determined. Function-oriented navigation is based on a functional decomposition of the application, which is often represented as a menu hierarchy.

function overloading having more than one function of the same name, the appropriate function being chosen by matching the types of the actual parameters to the types declared in the definitions of the function, and possibly by matching the result type determined by the context of the call with the result types in the definitions of the function. *See also* polymorphism.

function point a software metric. Function points were originally proposed as an early lifecycle estimation measure. Each *function point* corresponds to one end-user business function and, for data-intensive MIS applications, this works reasonably well.

function point analysis the use of function points to help forecast effort from an early analysis stage. However, even in that restricted environment, many studies have shown that a number of well-trained FP counters do not come up with identical FP counts despite the careful preparation of the Function Point Counting Practices Manual, created by the International Function Point Users' Group.

function test a check for correct device operation generally by truth table verification.

fundamental design quality quality represented by the core parts of software design, e.g., software architecture, software structure, user interface design strategy.

fundamental measure any measure that is independent of the evaluation items and their software quality characteristics, e.g., operation time, number of function items, lines of code.

fundamental metric a rule and method by which a target entity is measured to get the value of a fundamental data element.

fuse link used in non-erasable programmable memory devices. Each bit in the memory device is represented by a separate fuse link. During programming the link is either "blown" or left intact to reflect the value of the bit.

fusible link ROM read only memory using fuse links to represent binary data. *See also* fuse link.

Futurebus a bus specification standardized by the IEEE, originally defined for CPU-memory data transfers.

fuzzification a procedure of transforming a crisp set or a real-valued number into a fuzzy set.

fuzzifier a fuzzy system that transforms a crisp (nonfuzzy) input value in a fuzzy set. The most used fuzzifier is the *singleton fuzzifier* that interpretes a crisp point as a fuzzy singleton. It is normally used in fuzzy control systems. *See also* fuzzy inference system, fuzzy singleton.

fuzziness the degree or extent of imprecision that is naturally associated with a property, process, or concept.

fuzzy aggregation network artificial neural network which can be trained to produce a compact set of fuzzy rules with conjunctive and disjunctive antecedents. An aggregation operator is used in each node.

fuzzy algorithm an ordered set of fuzzy instructions which, upon execution, yield an approximate solution to a specified problem. Examples of fuzzy algorithms include fuzzy c-mean clustering, fuzzy-rule-based classification, etc.

fuzzy AND *See* triangular norm.

fuzzy associative memory a look-up table constructed from a fuzzy IF-THEN rule base and inference mechanism to define a relationship between the input and output of a fuzzy controller or a fuzzy system. *See also* fuzzy inference system.

fuzzy automata based on the concept of fuzzy sets, a class of fuzzy automata is formulated similar to Mealy's formulation of finite automata.

fuzzy basis functions a set of fuzzy membership functions or their combinations in a fuzzy system which forms a basis of a function space, normally for function approximation.

fuzzy behavioral algorithm a relational algorithm that is used for the specific purpose of approximate description of the behavior of a system.

fuzzy C the C programming language incorporating fuzzy quantities and fuzzy logic operations.

fuzzy clustering a method used to cluster data into subsets based on a distance or similarity measure that incorporates fuzzy membership functions.

fuzzy c-means (FCM) a fuzzy version of the commonly used K-means clustering algorithm. The main feature of FCM is that a type of membership function is utilized in computing a distance measure. *Alias*: fuzzy ISODATA.

fuzzy cognitive map (FCM) nonlinear dynamical systems whose state trajectories are constrained to reside in the unit hypercube $[0, 1]^n$. The components of the state vector of an FCM stand for fuzzy sets or events that occur to some degree and may model events, actions, goals, etc. The trajectory of an FCM can be viewed as an inference process. For example, if an FCM's trajectory starting from an initial condition $\mathbf{x}(0)$ converges to an equilibrium state $\mathbf{A}$, then this can be interpreted as the FCM providing an answer to a question "What if $\mathbf{x}(0)$ happens"? In this sense it can be interpreted that the FCM stores the rule "IF $\mathbf{x}(0)$ THEN the equilibrium $\mathbf{A}$". FCMs were introduced by B. Kosko.

fuzzy complement the complement of a fuzzy set A is understood as NOT (A).

fuzzy concentration the concentration of a fuzzy set produces a reduction by squaring the membership function of that fuzzy set.

fuzzy conditional statement an IF-THEN statement of which either the antecedents or consequent(s) or both may be labels of fuzzy sets. Also know as fuzzy rule, linguistic fuzzy model, linguistic rules. *See* fuzzy IF-THEN rule.

fuzzy control the application of fuzzy logic and fuzzy inference rules utilizing knowledge elicited from human experts in generating control decisions for the control of processes. More specifically, a means of expressing an operator's knowledge of controlling a process with a set of fuzzy IF-THEN rules and linguistic variables.

fuzzy controller devices for implementing fuzzy control, normally including the following components: a fuzzifier to transform a crisp real-valued number to a fuzzy set; a fuzzy rule base to define IF-THEN control rules; a fuzzy inference engine to combine the IF-THEN rules; and a defuzzifier to transform a fuzzy set to a crisp real-valued number.

fuzzy controller design a process of determining a fuzzy controller including a fuzzifier, a fuzzy rule base, a fuzzy inference engine, and a defuzzifier.

fuzzy control system stability stability of a fuzzy control system which normally includes a plant to be controlled and a fuzzy controller.

fuzzy decisional algorithm a fuzzy algorithm which serves to provide an approximate description of a strategy or decision rule.

fuzzy decision rule a decision rule with fuzzified antecedents.

fuzzy decision tree a decision tree with fuzzy decision functions.

fuzzy decision tree algorithm an algorithm to generate a fuzzy decision tree, such as the branch-bound-backtrack algorithm.

fuzzy definitional algorithm a finite set of possible fuzzy instructions which defines a fuzzy set in terms of other fuzzy sets (and possibly itself, i.e., recursively).

fuzzy digital topology an extension of the topological concepts of connectedness and surroundness in a digital picture using fuzzy subsets.

fuzzy dilation an operator that increases the degree of belief in each object of a fuzzy set by taking the square root of the membership function.

fuzzy expert system a rule-based system for approximate reasoning in which rules have fuzzy conditions and their triggering are driven by fuzzy matching with fuzzy facts. *See also* fuzzy inference system.

fuzzy generational algorithm a fuzzy algorithm serves to generate rather than define a fuzzy set. Possible applications of generational algorithms include generation of handwritten characters and generation of speech.

fuzzy geometry a way to describe an image by fuzzy subsets.

fuzzy grammar an attributed grammar that uses fuzzy primitives as the syntactic units of a language.

fuzzy hierarchical systems fuzzy systems of hierarchical structures. A typical example is a two-level fuzzy system with a higher level of fuzzy inference rules and lower level of analytical linear models.

fuzzy identification a process of determining a fuzzy system or a fuzzy model. A typical example is identification of fuzzy dynamic models consisting of determination of the number of fuzzy space partitions, determination of membership functions, and determination of parameters of local dynamic models.

fuzzy IF-THEN rule rule of the form

if x is A then y is B

where A and B are linguistic values defined by fuzzy sets on universe of discourse X and Y, respectively (abbreviated as $A \longrightarrow B$). "x *is* A" is called the *antecedent* or *premise,* while "y *is* B" is called the *consequence* or *conclusion.* A fuzzy IF-THEN rule can be defined as a binary *fuzzy relation.* The most common definition of a fuzzy rule $A \longrightarrow B$ is as *A coupled with B*, i.e.,

$$R = A \longrightarrow B = A \times B$$

or

$$\mu_{A \times B}(x, y) = \mu_A(x) \star \mu_B(y) ,$$

where $\star$ is a triangular norm.

See also fuzzy relation, fuzzy set, linguistic variable, triangular norm.

fuzzy implication *See* fuzzy IF-THEN rule.

fuzzy inference a fuzzy logic principle of combining fuzzy IF-THEN rules in a fuzzy rule base into a mapping from a fuzzy set in the input universe of discourse to a fuzzy set in the output universe of discourse. A typical example is a composition inference.

fuzzy inference engine a device or component carrying out the operation of fuzzy inference, that is, combining fuzzy IF-THEN rules in a fuzzy rule base into a mapping from a fuzzy set in the input universe of discourse to a fuzzy set in the output universe of discourse. *See also* approximate reasoning, fuzzy inference system, fuzzy rule.

fuzzy inference system a computing framework based on fuzzy set theory, fuzzy IF-THEN rules, and approximate reasoning. There are two principal types of fuzzy inference systems:

1. fuzzy inference systems mapping fuzzy sets into fuzzy sets (pure fuzzy inference systems) that are composed of a knowledge base which contains the definitions of the fuzzy sets and the database of fuzzy rules provided by experts; and a fuzzy inference engine which performs the fuzzy inferences.

2. fuzzy inference systems performing nonlinear mapping from crisp (nonfuzzy) input data to crisp (nonfuzzy) output data. In the case of a Mamdani fuzzy system, in addition to a knowledge base and a fuzzy inference engine, there is a fuzzifier which represents real valued inputs as fuzzy sets, and a defuzzifier which transforms the output set to a real value. In the case of a Sugeno fuzzy system, special fuzzy rules are used giving crisp (nonfuzzy) conclusions and the output of the system is given by the sum of those crisp conclusions, weighted on the activation of the premises of rules. Some fuzzy systems of this type hold the universal function approximation property.

See also defuzzifier, fuzzifier, fuzzy set, fuzzy IF-THEN rule, fuzzy reasoning, Sugeno fuzzy rule.

fuzzy input-output model input-output models involving fuzzy logic concepts. A typical example is a fuzzy dynamic model consisting of a number of local linear transfer functions connected by a set of nonlinear membership functions.

fuzzy integral an aggregation operator used to integrate multi-attribute fuzzy information. It is a functional defined by using fuzzy measures, which corresponds to probability expectations. Two commonly used fuzzy integrals are Sugeno integral and Choquet integral.

fuzzy intensification an operator that increases the membership function of a fuzzy set above the crossover point and decreases that of a fuzzy set below the crossover point.

fuzzy intersection the fuzzy intersection is interpreted as "A AND B", which takes the minimum value of the two membership functions.

fuzzy logic a branch of artificial intelligence that is applied to fuzzy sets where membership in a fuzzy set is a probability, not necessarily 0 or 1. Introduced by Zadeh (1973), it gives us a system of logic that deals with fuzzy quantities, the kind of information humans manipulate which is generally imprecise and uncertain. Non-fuzzy logic manipulates outcomes that are either true or false. Fuzzy logic needs to be able to manipulate degrees of "maybe" in addition to true and false. Fuzzy logic's main characteristic is the robustness of its interpolative reasoning mechanism. *See also* approximate reasoning, linguistic variable, modifier, generalized modus ponens.

fuzzy measure a subjective scale used to express the grade of fuzziness similar to the way probability measure is used to express the degree of randomness. An example of fuzzy measure is g_λ-fuzzy measure proposed by Sugeno.

fuzzy membership function function of characterizing fuzzy sets in a universe of discourse which takes values in the interval [0, 1]. Typical membership functions include triangular, trapezoid, pseudo-trapezoid, and Gaussian functions.

fuzzy minus the operation of minus applied to a fuzzy set gives an intermediate effect of concentration.

fuzzy modeling combination of available mathematical description of the system dynamics with its linguistic description in terms of IF-THEN rules. In the early stages of fuzzy logic control, fuzzy modeling meant just a linguistic description in terms of IF-THEN rules of the dynamics of the plant and the control objective. Typical examples of fuzzy models in control application include Mamdani model, Takgi-Sugeno-Kang model, and fuzzy dynamic model.

fuzzy neural control a control system that incorporates fuzzy logic and fuzzy inference rules together with artificial neural networks.

fuzzy neural network artificial neural network for processing fuzzy quantities or variables with some or all of the following features: inputs are fuzzy quantities; outputs are fuzzy quantities; weights are fuzzy quantities; or the neurons perform their functions using fuzzy arithmetic.

fuzzy neuron a McCulloch-Pitts neuron with excitatory and inhibitory inputs represented as degrees between 0 and 1 and the output as a degree to which it is fired.

fuzzy number a convex fuzzy set of the real line such that:

1. it exists exactly one point of the real line with membership 1 to the fuzzy set;

2. its membership function is piecewise continuous.

In fuzzy set theory, crisp (nonfuzzy) numbers are modeled as fuzzy singletons.

See also convex fuzzy set, fuzzy singleton.

fuzzy operator logical operator used on fuzzy sets for fuzzy reasoning. Examples are the complement (NOT), union (OR), and the intersection (AND).

fuzzy OR *See* triangular conorm.

fuzzy parameter estimation a method that uses fuzzy interpolation and fuzzy extrapolation to estimate fuzzy grades in a fuzzy search domain based on a few cluster center-grade pairs. An application of this method is estimation of mining deposits.

fuzzy pattern matching a pattern matching technique that applies fuzzy logic to deal with ambiguous or fuzzy features of noisy point or line patterns.

fuzzy plus the operation of plus applied to a fuzzy set gives an intermediate effect on dilation.

fuzzy PROLOG the PROLOG programming language incorporating fuzzy quantities and fuzzy logic operations.

fuzzy proposition a proposition in which the truth or falsity is a matter of degree.

fuzzy reasoning approximate reasoning based on fuzzy quantities and fuzzy rules. *See also* approximate reasoning.

fuzzy relation a fuzzy set representing the degree of association between the elements of two or more universes of discourse. *See also* fuzzy set.

fuzzy relational algorithm a fuzzy algorithm that serves to describe a relation or relations between fuzzy variables.

fuzzy relational matrix a matrix whose elements are membership values of the corresponding pairs belonging to a fuzzy relation.

fuzzy relaxation a relaxation technique with fuzzy membership functions applied.

fuzzy restriction a fuzzy relation which places an elastic constraint on the values that a variable may take.

fuzzy rule *See* fuzzy IF-THEN rule.

fuzzy rule bank a collection of all fuzzy rules which are arranged in n-dimensional maps.

fuzzy rule base a set of fuzzy IF-THEN rules. It is a central component of a fuzzy system and defines major functions of a fuzzy system.

fuzzy rule-based system system based on fuzzy IF-THEN rules, or another name for fuzzy systems. *See* fuzzy inference system.

fuzzy rule minimization a technique to simplify the antecedent and consequent parts of rules and to reduce the total number of rules.

fuzzy rules of operation a system of relational assignment equations for the representation of the meaning of a fuzzy proposition.

fuzzy set introduced by L. Zadeh (1965). A fuzzy set A in a universe of discourse X is characterized by a *membership function* μ_A which maps each element of X to the interval $[0, 1]$. A fuzzy set may viewed as a generalization of a classical (crisp) set whose membership function only takes two values, zero or unity. *See also* crisp set, membership function.

fuzzy singleton a fuzzy set of membership value equal to 1 at a single real-valued point and 0 at all other points in the universe of discourse. *See also* membership function, fuzzy set.

fuzzy sliding mode control a combination of available mathematical descriptions of the system dynamics with its linguistic description in terms of IF-THEN rules. In the early stages of fuzzy logic control, fuzzy modeling meant just a linguistic description in terms of IF-THEN rules of the dynamics of the plant and the control objective. *See also* TSK fuzzy model, fuzzy system.

fuzzy space the region containing the fuzzy sets created by set theoretic operations, as well as the consequent sets produced by approximate reasoning mechanisms.

fuzzy system a fuzzy system is a set of IF-THEN rules that maps the input space, say X, into the output space Y. Thus, the fuzzy system approximates a given function $F : X \rightarrow Y$ by covering the function's graph with patches, where each patch corresponds to an IF-THEN rule. The patches that overlap are averaged. *See also* fuzzy inference system.

fuzzy union an operator that takes the maximum value of the membership grades.

G

G (giga) a prefix indicating a quantity of 10^9. For instance, a gigabyte of storage is 1,000,000,000 (typically implemented as 2^{30}) bytes.

GA *See* genetic algorithm.

gamma camera a device that uses a scintillation crystal to detect gamma photons for radionuclide imaging. Also known as a scintillation camera.

gamma correction the correction of nonlinearities of a monitor so that it is possible to specify a color in linear coordinates.

Inside a CRT, the luminance of a pixel is represented by a voltage V; however, the luminance of the ray produced by that voltage is not proportional to V, but rather to V^γ, where γ is a device-dependent constant between 2 and 2.5. It is thus necessary to compensate for this effect by transforming the luminance L into $L^{1/\gamma}$; this is called gamma correction. Some CCD cameras implement gamma correction directly on the signal that they generate. Some computer monitors allow modification of the factor γ or gamma correction through software.

gamma function a function of n, written $\Gamma(n)$, is $\int_0^\infty e^{-x}x^{n-1}dx$. Recursively $\Gamma(n+1) = n\Gamma(n)$. For non-negative integers $\Gamma(n+1) = n!$. *See also* Stirling's approximation.

gamut normally refers to the full range of colors available in a color space. The gamut varies with resource: photographic film, printing inks, color displays, etc. A 24-bit color system has a gamut of $2^{24} = 16$ million different colors. Moving between systems with different color gamuts will require quantization.

gang scheduling the scheduling of a set of related threads to run on a set of processors at the same time, on a one-to-one basis.

Gantt diagram a horizontal histogram for the showing of a project scheduling. Generally, a Gantt diagram represents a temporal scheduling. Also, if often it is completed by logical relationships among project tasks and milestones of the project, in this case, it integrates relationships typical of Pert Diagrams. Gantt diagrams were developed during World War I by Henry Gantt.

garbage an object or a set of objects that can no longer be accessed, typically because all pointers that direct accesses to the object or set have been eliminated.

garbage collection a storage allocation technique in which unused portions of memory are deallocated.

Relying on garbage collection to manage memory simplifies the interfaces between program components. Two main methods used to implement garbage collection are reference counting and mark-and-sweep.

garbage collector a software run-time system component that periodically scans dynamically allocated storage and reclaims allocated storage that is no longer in use (garbage).

gate (1) a logical or physical entity that performs one logical operation, such as AND, NOT, or OR.

(2) the terminal of a FET that controls the flow of electrons from source to drain. It is usually considered to be the metal contact at the surface of the die. The gate is usually so thin and narrow that if any appreciable current is allowed to flow, it will rapidly heat up and self-destruct due to I-R loss. This same resistance is a continuing problem in low noise devices, and has resulted in the creation of numerous methods to alter the gate structure and reduce this effect.

gate array a semicustom integrated circuit (IC) consisting of a regular arrangement of gates that are interconnected through one or more layers of metal to provide custom functions. Generally, gate arrays are preprocessed up to the first interconnect level so they can be quickly processed with final metal to meet customer's specified function.

0-8493-2691-5/01/$0.00+$.50

Gaussian distribution a probability density function characterized by a mean μ and covariance Σ:

$$f(x) = (2\pi)^{-N/2}|\Sigma|^{-1/2} \exp\left(-(x-\mu)^T\Sigma^{-1}(x-\mu)/2\right),$$

where $|\Sigma|$ represents the determinant of Σ and N represents the dimensionality of x. If x is scalar, then the above function simplifies considerably to its more familiar form:

$$f(x) = \frac{1}{\sqrt{2\pi}\sigma}\exp\left(-\frac{1}{2}\frac{(x-\mu)^2}{\sigma^2}\right).$$

The Gaussian distribution is tremendously important in modeling signals, images, and noise, due to its convenient analytic properties and due to the central limit theorem. *See also* probability density function, mean, Cauchy distribution, exponential distribution.

Gaussian noise a noise process that has a Gaussian distribution for the measured value at any time instant.

Gaussian process a random (stochastic) process $x(t)$ is Gaussian if the random variables $x(t_1), x(t_2), \ldots, x(t_n)$ are jointly Gaussian for any n.

Gaussian smoothing convolution of a signal with a Gaussian function in order to remove high spatial frequency changes from the image.

gaze direction view direction is specified as a target object rather than the more usual vector form from a camera or eye position and direction.

gender the orientation on a cable connector either female (holes) or male (pins).

gene a unit of information on the chromosome which describes the synthesis of a single type of protein. In genetic algorithm, the term also refers to a single symbol of a particular symbol string in the population.

general cell layout a style of layout in which the components may be of an arbitrary height and width and functional complexity.

generalization (1) a type of abstraction in which a more general concept is derived from a more specific one.

(2) another name for the specialization relationship which represents the is-a-kind-of relationship used in object technology. Sometimes this is known as generalization/specialization.

generalized cone data structure for volumetric representations, generated by sweeping an arbitrarily shaped cross-section along a 3D line called the "generalized cone axis".

generalized delta rule the weight update rule employed in the backpropagation algorithm.

generalized Lloyd algorithm (GLA) a generalization of the Lloyd (or Lloyd-Max) algorithm for scalar quantizer design to optimal design of vector quantizers. *See also* K-means algorithm.

generalized modus ponens generalization of the classical modus ponens based on the compositional rule of inference.

Let a fuzzy rule $A \longrightarrow B$, that can be interpreted as a fuzzy relation R, and a fuzzy set A', then the compositional rule of inference maps a fuzzy set:

$$B' = A' \circ R = A' \circ (A \longrightarrow B)$$

that can be interpreted as

premise 1: x is A'
(fact)
premise 2: if x is A then y is B
(fuzzy rule)
consequence: y is B'
(conclusion)

For example, if we have the fuzzy rule *"If the tomato is red then it is ripe"*, and we know *"The tomato is more or less red"* (fact) then the generalized modus ponens can infer *"The tomato is more or less ripe"*.

See compositional rule of inference, fuzzy inference system, fuzzy relation, linguistic variable, modifier.

generalized polygon a planar shape that is constructed using an ordered number of vertices that are connected to form an enclosed polygonal area. It is a graphics engine's most abstract internal representation of a shape. Specific shapes such as square and triangle have a fixed number of vertices (3 and 4 in this example) and can be generalized using a *generalized polygon.* A generalized polygon may have holes or be concave.

Other shapes such as circle and ellipse have an infinite number of vertices. A generalized polygon can provide the graphics engine with an approximate representation by using a large number of vertices.

general-purpose register a register that is not assigned to a specific purpose, such as holding condition codes or a stack pointer, but which may be used to hold any sort of value. *General-purpose registers* are typically not equipped with any dedicated logic to operate on the data stored in the register.

general register a machine register that can be used either for general arithmetic computations or in address computations. *See* general-purpose register.

generated borrow a borrow that occurs in the subtraction of two one-bit numbers without there being any borrow-in, i.e., from the subtraction of 1 from 0. *See also* propagated borrow.

generated carry a carry that is produced from the addition of two one-bit numbers without there being any carry-in, i.e., from the addition of 1 and 1. *See also* propagated carry.

generated error the error that is produced because the machine operations are only approximations of the true mathematical operation.

generation in evolutionary computation, a single iteration of the algorithm during which the original population is transformed into a new population through the processes of recombination, mutation, competition, and selection.

generator a function that provides guaranteed unique identifiers from outside of transaction scope.

generic from the Ada language, a specification of a program component in which not all of the type information is bound at the point of definition. This allows the specifications of families of components that differ only by the type binding. *See* generic instantiation. A limited form of polymorphism.

generic class *See* abstract class.

generic function (1) a function that can operate on objects of more than one data type.

(2) from the Ada language, a function containing one or more unbound type specifications. See also generic instantiation. *See* virtual method.

generic instantiation from the Ada language, a declaration that is based on a generic prototype, but which binds actual types to formal types.

genetic algorithm (GA) an optimization technique that searches for parameter values by mimicking natural selection and the laws of genetics. A representative algorithm in the field of evolutionary computation proposed by J. Holland, individual potential solutions in the population are represented as strings of symbols from a particular alphabet. In most cases the binary alphabet {0, 1} is adopted for this purpose. Adopting terms from the field of genetics, each symbol string is known as a chromosome and individual symbols are referred to as genes. New symbol strings in the population are generated using crossover, which is a specific example of the recombination operation, and mutation. In other words, the primary emphasis of GA is on the role of genotypic changes to bring about fitness improvement in the population. Between the above two operations, crossover serves as the dominant operation to allow exchange of symbol sub-strings of high fitness between different chromosomes such that longer segments with desirable characteristics can be constructed.

On the other hand, mutation only serves as a background operator to allow introduction of new variants while preventing excessive disruption to the current gene pool. When applied to optimization problems, each potential solution is transformed into a symbol string using a suitable encoding scheme, and the fitness function is derived from the corresponding optimization cost function through a scaling operation to result in positive fitness values for the selection process. In each generation, after the construction of new symbol strings through crossover and mutation, the associated fitness value of each new string is evaluated and normalized with respect to the total fitness value sum in the population. This is then interpreted as a selection probability by the stochastic proportional selection method to allow incorporation of these modified symbol strings into a new population according to this normalized fitness value. Under this selection mechanism, higher fitness values thus translate into higher probabilities of survival into the next generation.

A genetic algorithm takes a set of solutions to a problem and measures the "goodness" of those solutions. It then discards the "bad" solutions and keeps the "good" solutions. Next, one or more genetic operators, such as mutation and crossover, are applied to the set of solutions. The "goodness" metric is applied again and the algorithm iterates until all solutions meet certain criteria or a specific number of iterations has been completed. *See also* optimization, schema theorem.

genetic drift random variations in the frequencies of particular genes appearing in the gene pool, which is the result of sampling from a finite population.

genetic program a form of genetic algorithm that uses a tree-based representation. The tree represents a program that can be evaluated, for example, as an S-expression.

genetic programming (GP) a special form of genetic algorithm where the individuals in the population are represented as executable programs. In other words, rather than interpreting each individual as a potential solution to a particular problem, GP allows the evolution of potential algorithms for solving the problem. Each executable program is represented as a tree structure, where the leaves of the tree denote input variables and constants to the program, and the internal nodes represent specific operations on these inputs. Although the tree structure facilitates the interpretation of the individual nodes as LISP expressions, the operations performed are not necessarily restricted to those specified in the LISP language. The associated recombination process combines the characteristics of two parent program trees to form a new program tree by exchanging sub-trees at particular nodes of the two parents. On the other hand, the mutation operator generates new individuals by replacing the original sub-tree at particular nodes with a random sub-tree. The fitness of each individual is determined by executing the program for a certain task and evaluating the result. *See also* genetic algorithm.

genotype the totality of information embedded in the genes of an organism which together determines its various external traits. *See also* phenotype.

genus *See* Euler number.

geometrical modeling the process of developing a geometric model.

geometric distribution a discrete probability density function of a random variable $\mathbf{x}$ that has the form

$$p\{\mathbf{x} = k\} = p(1-p)^k k = 0, 1, 2, \dots .$$

This is the probability that k independent trials, each with probability of success p and failure $1 - p$, fail before one succeeds. *See also* probability density function.

geometric hashing a model-based recognition technique using a highly redundant representation that is invariant to certain geometric transformations.

geometric invariant a property of a geometric model (e.g., a set of points) that remains unchanged under specified types of geometric transformations (e.g., distances between points under rigid motion).

geometric model a mathematical model of the geometry of a system or structure.

geometric transformation transforms the pixel coordinates of an image to effect a change in the spatial relationships of elements in the image. The change often takes the form of a stretching or warping of the image.

gesture recognition human-computer interface technique in which the computer is able to recognize and assign meaning to human movements.

ghosting the formation of an image in which a ghost image (i.e., a similar, fainter, slightly displaced image) appears superimposed on the intended image; ghosts result from a form of crosstalk, and typically arise by radio transmission via an alternative path occurring as a result of random reflection.

Gibson mix an analysis of computer machine language instructions that concluded that approximately 1/4 of the instructions accounted for 3/4 of the instructions executed on a computer.

GIF *See* graphics interchange format.

gigaflop (GFLOP) 1000 million floating-point operations per second.

GKS *See* graphical kernel system.

GLA *See* generalized Lloyd algorithm.

glass-box testing testing strategy where test cases are selected solely on the bases of the code. Also called white-box testing or clear box-testing. *See also* black-box testing.

glitch an incorrect state of a signal that lasts a short time compared to the clock period of the circuit. The use of "glitch" in describing power systems is generally avoided. *See* hazard.

global coordinate system a coordinate system with respect to which all other coordinate systems in the definition of 3D scene are defined.

global memory in a multiprocessor system, memory that is accessible to all processors. *See also* local memory, distributed memory.

global optimization (**1**) generally, the process of searching for the set of global minima (maxima) for a function f in a minimization (maximization) problem. More precisely, if the function is defined on the set A, the process will try to locate the set of points $\mathbf{x}^*$ such that $\mathbf{x}^* \leq \mathbf{x}$ ($\mathbf{x}^* \geq \mathbf{x}$) for all $\mathbf{x}$ in A.

(**2**) in compilers, optimization involving one or more basic blocks. *Contrast with* local optimization. Global optimization may be limited in scope to a single function, or may analyze a context as large as an entire program with all its procedures.

global position sensor (GPS) a sensor using communication with orbiting satellites to provide the absolute position of the sensor with respect to the globe.

global recovery manager program construct used to coordinate recovery in distributed databases.

global routing when a layout area is decomposed into channels, global routing is the problem of choosing which channels will be used to make the interconnections for each net.

global variable a variable whose scope is the entire program and whose extent is the program's complete execution.

gloss a property of an image. An object is said to have a *gloss* surface when specular reflection is observed. This causes a highlight on the surface when a bright light is directed at the object.

glyph generally, a symbol used to represent information. For example, in visualizing vector fields, an arrow is often used to show the direction and magnitude of the vector value at each location.

GMQ *See* goal metric question.

gnoseological platform a computer-based framework for cognition.

goal a rule with an "empty" head.

Goals are used for formulating queries to databases DB; e.g., the goal G {brother (cain, B)} asks for all brothers of cain in the database

DB. A goal can also be perceived as an integrity constraint — more precisely: as a denial rule — that forbids $B_1, \ldots, B_n$ to hold simultaneously; e.g., the goal G $\{male(X) \wedge female(X)\}$ states that the same person X in DB cannot be both male and female.

goal metric question (GMQ) a widely used and recommended framework for constructing a metrics program. Initially, and before the collection of any specific metric is decided upon, the overall goal is identified, e.g., how to increase code quality. Then the questions that need to be asked in order to fulfill this goal are identified. Then, and only then, should the metrics be chosen which most readily measure the artifacts identified in the questions.

GOMS model a theoretical description of human procedural knowledge in terms of a set of goals, operators (basic actions), methods (sequences of operators that accomplish goals), and selection rules that select methods appropriate for goals. The goals and methods typically have a hierarchical structure. GOMS models can be thought of as programs that the user learns and then executes in the course of accomplishing task goals.

goniometer device for measuring joint angles.

good behavior a program that does not exhibit forbidden errors.

goto a language construct that unconditionally transfers control flow to a designated point in the program. In modern languages and under modern programming paradigms, often considered bad practice although many languages do not provide good alternatives to its use.

goto-free programming a style of programming in which the goto is never used, or used extremely rarely. Various languages make this more or less difficult depending on the power and sophistication of their control constructs. Historically, the only imperative language that succeeded totally in achieving this goal was the BLISS language, which had no goto statement. Functional languages such as ML or pure LISP are intrinsically goto-free.

Gouraud shading a technique that assigns an intensity for each vertex of a polygon using Lambert-law shading and then interpolates the computed intensities across the polygon by performing a bilinear interpolation of the intensities down and then across scan lines. It thus eliminates the sharp changes at polygon boundaries.

GP *See* genetic programming.

GPS *See* global position sensor.

GPSS acronym for General Purpose Simulation System, a discrete event simulation language developed by IBM.

graceful degradation the ability of a system to perform at a specified performance even in the presence of failures.

grade category or rank given to entities having the same functional use but different requirements for quality. *Grade* reflects a planned or recognized difference in requirements for quality. The emphasis is on the functional use and cost relationship. A high grade entity (e.g., luxurious hotel) can be of unsatisfactory quality and vice versa. Where grade is denoted numerically, the highest grade is usually designated as 1 with the lower grades extending to 2, 3, 4, etc. Where grade is denoted by a point score, such as a number of stars, the lowest grade usually has the fewest points or stars.

grade of service the part of quality of service that depends on network dimensioning. *See also* quality of service, network performance.

gradient a vector function denoted by ∇f or **grad** f, where f is a continuous, differentiable scalar function. For a 2D function $f(x, y)$, the gradient is

$$\nabla f = \begin{bmatrix} \frac{\partial f}{\partial x} \\ \frac{\partial f}{\partial y} \end{bmatrix} .$$

The magnitude of this gradient is

$$|\nabla f| = \sqrt{\left(\frac{\partial f}{\partial x}\right)^2 + \left(\frac{\partial f}{\partial y}\right)^2} .$$

In image processing, the term *gradient* often refers to the magnitude of the gradient. *See also* Sobel operator.

gradient descent a method for finding the minimum of a multidimensional function $f(x)$. The technique starts at some point and advances toward the minimum by iteratively moving in a direction opposite to that of the gradient:

$$x_{i+1} = x_i - \alpha \frac{\partial f}{\partial x} ,$$

where α is some scalar, usually set empirically. *See also* gradient, optimization.

gradient edge detector an edge detector that defines an edge to be present at a pixel only if the magnitude of the gradient at that pixel is greater than some threshold. *See also* edge, edge detection, gradient, Sobel operator.

gradient magnitude the magnitude of the gradient vector, or equivalently the square root of the sums of the squares of the local directional derivatives.

gradient motion detection differential motion detection algorithms that attach a vector to each point in a displacement image. The optical flow from this gradient field is then derived.

gradual underflow in computations in which results are usually normalized, this is the replacement of an underflowing value with a denormalized value.

grammar (**1**) a set of symbols, one of which is designated as an initial symbol, and a set of production rules that define how symbols may be organized into sentences specified by the grammar. Formally, it is an ordered 4-tuple ($G = T, V, P, S$), where T is said to be the set of terminals, V is the set of variables or non-terminals, P is a set of production rules, and $S \in V$ is called the start symbol. Terminals are the symbols with which the strings of the language are made. For example, T may be the English alphabet, the ASCII character set, or the set $\{0, 1\}$. Non-terminals are symbols that are replaced by a string of zeros or more terminals and non-terminals. The production rules specify which strings can be used to replace non-terminals. The symbol S is the first symbol with which every production starts. All the strings that can be generated from S using rules in P are said to be in $L(G)$, the language of G. Example: Let G be the grammar (T, V, P, S) such that

$$T = \{a, b\}$$
$$V = \{S\}$$

where S is the start symbol, and

$$P = \{S \rightarrow aSb, S \rightarrow \epsilon\}$$

where ϵ is the empty string. The language $L(G)$ is the set of all strings of the form $a^n b^n$. The $\rightarrow$ symbol signifies that S can be replaced with whatever follows the arrow. *See* phrase structure grammar.

(**2**) a specification of the syntax of a programming language.

Gram-Schmidt orthogonalization a recursive procedure for whitening (decorrelating) a sequence of random vectors. *See also* whitening filter.

GRANT an SQL command that allows the granting of privileges to a user. The GRANT command may be used to specify access rights to individual relations or may be used to specify rights at a database level.

grant option an option that may be used with the GRANT command. It allows the propagation of rights.

granularity Size or scope, particularly, of a transaction or lock.

granular lock a fine-grained lock used to implement predicate locking.

granulometric size distribution a distribution generated by counting the pixels in each succeeding filtered image using a granulometry.

granulometry a sequence of openings using structuring elements that are of increasing size. *See* size distribution.

graph a couple $G = (E, V)$ where V is a set of nodes and $E \subseteq V \times V$ is a set of edges. Graphs are widely used in modeling networks, circuits, and software. For example, they are often used in compilers to represent the program during intermediate stages of compilation. *See also* directed graph, undirected graph, biconnected graph, connected graph, weighted graph, isomorphic, homomorphic, graph drawing, diameter of a graph, degree.

graph-coloring algorithm an algorithm that attempts to color the nodes of a graph from a fixed set of colors such that no two nodes connected by a single arc are the same color. Commonly applied in compilers as a means of determining register allocation.

graph concentration contracting a graph by removing a subset of the vertices.

graph drawing a geometric representation of a graph. Usually, the representation is in either two- or three-dimensional Euclidean space, where a vertex v is represented by a point $p(v)$ and an edge (u, v) is represented by a simple curve whose endpoints are $p(u)$ and $p(v)$.

graphic adapter an adapter for interfacing the computer towards a monitor. *See* MDA, CGA, EGA, VGA.

graphical kernel system (GKS) a standard for computer graphics recognized by both ANSI and ISO. GKS defines the manipulation of graphic objects, including their visualization, print, etc. All manipulation is performed by regarding the graphic adaption as an independent device driver. In this way, GKS applications can be executed on different kinds of graphics adapters.

graphical preview control system involving rehearsal of robotic operations in a virtual environment prior to performing those operations in a real, remote world; rehearsals may be for practice or used to program later operations performed under supervisory control.

graphical programming composing programs by combining graphical elements from a toolbox, with each graphical element representing a pre-written subroutine or algorithm.

graphical user interface (GUI) human-computer interface using graphics, icons, and menus and other visual aids to facilitate and structure the user's actions; often coupled with a direct manipulation interface.

graphic controller *See* graphic adapter.

graphics the discipline dealing with the generation of artificial images by a computer. Its two main aspects are geometric modeling, whose subject is the computational representation of the geometry and topology of objects and scenes, and rendering, which studies the generation of images from the interaction of light and objects. Graphics has been generalized to the synthesis of animated image sequences, in which case one speaks of computer animation. Also called computer graphics, image synthesis.

graphics accelerator an auxiliary computer or peripheral board in a computer that handles graphical related computations, with the aim of improving frame rate or scene detail.

graphics interchange format (GIF) a popular image-file format that compresses the image with LZW coding. *See also* file format, image compression, Lempel-Ziv-Welch coding.

graphics library a set of computational functions for manipulating graphic images. *See also* application program interface.

graphics primitive a basic building block of an image, such as an edge or a surface.

graphics processor a computer or processor designed specifically for manipulating image information.

graphics workstation a computer system designed specifically for manipulating graphics or image programs.

graph isomorphism the problem of determining if two graphs are isomorphic. *See also* subgraph isomorphism, homomorphic.

graph isomorphism problem deciding if two given graphs are isomorphic to each other.

graph search an optimization technique used to find the minimum cost path from a starting point to a goal point, through a graph of interconnected nodes. Each link between nodes has an associated path cost, which must be selected based on the problem of interest. *See also* optimization.

grasp requirements the pattern of motions and forces required to securely grasp an object.

Gray code a code in which each of a sequence of code words differs by one bit from the preceding one, and the assigned value of each code word is one more than that of the preceding one. Such a code avoids glitches (i.e., sharp momentary unwanted spikes) when, in an electromechanical system, the sensors giving the code words are imperfectly aligned.

For example, one possible three-bit Gray code is:

Decimal digit					
0	1	2	3	4	5
Gray code					
000	001	011	010	110	111

gray level the individual numerical value corresponding to a particular degree of brightness in a digital image, often on an 8-bit gray scale consisting of 256 gray levels stretching from pure black to pure white.

gray level co-occurrence a means of measuring texture and other brightness variations in digital images by generating matrices which tabulate the frequencies with which different gray levels co-exist at different distance vectors from each other.

gray level saturation the restriction of image gray scale so that intensities above a certain level become fixed at the white level corresponding to the highest numerical value available within the current storage capacity, typically one byte, of each pixel.

gray scale a color space where colors are represented by their luminance values only, i.e., saturation and hue are zero.

gray-scale moving median filter a moving median filter for images that are not binary.

greedy algorithm a paradigm of algorithm design in which an optimization problem is solved by making locally optimum decisions.

grey level *See* gray level.

grey scale *See* gray scale.

grid drawing a grid drawing is a graph drawing in which each vertex is represented by a point with integer coordinates.

gripper an end effector used to grasp objects in a remote world.

grooved media on an optical disk, the embossment of the disk surface with grooves such that the disk tracks are either the grooves themselves or the regions between the grooves.

grounding strap special bracelet used to connect your wrist to the computer chassis at all times in order to prevent buildup of harmful static electricity.

ground instance for a formula, a new formula ϕ obtained by replacing all variable symbols in ϕ by terms from the Herbrand universe that are built using constants and function symbols only; example: father(adam, abel) is a ground instance of father(adam, Y).

group an association of users who have something in common; the user group associated with a file.

GROUP BY an SQL clause which specifies an attribute by which tuples are grouped.

group code a recording method used for 6250 BPI (bits per inch) magnetic tapes.

group decision support system a special decision support system used in support of groups of decision makers that might be geographically dispersed.

group ID an integer that uniquely identifies a user group.

groupware software that supports people to work across distances. *See* computer supported cooperative work.

groupware tools tools that allow multiple programmers or designers to work cooperatively. These tools are specifically designed to support geographically distributed cooperation for software development.

growing phase the phase in which locks are acquired by a transaction operating under the two-phase locking protocol.

guard band a design technique for a color CRT intended to improve the purity performance of the device by making the lighted area of the screen smaller than the theoretical tangency condition of the device geometry.

guard block a designated value placed on data structures to detect an out-of-bounds write access.

guard condition in parallel languages, a Boolean expression evaluated to permit a communication action (send or receive).

guard digit those extra digits of the significand in a floating-point operand which must be retained in order to allow correct normalization and rounding of the result's significand.

GUID a form of user ID which is guaranteed to be unique across all machines. Particularly important when an object-based system is used, to guarantee that an object has a unique identifier.

Gupta-Sproull algorithm a technique developed in the framework of raster graphics. It aims to reduce aliasing when doing line drawing.

H

Haar transform unitary transform mapping N samples $g(n)$ to N coefficients $G(k)$ in a way that corresponds to repeated two-point averaging and two-point differencing. The 2×2 Haar transform is

$$H_2 = \frac{1}{\sqrt{2}} \begin{pmatrix} 1 & 1 \\ 1 & -1 \end{pmatrix} .$$

The scaling factor allows the same matrix to be used for the inverse transform. The 4×4 Haar transform can be interpreted as follows: first apply the 2×2 transform to two independent pairs of samples; then apply the 2×2 transform to the two average coefficients just computed. Larger Haar transforms are constructed by continuing this process recursively. The Haar transform yields coefficients equal to the subband values generated by dyadic decomposition with the Haar wavelet. This transform has achieved rather less use than the other transforms in this family, such as the discrete cosine and Hadamard transforms.

Haar wavelet the orthonormal wavelet pair $(\frac{1}{\sqrt{2}}, \frac{1}{\sqrt{2}}), (\frac{1}{\sqrt{2}}, \frac{-1}{\sqrt{2}})$. Analysis and synthesis pairs are identical. This is the most compact wavelet pair. Dyadic subband decomposition with the Haar wavelets yields coefficients equal to those from the Haar transform.

hacker a person who explores computer and communication systems, usually for intellectual challenge. This term is commonly applied to those who try to circumvent security barriers (crackers).

Hadamard matrix a special matrix used in a lossy compression scheme. *See also* fast Hadamard transform.

Hadamard transform a unitary transform mapping N samples $g(n)$ to N coefficients $G(k)$ according to the transform matrix H_N, where

$$H_2 = \frac{1}{\sqrt{2}} \begin{pmatrix} 1 & 1 \\ 1 & -1 \end{pmatrix}$$

and larger arrays are formed by the recursive definition

$$H_N = \frac{1}{\sqrt{2}} \begin{pmatrix} H_{\frac{N}{2}} & H_{\frac{N}{2}} \\ H_{\frac{N}{2}} & -H_{\frac{N}{2}} \end{pmatrix} .$$

The inverse transform is identical. The Hadamard transform was formerly used for data compression because its entries are all 1 or -1, allowing computation without multiplications. In this context it is now superseded by the discrete cosine transform.

half adder a one-bit adder that takes two operand bits as input and produces two result bits, a summand bit and a carry bit, as output. A half adder has no carry input. *See also* full adder.

half-adjust rounding *See* round-to-nearest-up.

half-size card short card that plugs into half-slots on the motherboard. Also called half card.

half subtracter a logic circuit that provides the difference and borrow outputs for two input signals. A *half subtracter* has no borrow input.

halftone technique of simulating continuous tones by varying the amount of area covered by the colorant. Typically accomplished by varying the size of the printed dots in relation to the desired intensity.

halfword a unit of information in a computer, representing a data item whose representation is half of what the machine calls a word. Traditionally, for machines with 32-bit words, a *halfword* is 16 bits, but for machines that have a 16-bit word, this term is not used; the more common term byte represents half of a 16-bit word. The halfword data type is frequently directly reflected in a programming language as a specific data type. *See also* byte, word, longword, quadword.

0-8493-2691-5/01/$0.00+$.50

Halstead's metrics a set of metrics used to measure various aspects of information content in a collection of software.

A modified form of *Halstead's metrics* can be computed as follows: Let n_1 be the number of logical distinct BEGIN-END pairs, known as operators. Let n_2 be the number of distinct statements (e.g., lines terminated by semicolons in C) known as operands. Let N_1 be the total number of occurrences of distinct operators in the program. Let N_2 be the total number of occurrences of operands in the program.

Then the program vocabulary is defined as

$$n = n_1 + n_2 ,$$

program length is defined as

$$N = N_1 + N_2 ,$$

program volume is defined as

$$V = N \log_2 n ,$$

potential volume is defined as

$$V^* = (2 + n_2) \log_2 (2 + n_2) ,$$

and program level is defined as

$$L = V^* / V .$$

Effort is defined as

$$E = V / L .$$

halting problem the problem of deciding whether a given program (or Turing machine) halts on a given input. The *halting problem* is undecidable.

halt instruction an instruction (typically privileged) that causes a microprocessor to stop execution.

Hamming code an encoding of binary numbers that permit error detection and correction first discovered by Richard Hamming at Bell Laboratories.

A Hamming code is a perfect code with code word length $n = 2^m - 1$ and source word length $k = n - m$ for any $m > 2$. A Hamming code can correct an error involving a single bit in the binary number. It can also detect an error involving two bits. *See also* parity.

Hamming distance the number of digit positions in which the corresponding digits of two binary words of the same length differ. The minimum distance of a code is the smallest Hamming distance between any pair of code words. For example, if the sequences are 1010110 and 1001010, then the *Hamming distance* is 3.

Hamming net a pattern recognition network that has a set of prototype patterns stored in its weights. A given input pattern is identified with the prototype whose Hamming distance from the input pattern is least.

Hamming weight the number of nonzero symbols in a given sequence of symbols.

hand the distal end of the radius and ulna in humans; analogously, a gripper or more complex end-effector attached to the distal end of a robotic manipulator. *See* end-effector.

hand controller input device grasped in the user's hand or fingers.

hand-eye coordination the ability of the hands to work efficiently and safely in response to visual information, and of the eyes to seek out information required for manual tasks.

hand frame coordinate frame centered on the user's hand.

hand gesture hand posture or pattern of hand movements to which meaning may be assigned and which may be used to communicate with other persons or machines.

hand-hand coordination the ability of the hands to work efficiently and safely together to perform manual tasks.

hand-held computer a small lightweight computer that performs functions such as electronic mail, handwriting recognition for taking notes, and holding addresses and appointments. Also called a "palm-top".

handler (1) a section of code, often a function, designated to be invoked under some specified conditions. These may be asynchronous conditions (such as an interrupt, usage interrupt handler), synchronous but not necessarily determinate conditions (for example, a message handler that handles a specific type of message obtained by the programmer), or synchronous and determinate (for example, an error handler that is called by the programmer in response to some detected error condition). *See* interrupt handler.

(2) in a language that supports exceptions, the code which is designated to handle the exception.

hand master small master controller operated by the hand alone, i.e., without requiring arm movements.

handshaking I/O protocol in which a device wishing to initiate a transfer first tests the readiness of the other device, which then responds accordingly. The transfer takes place only when both devices are ready.

handwritten character recognition the process of recognizing handwritten characters that are clearly separated.

haptic display display communicating information to users through haptic perception (tactition, kinesthesia, or both).

haptic feedback information provided to users by tactition or kinesthesia, or a combination of both.

haptics the study of human haptic senses, including tactition, proprioception, and kinesthesia.

haptic sensor sensor used to detect object surface characteristics like stiffness or smoothness; somewhat analogous to human tactition and kinesthesia.

hard deadline a deadline that cannot be missed under any circumstances.

hard-decision decoding decoding of encoded information given symbol decisions of the individually coded message symbols. *Compare with* soft-decision decoding.

hard disk a rigid magnetic disk used for storing data. Typically a non-removable collection of one or more metallic disks covered by a magnetic material that allows the recording of computer data. The hard disk spins about its spindle while an electromagnetic head on a movable arm stays close to the disk's surface to read from or write to the disk. Each disk is read and written on both above and below. N disks are read/written by using $2N$ heads. The information is stored by cylinders, circular segments of the collection of the disks. Cylinders are divided in sectors as a pie. The mean time to access at data is typically close to 10 msec.

Generally, hard disks are the backing memory in a hierarchical memory. *See also* floppy disk.

hard link a directory entry that refers to a file via its inode; a hard link provides direct access to a file.

hard real-time system a real-time system in which missing even one deadline results in system failure. *Compare with* soft real-time system, firm real-time system.

hard real-time task a task with precise timing constraints that cannot be missed without creating damage.

hardware the physical part of a system: wires, connectors, relays, silicon devices, etc.

hardware accelerator a piece of hardware dedicated to performing a particular function (such as image convolution or matrix-vector products) which would otherwise be performed in software. Although much less flexible, dedicated hardware implementations can give significant speed improvements over software, and are especially useful for real-time applications. *See also* real-time system.

hardware design the logic of the construction of the hardware of a system.

hardware interrupt an interrupt generated by a hardware device; for example, keyboard, the DMA, PIC, the serial adapter, the printer adapter, etc. Other hardware interrupts can be generated by the control unit or by the ALU; for example, for the presence of a division per zero, for attempting to execute a unknown instruction. This last class of hardware interrupts is called internal exception.

hardware software interface (HSI) the interface between a hardware device and the software which is executed within it (e.g., in a microprocessor), or which accepts input from it (e.g., from a sensor), or sends output to it (e.g., to an actuator).

Harvard architecture a computer design feature where there are two separate memory units: one for instructions and the other for data. The name originates from an early computer development project at Harvard University in the late 1940s. *Compare with* Princeton architecture.

hash collision in a hash table, the consequence of a hash function producing a hash key which selects a non-empty equivalence class as the place to put the new value. When a non-empty equivalence class is discovered, either the value will be placed in it, or a new equivalence class can be selected by using a secondary hash.

hashed page table a page table where the translation of each virtual page number is stored in a position determined by a hash function applied to the virtual page number. This technique is used to reduce the size of page tables.

hash field a field of a record chosen on which to apply the hash function.

hash field space the range of potential hash field values.

hash file a type of file organization, in which the file is viewed as a collection of addressable slots capable of holding records. A hash function is used to calculate the file address based on the values of the hash field of the record. This address is used to insert and retrieve records.

hash function a function that is applied to the hash field value of a given record and returns the address at which the record is stored. The address may be a memory location or a disk block.

hashing a data mapping process where a function maps a database key to the location (address) of the record having that key. (A secondary hashing method maps the database key to the location containing the address of the record.)

hash join an implementation of the join operator. To implement $R \bowtie S$, the records of both relations are hashed, on the join attribute, to the same file using the same hash function.

hash key a computed value based on the contents of an entry in a hash table (or about to be inserted in a hash table) that is used to initiate search in or insertion to the table. *See also* secondary hash.

hash partitioning a technique for partitioning a relation across a number of disks where tuples are hashed to disks via a hash function on the record's partitioning attributes.

hash structure a tree structure whose root is a function, called a hash function. Given a key, the hash function returns a page that contains pointers to records holding that key or is the root of an overflow chain. It should be used when selective equality queries and updates are the dominant access patterns.

hash table a data structure characterized by a mapping function from a value to a hash key which has the property that the computed key uniquely identifies an equivalence class of entries which is a subset of the entire set of entries. The location of a specified value within the equivalence class is typically very efficient, and consequently both insertion and search in the table are more efficient than many alternative representations. *Hash tables* are used for many memory and name mapping functions, such as symbol tables in assemblers and compilers. *See also* perfect hash.

Hausdorff distance an important distance measure, used in fractal geometry, among other places. Given a distance function d defined on a Euclidean space E, one derives from it the Hausdorff distance H_d on the family of all compact (i.e., bounded and topologically closed) subsets of E; for any two compact subsets K, L of E, $H_d(K, L)$ is the least $r \geq 0$ such that each one of K, L is contained in the other's dilation by a closed ball of radius r, that is:

$$K \subseteq \bigcup_{p \in L} B_r(p) \quad \text{and} \quad L \subseteq \bigcup_{p \in K} B_r(p) ,$$

where

$$B_r(p) = \{q \in E \mid d(p, q) \leq r\} .$$

HAVING an SQL clause used in conjunction with the GROUP BY clause to select groups that satisfy certain conditions. Only groups that satisfy the HAVING clause are returned as part of the query collection.

Hayes-compatible modem refers to a modem when it is capable of responding at the commands of modems made by Hayes Microcomputer Products. The Hayes set of commands represents a sort of standard for modems.

hazard a momentary output error that occurs in a logic circuit because of input signal propagation along different delay paths in the circuit.

hazardous environment An area that poses health threats to humans, should they enter, either by the presence of harmful substances or by the presence of dangerous events or conditions inherent to the area (e.g., battlefields or the deep ocean).

hazardous state a state of an item which, in conjunction with certain environmental conditions, could lead to an accident. A physical situation, often following from some initiating event, that can lead to an accident.

hazard probability probability of a system getting into a hazardous state within a given time interval under given conditions. It is assumed that the system is not in a hazardous state at the start of the interval.

HCI *See* human computer interface.

HDA *See* head disk assembly.

head an electromagnet that produces switchable magnetic fields to read and record bit streams on a platter's track.

head disk assembly (HDA) collection of platters, heads, arms, and actuators, plus the airtight enclosing of a magnetic disk system.

header **(1)** with respect to a binary executable file, it describes the file so that the loader can properly transfer it to memory; with respect to C language source code, it contains function prototypes, data structure and symbol definitions, and code macros.

(2) a section of an image file, usually of fixed size and occurring at the start of the file, that contains information about the image, such as the number of rows and columns and the size of each pixel.

header file a file containing definitions which are read as part of a compilation process. Also known as an include file.

head-medium gap the distance between the read-write head and the disk in magnetic or optical disk memory devices.

head motion movements of the head; head motion during virtual reality use may be a contributor to simulator sickness.

head-mounted display (HMD) displays worn by the user, on the head, as a helmet or cap; most commonly visual displays.

head-of-line blocking a blocking phenomenon in a switching system in which a queued request cannot be delivered to its output port because a request waiting in front of it is blocked.

headphones auditory displays worn by the user on the ears.

head related transfer function a mathematical relationship providing an estimate of the response of the head to external stimuli.

head tracker tracker used to detect the position and orientation of a user's head.

head-up display (HUD) display system used to present visual information to the user by projecting it within the primary visual cone, so that the user need not look down or away to acquire that information; for example, projection of flight parameters onto the cockpit canopy during piloting.

heap (**1**) an area of memory which is organized to support the allocation of blocks of memory on demand. Also known as free storage, free storage pool, storage pool, or a collection.

(**2**) data storage structure that accepts items of various sizes and is not ordered.

(**3**) a tree in which parent-child relationships are consistently "less than" or "greater than".

heap file a file in which new records are appended at the end of the file.

heapify to rearrange an array in-place so that each node has a key more extreme (greater or less) than or equal to the key of its parent. *See also* binary heap.

heaviest common subsequence problem *See* longest common subsequence problem.

Hebbian algorithm in general, a method of updating the synaptic weight of a neuron w_i using the product of the value of the ith input neuron, x_i, with the output value of the neuron y. A simple example is:

$$w_i(n+1) = w_i(n) + \alpha y(n) x_i(n) ,$$

where n represents the nth iteration and α is a learning-rate parameter.

Hebbian learning a method of modifying synaptic weights such that the strength of a synaptic weight is increased if both the presynaptic neuron and postsynaptic neuron are active simultaneously (synchronously) and decreased if the activity is asynchronous. In artificial neural networks, the synaptic weight is modified by a function of the correlation between the two activities.

hedge a special linguistic term by which other linguistic terms are modified. Examples are very, fairly, more or less, and quite.

hedgehog a visual representation of a surface in which the surface normals are rendered like pins sticking out of the surface.

height the height of a tree is the maximum distance of any leaf from the root. If a tree has only one node, the root, the height is 1. The height of the figure at the definition of the tree is 3.

height-balanced binary search tree a data structure used to support membership, insert/delete operations each in time logarithmic in the size of the tree. A typical example is the AVL tree or red-black tree.

height defuzzification *See* centroid defuzzification.

helical scan tape a magnetic tape in which recording is carried out on a diagonal to the tape, by a head that spins faster than the tape movement. Popular for VCRs, camcorders, etc., as it improves recording density and tape speed.

hemicube an algorithmic device that enables an efficient calculation of the form-factor values in the radiosity method. A *hemicube* is an efficient approximation to a hemisphere.

henceforth operator a temporal logic operator used to state that for all time instants from the evaluation instant (in the future or in the past) a formula is true. Usually there are two operators, one for the future and one for the past, and they are represented by a box (white for future and black for past). For example, formula $(\Box P) \wedge (\Box Q)$ is true if for all time instants in the future, P and Q are true.

Herbrand interpretation a concept in Boolean logic. Herbrand interpretation I is given by a set of ground atoms. It corresponds to a Boolean function on ground atoms A, where I(A) = true, if and only if $A \in I$. I satisfies a disjunction $A_1 \vee \ldots \vee A_k$ (a conjunction $B_1 \wedge \ldots \wedge B_n$) of ground atoms if at least one atom A_i is true (all atoms B_i are true) w.r.t. I.

Herbrand model a database model. A Herbrand model M of a database DB is a Herbrand interpretation that satisfies all ground instances of the rules $r \in DB$. M satisfies a ground rule $r\alpha\beta$ either if (1) M does not satisfy the body β, or if (2) M satisfies both β and the head α. For definite rules $AB_1 \wedge \ldots \wedge B_n$ this means that $B_1, \ldots, B_n \in M$ implies $A \in M$.

Herbrand universe the set of all terms that can constructed by combining the terms and constants that appear in a logic formula.

Hermitian matrix a square matrix that equals its conjugate transpose.

heterogeneous having dissimilar components in a system; in the context of computers, having different types or classes of machines in a multiprocessor or multicomputer system.

heterogeneous array an array that does not require that each value be the same data type. *Compare with* homogeneous array.

heterogeneous list a list containing objects of different types. *Compare with* homogeneous list.

heterogeneous network a network composed of systems of more than one architecture. *Compare with* homogeneous network.

heuristic a rule-of-thumb or other device or simplification that allows its user to draw conclusions without being certain.

heuristic approximation a technique commonly employed in writing compilers to handle problems that are computationally difficult to solve completely, such as the register allocation problem.

heuristic search a search that uses problem-specific knowledge in the form of heuristics in an effort to reduce the size of the search space or speed convergence towards a solution. Heuristic searches sacrifice the guarantee of an optimal solution for reduced search time.

hexadecimal *See* hexadecimal number system.

hexadecimal notation expressing numbers in base-16 format. *See* hexadecimal number system.

hexadecimal number system a numerical system of radix 16. Convenient for expressing computer value representations because three bits make up one hexadecimal digit, and two hexadecimal digits represent a single 8-bit byte. The traditional representation of hexadecimal values beyond the value 9 are the letters A, B, C, D, E, and F representing, respectively, the values 10, 11, 12, 13, 14, and 15.

hexagonal pixel *See* pixel.

hidden bit in a floating-point representation if the radix is 2, the most significant bit of a normalized significand is always 1 and therefore need not be explicitly represented; if the bit is implicit, then it is said to be hidden.

hidden file in a PC system, files that cannot be seen with an ordinary DOS dir (directory) command.

hidden layer a layer of neurons in a multilayer perceptron network which is intermediate between the output layer and the input layer.

hidden line removal a technique used in wireframe rendering (which is when one draws the straight line boundaries of the polygonal patches, or polyhedral solids that define the scene). If all boundaries are drawn, this is as if all surfaces and objects are transparent. If all surfaces and objects are opaque, then some boundaries would not be visible because they are hidden by closer surfaces. Removing the obscured or occluded portions of the boundaries is called *hidden line removal.*

hidden Markov model (HMM) statistical model of temporal processes based on the Markovian assumption. HMMs are discrete-time, discrete-space, dynamical systems governed by a Markov chain that emits a sequence of observable outputs, usually one output (observation) for each state in a trajectory of such states. More formally, an HMM is a five-tuple $(\Omega_X, \Omega_O, A, B, \pi)$, where Ω_X is a finite set of possible states, Ω_O is a finite set of possible observations, A is a set of transition probabilities, B is a set of observation probabilities, and π is an initial state distribution.

hidden surface problem sometimes called visible-surface determination or hidden-surface removal. It is the problem of only displaying the parts of a surface in a scene which are visible to the user. For a scene to make sense to a user, any surface that is obscured by an opaque surface must not be rendered. For raster graphics, an example of a rendering algorithm that solves the hidden-surface problem is the z-buffer algorithm.

hidden-surface removal the general algorithm that removes hidden surfaces and deals with such cases where one object partially obscures another. A hidden surface algorithm will, in general, eliminate those fragments of a polygon that are not visible because they are behind another object.

hierarchical clustering a clustering technique that generates a nested category structure in which pairs or sets of items or categories are successively linked until every item in the data set is connected. Hierarchical clustering can be achieved agglomeratively or divisively. The cluster structure generated by a hierarchical agglomerative clustering method is often a dendrogram but it can also be as complex as any acyclic directed graph. All concept formation systems are hierarchical clustering systems including self-generating neural networks. *See* clustering. *See also* concept formation, self-generating neural network.

hierarchical coding (**1**) coding a signal at several resolutions and in order of increasing resolution. In a hierarchical image coder, an image is coded at several different sizes and in order of increasing size. Typically, the smaller sized images are used to encode the larger size images to obtain better compression.

(**2**) coding where image data are encoded to take care of different resolutions and scales of the image. Additional data is transmitted from the coder to refine the image search. *See* progressive transmission.

hierarchical decomposition a decomposition approach by means of a system is decomposed down into a hierarchy of sub-components via a top-down process. *See* decomposition, system decomposition.

hierarchical diagram used to describe hierarchical schemas. The hierarchical diagram contains boxes and lines which correspond to record types and links, respectively.

hierarchical encoding *See* hierarchical coding.

hierarchical feature map a hybrid learning network structure also called counter-propagation network which consists of an unsupervised (hidden) layer and a supervised layer. The first layer uses the typical (instar) competitive learning, the output will be sent to the second layer, and an outstar supervised learning is performed to produce the result. *See also* outstar training.

Kohonen calls a feature map which learns hierarchical relations a *hierarchical feature map.*

hierarchical interpolation a technique of forming image pyramids. In this method pixels are interpolated to higher levels in the hierarchy. Thus, only the interpolated pixels need a residual in order to be reconstructed exactly since the subsampled pixels are already correct.

hierarchical locking where coarse grained locks are acquired initially and then finer grained locks are acquired.

hierarchical memory the organization of memory in several levels of varying speed and capacity such that the faster memories lie close to the processor and slower memories lie further away from the processor. Faster memories are expensive, and therefore are small. The memories that lie close to the processor store the current instruction and data set of the processor. When an object is not found in the memories close to the processor, it is fetched from the lower levels of the memory hierarchy. The top levels in the hierarchy are registers and caches. The lowest level in the hierarchy is the backing memory which is usually a disk. Also known as multi-level memory.

hierarchical model a data model, in which the data is represented as collections of records and the relationships are represented by links. The records and links are organized as collections of trees.

hierarchical schema a collection of hierarchical diagrams which define the logical structure of the data in the database.

hierarchical subsystem structure a structure that is formed by imposing equivalence relations on the resource-flow graphs of the source code among subsystems.

hierarchy a tree-like or forest-like structure in which there are subordinate nodes. Often used to represent inheritance relationships and sometimes aggregation structure.

high byte the most-significant byte of a multi-byte numeric representation.

higher-order function a function that takes another function as a parameter or returns a function as its result.

higher-order language (HOL) generally any language beyond the second generation. Higher-order languages are compiled or interpreted and include languages as "ancient" as FORTRAN. Also known as high level language (HLL).

higher-order unit a neural unit whose input connections provide not the usual weighted linear sum of the input variables but rather a weighted polynomial sum of those variables.

high-fidelity prototype a prototype constructed in a medium and environment very similar to that of the eventual product. Often this takes the form of running code. *See also* low-fidelity prototype.

high-impedance state the third state of tri-state logic, where the gate does not drive or load the output line. *See* tri-state circuit.

high-level language (HLL) *See* higher-order language.

high-level vision the highest stage of vision, leading to the full extraction of the 3D information and to its exploitation for the description and understanding of the scene.

highlight the area of a glossy object over which specular reflection can be viewed. It is normally the color of the light source, not of the object.

high order interleaving in memory interleaving, using the most significant address bits to select the memory module and least significant address bits to select the location within the memory module.

high order logic (HOL) an extension of first order logic where quantification can be done over predicates and not only variables as for FOL. For example, $\forall P . \exists x . P(x)$.

high-pass filter a filter that prevents signals below the nominal frequency from propagating.

high-speed carry in an arithmetic logic unit, a carry signal that is generated separately from the generation of the result and is therefore faster; a lookahead carry.

highway *See* bus.

Hilbert space an inner product space with the additional property that is a complete metric space. An inner product space is a linear space on which an inner product is defined, while completeness means that there are no "missing" vectors arbitrarily close to but not included in the space.

Hilbert transform a transform that relates the real and imaginary parts of a complex quantity, such as the index of refraction. For f, a function of one real variable, the transform $\mathcal{H}(f)$ defined as follows:

$$\mathcal{H}(f)(x) = \lim_{\varepsilon \to 0} \frac{1}{\pi} \int_{|t| \geq \varepsilon} \frac{f(x-t)}{t} dt .$$

An alternate formula, which coincides with the previous one, except possibly on a zero-measure set of points of discontinuity of f, is:

$$\mathcal{H}(f)(x) = \lim_{\varepsilon \to 0} \frac{1}{\pi} \int_{-\infty}^{+\infty} f(x-t) \frac{t}{t^2 + \varepsilon^2} dt .$$

When f is square-integrable, the Fourier transforms $\mathcal{F}(f)$ and $\mathcal{F}(\mathcal{H}(f))$ of f and $\mathcal{H}(f)$ are related by:

$$\mathcal{F}(\mathcal{H}(f))(u) = -i \operatorname{sign}(u) \mathcal{F}(f)(u)$$

almost everywhere; thus for positive frequencies, the Fourier phase is shifted by $-\pi/2$; in particular, f and $\mathcal{H}(f)$ have the same L^2 norm.

histogram (**1**) graphical representation of information in vertical bars proportional to the value of the variable displayed. For example, the histogram of people having a PC in the last 10 years will show 10 vertical bars, each of them having height proportional to the number of people that have a PC in the selected year.

(**2**) in image processing, the distribution of the gray levels in an image, typically plotted as the number or percentage of pixels at a certain gray level vs. the gray levels. If the ordinate is the ratio of the number of pixels at a gray level to the total number of pixels, the histogram is an approximation of the probability density function of the gray levels. *See also* probability density function.

histogram equalization a technique for computing an image gray level transformation to redistribute the pixel intensity values and "flatten" the pixel intensity histogram. Histogram equalization can be used to enhance the contrast of the image. Also called histogram leveling, histogram flattening. *See also* contrast enhancement, histogram.

histogram modeling (**1**) making the histogram of an image have another shape, for example, the flat line produced by histogram equalization. This procedure is usually not automatic, i.e., the shape is usually specified by a person. Also called histogram modification or histogram specification. *See also* histogram equalization.

(**2**) the fitting of a function to a histogram in order to obtain an analytical expression that approximates the histogram.

histogram sliding the addition of a constant to all pixels in an image to brighten or darken the image. A positive constant makes the histogram slide (translate) to the right; a negative constant makes it slide to the left. *See also* histogram.

histogram stretching expansion or contraction of the gray-level histogram of an image in order to enhance contrast. Usually performed by multiplying or dividing all pixels by some constant value. *See also* contrast enhancement, histogram.

history (**1**) record of changes to all, or part of, a database.

(**2**) a schedule.

hit the notion of searching for an address in a level of the memory hierarchy and finding the address mapped (and the data present) in that level of the hierarchy. Often applies to cache memory hierarchies.

hit-miss ratio in cache memory, the ratio of memory access requests that are successfully fulfilled within the cache to those which require access to standard memory or to an auxiliary cache.

hit-miss transform *See* hit-or-miss transform.

hit-or-miss transform a class of transforms in image processing that locates objects larger than some specific size, S_1, and smaller than some size S_2.

hit rate the percentage of references to a cache which find the data word requested in the cache. Also the probability of finding the data and given by

$$\frac{\text{number of times required word found in cache}}{\text{total number references}}.$$

Also known as hit ratio.

hit ratio *See* hit rate.

hit time the time to process a cache access that results in a hit.

HLL high level language. *See* higher-order language.

HMD *See* head-mounted display.

HMM *See* hidden Markov model.

HOL *See* high order logic, higher-order language.

holding time the time for which a request occupies a server resource. In the case of circuit switching, the holding time of a circuit is equal to the holding time of the corresponding connection.

Hollerith card a punched card used for data storage. Developed by Hermann Hollerith for use in the 1890 U.S. census, these cards were widely used for early computer data storage.

The card size, which has been standardized since 1890, was based on the size of the one dollar bank note of the time — a decision made because of the availability of standard cabinetry to hold the millions of punched cards needed for the survey. The punched card is now obsolete.

Hollerith code an encoding of data where the data is represented by a pattern of punched holes in a card.

Hollerith, Herman (1860–1929) Born: Buffalo, New York

Hollerith is best known for his development of a tabulating machine using punched cards for entering data. This machine was used to tabulate the 1890 U.S. census. The success of this machine spurred a number of other countries to adopt this system. The company Hollerith formed to develop and market his ideas became one of several companies that merged to form IBM, International Business Machines. Hollerith's initial interest in automatic data processing probably stems from his work as an assistant to several census preparers at Columbia University. The success of Hollerith's work gives him claim to the title of the father of automated data processing.

hollow fill a three-dimensional object whose internal volume (defined as the space enclosed by the object's skin) is not rendered. Such three-dimensional solid objects are frequently used in virtual environments and are constructed using infinitely thin polygons to form the skin of the object. If the user's viewpoint is positioned within the skin of the object, the reverse of the surface will be rendered if there is sufficient illumination.

holographic display a display that provides three-dimensional or holographic images.

holonomic constraints in which it is possible to find a set of independent generalized coordinates such that there are the same number of coordinates as degrees of freedom.

HOL scheduling a non-preemptive priority scheduling discipline in which customers form a separate first-come-first-serve queue for each class and at each start of service event the server selects the first customer from the highest priority queue that is non-empty.

home directory a directory, specified in the password file, that is a user's current directory during and immediately following the login sequence.

homeomorphic *See* homomorphic.

homogeneity a property stating that in a given set all the components have common features. For example, they have the same type or class.

homogeneous array an array that requires that all of its values be of the same data type.

homogeneous coordinates normally, the transformations for scaling, rotation, and translation are treated differently. Scaling and rotation use matrix multiplication, whereas translation uses vector addition. When the homogeneous coordinate system is used, all three transformations can be performed using matrix multiplication. This representation is commonly used in graphics systems because of its simplicity of representation and use.

A homogeneous coordinate is expressed with an additional coordinate to the point. So, a two-dimensional point $(x, y)^T$ is represented as a homogeneous coordinate by a triple $(x, y, W)^T$. Two sets of homogeneous coordinates are equivalent if one is a multiple of the other. If the W coordinate is non-zero, we can divide each coordinate by W, transforming $(x, y, W)^T$ into $(x/W, y/W, 1)^T$; the numbers x/W and y/W are the Cartesian coordinates of the homogeneous point. If W is zero, the homogeneous coordinate is a point at infinity.

homogeneous list a list containing objects of different types. *Contrast with* heterogeneous list.

homogeneous network a network composed of systems of only one architecture. *Compare with* heterogeneous network.

homogenous transformation matrix a coordinate transformation in which both the translational and rotational transformations are given by a single matrix.

homomorphic a mathematical property between two relations. To illustrate, two graphs are *homomorphic* if they can be made isomorphic by inserting new vertices of degree 2 into edges.

homomorphic filter an image enhancement technique based upon the illumination-reflectance model. The homomorphic filter assumes the image function $f(x, y)$ is the combination of two functions $i(x, y)$ and $r(x, y)$. By taking the natural log of the images, the components are separated and can be processed separately. The reflectance of an image usually contains high frequency components, while the illumination component tends to vary slowly (the low frequencies); thus, the filter applied to the logarithm image should affect the low and high frequency components differently. The most common use for the filter is to enhance images by compressing the brightness range while increasing contrast; the filter applied in this case is a circularly symmetric high pass filter with the stop band magnitude < 1 and pass band magnitude > 1.

homotopy method a technique for solving nonlinear algebraic equations $F(x) = 0$ based on higher-dimensional function embedding and curve tracing. The idea is to construct a parameterized function such that at one parameter value, say $\lambda = \lambda_0$, the system of equations is easy to solve or has one or more known solutions, and at another parameter value, say $\lambda = \lambda_f$, the system of equations is identical to that of the system of interest, $F(x) = 0$. A homotopy method may then be interpreted as a geometric curve following through solution space from the known solutions of the "easy" problem to the unknown solutions of $F(x) = 0$.

homunculus in literature and art, the little person who sits inside our heads and operates the controls; the operator of a teleoperator is a sort of homunculus for the robot.

hop the act of a packet being forwarded by a router.

Hopfield memory *See* Hopfield model.

Hopfield model a neural algorithm capable of recognizing an incomplete input. Also known as Hopfield memory.

Hopfield suggested that an incomplete input can be recognized in an iterative process, in which the input is gradually recognized in every cycle of the iteration. The iteration is completed when the input finally matches with a stored memory. The Hopfield model is a sort of associative memory. A hologram can also be directly used as associative memory. The main difference of the non-neural holographic associative memory and the Hopfield model is as follows. The direct holographic associative memory is one-step and its signal-to-noise ratio depends on the incompleteness of the input that cannot be improved. The Hopfield model is an iterative process involving a nonlinear operation, such as thresholding in which the signal-to-noise ratio of the input can be improved gradually during the iterative process. A large number of optical systems have been proposed to implement the Hopfield model, including the first optical neural networks. Those optical implementations are primarily based on optical matrix-vector or tensor-matrix multiplication.

Hopfield network a recurrent, associative neural network with n processing elements. Each of the processing elements receives inputs from all the others. The input that a processing element receives from itself is ignored. All of the processing element output signals are bipolar. The network has an energy function associated with it; whenever a processing element changes state, this energy function always decreases. Starting at some initial position, the system's state vector simply moves downhill on the network's energy surface until it reaches a local minimum of the energy function. This convergence process is guaranteed to be completed in a fixed number of steps. *See* discrete Hopfield network.

Hopper, Grace Murray (1906–1992) Born: New York, New York

Hopper is best known as the author of the first compiler. Hopper began her career as a mathematics professor at Vassar College. During WWII she volunteered for service in the U.S. Navy and was assigned to work at Harvard with Howard Aiken on the Mark I computer. She later joined J. Presper Eckert and John Mauchly working on the UNIVAC computer. It was at this time she wrote the first compiler. Her compiler and her views on computer programming significantly influenced the development of the first "English-like" business computer language COBOL.

horizontal fragmentation in object oriented databases, fragmenting the attributes of an object into more than one storage structure.

horizontal microinstruction a microinstruction made up of all the possible microcommands available in a given CPU. In practice, some encoding is provided to reduce the length of the instruction.

horizontal propagation used to limit the propagation of privileges. A user granted privileges with horizontal propagation limited to N enforces that that user may not grant the privilege to more than N users.

Horn clause also known as a definite clause, it is a clause with one positive literal. It is written as a disjunction $\mathsf{A} \vee \neg\mathsf{B}_1 \vee \ldots \vee \neg\mathsf{B}_n$ or $\neg\mathsf{B}_1 \vee \ldots \vee \neg\mathsf{B}_n$ with $\mathsf{k} \leq 1$ positive atomic formulas. Horn clauses correspond to definite rules (for $\mathsf{k} = 1$) or goals (for $\mathsf{k} = 0$), respectively. Horn clauses can express a subset of statements of first order logic. *See* clausal form.

Horner's rule a rule that states that a polynomial $A(x) = a_0 + a_1x + a_2x^2 + a_3x^3 + \ldots$ may be written as $A(x) = a_0 + x(a_1 + x(a_2 + x(a_3 + \ldots)))$.

Horner's scheme an efficient and optimal method for the evaluation of polynomials. The polynomial $c_0 + c_1x + c_2x^2 + \cdots + c_nx^n$ is evaluated as $c_0 + x(c_1 + x(c_2 + x(\cdots + (c_{n-1} + c_nx)\cdots)$, which requires n steps, each consisting of one multiplication and one addition.

horopter in binocular vision, the curved surface occupied by points having no retinal disparity; objects closer than the horopter have crossed disparity, and objects beyond the horopter have uncrossed disparity.

host (**1**) a computer that is the one responsible for performing a certain computation or function.

(**2**) computer on which the command processor is running, and on which the user is logged on.

(**3**) in networking, a computer, printer, switch, etc. that can act as a source or sink of data.

host system In reference to a compiler, the system on which the compiler executes. This is usually used to distinguish the system on which the compiler runs from the target system on which its output is intended to run. *See also* cross-compiler.

hot 1 in 2's complement notation, the (representation of the) negation of a number is computed by forming a 1's complement and then adding a 1; the 1 is known as a hot 1.

hot backup (**1**) to backup a live running database.

(**2**) copy of a database, immediately available for use, should the primary version fail.

hot spot a frequently accessed database object, likely to cause deadlock or blocking.

Hough transform a transform which transforms image features and presents them in a suitable form as votes in a parameter space, which may then be analyzed to locate peaks and thereby infer the presence of desired arrangements of features in the original image space: typically, Hough transforms are used to locate specific types of objects or shapes in the original image. Hough transform detection schemes are especially robust.

Householder transformation a matrix Q that maps each vector to its reflection through a defined hyperplane; specifically,

$$Q = I - 2\frac{uu^T}{u^T u}$$

reflects through the plane having normal vector u.

housekeeping an activity that establishes a set of initial conditions to facilitate the execution of a computer program.

HSI *See* hardware software interface, hue-saturation-intensity.

HSL *See* hue-saturation-lightness.

HSV *See* hue-saturation-value.

HTML *See* hypertext markup language.

HTTP usually written as http. *See* hypertext transport protocol.

http *See* hypertext transport protocol.

HUD *See* head-up display.

hue a perceptual term referring to the colorimetry quantity "dominant wavelength" of a color. *Hue* can be used together with saturation and luminance to define the HSL color space.

hue-saturation-intensity (HSI) *See* hue-saturation-lightness.

hue-saturation-lightness (HSL) a color model based on the specification of hue (H), saturation (S), and intensity (I). A useful and convenient property of this model is the fact that intensity is separate from the color components. The hue and saturation components (together referred to as chromaticity) relate closely to color perception in humans; however, the HSI intensities are linear and do not correspond to those observed by the eye.

HSL is based on polar coordinates, while the RGB color space is based on a three-dimensional Cartesian coordinate system. Intensity is the vertical axis of the polar system, hue is the relative angle, and saturation is the planar distance from the axis. The conversion from RGB to HSI is as follows:

$$\begin{aligned} H &= \arccos(0.5((R-G)+(R-B))/ \\ &\quad \sqrt{((R-G)^2+(G-B)*(R-B))}\,, \\ S &= 1-(3/(R+G+B))*(\min\{R,G,B\})\,, \\ I &= (R+G+B)/3\,. \end{aligned}$$

See also RGB color model, color space.

HSL is thought to be more intuitive to manipulate than RGB space. For example, in the HSL space, to change red to pink requires only changing the saturation parameter.

Also known as HSI (hue-saturation-intensity).

hue-saturation-value (HSV) a color space that describes color using three basis components: hue, saturation, and brightness. *See also* HSL, Munsell color system.

Huffman coding a variable length coding scheme whose codewords are generated from the probability distribution of the source. The method is optimal in the sense that no other method can give a higher compression rate. It is capable of achieving the bound on compression given by the source coding theorem. It is due to D.A. Huffman (1952).

Decoding a Huffman codeword corresponds to traversing an unbalanced binary tree according to the value (0 or 1) of each bit in the word; the leaves of the tree are the source symbols, with the most probable ones being the shortest distance from the root of the tree. *Huffman coding* achieves an average code rate equal to the source entropy if and only if all the probabilities are negative powers of 2. In general, it achieves less compression than arithmetic coding but is easier to implement.

See also k-ary Huffman encoding.

Huffman encoding *See* Huffman coding.

human-centered automation a system that interacts with human users and that features automation of tasks or sub-tasks; usually used to refer to an automated system that is not a robot.

human-centered design a design philosophy which starts with the user and the user's mission, and strives to adapt hardware and software to the needs of the former in the context of the latter.

human computer interface (HCI) the interface between a (computer) system and its human users. A well-defined HCI is essential to ensure the usability of a system. Also referred to in the past as man-machine interface.

human error human action or failure to act that leads to an undesirable system state during performance of a task or during the operation, programming, or design of a system.

human factors discipline devoted to the study of human-machine interaction and human-centered design.

human interface *See* human-machine interface.

human-machine interaction reciprocal interaction of humans with machines, whether the human is controlling the machine, providing logistical support for the machine, serving as the object of machine actions, or simply in the presence of the machine.

human-machine interface the subsystem serving to accept inputs from the user and display information to the user; an expanded definition would include equipment designed to house and support the user, such as chairs; heating, ventilating, and air conditioning; lighting; etc.

human oriented concepts an approach to knowledge-based concept assignment. Human oriented concepts are often rather informal or semi-formal. They cannot always be expressed in structural information alone. Therefore, a pure parsing approach that depends on structural plans fails to assign human oriented concepts.

human resources people involved in a project. This term distinguishes them from general resources, which are the set of facilities and tools that will be used for a project or are available in a plant.

human-robot synergy concept of systems that are more than robots and more than people because they combine and take advantage of the best elements of each.

human senses human tissue devoted to the translation of physical energy into nervous system signals; traditionally sight, hearing, touch, taste, and smell, but actually a much more complex set of systems designed to search for information in the environment.

human tasks those parts of the mission of a human-robot synergistic system that must be done by the human.

human transfer function mathematical relationship describing the human output made in response to external stimuli.

human visual system (HVS) the collection of mechanisms in humans which process and interpret the visual world. These mechanisms include the eye, the retina, the optic nerve, the visual cortex, and other parts of the brain.

Hungarian notation a convention of programming in the C language, invented by Hungarian programmer Charles Simonyi at Microsoft, and promulgated as a means of simplifying type checking. It consists of encoding a sequence of lower-case characters at the start of every identifier name to indicate its type, for example "nValue" meaning an integer value and "lpstrText" meaning a string pointer. The use of these prefixes is totally redundant with the type declaration.

Hurst parameter indicates the degree of self-similarity of a time series. For an autocorrelation function of $k^{-\beta}$, the Hurst parameter $H = 1 - \beta/2$ indicates that the series is self-similar if $1/2 < H < 1$. *See* self-similar stochastic process.

HVS *See* human visual system.

hybrid computer a computer based on at least two different technologies. For instance, a computer presenting both digital and analog circuits and signals.

hybrid control control strategy that provides both force and motion control in orthogonal directions.

hybrid fragmentation in distributed databases, the subdivision of a relation combining horizontal and vertical fragmentation.

hybrid hash join a variation of the hash join. Similar to the hash join, but differs when the hashed records do not fit in memory. In this case, part of each hash bucket is stored on disk. Each hash bucket is allowed one block in memory; the remainder is stored on disk.

hybrid model *See* dual model, formal model.

hybrid specification models a specification technique for modeling linear continuous properties of a system. Hybrid automata are extensions of finite state automata to continuous quantities. They allow continuous properties of the operating environment to be specified and modeled directly. Finite state automata provide a mathematical foundation for reasoning about systems in terms of their discrete properties. In hybrid automata, state transitions may be triggered by functions on continuous variables.

hydraulic actuator an actuator that uses hydraulics as the medium for power.

hydraulics the science of liquid flow.

hyperalternative in a W-grammar, a possibly empty sequence of hypernotions separated by "," and followed by a ".".

hypercube processor a processor configuration that is similar to the linear array processor except that each processor element communicates data along a number of other higher dimensional pathways. There are typically 2^N processing elements in a hypercube, where N is a positive integer.

hypermedia presentation of information in a nonlinear and hierarchical structure. Hypermedia are an extension of hypertext that include graphics, sound, video, and other kinds of data.

hypernotion in a W-grammar, a possibly empty sequence of metanotions and protonotions.

hyper rules one of the two sets of context-free grammars that define a W-grammar. A *hyper rule* is written as a nonempty hypernotion, followed by a ":", followed by a sequence of hyperalternatives separated by ";", followed by ".". A hyper rule represents one or more production rules. *See also* metaproduction rules.

hypertext a term coined by Ted Nelson around 1965 for a collection of documents (or nodes) containing cross-references or links which, with the aid of an interactive browser program, allow the reader to move easily from one document to another following links that can be defined on words, figures, etc.

hypertext markup language (HTML) a hypertext document format used on the World-Wide Web. Built on top of SGML. "Tags" are embedded in the text. A tag consists of a "<", a "directive", zero or more parameters, and a ">". Matched pairs of directives, like "<title>" and "</title>" are used to delimit text which is to appear in a special place or style.

hypertext transport protocol (HTTP) the client-server TCP/IP protocol used on the World Wide Web for the exchange of HTML documents. It conventionally uses port 80. Version 1.0 is the current standard.

I

IC *See* integrated circuit.

ICE *See* in-circuit emulator.

ICON a string-pattern-matching language which was developed as a successor to SNOBOL.

icon a picture having meaning in the context of a human-computer interface.

ID *See* index of difficulty, identifier.

IDE *See* integrated device electronics, integrated development environment.

idempotency a property possessed by operations for which the effect of executing the operation many times is equivalent to executing it once.

idempotent an operator where applying it twice gives the same result as applying it only once; mathematically speaking, if ψ is the operator and X the object to which it is applied, this means that $\psi(\psi(X)) = \psi(X)$.

identification a mechanism to guarantee the identity of the user.

identifier the name bound to an abstraction.

identifying relationship a relationship between a strong entity type and a weak entity type.

IDL *See* interface definition language.

idle pertaining to a system or component that is fully functional but currently not in use.

idle CPU the amount of time a processing platform's processor spends not executing instructions associated with system or user applications.

idle period for a device, this is the period during which the device has no workload to process.

IEEE *See* Institute of Electrical and Electronics Engineers.

IEEE-754 *See* IEEE Standard 754.

IEEE 802 address the 48-b address defined by the IEEE 802 committee as the standard address on 802 LANs.

IEEE floating point a representation of floating point values established by the IEEE as IEEE standard number 754. Not only does the standard fix the representation, but it also specifies the behavior in case of overflow or underflow.

IEEE Standard 754 the specification that defines a standard set of formats for the representation of floating-point numbers.

IGES *See* International Graphics Exchange Specification.

IID *See* independent and identically distributed.

ill-posed problem a problem whose solution may not exist, may not be unique, or may depend discontinuously on the data. (A problem is well posed if it can be shown that its solution exists, is unique, and varies continuously with perturbation of the data. If any of the three conditions does not hold, the problem is ill posed.).

ill-typed a program fragment that does not comply with the rules of a given type system.

illuminance the amount of light falling into a patch of unit surface area. It is measured in lux.

illuminant a source of illumination.

illumination photometric measure of the amount of light radiated onto a surface.

image (1) information stored digitally in a computer, often representing a natural scene or other physical phenomenon, and presented as a picture. The brightness at each spot in the pic-

0-8493-2691-5/01/$0.00+$.50

ture is represented by a number. The numbers may be stored explicitly as pixels or implicitly in the form of equations. *See also* bitmapped image, image acquisition, image processing, imaging modalities, pixel, vector image.

(**2**) the representation of a computer program and its related data as they exist at the time they reside in main storage.

image acquisition the conversion of information into an image. Acquisition is the first stage of an image processing system and involves converting the input signal into a more amenable form (such as an electrical signal), and sampling and quantizing this signal to produce the pixels in the image. Hardware, such as lenses, sensors, and transducers are particularly important in image acquisition. *See also* analog-to-digital converter, digitization, image, pixel, quantization, sample.

image analysis the extraction of information from an image by a computer. The results are usually numerical rather than pictorial and the information is often complex, typically including the recognition and interpretation of objects in or features of the image.

image-based rendering an approach to rendering in which objects and environments are modeled using image data instead of geometric primitives.

image classification the division by a computer of objects in an image into classes or groups. Each class contains objects with similar features. Also, the division of the images themselves into groups according to content.

image coding any of a class of techniques for recoding images to achieve the high rates of data compression required to cope with the storage and transmission of prodigious amounts of image data. For methods, *see* still image coding and video coding.

image coding for videoconferencing this is also called coding at primary rates for videoconferencing. In North America and Japan, one of the most plentiful is the DSI rate of 1.544 Mb/s. In Europe, the corresponding rate is 2.044 Mb/s. The main difference is that in videoconferencing the video camera is fixed and picture data is produced only if there are moving objects in the scene. Typically, the amount of moving data is much smaller than the stationary area.

image compression representation of still and moving images using fewer bits than the original representation. For methods, *see* still image coding and video coding.

image enhancement processing an image to improve its appearance, or to make it more suited to human or machine analysis. Enhancement is based not on a model of degradation (for which, *see* restoration), but on qualitative or subjective goals such as removing noise and increasing sharpness. Contrast stretching, edge emphasis, and smoothing are examples of enhancement techniques.

image feature an attribute of a block of image pixels.

image file format a representation (usually binary) used by a computer system as an agreed format to store an image. Examples of image file formats include the Graphics Interchange Format (GIF) and Tagged Image File Format (TIFF).

image fusion the merger of images taken with different imaging modalities or with different types of the same modality, for example, different spectral bands. *See also* multispectral image.

image motion field the two-dimensional projection of the velocity field onto the image plane.

image noise unmodeled and arbitrary data and/or signals on an image.

image plane the collection of two-dimensional coordinates from an image.

image processing the manipulation of digital images on a computer, usually done automatically (solely by the computer) but also semi-automatically (the computer with the aid of a person). The goal of the processing may include compressing the image, making it look better, extracting information or measurements from it, or getting rid of degradations. *See also* image acquisition, image analysis, image classification, image compression, image enhancement, image recognition, image restoration.

image recognition the identification and interpretation of objects in an image. Image recognition is a higher-level process that requires more intelligence than other parts of image processing such as image enhancement. Artificial intelligence, neural nets, or fuzzy logic are often used. *See also* artificial intelligence, artificial neural network, fuzzy logic, classifier, image classification, image understanding, pattern recognition.

image reconstruction (**1**) the conversion of a digital image to a continuous one suitable for display. *See also* analog-to-digital converter, interpolation.

(**2**) in tomography, the reconstruction of a spatial image from its projections. *See also* computed tomography, Radon transform.

image registration the spatial alignment of a pair of images, which may involve correcting for differences in translation, rotation, scale, deformation, or perspective. Especially important in binocular vision.

image regularization the process of shaping the raw image data (natural image) into a format suitable for image analysis or visualization.

image resolution a measure of the amount of detail that can be present in an image. Spatial resolution refers to the amount of a scene's area that one pixel represents, with a smaller area being a higher resolution. Brightness resolution refers to the amount of luminance or intensity that each gray level represents, with a smaller amount being a higher resolution.

image restoration the removal of degradations in an image in order to recover the original image. Some typical image degradations and possible causes are: blurring (intervening atmosphere or media, optical aberrations, motion), random noise (photodetectors, electronics, film grain), periodic noise (electronics, vibrations), and geometric distortions (optical aberrations, angle of image capture). *See also* motion compensation, noise

image segmentation the division of an image into distinct regions, which are later classified and recognized. The regions are usually separated by edges; pixels within a region often have similar gray levels. *See also* edge detection, image classification, image recognition.

image smoothing the reduction of abrupt changes in the gray levels of an image. Usually performed by replacing a pixel by the average of pixels in some region around it. Also called image averaging. *See also* smoothing.

image transform (**1**) the conversion of the image from the spatial domain to another domain, such as the Fourier domain. It is often easier and more effective to carry out image processing such as image enhancement or image compression in the transformed domain.

(**2**) transform involving the subdivision of $N \times N$ image into smaller $n \times n$ blocks and performing a unitary transform on each subimage. Any reversible liner transform where the kernal describes a complete orthogonal discrete basis function is called a unitary transform.

See also block transform, discrete Fourier transform (DFT), discrete Hadamard transform, discrete cosine transform (DCT), discrete sine transform (DST), discrete wavelet transform (DWT), distance transform, Fourier transform, Haar transform, Hadamard transform, Hilbert transform, Hough transform, Karhunen-Loeve transform (KLT), medial axis transform (MAT), Radon transform, slope transform, top hat transform, transform, Walsh transform, wavelet transform.

image understanding the interpretation by a computer of the contents of an image. The process seeks to emulate people's ability to intelligently extract information from or make conclusions about the scene in an image. Also called image interpretation.

imaginary part in a complex number, the component that consists of a real number (a number found on the number line) times the positive square root of -1, denoted i.

imaging modalities the general physical quantity that the pixels in an image represent; the type of energy an image processing system converts during image acquisition. The most common modality is visible light, but other modalities include invisible light (infrared or X-ray), sound (ultrasound), and magnetism (nuclear magnetic resonance). *See also* computed tomography (CT), fluoroscopy, image acquisition, magnetic resonance imaging (MRI), multispectral image, positron emission tomography (PET), tomography, X-ray image.

immediate addressing an addressing mode where the operand is specified in the instruction itself. The address field in the instruction holds the data required for the operation.

immediate mode a memory addressing scheme that involves an integer operand that is usually contained in the next address after the instruction.

immediate operand a data item contained as a literal within an instruction.

immediate update protocol where data updates are written to the database immediately. The database is updated before the transaction reaches its commit point.

immersion the degree to which a synthetic environment controls sensory inputs to the user.

immersion display a display in which the user is immersed in an artificial reality. The everyday environment is blocked out.

immersive synthetic environment which largely controls sensory inputs to users.

immersive virtual reality a system where a user's field of view is completely filled by the display medium and the user can interact with the visualization in a natural way such as pointing, grabbing, head movement to change view, etc. The user is also shielded from external factors such that the overall perception is one of being immersed within the visualization.

impact analysis the activity aimed at determining the impact of a change before its actual implementation, in order to anticipate the extent of its effects, and thus the associated risk and cost.

impedance the relationship between force and motion.

impedance control a robotic manipulator control strategy based on modifying the impedance of a manipulator or controlling the response a robot has to external forces.

imperative language a programming language which achieves its primary effect by changing the state of variables by assignment.

implementation the transformation of a design into a more detailed form, or its realization as a working product. The implementation may be hardware or software end product, which actually performs the required functions, or it may be an intermediate product at a lower level of stepwise refinement. The goal of abstraction is to provide mechanisms by which functionality may be made visible without the user of that functionality needing to be aware of the details of how that functionality is achieved, that is, its *implementation.*

implementation-dependent behavior of a construct in a programming language whose behavior is not specified by a formal description of the language. The decisions that are implementation-dependent may be as simple as choice of representation for primitive values (for example, choices for "true" and "false"), or decisions that can affect the fundamental behavior of the program. *See* portable.

implementation design description a complete description of all the information needed to produce a concrete system from one or more functional designs. It describes which components of a generic system family must be used, and which tools must be invoked with which parameters. The *implementation design descrip-*

tion acts as a meta-description referring to other descriptions.

implementation inheritance the use of inheritance to access code in the superclass at the likely expense of destroying the inheritance of the specification or interface. Usually negates/prevents the use of polymorphism.

implementation model a model that consists of the code files and the used work structure. It includes the application software description as well as the support software description. While the design model is a more abstract view, the implementation model contains the full information necessary to build the system.

implemented software quality quality of an intermediate software product just after code implementation, but before testing. Implemented quality reflects fundamental design quality and the quality of the code implementors and should be evaluated on the basis of internal and external attributes.

implication the Boolean implies function. *See also* conjunction, disjunction.

implicit declaration a mechanism specified by many programming languages in which the use of a name defines the name. *Contrast with* explicit declaration. Programming languages that allow implicit declarations usually specify the way in which the characteristics of the name are determined, for example, the FORTRAN rule for variables starting with the letters I-O, the BASIC rule that a string variable or function ends in $, or the C rule that an unnamed function always returns an integer result.

implicit lock *See* automatic lock.

implicitly typed language a typed language where types are not part of the syntax.

implicit parallelism the assertion that, in genetic algorithms, each symbol string in the population contains information about every schema to which the string belongs. As a result, when a particular string in the population is processed, we are in effect performing searches on the hyperplanes of all schemata associated with that string. It was estimated by J. Holland that, for a population with μ individuals, the number of schemata processed in a generation is on the order of μ^3. *See also* schema.

implicit surface a surface defined by using an implicit equation given by $f(x, y, z) = 0$ for some function $f(x, y, z)$. The equation restricts the interaction of x, y, and z to ensure that the point (x, y, z) is confined to the surface.

implied mode a memory addressing scheme that involves one or more registers that are implicitly defined in the operation determined by instruction.

implied needs needs that may not have been stated but are actual needs when the entity is used in particular conditions. These are real needs that have not been documented.

implies the Boolean function of implication: $0 \to 0 = 1, 0 \to 1 = 1, 1 \to 0 = 0, 1 \to 1 = 1$. *See also* AND, OR, NOT, XOR.

import to create part or all of a database from a portable format.

imprecise computation techniques in which accuracy is sacrificed in order to meet deadlines. These techniques involve early truncation of a numerical series. Also called approximate reasoning.

imprecise interrupt an implementation of the interrupt mechanism in which instructions that have started may not have completed before the interrupt takes place and insufficient information is stored to allow the processor to restart after the interrupt in exactly the same state. This can cause problems, especially if the source of the interrupt is an arithmetic exception. *See* precise interrupt.

impulse noise refers to a noise process with infrequent, but very large, noise spikes; it is also known as shot noise or Salt and Pepper noise. The phrase "impulsive noise" is frequently used to characterize a noise process as being fundamentally different from Gaussian white noise, in being derived from a probability density function with very heavy (long) tails. Applying impulsive noise to an image leaves most pixels unaffected, but with some pixels very bright or dark.

inbetweening the generation of intermediate transition positions from a given start and end point or keyframes. This technique is often used in animation, where a lead artist generates the beginning and end keyframes of a sequence (typically 1 second apart), a breakdown artist does the breakdowns (typically 4 frames apart), and an "inbetweener" completes the rest.

incident an event during operation of an item which may indicate that a failure has occurred. The true nature of an *incident* may not become apparent until after diagnosis. Some incidents will turn out to be due to mistaken observation. Others will turn out to be "genuine" and relevant to the measurement of one or more dependability attributes.

in-circuit emulator (ICE) a device that replaces the processor and provides the functions of the processor plus testing and debugging functions.

inclinometer a sensor that provides a measurement of the angle of inclination in a gravity field.

include file a header file. The name comes from the C/C++ preprocessor directive `#include`.

inclusion-exclusion principle a rule that allows computation of the probability of exactly r occurrences of events $A_1, A_2, \dots, A_n$.

inclusive or disjunction, otherwise known as the OR function: 0 OR 0 = 0, 0 OR 1 = 1, 1 OR 0 = 0, 1 OR 1 = 1. *See also* XOR, AND, NOT.

incompressible string a string whose Kolmogorov complexity equals its length, so that it has no shorter encodings.

inconsistent backup a backup of a database in an inconsistent state. Specifically, a backup containing incomplete and pending transactions.

incorrect summary problem an incorrect aggregate calculation resulting from the aggregated data set being updated during the calculation.

increasing Boolean function *See* positive Boolean function.

increment to add a constant value (usually 1) to a variable or a register. Pointers to memory are usually incremented by the size of the data item pointed to.

incremental backup a backup comprising only changes made since the last backup.

incremental development a software process which proceeds by producing progressively larger increments of the desired product. At each stage, the completed increments form a subset of the final product. Each increment provides additional functionality and brings the currently available subset closer to the desired one.

incremental encoder similar to an absolute encoder, except there are several radial lines drawn around the disc, so that pulses of light are produced on the light detector as the disc rotates. Thus, incremental information is obtained relative to the starting position. A larger number of lines allows higher resolution, with one light detector. A once per rev counter may be added with a second light detector. *See also* encoder.

incremental garbage collection a form of garbage collection in which the garbage collection proceeds incrementally, often as a background task, rather than interrupting the program execution for long periods while a traditional mark-and-scan garbage collection is performed. Considered essential when there is a large virtual memory.

incremental model a process model that results in delivery of versions of an application that provide increasingly greater functionality.

incremental testing a test approach performed in an iterative manner with the growth of the system occurring by following an evolutionary life cycle.

indefinite number *See* not-a-number.

independence a complete absence of any dependence between statistical quantities. In terms of probability density functions (PDFs), a set of random quantities are independent if their joint PDF equals the product of their marginal PDFs:

$$p_{x_1,x_2,\ldots}(\mathbf{x}_1, \mathbf{x}_2, \ldots) = p_{x_1}(\mathbf{x}_1) \cdot p_{x_2}(\mathbf{x}_2) \cdot \ldots .$$

Independence implies uncorrelatedness. *See also* correlation, probability density function.

independent and identically distributed (IID) a term to describe a number of random variables, each of which exhibits identical statistical characteristics, but act completely independent, such that the state or output of one random variable has no influence upon the state or output of any other.

independent event event with the property that it gives no information about the occurrence of the other events.

independent identically distributed process a random process $\mathbf{x}[i]$, where $\mathbf{x}[i]$ and $\mathbf{x}[j]$ are independent for $i \neq j$, and where the probability distribution $p(\mathbf{x}[i])$ for each element of the process is not a function of i. *See also* independence, probability density function, random process.

independent increments process a random process $\mathbf{x}(t)$, where the process increments over non-overlapping periods are independent. That is, for $t_i < s_i, s_i \leq t_{i+1}$, then

$$(\mathbf{x}(s_1) - \mathbf{x}(t_1)), (\mathbf{x}(s_2) - \mathbf{x}(t_2)), \ldots$$

are all independent. *See also* independence, random process.

independent joint control control of a single joint of a robot while all the other joints are fixed.

independent parallelism a state achieved by executing in parallel operations within a query that are not dependent on each other.

independent set a set of vertices in a graph such that for any pair of vertices, there is no edge between them. *See also* maximal independent set.

independent test group (ITG) a group of people whose primary responsibility is software testing. The most legendary testing team was the Black Team.

independent testing a test done by a company other than the developing company.

indeterminate a specification of the semantics of a language which indicates that under some specified set of conditions the meaning of a construct in the program cannot be determined by reading the language specification. This usually applies to execution semantics of the program, and may deal with situations in which the execution semantics are violated in a way the language does not detect, or to situations in which the language specification leaves the behavior unspecified.

index **(1)** an indication of a position of a value in a sequence of values. An index is often represented by an integer value. By implication, for a given value n for an index, value $(n-1)$ precedes value n and value $(n+1)$ follows value n. Synonym, subscript.

(2) a data structure that maps between object designators and the object representation. The index and the objects may be partially or wholly contained on a secondary storage medium such as disk, with only portions residing in the computer's main memory, but this is an implementation detail.

(3) an access structure used for efficient retrieval of records from a file given a search condition.

(4) that part of memory address used to access the locations in the cache, generally the next most significant bits of the address after the tag.

index authorization authorization to create and delete indexes.

indexed addressing an addressing mode in which an index value is added to a base address to determine the location (effective address) of an operand or instruction in memory. Typically, the base address designates the beginning location of a data structure in memory such as a table or array and the index value indicates a particular location in the structure.

indexed color image an image that consists of a set of references to values stored in a color table or palette. The palette, which is often contiguous in an image file, lists all the colors as sets of coordinates in color space. An indexed 16-color image contains a palette with 16 color entries (4 bits), whereas in an indexed 256-color image 256 colors are listed (8 bits).

indexed nested loop join similar to the nested loop join algorithm. This is applicable where an index exists on the join attribute of at least one of the participating relations. In this case, the index is used to retrieve records instead of executing a linear scan of the relations.

indexed sequential a file organization in which a multi-level primary index is maintained on the ordering key field.

index entry an individual entry in an index. An entry usually comprises two fields: a search value and a pointer or reference to the actual record in a file.

indexing changing the geometric relationship between a master controller and a remote manipulator to provide a more comfortable working posture or to change the manipulator work envelope. *See also* indexed addressing.

index lock a lock applied to an index rather than a database object.

index of difficulty (ID) in Fitts' Law, a measure of task difficulty calculated by $ID = \log_2 (A/W)$, where A is movement distance or amplitude and W is target width.

index of fuzziness an index to measure the degree of ambiguity of a fuzzy set by computing the distance between the fuzzy set and its nearest ordinary set.

index page a disk page in an access method or indexing method which does not contain any data records.

index register register used to index a data structure in memory; the starting address of the structure will be stored in a base register. Used in indexed addressing. *See* register.

indicator a measure that can be used to estimate or predict another measure. To infer the estimated or predicted value requires a conceptual model linking the attributes. Indicators may be measured early in the software development cycle in order to predict system attributes that can only be measured later, when the system is operating. Examples: "specification size" may be an indicator of "development effort"; "code complexity" is an indicator of reliability. *See also* indirect measure, flag.

indicator measure a measure (or metric) used to quantify an indicator.

indifferent a point that under iteration acts as neither an attractor nor a repelling point.

indirect addressing an addressing mode in which an index value is added to a base address to determine the location (effective address) of an operand or instruction in memory. Typically, the base address designates the beginning location of a data structure in memory such as a table or array and the index value indicates a particular location in the structure. Also known as deferred addressing.

indirect block a file system data structure composed of an array of pointers to disk blocks used to locate the data blocks associated with a file.

indirect input device a device that the user operates by moving a control that is located away from the screen or other display to be controlled, such as a mouse or trackball.

indirect measure a measure of a feature which involves the measurement of one or more other features/attributes. *See* indirect measure, flag.

indirect measurement a measurement which is made by calculating the value of a function of one or more other measures. The function may be simple (e.g., taking the ratio of two quantities) or complex (e.g., estimation of reliability using a reliability growth model). An example of indirect measurement is the calculation of "productivity" as the ratio of "lines of code" to "total hours worked".

indirect mode a memory addressing scheme in which the operand of the instruction is a memory address containing the effective address of the address.

indiscernibility relation in Pawlak's information system $S = (U, A)$ whose universe U has n members denoted $x, y, z, \ldots$ and the set A consisting of m attributes $\mathbf{a}_j$, the indiscernibility relation on U with respect to the set $B \subseteq A$ defines the partition U/B of U into equivalence classes such that $x, y \in U$ belong to the same class of U/B if for every attribute $\mathbf{a} \in B$,

$$\mathbf{a}(x) = \mathbf{a}(y) .$$

In terms of negotiations, the partition U/B may be interpreted as the partition of the universe U into blocks of negotiators that have the same opinion on all of the issues of B.

individual productivity the productivity of actual software developers. This is a number which presents a lot of uncertainty. In many controlled experiments, a ratio of 25 to 1 between the fastest and slowest has been measured.

induction (1) a technique that infers generalizations from the information in the data.

(2) in learning, the process of acquiring general concepts from specific examples. Very often the specific examples will be historical data of some form, such as production data for some product, the past performance of a share price against various indicators, or a customer's purchasing history and mailshot response.

(3) the proof technique of mathematical induction.

induction variable a variable that every time its value changes, it is incremented or decremented by some constant.

inductive bias any preference not due to the training set which is exhibited by a concept acquisition algorithm for one concept descriptor over another.

industrial benefit in project management, any benefit that will be gained by the industry once a project is completed. These are included in the project objectives and results and are typically quantified in terms of percentages or direct values.

industrial inspection the use of computer vision in manufacturing, for example, to look for defects, measure distances and sizes, and evaluate quality. *See also* computer vision, inspection system.

industrial robot a mechanical device or manipulator, used for production and service, that can be programmed to perform a variety of tasks of manipulation and locomotion under automatic control. Robots may be used as a means of production on assembly lines, performing repetitive tasks or tasks that benefit from the accuracy and dependability of robots.

Robots may take the form of a sequence of rigid body (links) connected by means of articulation joints. A manipulator is characterized by an arm, a wrist, and an end-effector. Motion of the manipulator is imposed by actuators through actuation of the joints. Control and supervision of manipulator motion is performed by a control system. The manipulator is equipped with sensors that measure the status of the manipulator and the environment.

inertia the measure of a body's resistance to changes in velocity.

inertia tensor a matrix providing the distributed mass of a body with respect to a set of generalized coordinates.

inference (1) an inference in a deductive database is based on a rule saying that A should be inferred, if all of $B_1, B_2, \ldots, B_n$ have been inferred already.

(2) the process of drawing a conclusion from given evidence, or to reach a decision by reasoning.

inference engine a reasoning mechanism for manipulating the encoded information and rules in a knowledge base to form inferences and draw conclusions.

infinite light a light source taken to be infinitely far from the model being illuminated so that all of its rays reaching the model can be considered to be parallel.

infinite servers a model for a delay line. In a queueing model, a delay line can be represented by an infinite number of deterministic (D) servers in parallel, so that each request is delayed by the same amount of time.

infinite source model in networking, a technique for simplifying the analytical models for medium access control protocols by ignoring queueing effects at the individual hosts. More formally, we assume a symmetric system in which the number of hosts is so large (and hence the arrival rate per host is so small) that the probability of a second packet arriving to a host that is still attempting to transmit its previous packet is negligible.

infinite tree a tree that can be unified by special unification algorithms which bypass the occur-check. These trees constitute a new domain, different from that of usual PROLOG trees.

infinite type a data type whose value set is not enumerable. Traditionally, integer and floating-point values are considered to be infinite types, although on any particular machine or compiler the value set may actually be fully enumerable and hence finite.

infinity in computer arithmetic, a unique pattern that is used to denote the corresponding real number. It is useful in handling overflow, e.g., when dividing a non-zero number by zero.

infix a mathematical notation in which binary operators appear between their associated operands.

influential factors (1) the set of 14 factors that create a multiplier for complexity to calculate function points.

(2) the set of factors that influence the estimation of indirect metrics.

informal the contrary of formal. A non-structured and formalized way of expressing data, descriptions, etc.

informal reasoning an approach to knowledge-based concept assignment. *Informal reasoning* is based on human oriented concepts, but also takes into account knowledge like natural language comments or grouping. The needed base of knowledge about the problem for the informal reasoning is called the domain model. It is assumed that the domain model (problem, program, and application) knowledge can be usefully represented as patterns of informal and semi-formal information, which are called conceptual abstractions.

informal testing a test performed disregarding the formal test plan and procedure.

information a mathematical model of the amount of surprise contained in a message. For a discrete random source with a finite number of possible symbols or messages, the information associated with the symbol x_k is

$$I_k = -\log_2(p_k) \ ,$$

where p_k is the probability of the symbol x_k. The expected value of the information of the symbols is the first order entropy of the source.

information base the main repository of information about the software. It can be created by decomposing any number of views of a system.

information gain for an attribute in a set of objects to be classified is a measure of the importance of the attribute to the classification. The information gain G_i of the ith attribute A_i of a set of n objects S in the classification is defined as

$$G_i = I(S) - E_i ,$$

where $I(S)$ is the expected information (or entropy) for the classification and E_i is the expected information required for the value of A_i to be known. $I(S)$ is defined as

$$I(S) = -\sum_{c=1}^{N_c} \frac{n_c}{n} \log_2 \frac{n_c}{n} ,$$

where N_c is the total number of classes in the classification, and n_c is the number of objects in the cth class C_c. E_i is defined as

$$E_i = \sum_{k=1}^{N_i} \frac{n_{ik}}{n} I(S_{ik}) ,$$

where S_{ik} is the subset of S in which A_i of all objects takes its kth value, N_i is the number of values A_i can take, n_{ik} is the number of objects in S_{ik}, and the information required in S_{ik} is

$$I(S_{ik}) = -\sum_{c=1}^{N_c} \frac{n_{ikc}}{n_{ik}} \log_2 \frac{n_{ikc}}{n_{ik}} ,$$

where n_{ikc} is the number of objects in S_{ik} belonging to class C_c.

information hiding a programming methodology that conceals the details of the implementation of a program unit, such as a module or a class. A user of the module or class works solely from a formal specification of the behavior of the module.

Modularization produces chunks in code (subroutines, functions, objects). These chunks have interfaces (to the rest of the system). When information hiding is practiced, all information is hidden inside these chunks and is not visible from outside, i.e., it is not visible in the interface and there is no use of any global visibility concept (as typically used in procedural languages such as COBOL, FORTRAN, and C). Information includes data and behavior. Information thus extends the notion of data hiding to objects. *See* data hiding.

information overload situation in which more information is presented to a user than can be processed.

information retrieval refers to the range of techniques used to retrieve relevant documents from a collection of documents. The documents are largely unstructured natural language text. The most relevant documents, according to some criteria, are returned.

information system system that provides some kind of data manipulation producing results in several formats: reports, data, messages, etc. Examples are general ledger systems, sales and lead systems, accounting systems, etc.

information theoretic bounds lower bounds based on the rate at which information can be accumulated.

information theory theory relating the information content of a message to its representation for transmission through electronic media. This subject includes the theory of coding, and also topics such as entropy, modulation, rate distortion theory, redundancy, and sampling.

inheritance a method in object-oriented languages for creating new classes, called subclasses or derived classes. A class from which another class has been derived is said to be a superclass of that class. A subclass will inherit some or all of the methods of its superclass, and may selectively override these methods to modify the subclass's behavior. A subclass may contain additional methods not in its superclass. If the language constrains the subclass to be derived from no more than one class, then this is called single inheritance. If a subclass can be derived directly from two or more superclasses, inheriting data and methods from each, the language is said to allow multiple inheritance.

inherited attribute an attribute of a class that has been inherited from a superclass. *See* attribute grammar, synthesized attribute.

inhibit to prevent an action from taking place; a signal that prevents some action from occurring. For example, the READY signal on a memory bus for a read operation may inhibit further processing until the data item has become available, at which time the READY signal is released and the processor continues.

inhibit sense multiple access (ISMA) protocol multiple access protocol attempting to solve the hidden station problem. In ISMA, the base station transmits an "idle" signal at the beginning of a free slot. Each user has to monitor this signal before it attempts to access the base station. ISMA is sometimes called busy tone multiple access (BTMA).

initialize (1) to set a variable, register, or other storage location to a starting value.

(2) to place a hardware system in a known state, for example, at power up.

initial query tree the query tree prior to the application of heuristic optimization.

initial symbol in a phrase structure grammar, a designated nonterminal symbol from which all sentences in the language are generated.

inline function a function written as a subroutine call, but which is actually compiled directly at the call site. This may be the result of explicit specification by the programmer, or done automatically by the compiler. *See also* intrinsic function.

inner product space over the field F, $\mathcal{H}$ is an inner product space if it is a linear space together with a function $\langle , \rangle : \mathcal{H} \times \mathcal{H} \to F$, such that

1. $\langle x, y\rangle = \overline{\langle y, x\rangle}, \quad \forall x, y \in \mathcal{H}.$
2. $\langle \lambda x+\mu y, z\rangle = \lambda\langle x, z\rangle+\mu\langle y, z\rangle, \quad \forall x, y, z \in \mathcal{H}, \forall\lambda, \mu \in F.$
3. $\langle x, x\rangle \geq 0.$
4. $\langle x, x\rangle = 0$ if and only if $x = 0$.

The field F could be either the real or the complex numbers.

inode in Unix systems, a system table that contains a file's disk layout and other vital information.

IN OUT parameter a parameter that can be implemented by call-by-reference or call-by-value-result.

IN parameter a parameter that is implemented using call-by-value.

input data conveyed to a computer system from the outside world. The data is conveyed through some input device.

input buffer a temporary storage area where input is held. An *input buffer* is necessary when the data transfer rate from an input device is different from the rate at which a computer system can accept data. Having an input buffer frees the system to perform other tasks while waiting for input.

input device a peripheral unit connected to a computer system, and used for transferring data to the system from the outside world. Examples of input devices include keyboard, mouse, and light pen.

input layer a layer of neurons in a network that receives inputs from outside the network. In feedforward networks, the set of weights connected directly to the input neurons is often also referred to as the *input layer.*

input neuron a neuron in a network that receives inputs from outside the network. In feedforward networks, the set of weights connected directly to the input neurons is often also referred to as the input layer.

input/output (I/O) devices or subsystems that provide for the input and output of data to and from the computing subsystem. *See all terms beginning with "I/O".*

input product a type of information. An *input product* is a class of input to a process, and is transformed into an output by the process, or used as a signal or parameter to control the process. It is also an output product of a previous or an external process. An intermediate product which is input to some phase of the development process.

input register a register to buffer transfers to the memory from the I/O bus. The transfer of information from an input device takes three steps:

1. from the device to its interface logic,
2. from the interface logic to the input register via the I/O bus, and
3. from the input register to the memory.

input routine a function responsible for handling input and transferring the input to memory.

input string a sequence of zero or more input symbols which are processed by an automaton.

input symbols a finite set of symbols which form the alphabet that is processed by an automaton. A sequence of *input symbols* is designated as a string.

input vector a vector formed by the input variables to a network.

inquiry a question or explanatory information that users of a software system can request. These are used for calculating function points as well as other questions based on metrics.

insert authorization authorization to insert tuples into a relation.

insertion anomaly phenomenon that occurs when the insertion of data into a relation causes inconsistencies in the data or when a valid insertion requirement cannot be satisfied.

insertion sort the family of sorting algorithms where one item is analyzed at a time and inserted into a data structure holding a representation of a sorted list of previously analyzed items.

inspection activity such as measuring, examining, testing, or gauging one or more characteristics of an entity and comparing the results with specified requirements in order to establish whether conformity is achieved for each characteristic.

inspection system an image processing system, usually automatic or semi-automatic, that performs industrial inspection.

instability the condition where a system does not reach some sort of equilibrium or steady-state behavior over the long term. For example, the system may have a transient state distribution because the arrival rate exceeds the departure rate, say, or the system may be oscillating chaotically among multiple states.

install to introduce a computer program on a disk or in computer memory to have it ready for invocation.

installability attributes of software that bear on the effort needed to install the software in a specified environment. The capability of the software product to be installed in a specified environment. If the software is to be installed by an end user, *installability* can affect the resulting suitability and operability.

installation manual the manual or report in which the process of software or hardware installation is described.

instance (**1**) in software engineering, a single copy of a software product in operation or on test on a single installation.

(**2**) in object-oriented programming, an object produced from a class.

(**3**) in artificial intelligence, a single observation or datum described by its values of the attributes.

instance diagram *See* object diagram.

instance variable a variable of a class that is private to each instance of an object of that class. *Contrast with* class global member, static member.

instantaneous description a string that encodes the current state, head position, and (work) tape contents at one step of a Turing machine computation.

instantiation the creation of a new instance. In object-oriented programming, the process to produce and object from a class. It can be static and dynamic.

instar configuration a term used for a neuron fed by a set of inputs through synaptic weights. An instar neuron fires whenever a specific input pattern is applied. Therefore, an instar performs pattern recognition. *See* instar training. *See also* outstar configuration.

instar training situation in which an instar is trained to respond to a specific input vector and to no other. The weights are updated according to

$$w_i(t+1) = w_i(t) + \mu\,(x_i - w_i(t))\ ,$$

where x_i is the ith input, $w_i(t)$ the weight from input x_i, μ is the training rate starting from about 0.1 and gradually reduced during the training. *See* instar configuration. *See also* outstar training.

Institute of Electrical and Electronics Engineers (IEEE) a professional organization of electrical engineers and computer scientists. The world's largest professional organization. Also one of the major standards-setting bodies in the computer industry. *See also* ANSI, NIST, ISO.

instruction specification of a collection of operations that may be treated as an atomic entity with a guarantee of no dependencies between these operations.

instruction access fault a fault, signaled in the processor, related to abnormal instruction fetches.

instruction cache *See* code cache.

instruction cache unit The instruction cache unit (ICU) is that component of the processing platform's processor responsible for parsing instructions to the fixed-point and floating-point units for execution. The total number of instructions processed by a processing platform within a given CPU cycle is typically defined as the sum of the instruction cache, fixed-point, and floating-point instructions. This sum is usually referred to as the number of executed instructions per clock cycle.

instruction counter the memory register within a computer that maintains the location of the next instruction that will be fetched for execution. This register is also called a program counter.

instruction decoder the part of a processor that takes instructions as input and produces control signals to the processor registers as output. All processors must perform this function; some perform it in several steps, with part of the decoding performed before instruction issue and part of the decoding performed after.

instruction decoder unit in modern CPU implementations, the module that receives an instruction from the instruction fetch unit, identifies the type of instruction from the opcode, assembles the complete instruction with its operands, and sends the instruction to the appropriate execution unit, or to an instruction pool to await execution.

instruction fetch unit in modern CPU implementations, the module that fetches instructions from memory, usually in conjunction with a bus interface unit, and prepares them for subsequent decoding and execution by one or more execution units.

instruction field a portion of an instruction word that contains a specified value, such as a register address, a 16-bit immediate value, or an operand code.

instruction format the specification of the number and size of all possible instruction fields in an instruction-set architecture.

instruction issue the sending of an instruction to functional units for execution.

instruction-level parallelism the concept of executing two or more instructions in parallel (generally instructions taken from a sequential, not parallel, stream of instructions).

instruction pipeline a mechanism of a computer by which more than one instruction is active in the processor. Instructions may concurrently be in various stages of decode and execution, with the operands in various stages of being fetched or stored. An instruction pipeline often has the property that "short" control transfers, those within the limits of the pipeline, do not require the instructions be re-fetched and re-decoded. The characteristics of an instruction pipeline can be very important to a compiler writer.

instruction pointer another name for program counter, the processor register holding the address of the next instruction to be executed. *See* program counter.

instruction pool in modern CPU implementations, a holding area in which instructions that have been fetched by an instruction fetch unit await access to an execution unit.

instruction prefetch the technique of improving total execution time for a sequence of code by prefetching instructions from memory so that they are available to the processor (whether in cache or in an instruction pipeline) before they are needed. *See* memory caching.

instruction prefix a field within a program instruction word used for some special purpose. Found only rarely. The Intel X86 architecture occasionally uses *instruction prefixes* to override certain CPU addressing conventions.

instruction reordering a technique in which the CPU executes instructions in an order different from that specified by the program, with the purpose of increasing the overall execution speed of the CPU.

instruction repertoire *See* instruction set.

instruction set the collection of all the machine-language instructions available to the programmer. Also known as instruction repertoire.

instruction set processor (ISP) a formal notation used to describe machine instruction set architectures. ISP definitions have been used to describe machines to compiler automation systems.

instructions per clock cycle (IPC) the total number of floating-point, fixed-point, and instruction cache unit instructions which are processed by a given processing platform within a given clock cycle.

instructions per executed line of code (IPELOC) the total number of machine instructions required to execute a high-level language line of code. For example, the number of instructions required to execute the C-code segment X = A + B are four: Load A; Load B; Add A + B; Store sum in X.

instruction window for an out-of-order issue mechanism, a buffer holding a group of instructions being considered for issue to execution units. Instructions are issued from the *instruction window* when dependencies have been resolved.

integer a fixed-point, whole number value that is usually represented by the word size of a given machine.

integer linear program a linear program augmented with additional constraints specifying that the variables must take on integer values. Solving such problems is NP-hard.

integer multi-commodity flow a multi-commodity flow in which the flow through each edge of each commodity is an integral value. The term is also used to capture the multi-commodity flow problem in which each demand is routed along a single path.

integer polyhedron a polyhedron, all of whose extreme points are integer valued.

integer unit in modern CPU implementations, a type of execution unit designed specifically for the execution of integer-type instructions.

integrated application an application that agrees on data formats with other applications so they can use each other's outputs.

integrated circuit (IC) an assembly of miniature electronic components simultaneously produced in batch processing, on or within a single substrate, which performs an electronic circuit function.

integrated development environment (IDE) a style of programming environment in which such features as the text editor, compiler, linker, debugger, and so on form a seamless (or nearly seamless) operating environment.

integrated device electronics (IDE) an implementation and protocol for communication between a disk controller and a disk.

integrated display a display integrating two or more variables into a single representation as, for example, when a cursor leaves a trail as it is moved to indicate the direction (by trail vector) and velocity (by trail length) of mouse movement.

Integrated Services Digital Network (ISDN) a network that provided end-to-end digital connectivity to support a wide range of services, including voice and nonvoice services, to which users have access by a limited set of standard multipurpose user-network interfaces.

integration test a type of test in which the interactions among components (hardware and software) are tested.

integrity (1) a belief in the truth of the information represented by a set of data.

(2) the degree to which a component or a system reacts and provides measures against the unauthorized access for performing changes and manipulations to data and code.

(3) a condition stating that the information in a set of data does satisfy a set of logical constraints (the integrity constraints).

integrity constraint a constraint imposed on the database by the database management system that helps maintain the integrity of the data.

intelligence (1) ability to deal with abstract concepts and form complex pictures of the outside world such as creativity, ability with spoken and written language, etc.

(2) in the military sense, the aggregated and processed information about the environment, including potential adversaries, available to commanders and their staff.

intelligent agent *See* autonomous agent.

intelligent control a sensory-interactive control structure incorporating cognitive characteristics that can include artificial intelligence techniques and contain knowledge-based constructs to emulate learning behavior with an overall capacity for performance and/or parameter adaptation. Intelligent control techniques include adaptation, learning, fuzzy logic, neural networks, and genetic algorithms, as well as their various combinations.

intention lock a lock acquired when the intention is to perform an operation.

interaction diagram a depiction of interactions. An interaction is a dynamic event in which a message is exchanged between two objects in a collaborative embrace. Thus, for one collaboration we may have very many interactions.

interactive pertaining to a mode of operation of a computer system in which a sequence of alternating entries and responses takes place in the form of a dialog.

interactive graphics data from sensors is integrated with a graphics model to provide a closer fit between synthetic and real environments.

interactive proof system a protocol in which one or more provers try to convince another party called the verifier that the provers possess certain true knowledge, such as the membership of a string x in a given language, often with the goal of revealing no further details about this knowledge. The provers and verifier are formally defined as probabilistic Turing machines with special "interation tapes" for exchanging messages.

interactive steering the practice of dynamically modifying the parameters of a running simulation, guided by a real-time visualization of the simulation's progress.

interactive system a system in which the user interacts with the system through a series of query-response transactions in real time.

interactive tools tools for the generation of programs by a human with the assistance of an instruction database.

interactivity the degree to which a computer simulation allows people to change or affect the simulated world and objects within it.

interarrival time the time between two successive arrivals of jobs or requests at a system or device.

interconnection network the combination of switches and wires that allow multiple processors in a multiprocessor or multicomputer to communicate.

interdeparture time the time between two consecutive departure events.

interface (**1**) the boundary between a system and its environment, across which interaction occurs by the passing of information.

(**2**) the externally visible features or characteristics (of an object, use case, subroutine, etc.). This term is used in the languages supporting the distinction between interfaces and classes such as C++.

interface definition language (IDL) a way to accomplish interoperability across languages and tools. An IDL is an object model that specifies standards for defining application interfaces in terms of an independent language — an *interface definition language.* Interface definitions are typically stored in a repository which clients can query at run-time.

interface device a device that is used to interconnect multiple system components for resolving any differences in data or control signals that are exchanged by these components.

interface inheritance the exploitation of the inheritance mechanism between interfaces. An interface can be derived/specialized from another interface. *See* specification inheritance.

inter-feature distance the distance between a pair of feature points in an image, often used to help identify the object, or (in conjunction with other measures of this type) to infer the presence of the object.

interframe referring to the use of more than a single image frame (e.g., an algorithm which differentiates successive image frames or detects frame-to-frame motion). *See also* intraframe.

interframe coding the process of coding the difference of consecutive video frames based on motion estimation instead of the frames themselves.

interframe coding-3D in situations where motion picture film or television images need to be coded, three-dimensional transform coding is used. In this approach the frames are coded L at a time and partitioned into $L \times L \times L$ blocks per pixel prior to transformation.

interframe hybrid transform coding $L \times L$ blocks of pixels are first 2D-transformed. The resulting coefficients are then DPCM (interframe) coded and transmitted to the receiver. This is then reversed at the receiver for reconstruction. In the interframe hybrid transform coding, the DC components are determined exclusively by the amount of spatial detail in the picture.

interframe prediction this technique uses a combination of pixels from the present field as well as previous fields for prediction. For scenes with low detail and small motion, field difference prediction appears to perform better than frame difference prediction.

interior-based representation a representation of a region that is based on its interior (i.e., the cells that comprise it).

interlaced *See* interlaced scanning.

interlaced display a technique for displaying images at a higher resolution than the monitor. Two images consisting of every second row of pixels are alternately displayed during every screen refresh (e.g., every fiftieth of a second). There is hence a flickering artifact.

interlaced scanning a system of video scanning whereby the odd- and even-numbered lines of a picture are transmitted consecutively as two separate interleaved fields.

interlacing a technique used in video monitors where alternating rows of the screen are updated to speed refresh. Its opposite is noninterlaced.

interleave (**1**) a technique used to prevent data loss during transfer of data to/from a disk. Data is written to nonconsecutive sectors in order to allow the data transfer rate to keep up with the rotation speed of the disk.

(**2**) to arrange parts of one sequence of events so that they can alternate with parts of one or more other sequences.

interleaved data ordering data ordering in which different components are combined into so-called Minimum Coded Units (MCUs).

interleaved memory a memory system consisting of several memory modules (or banks) with an assignment of addresses that makes consecutive accesses fall into different modules. This increases the effective memory bandwidth, as several memory requests can be satisfied (concurrently by several modules) in the same time as it would take to satisfy one memory request. For example, a simple arrangement (sequential interleaving), which favors sequential access, is to assign address k to memory bank k mod N, where N is the number of modules employed: the reference stride is 1, so consecutive requests are to different banks.

There are other schemes for assigning addresses to banks, all of which aim to (for non-unit strides) give the same performance as with a stride of 1. These include: skewed addressing, which is similar to sequential interleaving but with a fixed displacement (the skew) added to the chosen module number; dynamic skewed addressing, in which the skew is variable and determined at run-time; pseudo-random skewing, in which the module number is chosen on a pseudo-random basis; prime-number interleaving, in which the number of modules employed is prime relative to the degree of interleaving; and superinterleaving, in which the number of modules exceeds the degree of interleaving.

interleaved storage the notion of breaking storage into multiple pieces that may be accessed simultaneously by the same request, increasing the bandwidth from the storage.

interleaving (**1**) regular permutation of symbols usually used to split bursts of errors caused by channels with memory (e.g., fading channels) making a coding scheme designed for memoryless channels more effective. The two main categories are *block interleaving* and *convolutional interleaving*. The inverse process is called deinterleaving.

(**2**) a characteristic of a memory subsystem such that successive memory addresses refer to separately controlled sections or banks of memory. Sequential accesses to interleaved memory will activate the banks in turn, thus allowing higher data transfer rates.

intermediate-level vision the set of visual processes related to the perception/description of surfaces and of their relationships in a scene.

intermediate product an item that is produced during some phase of the software development process, and is an input product to a later phase, but is not provided to the user. Examples of intermediate products are "requirements specification", "design documentation," "test report", etc. Some products may be either intermediate products or end products, depending on the circumstances, e.g., source code may be sold to the customer in some cases, but not in others.

intermediate product quality the quality of an intermediate software product that can be evaluated on the basis of internal quality attributes.

intermediate representation a representation of a program used within a compiler. A compiler may have many intermediate representations, such as a parse tree or tuples, a flow graph, sequences of abstract machine instructions, sequences of actual machine instructions, and the like, as appropriate for the stage in processing.

intermittent fault a fault of an item that persists for a limited time duration following which the item recovers the ability to perform a required function without being subjected to any action of corrective maintenance. Such a fault is often recurrent.

internal attribute an attribute describing a system considered in isolation from its environment. An internal attribute can be measured by observing the properties of a system, without regard to how it performs in its environment. Examples of internal attributes are "code size", "complexity", etc.

internal blocking a phenomenon in certain types of switching systems where data cannot pass through a switching network from its input port to a free output port because the switch lacks sufficient internal paths to support arbitrary permutations of its inputs.

internal bus a bus used to connect internal components of a computer system, such as processors and memory devices.

internal command a Unix command that is an internal part of the shell. *Compare with* command program.

internal containment a class is subject to *internal containment* each time the data members are referenced by an is-part-of relationship and, therefore, are entirely contained in the memory space of the object. *Compare with* external containment.

internal forwarding a mechanism in a pipeline which allows results from one pipeline stage to be sent directly back to one or more waiting pipeline stages. The technique can reduce stalls in the pipeline.

internal fragmentation in paging, the effect of unused space at the end of a page. On average the last page of a program is likely to be 50% full. Internal fragmentation increases if the page size is increased.

internal hashing a form of hashing, where the value returned from the hash function is a memory address or an array location.

internal memory main memory of a computer.

internal node a node in a tree which has the property that it has one or more arcs leading to other nodes, and one arc leading into it. *Compare with* root node, leaf node.

internal schema a description of the physical storage of the database.

internal software software created for inhouse usage only. As opposed to external software, meant for usage outside a development organization.

internal sorting the situation when the file to be sorted is small enough to fit in main memory.

internal specification specification of a subpart of the system, the subpart interacts with other parts or with the environment using an external interface. *See* external specification.

International Graphics Exchange Specification (IGES) a data exchange standard for graphics.

internationalization the problem of creating systems that are not ethnocentric to a particular culture. Typical problems in internationalization include representations of dates and times, collating sequence, and the mapping between the numeric representation of character values and the printed graphics (for example, currency symbols whether for Pounds Sterling or Dollars may have the same character representation). In language design, the notion of keywords and reserved words can present problems in internationalization (although there seems to be general acceptance of English keywords). Icons or colors that in one culture have accepted benign meanings (pause, caution), may in another culture have entirely different meanings (obscene gesture, death). *See also* unicode.

International Standards Organization (ISO) (**1**) as a standards committee, establishes the standards in many areas including programming languages and character sets.

(**2**) sometimes used to refer to the character sets defined by the ISO under ISO standard 8859.

internet generally speaking, any set of networks interconnected with routers. The Internet is the biggest example of an *internet.* The Internet is the largest internet in the world. It is a three-level hierarchy composed of backbone networks (e.g., ARPAnet, NSFNet, MILNET), mid-level networks, and stub networks. These include commercial (.com or .co), university (.ac or .edu) and other research networks (.org, .net), and military (.mil) networks and span many different physical networks around the world with various protocols including the Internet Protocol.

Internet service provider (ISP) any entity, such as a company, university, or organization, that provides Internet access to individual users, businesses, and other customers or constituents. America Online (AOL) is the world's largest ISP.

interoperability the capability of the software product to interact with one or more specified systems. *Interoperability* is used in place of compatibility in order to avoid possible ambiguity with replaceability.

interpenetration the surface of one object passing through the surface of another.

interpixel redundancy the tendency of pixels that are near each other in time or space redundancy to have highly correlated gray levels or color values. Reducing *interpixel redundancy* is one way of compressing images. *See also* image compression.

interpolation the approximation of a real function that consists of approximating the value of the function between two points where the value is known. Polynomials are commonly used for such approximations.

interpolative coding coding schemes that involve interpolation.

interpolative vector quantization (IRVQ) technique in which a subsample of the interpolated original image is used to form the predicted image and then a residual image is formed based upon this prediction. This approach reduces blocking artifacts by using a smooth prediction image.

interpret to translate and execute each statement or construct of a computer program before translating and executing the next.

interpretation in artificial intelligence, a function from the atom set to the set of truth values.

interpretation tree a model-based recognition technique that uses a pruned exponential tree search to find corresponding sets of model and image features.

interpreter a program that sequences through a program representation and performs the computations it specifies. Some interpreters work directly on the source representation, others work on compiled representations such as a byte code. While strictly speaking a computer acts as in interpreter for its instruction set, this term is generally reserved to indicate that the program being executed requires at least one level of isolation between its representation and the underlying machine performing the execution.

interprocess communication (IPC) (**1**) the class of mechanisms used for concurrent processes to synchronize and communicate.

(**2**) the transfer of information between two cooperating programs. Communication may take the form of signal (the arrival of an event) or the transfer of data.

interpupilary distance separation between the pupils when the line of gaze is horizontal and straightforward; approximately 63 mm (2.5 inches) on average for human adults.

interquery parallelism parallelism achieved by executing numerous queries in parallel.

inter-record gap the space between two records of data stored on a magnetic medium such as a tape. This space helps prevent interference between the two records. It can also contain markers marking the beginning and end of the records.

interreflection a phenomenon that occurs when a surface reflects light from other surfaces in its environment. The effects range from more or less sharp specular reflections that change with the viewer's position to diffuse reflections that are insensitive to viewer's position. Also called mutual illumination.

interrupt (**1**) a means of altering control flow in an asynchronous fashion, transferring control from an arbitrary context to a function, the interrupt handler, which performs an appropriate computation and returns to the main execution flow. Many languages support interrupts as basic control mechanisms.

(**2**) a suspension of an execution of a computer program, caused by an event external to that program, and performed in such a way that the execution can be resumed.

interrupt controller a device that provides additional interrupt handling capability to a CPU.

interrupt descriptor table the *interrupt descriptor table* is used in protected mode to store the gate descriptors for interrupts and exceptions.

interrupt disable (**1**) a state in which interrupts to the CPU are held, but not processed. In most systems it is possible to disable interrupts selectively, so that certain types of interrupts are processed, while others are disabled and held for later processing.

(**2**) the operation that changes the interrupt state of the CPU from enabled to disabled.

interrupt-driven I/O an input/output (I/O) scheme where the processor instructs an I/O device to handle I/O and proceeds to execute other tasks; when the I/O is complete, the I/O device will interrupt the processor to inform completion. An interrupt-driven I/O is more efficient than a programmed I/O where the processor busy-waits until the I/O is complete.

interrupt enable (**1**) a state in which interrupts to the CPU can be processed.

(**2**) the operation that changes the interrupt state of the CPU from disabled to enabled.

interrupt handler a piece of code designated to handle an interrupt, which is usually an asynchronous request to process an event generated outside of the control of the program (for example, the completion of an I/O operation). Although an interrupt handler often has many of the characteristics of a function, the exact linkage to the interrupt handler is usually determined by the underlying hardware and can be substantially different from the linkage used for ordinary function calls. Many languages support the necessary linkages for writing an interrupt handler within the language.

interrupt latency the delay between a computer system's receipt of an interrupt request and the beginning of its handling of this request.

interrupt line a wire carrying a signal to notify the processor of an external event that requires attention.

interrupt mask a bit or a set of bits that enable or disable the interrupt line to be transmitted at the interrupt detector circuit (inside or outside the CPU). The mechanisms of masking are typically implemented into interrupt controllers.

interrupt priority a value or a special setting that specifies the precedence to serve the corresponding interrupt with respect to other interrupt signals.

interrupt register contains a bit map of all pending (latched) interrupts.

interrupt request a signal or other input requesting that the currently executing program be suspended to permit execution of a more urgent code.

interrupt service the steps that make up the operation by which the CPU processes an interrupt. Briefly, the CPU suspends execution of its current program and branches to a special program known as an interrupt handler or interrupt routine to take appropriate action. Upon completion of the interrupt service, the CPU will take one of a number of actions, depending on circumstances: it can

1. return to the previous task if conditions permit,
2. process other pending interrupts, or
3. request further action from the operating system.

interrupt service routine a routine that is executed in response to an interrupt request.

interrupt signal a hardware signal generally used to initiate an asynchronous process or to handle an anomalous program condition.

interrupt vector in an interrupt controller, a register that contains the identity of the highest-priority interrupt request.

intersect (**1**) in operations on mathematical sets, an operation that produces the set intersection, a set of values that represent elements that are in both sets.

(**2**) in database systems, an operation that computes the relational intersect of two relational tables.

intersection given the sets A and B, A ∩ B the set comprising all elements that exist in both A and B.

intersection of fuzzy sets a logic operation, corresponding to the logical AND operation, forming the conjunction of two sets. In a crisp (non-fuzzy) system, the intersection of two sets contains the elements that are common to both. In fuzzy logic, the intersection of two sets is the set with a membership function which is the smaller of the two.

Denoted ∩, fuzzy intersection is more rigorously defined as follows. Let A and B be two fuzzy sets in the universe of discourse X with membership functions $\mu_A(x)$ and $\mu_B(x)$, $x \in X$. The membership function of the intersection $A \cap B$, for all $x \in X$, is

$$\mu_{A \cap B}(x) = \min\{\mu_A(x), \mu_B(x)\} \ .$$

See also fuzzy set, membership function.

intersymbol interference (ISI) the distortion caused by the overlap (in time) of adjacent symbols.

interval logic an extension of a logic (propositional or first order) with specific operators to express relationships between time intervals (e.g., meet, before, after, ...) and/or operators for combining intervals (e.g., chop) and operators to specify interval boundaries on the basis of the truth of predicates. Generally are useful to express relationships among events but they cannot express temporal constraints as needed by real-time systems. *See* temporal interval logic.

intractable a problem for which there is no closed form solution. A queueing model is intractable if there does not exist an exact analytical solution. A problem for which the execution times for all known solution algorithms grow faster than any polynomial function as the problem size increases, so that only small examples can be solved in a reasonable period of time.

intraframe image or video processing method that operates within a single frame without reference to other frames. *See also* interframe.

intraframe coding image coding schemes that are based on an intraframe restriction; that is, to separately code each image in a sequence.

intranet a network that provides similar services within an organization to those provided by the Internet outside it but which is not necessarily connected to the Internet. The common example is a company that sets up one or more World Wide Web servers on an internal network for distribution of information within the company.

intraoperation parallelism the execution of one relational operation as many suboperations.

intraquery parallelism parallelism achieved by parallel execution of a single query.

intrinsic function a function that is recognized by the compiler. In some languages this means only that the function need not be explicitly declared, and in some languages or compilers it may also imply that the function may be compiled as an inline function.

intrinsics the layer of a toolkit on which different widgets are implemented.

introspective sort a variant of quicksort which switches to heap sort for pathological inputs, that is, when execution time is becoming quadratic.

invalid defect a defect that looks like a defect in some software product but in reality it is not a defect. Either it is not a defect, or it is a defect in some other software product. It is common to receive about 15% of invalid defects for software vendors. Possible causes are user errors, hardware errors, and errors in other products.

invalid entry an entry in a table, register, or module that contains data that is not consistent with the global state. Invalidity is used in cache coherence mechanisms.

invariant (**1**) an assertion that is true at all times. In object-oriented programming, an invariant acts as a post-condition for all class methods.

(**2**) an unchanging relationship between actions and information within a world, e.g., unsupported objects accelerate towards the surface of the world.

inverse for a real function f, if its inverse is f^{-1} then $f^{-1}(f(x)) = x$.

inverse Jacobian the mathematical inverse of the Jacobian matrix. The transformation from a Cartesian velocity or to a joint velocity vector or joint torque to tip force.

inverse kinematics identification of possible robot joint configurations given desired Cartesian task coordinates.

inverse translation buffer a device to translate a real address into its virtual addresses to handle synonyms. Also called a reverse translation buffer and an inverse mapper.

inverse wavelet transform a computation procedure that calculates the function using the coefficients of the wavelet series expansion of that function.

inversion a mutation operator in which a random sub-string is chosen from a chromosome and then re-inserted into the same location in reverse order.

inverted index an index into a set of texts of the words in the texts. The index is sorted alphabetically, and each index entry gives the word and a list of texts, and possibly locations within the text, where the word occurs.

inverted page table the number of entries in a conventional page table in a virtual memory system is the number of pages in the virtual address space. In order to reduce the size of the page table, some systems use an inverted page table that has only as many entries as there are pages in the physical memory, and a hashing function to map virtual to physical addresses in nearly constant time.

I/O *See* input/output.

I/O bandwidth the data transfer rate into or out of a computer system. Measured in bits or bytes per second. The rate depends on the medium used to transfer the data as well as the architecture of the system. In some instances the bandwidth average rate is given and in others the maximum rate is given. *See* I/O throughput.

I/O bound pertaining to a process that spends most of its time performing input/output.

I/O buffer a temporary storage area where input and output are held. Having I/O buffers frees a processor to perform other tasks while the I/O is being done. Data transfer rates of the processor and an I/O device are, in general, different. The *I/O buffer* makes this difference transparent to both ends.

I/O bus a data path connecting input or output devices to the computer.

I/O card a device used to connect a peripheral device to the main computer; sometimes called an I/O controller, I/O interface, or peripheral controller.

I/O channel input/output subsystem capable of executing a program, relieving the CPU of input/output related tasks. The channel has the ability to execute read and write instructions as well as simple arithmetic and sequencing instructions that give it complete control over input/output operations (IBM terminology). *See also* direct memory access, selector channel, multiplexor channel, subchannel I/O.

I/O command a command that controls the transfer of data between peripheral devices and the computer systems. *I/O commands* are typically used to initiate I/O operations, sense completion of commands, and transfer data. These commands typically contain the ID of the device being addressed, the operation to be performed on the device, and parameters for the operation. Not all computers have I/O commands. *See* memory mapped I/O.

I/O controller (**1**) the logic that controls the input and output activities on the I/O bus.

(**2**) the software subroutine used to communicate with an I/O device. *See* I/O routine.

I/O data stream a program's input or output channel; conceptualized and handled as a file.

I/O device a physical mechanism attached to a computer that can be used to store output from the computer, provide input to the computer, or do both.

I/O interface the hardware required to connect input or output devices to a computer system. This hardware, together with the software drivers, is responsible for accommodating the operational differences, such as speed and data type or format, between the devices and the computer system.

I/O interrupt a signal sent from an input or output device to a processor, which states the status of the I/O device has changed. An interrupt usually causes the computer to transfer program control to the software subroutine that is responsible for controlling the device. *See* interrupt-driven I/O.

I/O port (**1**) a form of register designed specifically for data input-output purposes in a bus-oriented system.

(**2**) the place from which input and output occurs in a computer. Examples include printer port, serial port, and SCSI port.

I/O processor a processor dedicated to performing I/O operations exclusively, thus alleviating the burden on the CPU. It usually has a separate local bus for I/O operations data traffic, thus permitting the CPU to access memory on the main system bus without interruption. For example, the Intel 8086 CPU has an 8089 I/O processor associated with it. Both can operate simultaneously, in parallel.

I/O redirection a shell mechanism for redefining a process's I/O data streams, in particular standard input, output, and error; occurs on a process-by-process basis.

I/O register a special storage location used specifically for communicating with input/output devices.

I/O routine a function responsible for handling I/O and transferring data between the memory and an I/O device. *See* I/O controller.

I/O system the entire set of input/output constructs, including the I/O devices, device drivers, and the I/O bus.

I/O throughput the rate of data transfer between a computer system and I/O devices. Mainly determined by the speed of the I/O bus or channel. Throughput is typically measured in bits/second or bytes/second. In some instances, the throughput average rate is given and in others the maximum rate is given. *See* I/O bandwidth.

I/O trunk *See* I/O bus.

I/O unit the equipment and controls necessary for a computer to interact with a human operator or to access mass storage devices or to communicate with other computer systems over communication facilities.

IPC *See* instructions per clock cycle, interprocess communication.

IPELOC *See* instructions per executed line of code.

irradiance a measure of the amount of light energy incident on a unit area of surface per unit time. Measured in Watts per square meter.

IRVQ *See* interpolative vector quantization.

is-a a relationship in object-modeling. A is-a B means that an instance called A is an instance (object) created from class/type B. In some cases, it is also used as a shorter version of is-a-kind-of relationship.

is-a-kind-of a relationship used in object modeling. It often represents a combination of a taxonomic viewpoint and a subtyping viewpoint.

ISDN *See* Integrated Services Digital Network.

ISI *See* intersymbol interference.

ISMA *See* inhibit sense multiple access protocol.

ISO *See* International Standards Organization.

isodata a special elaboration of K-Means used for clustering.

isolated I/O an I/O system that is electrically isolated from where it is linked. *Isolated I/O* is often found in embedded systems used for control.

isolation the extent to which a transaction can access the updates made by other concurrent transactions.

isometric projection this is a form of orthographic projection in which the direction of projection and surface normal of the image plane are parallel to one of these eight directions { (1,1,1), (1,1,-1), (1,-1,1), (1,-1,-1), (-1,1,1), (-1,1,-1), (-1,-1,1), (-1,-1,-1) }.

isomorphic a relationship between two graphs. Two graphs are isomorphic if there is a one-to-one correspondence between their vertices and there is an edge between two vertices of one graph if and only if there is an edge between the two corresponding vertices in the other graph. *See also* homomorphic.

isomorphism a one-to-one correspondence between two sets of data.

ISO/OSI *See* open systems interconnection model.

isosurface a technique used in three-dimensional data visualization where a surface is drawn around points in three-dimensional space that represent the same data value. For example, the set of points $\{ (x, y, z) : f(x, y, z) = c \}$ where c is a given constant. *See* implicit surface.

ISP *See* instruction set processor, Internet service provider.

is-part-of in object-modeling, another name for a whole-part relationship.

is-referred-by in object-modeling, the relationship between two classes in which a class contains an attribute that is a reference or a pointer to an object of another class.

issue the act of initiating the performance of an instruction (not its fetch). Issue policies are important designs in systems that use parallelism and execution out of program order to achieve more speed.

ITB *See* inverse translation buffer.

item any part, component, device, subsystem, functional unit, equipment, or system that can be individually considered. Synonymous with entity.

iterated function system a finite collection of affine mappings in the plane which are combinations of translations, scalings, and rotations. Each mapping has a defined probability and should be contractive, that is, scalings are less than 1. *Iterated function systems* can be used for the generation of fractal objects and image compression.

iteration repetition. Most languages support an iteration construct, typically a counted iteration such as for, as well as conditional iteration such as while, do, or until.

iterative coordination coordination process concerned with such exchange of information between the coordinator and the local decision (control) units, which — after a number of iterations — leads to the local solutions being satisfactory from the overall (coordinator) point of view; iterative coordination is used in multilevel optimization, in particular in the direct method and the price method.

iterative decoding decoding technique that uses past estimates to provide additional information to improve future estimates. Used in the decoding of Turbo–codes.

iterative selection a thresholding algorithm based on a simple iteration. First select a simple threshold (say, 128) and segment the image, computing the mean of the black and white pixels along the way. Next, calculate a new threshold using the mid-point between the black and white pixels as just computed. Next, repeat the entire process, using the new threshold to perform the segmentation. The iteration stops when the same threshold is used on two consecutive passes.

iterator an iteration construct in a language. *See also* enumerator. Note that all enumerators are iterators, but not all iterators are enumerators (for example, do-while is an iterator but not an enumerator).

iterator class used to traverse and retrieve information from a set of objects managed by a container class.

ITG *See* independent test group.

J

Jackson network a class of queueing networks studied by J. R. Jackson in the 1950s that consists of N nodes, where the ith node consists of m_i exponential servers each with parameter μ_i. Arrivals come to the ith node from outside the system in the form of a Poisson process, and departures from the ith node go to the jth node with routing probability p_{ij}.

Jackson's theorem the result for Jackson networks that states that the joint distribution for all nodes factors into the product of each of the marginal distributions. That is, the solution for Markovian queues in equilibrium has a product form.

Jacobian a matrix that transforms joint velocity to tip velocity and tip force to joint torque.

Jacobian of the manipulator a matrix that maps the joint velocities into end-effector velocities.

jaggy *See* adaptive sampling.

jamming a rounding method that consists of first chopping and then forcing a 1 in the least significant bit retained. Also known as force-1 rounding or von Neumann rounding.

Java an object-oriented language based approximately on the syntax and semantics of the C language, designed to be portable across many operating platforms by using a byte code to represent the object program, said byte code being executed by an interpreter.

JavaScript a subset of the Java language which is limited in what it can do, so that its access to the operating environment is limited. *See* sandbox.

Java Virtual Machine (JVM) a virtual machine that executes the Java byte code.

JCL *See* job control language.

jerk time rate of change of acceleration.

JIT *See* just-in-time.

jitter the variability in the end-to-end packet delay in a network of queues. Jitter arises due to

1. variable length of packets
2. random transmission and queueing delays in packet switched networks.

Jitter is an important performance measure, e.g., in systems with packetized voice due to the real-time nature of voice traffic.

jittering a technique that adds noise to a rendered image; the advantage of jittering is that the human eye tolerates noise more easily than it tolerates aliasing artifacts. Consequently, humans perceive the jittered image as being of a higher quality.

Jittering is performed by displacing sample locations that are initially spaced regularly. Typically, this involves randomly shifting uniformly positioned sample points horizontally and vertically. Such a sample point is usually in the center of the pixel which is perturbed to some other location within it.

JK flip-flop device that uses two inputs (J and K) to control the state of its Q and Q' outputs. A negative edge sensitive clock input samples J and K and changes the outputs accordingly. The simplified truth table for the JK flip-flop is as follows:

J	K	Q	Q'
0	0	0	0
0	0	1	1
0	1	0	0
0	1	1	0
1	0	0	1
1	0	1	1
1	1	0	1
1	1	1	0

JND *See* just noticeable difference.

job (1) a unit of work to be accomplished by an operating system.

0-8493-2691-5/01/$0.00+$.50

(**2**) an executable entity such as a program that can give rise to multiple CPU, memory, and I/O operations.

job control language (JCL) a language used to identify a sequence of jobs, describe their requirements to an operating system, and control their execution.

job stream a sequence of programs or jobs set up so that a computer can proceed from one to the next without operator intervention.

jog in a rectilinear routing model, a vertical segment in a path that is generally running horizontally, or vice versa.

join (**1**) a control flow point where two separate execution paths that are executing in parallel (or conceptually executing in parallel) must synchronize. *See* fork, rendezvous.

(**2**) in languages operating on databases, an operator used to combine related tuples from two relations into single operations. Only operations that satisfy the join condition appear in the result of the operation. It is a composite operator comprising a Cartesian product, a select, and a project. In SQL, a JOIN TABLE allows users to specify a table resulting from a join operation in the FROM clause of a query. *See also* union, intersect.

join attributes attributes with common domains on which two relations may be joined.

join dependency a relation, denoted $*(R_1, R_2, \ldots R_N)$, specified on relation schema R where every instance r of R has a lossless decomposition into $R_1, R_2 \ldots R_N$.

joint (**1**) connecting point between links of a limb. A lower-pair joint connects two links and it has two surfaces sliding over one another while remaining in contact. Six different lower-pair joints are possible: revolute (rotary), prismatic (sliding), cylindrical, spherical, screw, and planar. Of these, only rotary and prismatic joints are common in manipulators. Each has one degree of freedom.

(**2**) site at which the force supplied by actuators is exerted to move a link.

(**3**) the origin of link motion.

joint angle the absolute or relative displacement of a manipulator's joint.

joint-by-joint control control scheme requiring independent activation of each joint of a robotic manipulator.

joint controller a controller designed to track a desired joint position, velocity, or torque.

joint motor a motor located at the joint of a manipulator that provides a desired velocity, position, or torque.

Joint Photographic Experts Group (JPEG) a lossy compression technique that is an industry standard for image information storage and retrieval.

joint sensor a sensor that provides a signal proportional to the position, velocity, or torque at a joint.

joint space the mathematical space related to the joints of a manipulator. The number of dimensions of this space is the same as the degrees of freedom.

A position and orientation of a manipulator of n degrees of freedom is specified as a result of the forward kinematics problem. This set of variables is called the $n \times 1$ vector. The space of all such joint vectors is referred to as *joint space.*

joint space control depicted in the figure. The joint space control scheme consists of the trajectory conversion which recalculates the desired trajectory from the operational space $(X_d, \dot{X}_d, \ddot{X}_d)$ into the joint space $(q_d, \dot{q}_d, \ddot{q}_d)$. Then, a joint space controller is designed that allows tracking of the reference inputs. Notice that only joint inputs are compared with joint positions, velocities, and accelerations. The joint space control does not influence the operational space variables which are controlled in an open-loop fashion through the manipulator mechanical structure. *Joint space control* is usually a single-input/single-output system; therefore a manipulator with n degrees of freedom has n such independent systems. A single-input/single-output system has PI (proportional-integral), PD (proportional-derivative), or PID (proportional-integral-derivative) structure. These schemes do not include dynamics of the manipulator. Dynamics of the manipulator can be included in, for example, computed torque control.

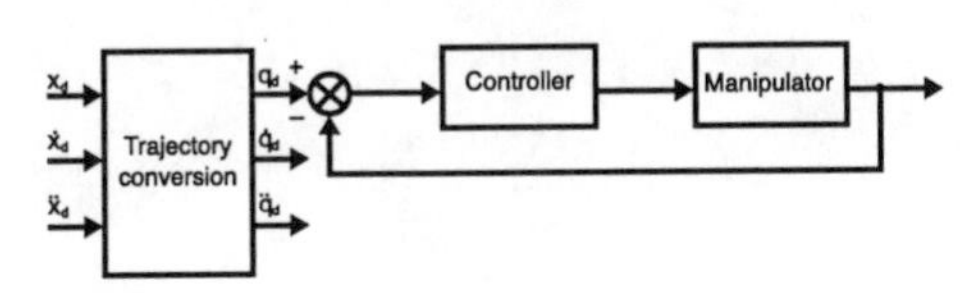

Joint space control scheme.

joint stiffness *See* stiffness of a manipulator arm.

joint torque the torque exerted at a joint due to an actuator or the effective torque experienced at a joint due to an external force applied to the system.

joint variable a scalar specifying position of each joint, one for each degree of freedom. The joint variable for a revolute joint is an angle in degrees; the *joint variable* for a prismatic joint is an extension in units of length.

journal a record of activity, specifically, a log of data before images and/or after images.

journaling a form of backward error recovery. A copy of the initial data (file, disk) is stored when the process begins. During process execution a record is made of all transactions that affect the data. If the process fails, its effect can be recreated by (re)running a copy of the backup data through the transactions after repair. The recovery time is equivalent to the time required for the initial attempt.

joystick an input device in the form of a control lever that transmits its movement in two dimensions to a computer. Joysticks are often used in games for control. They may also have a number of buttons whose state can be read by the computer.

JPEG *See* Joint Photographic Experts Group.

judgement a formal assertion relating entities such as terms, types, and environments. Type systems prescribe how to produce valid judgements from other valid judgements.

Julia set in fractal geometry, a complex function $f(z)$ that is the boundary of the set of points that escape.

jump/branch resolution an algorithm in a compiler that attempts to reduce code size by replacing control transfer instructions which use full machine addresses (in many architectures, called "jump instructions") with control transfer instructions which use shorter encodes, usually a relative offset from the instruction position (in many architectures, called "branch instructions").

jumper a plug or wire used for setting the configuration of system. It can be used for changing the hardware configuration by forcing some line to be high or low. It is used to change software configuration (especially on embedded systems) when the status of the jumper is read by the microprocessor.

jump instruction **(1)** an instruction that causes a control transfer.

(2) an instruction that conditionally causes a control transfer.

(3) in some computer architectures, a "jump" instruction is distinguished from a "branch" instruction in that a "branch" instruction is an instruction using a small number of bits, less than that required to represent a complete machine address, to encode a relative offset, and thus can cause a control transfer only a restricted distance from its occurrence. By contrast, a "jump" instruction uses a full machine address and can transfer control to any location. This problem gives rise to the code optimization feature in a compiler which is known as "jump/branch resolution".

just-in-time (JIT) a particular form of lazy evaluation in which certain complex computations are deferred until they are needed. Some languages such as Java have operating environments with just-in-time compilation, in which the byte codes that represent the executable program are translated to machine code the first time they are executed; thereafter, the machine code is executed, eliminating the overhead of the interpreter.

just noticeable difference (JND) the minimum difference between two stimuli in physical intensity which allows them to be perceived as different.

JVM *See* Java Virtual Machine.

K

Kalman filtering a digital averaging procedure for the enhancement of the signal-to-noise ratio in noisy signals. The Kalman filter is a recursive version of the true averaging procedure, and takes the form

$$y_i = (i-1)y_i - 1/i\ ,$$

in which the filter parameters are not constant but vary with the frame number i so that the latest averaged image y_n always equals $(x_1 + x_2 + \cdots + x_n)/n$. This gives a straightforward average over n frames, with a signal-to-noise ratio, without having to prespecify n.

Kanerva memory a sparse distributed memory developed for the storage of high-dimensional pattern vectors. Memory addresses are randomly generated patterns, and a given pattern is stored at the address that is closest in terms of Hamming distance. Patterns are usually bipolar and those stored at the same address are added together. Suitable thresholding of this sum allows a close approximation to individual patterns to be recalled, so long as the memory is not overloaded.

Karhunen-Loeve transform an optimal image transform in an energy-packing sense.

If the images are transformed by $\mathbf{y} = \mathbf{Ax}$, then the corresponding quadratic term associated with the covariance matrix becomes $\mathbf{x}'\mathbf{A}'$ $\mathbf{RAx}$. Hence the image uncorrelation is reduced to finding A such that

$$\mathbf{A}'\mathbf{RA} = \Lambda = \mathbf{diag}\,(\lambda_1, \ldots, \lambda_n)\ .$$

This equation can be solved by finding in the following three steps.

1. Solve equation $|\mathbf{R} - \lambda\mathbf{I}| = 0$ (find the $\mathbf{R}$'s eigenvalues).

2. Determine the n solutions of $(\mathbf{R} - \lambda_i\mathbf{I})\mathbf{a_i} = 0$ (eigenvectors).

3. Create the desired transformation $\mathbf{A}$ as follows

$$\mathbf{A} = [\mathbf{a_1}, \ldots, \mathbf{a_n}]\ .$$

In this transform a limited number of transform coefficients are retained. These coefficients contain a larger fraction of the total energy in the image. This transform is heavily dependent on the image features and requires a covariance function estimate for performing the transform. *See also* discrete cosine transform.

Karnaugh map a method for minimizing a Boolean expression, usually aided by a rectangular map of the value of the expression for all possible input values. Maximal rectangular groups which cover the inputs where the expression is true give a minimum implementation. *See also* Venn diagram.

Karp reduction a reduction given by a polynomial time computable transformation function. *See also* NP-complete, Turing reduction, Cook reduction, l-reduction, many-one reduction, polynomial time reduction.

k-ary Huffman encoding a minimal variable-length encoding based on the frequency of each character. Similar to a Huffman encoding, but joins k trees into a k-way tree at each step, and uses k symbols for each level.

KBS *See* knowledge-based systems.

KDC *See* key distribution center.

K-dominant match a match (i, j) having rank k and such that for any other pair $[i', j']$ of rank k either $i' > i$ and $j' \leq j$ or $i' \leq i$ and $j' > j$.

K-D Tree tree structure that caters to the storage and retrieval of data in k dimensions. It is an efficient approach to storing such data.

KE *See* knowledge engineering.

Kendall's classification a nomenclature for single stage service systems, describing the arrival process, the server process, number of servers, and the queue length in the form A/B/*n*/*s* where A is the type of arrival process, B is the type of server process, n is the number of servers, and s is the number of places in the system. A and B can be, e.g., D deterministic

0-8493-2691-5/01/$0.00+$.50

M negative exponential (Markov) E_k (Erlang-k) H_k (k-th order hyperexponential) G general GI-general independent $G^{[X]}$ general with batches the sizes of which have distribution type X.

Some authors use s for the number of queueing places in the system, not including the servers. The short notation A/B/n must be complemented by an indication if a waiting system or a loss system is meant. The number of sources can also be included in the notation, but the notation for this is not unique and therefore needs additional explanation.

Kerberos ticket-granting ticket (TGT) the cryptographic credential used to obtain credentials for other services.

K-Erlang the distribution of the sum of K exponential random variables. The K-Erlang density function with parameters (K, λ) is given by

$$f_K(t) = \frac{\lambda^K t^{K-1} e^{-\lambda t}}{(K-1)!} .$$

$K = 1$ corresponds to the exponential distribution.

kernel (1) a portion of an operating system that performs the basic functions related to process, memory, and I/O management, and is kept in main memory at all times (operates in privileged mode).

(2) the functional core of a Unix operating system.

kernel call a special instruction that switches execution from a user mode to a kernel mode. *Kernel calls* are used to create and manage processes, to access and manipulate the file system, and manage the system's resources.

kernel mode the mode of program execution providing protection for code and data.

key (1) in a table, the value that is used to select the desired entry (or entries). Formally, a set of attributes K, is a key if it is a superkey such that the removal of any attribute from the set will cause K to no longer be a key.

(2) in an access control system, a value held by a process to permit it to make access to certain objects within the system.

(3) in encryption, a value used to encrypt or decrypt a message according to the algorithm decrypt(key, encrypt(key, message)) = message.

(4) in accessing an associative array, the value which names an element of the array.

(5) in accessing any value set, the value which is used to name a value of the set, for example, a hash key.

key attribute any attribute that is a member of the primary key.

keyboard an input device with a set of buttons (or keys) through which characters are input to a computer. In addition to the keys for inputting characters, a *keyboard* may also have function keys and special keys, such as power-on or print-screen.

keyboard controller the device controller that processes keyboard input. Because of its importance, I/O from the *keyboard controller* is often handled differently from other I/O processes, with its own direct connection to the CPU.

key class in object-oriented design, the classes that collect the main functionalities of the system are called key classes. The number of key classes is directly related to the number of final classes of the system and the effort spent for development. Engine classes are typically the most important and big key classes of the system. *See* engine class.

key constraint a constraint that states that all tuples in a relation are unique.

key distribution center (KDC) a trusted third party in cryptographic protocols that has knowledge of the keys of other parties.

key field a field of a record whose value is unique for all records in the file.

keyframe an image that is stored in some way to be used as a reference point. Keyframes are often used in animation.

key point detection a technique usually employed in specialized linear or morphological filters designed for measuring gray-level changes in several directions. A key point is an isolated image point corresponding to a peculiar physical or geometrical phenomenon in the scene from which the image arises; it can be, for example, a corner, a line termination, a junction, a bright or dark spot, etc. Key points are distinguished from edges by two properties: they are sparse and display strong gray-level variations in two or more directions (while edge points are grouped into lines and have gray-level variations essentially in the direction normal to that line). *See* edge, edge detection. *See also* salient feature.

key punch a device with a keyboard used for storing data in paper cards or paper tapes by punching holes. The patterns of holes punched across these cards or tapes represent the data keyed in. Now obsolete.

key-range lock a lock acquired on a range of key values to prevent phantoms.

keyword a word (symbol) in a language which has meaning only in specific contexts. Unlike a reserved word, a keyword can usually be redefined by the programmer in any context other than the one its meaning is defined.

keyword parameter *See* named parameter.

kill character the terminal driver control character used to delete the current line.

kinematics the study of motion of mechanisms without the inclusion of dynamics.

kinesthesia human sense of forces acting on and applied by the limbs and body.

Kiviat graph a type of graph that displays the results of many changing variables simultaneously. The value corresponding to each variable is drawn considering its minimum, typical, and maximum values.

kixel (KInetic energy piXEL); fundamental unit of haptic displays, analogous to pixel.

Kleene star a notation used in the specification of a grammar to indicate zero or more replications of the starred component of the grammar. Most frequently used in the specification of regular expression.

KLT *See* Karhunen-Loeve transform.

KM *See* knowledge management.

K-means algorithm a clustering algorithm known from the statistical literature (E.W. Forgy, 1965). Relying on the same principles as the Lloyd algorithm and the generalized Lloyd algorithm.

Given a set of patterns and K, the number of desired clusters, K-means returns the centroid of the K clusters. In a Pascal-like language, the algorithm runs as follows:

function K-Means(Patterns:set-of-pattern):Centroids;
begin **repeat**
  for $h = 1$ **to** N **do**
    Assign point h to class k for which the distance $(x_h - \mu_k)^2$ is a minimum
  For $k = 1$ **to** K **do**
    Compute $\mu[k]$ (average of the assigned points)
  until no further change in the assignment
end

The algorithm requires an initialization that consists of providing K initial points, which can be chosen in many different ways. Each of the N given points is assigned to one of the K clusters according to the Euclidean minimum distance criterion. The average in the clusters is computed and the algorithm runs until no further re-assignment of the points to different clusters occurs.

knapsack problem a problem used in computational analysis. The problem is: given items of different values and volumes, find the most valuable set of items that fit in a knapsack of fixed volume. *See also* bin-packing.

K-nearest neighbor a classification method that classifies a point by calculating the distances between the point and points in the training data set. Then it assigns the point to the class that is most common among its k-nearest neighbors (where k is an integer). This is an extension of the nearest neighbor algorithm.

knot usually the join point between spline curve segments. (If the spline is a bundle of, e.g., cubics model, then a knot is the place where one cubic stops and another starts.)

knowledge a blanket term for any piece of information that can be applied to solve problems.

knowledge acquisition the extraction and formulation of knowledge derived from various sources, especially from experts.

knowledge base a collection of facts, rules, and procedures, organized into schemas, pertaining to a certain area. The information is usually obtained from experts in the area. A knowledge base forms part of a knowledge-based system.

knowledge-based concept assignment one of several approaches used in artificial intelligence. There are a few approaches to *knowledge-based concept assignment,* examples being: program plans, human oriented concepts, and informal reasoning. They differ in the kind of knowledge they use and which technique they apply to find concepts in the source code.

knowledge-based systems (KBS) the most mature, and still the most widely used, of the AI technologies. In a KBS, the knowledge is made explicit, rather than being implicitly mixed in with the algorithm. Also known as expert systems.

knowledge base management system a collection of programs that support the creation and maintenance of a knowledge base.

knowledge discovery the non-trivial process of identifying valid, novel, potentially useful, and ultimately understandable patterns in data.

knowledge engineer an artificial intelligence specialist responsible for the technical side of developing an expert system. The *knowledge engineer* works closely with the domain expert to capture the expert's knowledge in a knowledge base.

knowledge engineering (KE) the engineering discipline that involves integrating knowledge into computer systems in order to solve complex problems normally requiring a high level of human expertise.

Method for software developers to extract knowledge from domain experts.

knowledge management (KM) the distribution, access, and retrieval of unstructured information about "human experiences" between interdependent individuals or among members of a workgroup. Knowledge management involves identifying a group of people who have a need to share knowledge, developing technological support that enables knowledge sharing, and creating a process for transferring and disseminating knowledge.

knowledge representation formalism for expressing knowledge and reasoning with knowledge about many application domains. Knowledge representation paradigms include frames, production rules, semantic networks, first order logic, and systems of equations. Typical knowledge representation structures used in discovery systems are patterns such as trees, rules, equations, and contingency tables.

Knowlton's technique a pyramidal-type of hierarchical approach. In this approach, a reversible transformation takes adjacent pairs of k-bit pixel values and maps them into a k-bit composite value and a k-bit differentiator.

known segment table (KST) in the Multics system, a table listing all memory objects (segments) that are available to the process possessing the table. When a process desires access to a new segment, it requests permission to enter that segment in the KST, whence the system verifies the access rights for that process and then enters the segment and the appropriate access rights into the KST for that process.

Kogge-Stone adder a type of carry-lookahead adder; the organization is based on computing carries that are parallel prefixes.

Kohonen net a type of neural network that uses unsupervised learning to find patterns in data. In data mining it is employed for cluster analysis.

Kolmogorov complexity the minimum number of bits into which a string can be compressed without losing information. This is defined with respect to a fixed but universal decompression scheme, given by a universal Turing machine.

Kraft's inequality a theorem from information theory that sets a restriction on instantaneous codes (codes where no codeword is a prefix of any other codeword, i.e., a code containing 0 and 01 is not an instantaneous code). The Kraft inequality states that for an instantaneous code over an alphabet with size D and codeword lengths $l_1, l_2, l_3, \ldots, l_m$ the following must be true:

$$\sum_{i=1}^{m} D^{-l_i} \leq 1 ,$$

where m is the number of leaves in a binary tree and l_1 is the depth of leaf i.

Kripke structure a finite state machine, whose states are labeled with Boolean variables and whose next state is chosen nondeterministically. It may be extended with fairness constraints. *See also* model checking.

Kruskal's algorithm an algorithm for computing a minimum spanning tree. It maintains a set of partial minimum spanning trees, and repeatedly adds the shortest edge in the graph whose vertices are in different partial minimum spanning trees. *See also* Prim's algorithm.

KST *See* known segment table.

kv diagram *See* Karnaugh map.

L

L a suffix used in the C language to mean "long". For example, 23L is a 32-bit longword value. This was more important in the days when there were 16-bit implementations of C, when constants defaulted to 16-bit values. When this suffix is used on a floating point constant, it indicates the value should be an extended precision value.

label a construct used in a programming language, typically used to name a control transfer point in a program. Although in most languages the label must conform to the syntax of a name in the language, some languages such as FORTRAN and Pascal use integers as labels.

labeled transition system *See* finite state machine.

labelling (1) the computational problem of assigning labels consistently to objects or object components (segments) appearing in an image. One way to label an image involves appending to each pixel of an image the label number or index of its segment. Another way is to specify the closed contour of each segment and to use a contour filling technique to label each pixel within a contour.

lag the time elapsed between the onset of a signal and the onset of a response to that signal, as, for example, the time delay between transmission and receipt of a radio wave.

LALR(1) *See* lookahead LR(1).

Lamarckism a theory of evolution, proposed by J. Lamarck, which states that acquired physical and behavioral characteristic can be transmitted from parents to offspring. Although this does not represent the actual evolutionary process in which inheritable characteristics are only transmitted at the genotypic level, this principle has been adopted in some GA-based neural network design algorithms where individuals in the population represent potential network configurations. These individuals undergo alternate periods of evolution and training for fitness improvement. In other words, any desirable characteristics acquired through training can be reintroduced back into the gene pool.

lambda (λ) in the context of discussing automata, a metasymbol that is used to designate state transition that can take place without reading any input symbol. Often used as an artifice in the manipulation of the description of an automaton. Any automaton that is defined using a lambda transition can be replaced by a functionally identical automaton that does not require a lambda transition.

Lambda Calculus a notation designed by Alonzo Church that uses a functional notation to bind values to names. This notation was an early version of a functional language. The language LISP is an implementation of the notation of the *Lambda Calculus.*

lambda transition a state transition between two states of an automaton that takes place without the automaton reading a symbol. *See* lambda.

Lambertian diffuser a diffuser that spreads incoming light equally in all directions.

Lambertian reflection *See* diffuse shading.

Lambert's law a shading model in which the diffuse component of the brightness of a point on a surface is estimated as a scaled cosine of the angle between the surface normal and the direction from the point to a light source.

LAN *See* local area network.

landing zone a special track or cylinder where the disk heads are placed before shutting off power.

landmark a comparison point indicating a specific value achieved during a behavior, e.g., the successive heights reached by a partially elastic bouncing ball.

0-8493-2691-5/01/$0.00+$.50

language (**1**) a notation used to communicate with a computer.

(**2**) the set of symbol strings which can be generated by a grammar.

(**3**) the set of symbol strings that can be recognized by an automaton.

(**4**) any formalized system of symbols, concepts, and rules. For example, the language $\mathcal{L}$ of a deductive database is built from constant symbols, variable symbols, function symbols, predicate symbols, connectives, and quantifiers; *See also* term, atomic formula, well-formed formula.

language extension a syntactic or semantic enhancement to a language, typically to a language with a formal specification, to provide facilities not otherwise available in the standard language. The use of a language extension may destroy the portability of a program.

language-knowledgeable editor a text editor that has been specialized for editing programs in one or more source languages. The knowledge of the language is typically reflected in identation and in simple syntactic checks such as parenthesis balancing. *See also* structure editor.

laparoscopic surgery surgical procedure featuring very small cameras and cutting tools inserted into the body through small slits and operated by remote control.

laparoscopy *See* laparoscopic surgery.

Laplacian operator the second-order operator, defined in $\mathcal{R}^n$ as $\nabla^2 = \partial^2/\partial x_1^2 + \cdots + \partial^2/\partial_n^2$. The zero crossings of an image to which the Laplacian operator has been applied usually correspond to edges, as in such points a peak (trough) of the first derivative components can be found. Also simply called the Laplacian.

Laplacian pyramid a set of Laplacian images at multiple scales used in pyramid coding. An input image G_1 is Gaussian lowpass filtered and downsampled to form G_2. Typically G_2 is one quarter the size of G_1, i.e., it is downsampled by a factor of 2 in each direction. G_2 is upsampled and Gaussian lowpass filtered to form R_1 which is then subtracted from G_1 to give L_1. The process then repeats using G_2 as input. The sets of multiresolution images so generated are called "pyramids": $G_1 \ldots G_n$ form a Gaussian pyramid; $L_1 \ldots L_n$ form a Laplacian pyramid.

laptop a type of portable computer that is typified by a folding display and a battery-based power supply.

large object usually used to represent multimedia data, these objects require more than the normal unit of disk storage.

last-in-first-out (LIFO) a queueing discipline in which the customer that arrived most recently is serviced next.

Las Vegas algorithm a randomized algorithm that always produces correct reults, with the only variation from one run to another being in its running time.

latch (**1**) a bistable device that "holds" the value on an input wire to buffer upon occurrence of some event, such as a clock pulse or rising edge of a separate latch signal.

(**2**) a very short duration lock, used to control access to internal database structures.

latency in general, the time elapsed between issuing a request and completing it. For example: total time taken for a bit to pass through the network from origin to destination or the time between positioning a read/write head over a track of data and when the beginning of the track of data passes under the head. *See* lag.

latent fault an existing fault that has not yet been recognized. A fault that is present in a part of a system but has not yet contributed to a system failure.

latent type system a type system where types are associated with values, not variables. This usually requires run time type checking which is why latently typed languages such as SCHEME are often referred to as dynamically typed.

lateral inhibition in the human visual system, the inhibitory effect between nearby cells which acts to enhance changes (temporal or spatial) in the stimulus.

lattice (**1**) a directed acyclic graph which has one designated root node, the top node, and a designated bottom node. The bottom node has the property that no edges leave this node. Furthermore, any traversal that starts at the root node will always end at this bottom node. *See also* semi-lattice.

(**2**) a point lattice generated by taking integer linear combinations of a set of basis vectors.

lattice vector quantization a structured vector quantizer where the reproduction vectors are chosen from a highly regular geometrical structure known as a "lattice". The method is employed mainly because of the reduction in storage capacity obtained (compared to optimal vector quantization).

lattice VQ *See* lattice vector quantization.

L-attributed grammar an attribute grammar whose attributes may be computed by a left to right traversal of the source program. An attribute grammar must be L-attributed for the attributes to be computable during a parse that processes the input from left to right (as most parsers do). Synthesized attributes are always L-attributed.

layer assignment given a set of trees in the plane, each interconnecting the terminals of a net, an assignment of a routing layer to each segment of each tree so that the resulting wiring layout is legal under the routing model.

layered queueing model an extension of queueing models that allows reasoning about client/server architectures and the performance impacts of resource requests at different layers.

lazy evaluation an optimization technique applied to the execution of an algorithm by which the actual computation specified by the algorithm is deferred until the result is required. This can mean that many computations need never be performed at all. *See* eager evaluation.

LCD *See* liquid-crystal display.

LCFS *See* last-come-first-serve.

leading-zeros anticipator a hardware unit that predicts the position of the first non-0 digit in an expected sequence of digits and encodes this position as a number; the prediction may be slightly off and so may require a small adjustment. Most commonly used to predict the length of a normalization shift in a floating-point addition and hence to speed up the process by carrying the shifting concurrently with the significand addition. *See also* leading-zeros detector.

leading-zeros detector a hardware unit that detects the position of the first non-0 digit in a sequence of digits and encodes this position as a number. Most commonly used to determine the distance for a normalization shift following a floating-point addition. *See also* leading-zeros anticipator.

leading-zeros predictor *See* leading-zeros anticipator.

leaf node a node in a tree which has the property that there are no arcs out of it to other nodes. *Contrast with* tree node and internal node. *See also* root node. In a tree with only one node (a special case), the single node is both the root node and a leaf node.

leaf procedure a procedure that does not call another procedure.

learnability the capability of the software product to enable the user to learn its application.

learning (**1**) generally, any scheme whereby experience or past actions and reactions are automatically used to change parameters in an algorithm.

(**2**) in neural networks, the collection of learning rules or laws associated with each processing element. Each learning law is responsible for adapting the input-output behavior of the processing element transfer function over a period of time in response to the input signals that influence the processing element. This adaptation is usually obtained by modification of the values of variables (weights) stored in the processing element's local memory.

Sometimes neural network adaptation and learning can take place by creating or destroying the connections between processing elements. Learning may also be achieved by replacing the transfer function of a processing element with a new one.

learning classifier system an adaptive classifier system where the classifiers are subject to continuous modifications to allow improvement in task performance. Typically, new classifiers are generated from the original classifiers using genetic algorithm. In the Michigan approach, the condition/action pair of each classifier is regarded as a chromosome over the alphabet $\{0, 1, \#\}$, while for the Pitt approach, each chromosome represents a set of such rules for a complete classifier system. As a result, the crossover and mutation operators can be directly applied to each of these chromosomes for new rule generation. While the performance of each chromosome, as a complete classifier system, can be directly evaluated in the Pitt approach, we have to solve a credit assignment problem in the Michigan approach to determine the contribution of each individual rule to the achievement of the final goal. This is typically achieved using the bucket-brigade algorithm, which assigns a fitness value to each classifier based on its past capability in goal achievement and the specificity of the condition clause of its associated rule. *See also* classifier system.

learning law *See* learning rule.

learning rate a parameter in a learning rule which determines the amount of change for a parameter during the current iteration. While a large value may increase the rate of learning, it may also prevent a learning algorithm from converging.

learning rule in neural networks, an equation that modifies the connection weights in response to input, current state of the processing element, and possible desired output of the processing element.

learning vector quantization (LVQ) a supervised learning algorithm first prosposed by Kohonen that uses class information to move the Voronoi vectors slightly, so as to improve the quality of the classifier decision regions.

The training algorithm for LVQ is similar to its unsupervised counterpart with supervised error correction. After the winner is found, the weights of the winner and its neighbors will be updated according to the following rules:

if the class is correct,

$$w_i(t+1) = w_i(t) + \alpha \cdot (x_i(t) - w_i(t)) \ ;$$

otherwise

$$w_i(t+1) = w_i(t) - \alpha \cdot (x_i(t) - w_i(t)) \ .$$

See self-organizing system, self-organizing algorithm.

least fixpoint for a function $f : T \longrightarrow T$, an element $t \in T$ such that $f(t) = t$. A *least fixpoint* of f is the least element of the set of all fixpoints of f.

least mean square (LMS) algorithm in some cases of parameter estimation for stochastic dynamic systems, it is not feasible to use the least squares based algorithms due to the computational effort involved in updating and storing the $P(t)$ (probability) matrix. This is especially so when the number of parameters is large. In this case, it is possible to use a variant of the stochastic gradient algorithm — the least mean square (LMS) algorithm, which has the form

$$\begin{aligned}\hat{\theta}(t+1) &= \hat{\theta}(t) + \phi(t)(y(t+1) \\ &\quad - \phi^T(t+1)\hat{\theta}(t))\end{aligned}$$

where $\hat{\theta}$ is a parameter's estimates vector, ϕ is a regressor vector.

least recently used (LRU) a replacement algorithm based on program locality by which the choice of an object (usually a page) to be removed is based on the longest time since last use. The policy requires bookkeeping of essential information regarding the sequence of accesses, which may be kept as LRU bits or as an LRU stack.

Although most commonly used in operating systems to determine page replacement, it is also often used in caching systems, compilers, run time systems, and applications to improve performance.

least-significant bit (LSB) in a binary word, a bit with the lowest weight associated with it.

least-square-error fit an algorithm or set of equations resulting from fitting a polynomial or other type curve, such as logarithmic, to data pairs such that the sum of the squared errors between data points and the curve is minimized.

least squares an approach to determining the optimal set of free parameters $\vec{w}$ of an input-output mapping $\vec{y} = \vec{F}(\vec{x}, \vec{w})$, whereby the square of the difference between the output of the function $\vec{y}$ and the desired output $\vec{d}$ is minimized.

least-squares approximation an approximation in which the criteria for quality is the minimization of the square of the errors (at certain points) between the approximating and the approximated functions.

least squares solution the set of free parameters that satisfies the least squares criterion.

leave a construct in some programming languages that causes control to leave an execution context, such as transferring control to the end of a designated enclosing block. For example, in the C language, the break statement will cause control to leave the innermost enclosing loop or switch context.

LED *See* light emitting diode.

left-associative an interpretation of infix operators in which operations of equal operator precedence are interpreted as binding to the left; for example, A*B*C is interpreted as (A*B)*C. A language may be left-associative independent of its specified operator precedence, and some languages are left-associative with no operator precedence. *See also* right associative.

left outer join an outer join between two relations, whereby all the tuples from the leftmost operand (left relation) are included.

left-right model a hidden Markov model adopted in automatic speech recognition. The state diagram for these models is a directed acyclic graph; that is, there are no cycles apart from self-loops.

legacy system any application that is in a maintenance phase but not ready for retirement.

A legacy system can be 10 to 25 years old or even 1 day old if it has to be reused in a newer context with newer tools and programming languages. The maintenance for *legacy systems* is typically very expensive, and its integration with current or modern technology or software systems may be difficult or impossible. Legacy systems require reengineering to put them in a form where they may better suit modern requirements and may evolve more efficiently.

legal schedule a schedule that does not contain incompatible locks.

Lempel-Ziv-Welch (LZW) coding a variant of the dictionary-based coding scheme invented by Ziv and Lempel in 1978 (LZ78), where strings of symbols are coded as indices into a table. The table is built up progressively from the input data, such that strings already in the table are extended by one symbol each time they appear in the data.

lengthened code a code constructed from another code by adding message symbols to the codewords. Thus, an (n, k) original code becomes, after the adding of one message symbol, an $(n + 1, k + 1)$ code.

level-1 cache in systems with two separate sets of cache memory between the CPU and standard memory, the set nearest the CPU. Level-1 cache is often provided within the same integrated circuit that contains the CPU. In operation, the CPU accesses level-1 cache memory; if level-1 cache memory does not contain the required reference, it accesses level-2 cache memory, which in turn accesses standard memory, if necessary.

level-2 cach in systems with two separate sets of cache memory between the CPU and standard memory, the set between level-1 cache and standard memory.

leveling a process applied to an image in order to have global uniform illumination. There are many techniques for leveling. The simplest consists of subtracting from the original image the image of the background taken under the same conditions and then expanding the contrast of the difference.

level of automation degree to which a robot is computer controlled, i.e., functions autonomously.

level of control degree of human responsibility for system function; a continuum ranging from full control to strategic control.

level-of-detail blending when rendering models that are defined with levels-of-detail, artifacts can occur when one level-of-detail is replaced with another. This is known as "popping" and can be reduced by blending one level with the next when the transition takes place.

level of performance the degree to which the needs are satisfied, represented by a specific set of values for the quality characteristics.

level-sensitive pertaining to a bistable device that uses the level of a positive or negative pulse to be applied to the control input, to latch, capture, or store the value indicated by the data inputs.

levels of abstraction the degree of detail with which some representation of the software is presented.

level-triggered *See* level-sensitive.

Levenshtein distance the metric distance between two strings that counts the minium number of insertions, deletions, and substitutions of symbols to transform one string to the other.

lex a program that takes a description of a set of regular expressions defining different tokens and produces a lexical analyzer for these tokens that can then be integrated into the application. *See also* yacc.

lexeme a representation of an entity generally represented by a nonterminal symbol in a specification of lexical characteristics of a programming language.

lexer *See* lexical analyzer.

lexical analysis the processing of a sequence of characters to identify fundamental sequences of characters. Generally that component of a compiler that identifies names, operators, reserved words, and similar equivalence classes of lexical tokens (lexemes). *See also* syntactic analysis, semantic analysis.

lexical analyzer that component of a compiler which identifies fundamental lexical entities. Converts a string of input characters to a sequence of lexical tokens or lexemes. Also called a lexer.

lexical knowledge knowledge of words in a natural language. Includes knowledge of syntactic constructs in which the word(s) appears in the language, possible meanings of the word, pronunciation, part of speech, possible inflections, and so on.

lexical scope a name scope defined by the textual context, independent of any enclosing or potentially enclosing execution context. A name defined within a *lexical scope* is visible to all parts of the program contained within that lexical scope. *Compare with* dynamic scope.

lexicographically positive a vector $(x_1, x_2, \ldots, x_n)$ which is not all zeros and in which the first nonzero element is positive.

lexicographical order alphabetical or "dictionary" order.

library a set of precompiled routines that may be linked with a program at compile time or loaded at load time or dynamically at run time. *See* subroutine library, archive.

life-cycle all the steps or phases an item passes through during its useful life. The life may include: problem identification, requirements engineering, analysis, design, implementation, testing, assessment, risk analysis, maintenance, etc.

life-cycle model a framework containing the processes, activities, and tasks involved in the development, operation, and maintenance of a system, spanning the life of a system from the definition of its requirements to the termination of its use. A model that describes potential or prescribed sequencing of stages in the temporal evolution of a software product.

life-cycle process a process that describes the sequencing of activities and tasks within software development (software development process or software engineering process). A complete software engineering process takes into account not only the application development but also the resources (people, tools) that are used in the realization of the software development. Several distinct life-cycle processes have been defined: spiral, fountain, pin-ball, waterfall, etc.

lifetime (**1**) the time elapsed between the creation and destruction of an entity or object.

(**2**) the time during which a value is expected to be valid, that is, the interval from first assignment to last use. A value may have several disjoint lifetime intervals.

lifetime analysis the computation in a compiler that determines the lifetime of values. In compilers, *lifetime analysis* is used to prepare for tasks such as register allocation, because values with disjoint lifetimes can be stored in the same register.

LIFO acronym, last-in-first-out. *See* last-come-first-serve. *See also* first-in-first-out.

light emitting diode (LED) a forward-biased p-n junction that emits light through spontaneous emission by a phenomenon termed electroluminescence.

lifting lemma a lemma for proving completeness at the first-order level from completeness at the propositional level.

lighting quantity of light. Standard measurement unit: lumen/second.

lighting level brightness of an area judged relative to the brightness of a white or highly transmitting object/color which has similar illumination.

light pen an input device that allows the user to point directly to a position on the screen. This is an alternative to a mouse. Unlike a mouse, a light pen does not require any hand/eye coordination skills because the users point to where they look with the pen.

light source a source of visible electromagnetic radiation. *See also* local light source.

lightweight pertaining to a process that is designed to have a minimum of internal state and resources.

lightweight process a single-threaded subprocess which, unlike a thread, has its own process identifier and may also differ in its inheritance and controlling features. *See* thread.

likelihood ratio test a test using the likelihood ratio that can be used along with threshold information to test different information-content hypotheses. An example of a signal detection problem is the demodulation of a digital communication signal, for which the *likelihood ratio test* may be used to decide which of several possible transmitted symbols has resulted in a given received signal.

Linde Buzo Gray algorithm an iterative method for designing a codebook for a vector quantizer using a set of training data that is representative of the source to be coded. Otherwise known as the generalized Lloyd algorithm or K-means algorithm.

line a sequence of ASCII words terminated by a newline character.

linear (**1**) any function that is a constant times the argument plus a constant: $f(x) = c_1 x + c_0$.

(**2**) in complexity theory, the measure of computation, $m(n)$ (usually execution time or memory space), is bounded by a linear function of the problem size, n. More formally $m(n) = O(n)$. *See also* exponential, logarithmic, polynomial.

linear actuator an actuator that provides translational (instead of rotational) displacements.

linear approximation any technique used for the purpose of analysis and design of nonlinear systems. For example, one way of analyzing the stability of a system described by nonlinear differential equations is to linearize the equations around the equilibrium point of interest and check the location of the eigenvalues of the linear system approximation.

linear array processor a processor architecture that has one PE for each column in the image, and enough memory directly connected to each PE to hold the entire column of image data for several distinct images. Also known as vector processor.

linear block code a block coding scheme for which the mapping can be described by a linear transformation of the message block. The transformation matrix is referred to as the generator matrix.

linear code a forward error control code or line code whose code words form a vector space. Equivalently, a code where the element-wise finite field addition of any two code words forms another code word.

linear depth cueing a rendering technique used to give the effect of depth to an image. This is achieved by intensities in the image plane being reduced linearly with respect to the distance from the plane. *See* attenuation.

linear granulometries in a granulometry, the resulting opening sequences using vertical and horizontal line segments of increasing length.

linear hashing a technique that allows hashing to a dynamic file. The file size is 2^j, for some value j. Given H as the hash field value, the function $H \bmod 2^j$ is used as a hash function. Upon collision, (i) the record is inserted via chaining and (ii) a block is split with its contents redistributed according to a new hash function 2^{j+1}. This continues until all 2^j blocks are split. The blocks are split in a linear manner beginning with block zero. The process then begins a new iteration with j incremented.

linear interpolation a procedure for approximately reconstructing a function from its samples, whereby adjacent sample points are connected by a straight line.

linearity in the parameters a property of robot arm dynamics, important in adaptive controller design, where the nonlinearities are linear in the unknown parameters such as unknown masses and friction coefficients.

linear least squares estimator (LLSE) the linear estimator $\hat{\mathbf{x}} = K\mathbf{y} + c$, where matrix K and vector c are chosen to minimize the expected squared error $E\left[(\hat{\mathbf{x}} - \mathbf{x})^T(\hat{\mathbf{x}} - \mathbf{x})\right]$. The general LLSE solution to estimate a random vector $\mathbf{x}$ based on measurements $\mathbf{y}$ is given by

$$\hat{\mathbf{x}}(\mathbf{y}) = E[\mathbf{x}] + \text{cov}(\mathbf{x}, \mathbf{y}) \cdot \text{cov}(\mathbf{y}, \mathbf{y})^{-1} \cdot (\mathbf{y} - E[\mathbf{y}]),$$

where "cov" represents the covariance operation. *See also* least squares, minimum mean square estimator.

linear product for two vectors X and Y, and with respect to two suitable operations $\otimes$ and $\oplus$, a vector $Z = Z_0 Z_1 \ldots Z_{m+n}$ where $Z_k = \oplus_{i+j=k} X_i \otimes Y_j \ (k = 0, \ldots, m+n)$.

linear program optimization of a linear function subject to linear equality and inequality constraints.

linear programming a mathematical model for optimal solution of resource allocation problems.

linear scalar quantization *See* uniform scalar quantization. Also known as linear SQ.

linear scaleup sustained performance for a linear increase for a constant database size and linear increase in processing and storage power.

linear sequential model a process model that defines a set of linera activities for developing computer software. *See also* classic life-cycle, waterfall life-cycle.

linear SQ *See* linear scalar quantization.

linear system the systems in which the components exhibit linear characteristics, i.e., the principle of superposition applies. Strictly speaking, linear systems do not exist in practice; they are idealized models purely for the simplicity of theoretic analysis and design. However, the system is essentially linear when the magnitude of the signals in a control system are limited to a range in which the linear characteristics exist.

linear transformation a transformation operator A which satisfies superposition

$$A(x_1 + x_2) = Ax_1 + Ax_2 ,$$

and homogeneity

$$A(\lambda x_1) = \lambda A x_1 .$$

For a discrete linear transform A is a matrix and x_1 and x_2 are vectors. Any matrix transform is linear. *See* linear system, superposition.

linear waiting the situation in which a process making a request will be granted this request before any other process is granted the request more than once.

line clipping selecting the portion of a line segment that lies inside a clipping window. If the line intersects the window boundary, then the line is split into two or three segments, one of which moves with the clipping window and the other(s) remains unmoved.

line code modification of the source symbol stream in a digital communication system to control the statistics of the encoded symbol stream for purposes of avoiding the occurrence of symbol errors that may arise due to limitations of practical modulation and demodulation circuitry. Also called a recording codes or modulation codes.

line detection the location of lines or line segments in an image by computer. Often accomplished with the Hough transform.

line drawing an image created only from points connected by lines. It can be described using a series of end-point coordinate information, (x_1, y_1) and (x_2, y_2), for each connecting line. This can be combined with a weight (w) which denotes the thickness of the connecting line.

line-impact printer a printer that prints a whole line at a time (rather than a single character). An impact printer has physical contact between the printer head and the paper through a ribbon. A dot-matrix printer is an impact printer, whereas an ink-jet printer is not. A line-impact printer is both a line and impact printer.

line of gaze a line passing from the center of the fovea through the center of the lens; the aiming line of the eyes.

line of sight a wireless communication system passing signals along a straight line from transmitter to receiver and unable to communicate if the line is obstructed. *See* line of gaze.

lines-of-code metric (LOC) a software metric that counts the lines of code of a source in order to evaluate its size.

Ling adder a type of carry-lookahead adder.

linguistic variable variable whose values are not numbers, but words or sentences in a natural or artificial language. In fuzzy set theory, the linguistic values (or terms) of a *linguistic variable* are represented by fuzzy sets in a universe of discourse. *See also* membership function, fuzzy set.

linkage the propensity for a group of two or more genes at nearby loci on the same chromosome to be inherited together during reproduction. *See also* subroutine linkage.

linkage editor IBM-specific designation of a linker.

link count the number of (hard) links that refer to a particular file.

linked list a list implemented by each item having a link to the next item. *See also* list, doubly-linked list.

linker a computer program used to link. The linker takes one or more object files, assembles them into blocks which are to fit into particular regions in memory, and resolves all external (and possibly internal) references to other segments of a program and to libraries of precompiled program units. This prepares relocatable object code for execution, thus producing a binary executable file.

link inertial parameters for a manipulator arm, consists of six parameters of the inertia tensor, three parameters of its center of mass multiplied by mass of the link (more precisely three components of the first order moment), and mass of the link. Dynamic properties of each link are characterized by 10 inertial parameters. They appear in the dynamic equations of motion of the manipulator.

linking loader a program that performs the functions of a linker and in addition loads the program into main memory for execution.

linking record type used to represent M:N relationships and recursive relationships in the network model.

link time a designation usually applied to describe information that is known at the time a linking operation is performed. Usually contrasted to compile time and run time.

link vector an object containing information used to associate a name used, but not defined, in one program with a name defined in another program.

Linpack benchmark one of a variety of floating point benchmarks originally written in the Fortran programming language and suggested by Jack Dongarra of the University of Tennessee. The Linpack benchmark consists of a suite of matrix multiplication and related routines tailored to exercise the floating point processing capability of processing platforms. The result of executing this benchmark on a particular machine is an estimate of the MFLOPs (Millions of Floating Point Operations per Second) capacity of a particular machine. This benchmark is useful for comparing floating point capabilities between machines having similar hardware architectures for purposes of scaling between any two machines.

Linux a freeware, open source, version of Unix that was developed by literally thousands of code developers from around the world. Commercial versions are also available. Linux takes it name from its original developer, Linus Torvalds, a Finnish student who wanted to improve on MINIX.

liquid-crystal display (LCD) the screen technology commonly used in notebook and smaller computers.

LISP a programming language used in artificial intelligence that is based on List Processing. Developed at MIT by John McCarthy and his associates in the late 1950s. A language based on the lambda calculus and which was one of the first functional languages. Later generations of LISP included operations of imperative languages including assignment statements and goto control structures.

list an ordered set of values stored with the information required to sequence through the list. Typically implemented as a set of objects called list cells which hold a value or value reference and one or two pointers. A list with a single pointer is a singly linked list and defines only a single mapping to a successor cell, which can return a designator indicating the end-of-list. A list with two pointers is a doubly linked list and defines a mapping to a predecessor cell and a successor cell, which can return a value designating that the end of list (in the predecessor or successor direction) has been reached. *See also* circular list.

list cell a member of a list.

list contraction contracting a list by removing a subset of the nodes.

literal (1) a constant whose value is determined by its lexical denotation. For example, 1 is an integer constant, 1.2 is a floating point constant, and "Gray Cat" is a string constant. *See also* constant, symbolic constant.

(2) either an atomic formula A (positive literal) or its negation $\neg A$ (negative literal) or its default negation $\natural A$ (default negative literal).

(3) in artificial intelligence, an atom or negation of an atom.

little endian a memory organization whereby the byte within a word with the lowest address is the least significant, and bytes with increasing address are successively more significant. Opposite of big endian. Sometimes believed (with no merit) to be either the "right" or the "wrong" memory organization, hence the name (cf. Swift's *Gulliver's Travels*).

For example, in a 32-bit or four-byte word in memory, the most significant byte would be assigned address i, and the subsequent bytes would be assigned the addresses: $i-1, i-2$, and $i-3$. Thus, the least significant byte would have the lowest address of $i-3$ in a computer implementing the little endian address assignment. *See also* big endian, big endian/little endian problem.

little-o notation a theoretical measure of the execution of an algorithm, usually the time or memory needed, given the problem size n, which is usually the number of items. Informally saying some equation $f(n) = o(g(n))$ means $f(n)$ becomes insignificant relative to $g(n)$ as n approaches infinity. More formally it means for all $c > 0$, there exists some $k > 0$ such that $0 \leq f(n) < cg(n)$ for all $n \geq k$. The value of k must not depend on n, but may depend on c. *See also* big-O notation.

Little's formula a result formally proven by J.D.C. Little in 1961 that states that the average number of customers in a queueing system is equal to the average arrival rate of customers to that system times the average time spent in that system.

Little's law *See* Little's formula.

live insertion the process of removing and/or replacing hardware components (usually at the board level) without removal of system power and without shutting down the machine.

livelock (1) a situation in which referenced processes cannot make progress in execution except for a limited set of operations repeated in a loop. Also called process starvation.

(2) where a transaction cannot proceed while other transactions can proceed.

liveness conditions in the specification of a system, conditions expressing that something good will happen. For example, in a message passing system a liveness condition is as follows: If a message is sent it will be acquired by the receiver. Temporal logics are expressed with the formula

$$\Box(\forall m \text{ send } (m) \rightarrow \Diamond \text{ receive } (m)) .$$

See safeness condition.

liveness property a program property asserting that the program will eventually move into a good state during execution.

LL(k) a grammar whose rules contain no left-recursive nonterminal symbols. An LL(k) grammar can be parsed left-to-right, following a leftmost derivation, and requiring no more than k input symbols at each processing step. *Compare with* LR(k), LALR(k).

Lloyd-Max scalar quantization a scalar quantizer designed for optimum performance (in the minimum mean squared error sense). The method, and the corresponding design algorithm, are due to S.P. Lloyd (1957) and J. Max (1960). Also referred to as pdf-optimized quantization, since the structure of the scalar quantizer is optimized to "fit" the probability density function (pdf) of the source.

Lloyd-Max SQ *See* Lloyd-Max scalar quantization.

LLSE *See* linear least squares estimator.

l_m distance the generalized distance between two points.

In a plane with point p_1 at (x_1, y_1) and p_2 at (x_2, y_2), it is $(|x_1 - x_2|^m + |y_1 - y_2|^m)^{1/m}$.

See also Euclidean distance, rectilinear, Manhattan distance.

LMS algorithm *See* least mean square algorithm.

load (**1**) to read machine code into main memory in preparation for execution.

(**2**) the volume of work processed by the system. Load intensity is often expressed in terms of the proportion of time the system is busy processing.

load balancing the process of achieving an equal load at each computing node in an inter-connected multiple computer system through task transfers from one node to another.

load buffer a buffer that temporarily holds memory-load (i.e., memory-write) requests.

load bypass a read (or load) request that bypasses a previously issued write (store) request. Read requests stall a processor, whereas writes do not. Therefore, high-performance architectures permit *load bypass*. Typically implemented using write-buffers.

load dependent server a service center whose service rate is a function of the customer population at its queue.

loader a computer program used to load. The loader (typically an operating system), loads executable code into memory and may also perform computations (fixups) required to take into account the actual location into which the code is loaded.

load factor for a hash file, the number of records currently in the hash file divided by the number of records that currently could be stored, i.e., (number of records $\div$ (blocking factor $\times$ number of blocks)).

loading the operation of copying a program into memory for execution. *See also* loader.

load instruction an instruction that requests a datum from a virtual memory address, to be placed in a specified register.

load map a computer-generated list that identifies the location and size of all or parts of memory resident code and data.

load module a program unit that is the output of a linker, suitable for loading and execution.

load sharing the process of reducing the difference of load among the computing nodes in an inter-connected multiple computer system through task transfers from one node to another.

load/store architecture a system design in which the only processor operations that access memory are simple register loads and stores.

load/store unit a computer based on the load/store architecture.

LOC *See* lines-of-code metric.

local alignment the detection of local similarities among two or more strings.

local area network (LAN) a network of computers and connection devices (such as switches and routers) that are located on a single site. The connections are direct cables (such as UTP or optical fiber) rather than telecommunication lines. The computer network in a university campus is typically a *local area network.*

local attribute the attribute of a class. Opposite of the inherited attributes which are defined in the superclasses.

local bus the set of wires that connects a processor to its local memory module.

local coordinate system a coordinate system defined with respect to some reference coordinate system, usually used to define a part or subassembly within a scene definition.

local decision unit control agent or a part of the controller associated with a given subsystem of a partitioned system; *local decision unit* is usually in charge of the local decision variables and is a component of a decentralized or a hierarchical control system.

local decision variable control inputs associated with a given sub-process (sub-system) of the considered partitioned process (system); local decision variables can be either set locally by local decision unit or globally by a centralized controller.

locale a designation of mechanisms that customizes the behavior of a program to support local variation. Typically, these deal with internationalization issues, such as representations of dates and times, but in addition may extend to the character set permitted, the collating sequence used, and the like. Compilers and runtime systems often need to be either locale-independent or locale-aware.

local fault a fault that is present in a part of the system but has not yet contributed to a system failure, i.e., a latent fault.

locality one of two forms of program memory relationships.

1. Temporal locality is that if an object is being used, then there is a good chance that the object will be re-used soon.

2. Spatial locality is that when an object is being used, there is a good chance that objects which are in its neighborhood (with respect to the memory where these objects are stored) will be used.

These two forms of locality facilitate the effective use of hierarchical memory. Registers exploit temporal locality. Caches exploit both temporal and spatial locality. Interleaved memories exploit spatial locality. *See also* sequential locality.

locality-of-reference the notion that if you examine a list of recently executed program instructions, you will note that most of the instructions are confined to a small window of the actual code space.

local light source light source that directly (i.e., not by reflection or transmission) illuminates a point on a surface.

local memory memory that can be accessed by only one processor in a multiprocessor or distributed system. In many multiprocessors, each processor has its own local memory. *See also* global memory.

local optimization optimization whose scope is limited to a basic block, ignoring the context in which it executes. *Compare with* global optimization.

local reflection model a model that simulates the reflection of light incident directly on an object from a light source. Unlike global models, they take no account of (indirect) light reflected from another object.

local variable a variable whose scope is limited by its lexical context. Often used to designate variables that are allocated on the stack when the context is entered and which are deallocated when the context is exited. However, a local variable is not necessarily a stack-allocated variable; it may have permanent extent. When it is necessary to distinguish the allocation strategy, the local variable may be explicitly referred to as a stack local variable or an automatic variable.

location space the set of all hardware addresses of memory locations in RAM.

location transparency a system property by which how one refers to something depends on the location of neither the subject nor the object.

lock (**1**) generally, a state set on a resource restricting its allocation or use in some way. Locks are set in response to access or update of a resource. *See also* spin lock, page lock.

(**2**) to acquire a semaphore.

(**3**) in database systems, a standard method by which ACID transactions avoid isolation anomalies. A transaction attempting to read or update a datum must first be assigned a read or update lock to hold until commit. Two transactions cannot simultaneously hold locks in the same datum unless they are both read locks. Lock contention for popular data often leads to a bottleneck.

lock compatibility (**1**) whether a combination of particular lock types is allowed concurrently.

(2) whether a new lock is compatible with existing locks.

lock conversion the replacement of one lock by another lock.

lock downgrading replacement of a lock by one with fewer rights or privileges over the locked object, or over fewer objects.

lock duration the period of time a lock is held for.

lock escalation replacement of a lock by one with more rights or privileges over the locked object, or over more objects.

lock granularity the amount of data or number of objects covered by a lock.

locking (1) ensuring that a code segment will not be swapped from main memory. *See* bus locking.

(2) a method of concurrency control where locks are placed on database units (e.g., pages) on behalf of transactions that attempt to access them.

lock manager a component that controls the acquisition and release of locks.

lock table a table where locks are held.

lock-up-free cache *See* non-blocking cache.

lock upgrading *See* lock escalation.

log (1) in architecture, a record of a sequence of operations of a certain program unit.

(2) in database systems, a sequential file that stores information about transactions and the state of the system at certain instances.

logarithmic (1) any function that is the sum of constants times a logarithm of the argument: $f(x) = \Sigma_{i=0}^{k} c_i \log x$.

(2) in complexity theory, the measure of computation, $m(n)$ (usually execution time or memory space), is bounded by a logarithmic function of the problem size, n. More formally $m(n) = O(\log n)$.

(3) sometimes imprecisely used to mean polylogarithmic. *See also* linear, polynomial, exponential.

logging writing data to a file where existing data are never overwritten. The system thus modifies the file only by appending new data.

logging protocol the protocol which records, in a separate location, the changes that a transaction makes to the database before the change is actually made.

logical as "seen" by the microprocessor. *See* physical.

logical address space the maximum memory or address space accessible by the microprocessor. Various schemes are used to map the logical address space into the physical address space.

logical block the sequential fixed-sized pieces of a file. The *logical block* associated with a given byte offset in a file is calculated by dividing the offset by the file system block size. For example, byte 20,000 is located in the third logical block of a file residing on a file system with 8-kilobyte blocks.

logical cohesion a cohesion due to the presence of different tasks that work on the same module by performing the same logical actions. *See* cohesion, temporal cohesion, procedural cohesion.

logical complexity the complexity due to the logic part of the code: the logical expressions and their effects, that is the control flow. McCabe cyclomatic complexity can be considered a measure of the logic complexity since it considers the complexity of selection instructions. *See* computational complexity, complexity.

logical consequence a concept in artificial intelligence. A formula C is a *logical consequence* of $\mathcal{F}$ if every interpretation that satisfied $\mathcal{F}$ also satisfies C. *See also* interpretation.

logical data independence the ability to change the conceptual schema without affecting the application programs.

logical operation the machine-level instruction that performs Boolean operations such as AND, OR, and COMPLEMENT.

logical OR an operation that combines two values in the computer using the OR operation, interpreting each of the values according to two-valued logic rules. *See also* bitwise OR, exclusive OR. Depending on the programming language, the bitwise OR operation may or may not have different semantics from the logical OR operation. For example, in the C language the logical OR interprets 0 as "false" and any nonzero value as "true", so applying logical OR to two nonzero values does not necessarily produce the same result as when a bitwise OR is applied.

logical register *See* virtual register.

logical shift a shift in which all bits of the register are shifted. *See also* arithmetic shift.

logical trace an execution trace that reports only logical aspects of the execution: the jump, the selections. *See* execution trace.

logic analyser a machine that can be used to send signals to, and read output signals from, individual chips or circuit boards.

logic bomb destructive action triggered by some logical outcome.

logic database *See* deductive database.

logic error an error in which a program or function compiles and executes but produces an undesired result; that is, a program or function whose behavior is flawed even though its source code adheres to the rules of the source language.

logic programming the use of symbolic logic as a programming language.

login *See* logon.

login name identifies an individual as an authorized user of a Unix system; often called a login ID or simply a login.

login prompt a text string, issued during the login sequence, that indicates the system is ready to accept a user's login name.

logistics equation an equation first proposed as a model for population growth. It is given by

$$P(t+1) = aP(t) - aP(t)^2 ,$$

where $aP(t)$ is the previous year's population plus newborns, and $aP(t)^2$ is the death rate. Plotting the values of $P(t)$ over many years and over many values of a yields a bifurcation diagram.

log-normal distribution probability distribution with density

$$f(x) = 1/\left(\sqrt{(2\pi)}\sigma x\right) e^{-((\log x - \mu)^2/(2\sigma^2))} ,$$

where μ and σ are the mean and standard deviation of the logarithm.

logoff to end a session.

logon to start a session with a particular operating system, usually via a specified authentication procedure.

log sequence number (LSN) a number assigned to a log record, which serves to uniquely identify that record in the log. LSNs are typically assigned in monotonically increasing fashion so that they provide an indication of relative position.

long duration transaction a transaction that completes in a long period of time.

longest common subsequence problem the problem of finding a maximum-length (or maximum weight) subsequence for two or more input strings.

long integer an integer that has double the number of bits as a standard integer on a given machine. Some modern machines define long integers and regular integers to be the same size.

long-lasting transaction *See* long duration transaction.

longword a unit of information in a computer, representing a data item which is two words long. On machines with a 16-bit word length, or in languages with a history of a 16-bit word length, a longword represents 32 bits. *See also* byte, halfword, word, quadword.

lookahead for conditional branches, a strategy for choosing a probable outcome of the decision that must be made at a conditional branch, even though the conditions are not known yet, and initiates "speculative" execution of the instructions along the corresponding control path through the program. If the chosen outcome turns out to be incorrect, all effects of the speculative instructions must be erased (or, alternately, the effects are not stored until it has been determined that the choice was correct). Some lookahead mechanisms attempt execution of both possibilities from a conditional branch, and others will cascade lookahead choices from several conditional branches that occur in close proximity in the program flow.

look-ahead carry the concept (frequently used for adders) of breaking up a serial computation in which a carry may be propagated along the entire computation into several parts, and trying to anticipate what the carry will be, to be able to do the computation in parallel and not completely in series.

lookahead LR(1) [LALR(1)] the algorithm invented by DeRemer (1971). The algorithm used in many bottom-up parser generators, including yacc. A language is called LALR(1) if it can be parsed unambiguously by the LALR(1) algorithm.

lookaside list a data structure, often of small and limited size, used to expedite access to a larger structure. Often implemented as a cache of values.

lookbehind for an instruction buffer, a means of holding recently executed instructions in a buffer within the control unit of a processor to permit fast access to instructions in a loop.

look-up table an optimization technique that allows for the computation of continuous functions using mostly fixed-point arithmetic.

loop (**1**) a set of branches forming a closed current path, provided that the omission of any branch eliminates the closed path.

(**2**) a programming construct in which the same code is repeated multiple times until a programmed condition is met. *See also* iteration, do|while, do|until, while|do.

loop body the computations that are performed within a loop. In some languages, a *loop body* may have no statements at all, and hence be empty.

loop carried dependence a data dependence relation with a nonzero entry in the distance vector for some loop.

loop independent dependence a data dependence relation with all zero entries in the distance vector.

loop interchanging a loop restructuring transformation that switches two nested loops.

loop invariant in axiomatic semantics, a logical property of a while-loop that holds true no matter how many iterations the loop executes.

loop invariant removal an optimization technique that involves removing code that does not change inside a looping sequence.

loop jamming an optimization technique that involves combining two loops within the control of one loop variable.

loop testing a white box testing technique that exercises program loops.

loop unrolling a technique used by optimizing compilers to reduce execution times in a pipelined architecture. Deciding what to do next at the end of each loop interferes with prefetching (*See* branch prediction). Hence it is more efficient to duplicate the body of the loop to form a longer block of straight line code and reduce the number of iterations accordingly.

loosely coupled hardware a message passing multiprocessor.

loosely coupled multiprocessor a system with multiple processing units in which each processor has its own memory, and communication between the processors is over some type of bus.

lossless coding *See* lossless source coding.

lossless compression compression process wherein the original data can be recovered from the coded representation perfectly, i.e., without loss. Lossless compressors either convert fixed length input symbols into variable length codewords (Huffman and arithmetic coding) or parse the input into variable length strings and output fixed-length codewords (Ziv-Lempel coding). Lossless coders may have a static or adaptive probability model of the input data. *See also* entropy coding.

lossless encoding *See* lossy compression.

lossless predictive coding a lossless coding scheme that can encode an image at a bit rate close to the entropy of the mth-order Markov source. This is done by exploiting the correlation of the neighboring pixel values.

lossless source coding source coding methods (for digital data) where no information is "lost" in the coding, in the sense that the original can be exactly reproduced from its coded version. Such methods are used, for example, in computers to maximize storage capacity. *Compare with* lossy coding.

lossless source encoding *See* lossless source coding.

loss probability the probability for an arbitrary request or an ATM cell to be lost in a network or a network node.

lossy coding *See* lossy source coding.

lossy compression an image compression scheme where decompression yields an image that is not identical to the image that was compressed. *Compare with* lossless compression.

lossy encoding *See* lossy source coding.

lossy source coding refers to non-invertible coding, or quantization. In *lossy source coding,* information is always lost and the source data cannot be perfectly reconstructed from its coded representation. *Compare with* lossless source coding.

lossy source encoding *See* lossy source coding.

lost update problem the result of dirty writes.

low byte the least-significant 8 bits in a larger data word.

lower bound a function (or growth rate) below which solving a problem is impossible.

lower CASE a tool that is useful for supporting coding integration testing.

lower triangular matrix a matrix that is only defined at (i, j) when $i \leq j$. *See also* upper triangular matrix, ragged matrix.

lowest common ancestor the deepest node in a tree that is an ancestor of two given leaves.

low-fidelity prototype a prototype constructed in a medium and environment sharing only a few characterisitics with the eventual product. An example is the use of drawings of screen layouts to walk a user through the product design. *See also* high-fidelity prototype.

low-level formatting the first step in the process of preparing a disk to receive data.

low-level vision the set of visual processes related to the detection of simple primitives, describing raw intensity changes and/or their relationships in an image.

low order interleaving in memory interleaving, using the least significant address bits to select the memory module and the most significant address bits to select the location within the memory module.

low-pass filter (**1**) filter exhibiting frequency selective characteristic that allows low-frequency components of an input signal to pass from filter input to output unattenuated; all high-frequency components are attenuated.

(**2**) a filter that passes signal components whose frequencies are small and blocks (or greatly attenuates) signal components whose frequencies are large. For the ideal case, if $H(\omega)$ is the frequency response of the filter, then $H(\omega) = 0$ for $|\omega| > B$, and $H(\omega) = 1$ for $|\omega| < B$. The parameter B is the bandwidth of the filter. A filter whose impulse response is a low-pass signal.

l-reduction a Karp reduction that preserves approximation properties of optimization problems.

LR(k) a bottom-up parsing algorithm that processes the input from left to right, producing a rightmost derivation (in reverse) using k tokens of lookahead. The term can also be applied to a language that can be unambiguously parsed using this algorithm. *Contrast with* LL(k), LALR(k).

LRU *See* least recently used.

LRU bits a set of bits that records the relative recency of access among pairs of elements that are managed using an LRU replacement policy. If n objects are being managed, the number of bits required is $n(n-1)/2$. Upon each access, n bits are forced into certain states; to check one of the objects to determine whether it was the least recently accessed, n bits need to be examined.

LRU replacement *See* least recently used.

LRU stack a stack-based data structure to perform the bookkeeping for a least recently used (LRU) management policy. An object is promoted to the top of the stack when it is referenced; the object that has fallen to the bottom of the stack is the least recently used object.

LSB *See* least-significant bit.

LSN *See* log sequence number.

l-systems production-based approach to generate any concept of growth and life development in the resulting animation.

luminance (**1**) formally, the amount of light being emitted or reflected by a surface of unit area in the direction of the observer and taking into account the spectral sensitivity of the human eye.

(**2**) the amount of light coming from a scene. *See also* candela.

luminosity the relative quantity of radiation emitted by a light source.

LUV space a color space similar to the XYZ model, except that the components are scaled to be perceptually linear. The human eye perceives brightness on a logarithmic scale, and hence LUV components are logs of the corresponding XYZ values.

lvalue a value that can appear on the left-hand side of an assignment statement. Generally, an expression that evaluates to a location into which a value can be stored. *See also* rvalue.

LVQ *See* learning vector quantization.

LZ77 refers to string-based compression schemes based on Lempel and Ziv's 1977 method. An input string of symbols that matches an identical string previously (and recently) transmitted is coded as an offset pointer to the previous occurrence and a copy length.

LZ78 refers to string-based compression schemes based on Lempel and Ziv's 1978 method. A dictionary of prefix strings is built at the encoder and decoder progressively, based on the message. The encoder searches the dictionary for the longest string matching the current input, then encodes that input as a dictionary index plus the literal symbol which follows that string. The dictionary entry concatenated with the literal is then added to the dictionary as a new string.

LZA *See* leading-zeros anticipator.

LZD *See* leading-zeros detector.

LZP *See* leading-zeros predictor.

LZW coding *See* Lempel-Ziv-Welch coding.

M

M (mega) abbreviation for 1,048,576 (not for 1 million).

MAC *See* medium access control, mandatory access control.

MAC address synonym for an IEEE 802 address.

Mach band a perceived overshoot on the light side of an edge and an undershoot on the dark side of the edge. The *Mach band* is an artifact of the human visual system and not actually present in the edge. *See also* brightness, simultaneous contrast.

machine code (**1**) the native representation of a program for a specific machine architecture.

(**2**) source code in assembly language.

(**3**) an internal representation of the target machine instructions in a compiler. Often called machine language or object code.

machine epsilon the relative error when a number is rounded to the closest machine-representable number.

machine independent pertaining to software that can be executed on many platforms. *Compare with* portability.

machine interference the idle time experienced by any one machine in a multiple-machine system that is being serviced by an operator (or robot) and is typically measured as a percentage of the total idle time of all the machines in the systems to the operator (or robot) cycle time.

machine language the set of legal instructions to a machine's processor, expressed in binary notation. *See* machine code.

machine learning (**1**) In knowledge discovery, machine learning is most commonly used to mean the application of induction algorithms, which is one step in the knowledge discovery process. Machine learning is the field of scientific study that concentrates on induction algorithms and on other algorithms that can be said to "learn".

(**2**) the component of artificial intelligence that deals with the algorithms that improve with experience.

machine simulation that aspect of code generation that determines the preconditions for and postconditions of executing a particular instruction or instruction sequence.

machine translation translating a text in one natural language to another natural language by computer.

machine vision *See* robot vision.

macro a construct that specifies a source-to-source translation. A macro definition specifies the translation. When an instance of the macro occurs later in the program, it is expanded according to the definition. A macro definition may specify zero or more macro parameters that are replaced by text specified at the place the macro is used. Often the syntax and semantics of macros are substantially different from the syntax and semantics of the language in which they are used. The power of a system of macros may be as simple as straight textual substitution or as complex as a form of symbolic evaluation. *See also* macroprogram.

macro cycle the main repetitive set of activities that are performed in the main cycle of evolutionary life cycles, such as spiral.

macro development all the strategic activities related to the design of system architecture and system-level test.

macroinstruction (**1**) the lowest level of user-programmable computer instruction. *See* opcode.

(**2**) a shorthand for a number of language instructions or in integrated environments. In the latter, the definition of macros is a way to make shorter the execution and the writing of repeti-

0-8493-2691-5/01/$0.00+$.50

tive operations. In the first, it used for writing inline procedures.

Macro code is defined by the programmer, which the assembler or compiler will recognize and which will result in an inline insertion of a predifined block of code into the source code.

macrokernel a large operating system core that provides a wide range of services.

macroprogram a sequence of macroinstructions.

MAD *See* maximal area density.

magic cookie an opaque quantity, transmitted in the clear and used to authenticate access.

magic sets an evaluation technique for queries to deductive databases DB combines the nice properties of bottom–up reasoning (efficiency due to set–orientedness) with those of top–down reasoning (freedom of redundancy due to goal–orientedness). DB is rewritten into another deductive database magic(DB), such that the bottom–up evaluation of magic(DB) behaves like an efficient top–down evaluation of DB.

magnetic bubble memory a persistent memory consisting of a magnetic surface over which patterns of magnetic fields (bubbles) are transported by varying electromagnetic fields.

magnetic core memory a persistent, directly addressable memory consisting of an array of ferrite toruses (cores) each of which stores a single bit. Now obsolete.

magnetic disk a persistent, random-access storage device in which data is stored on a magnetic layer on one or both surfaces of a flat disk. *See also* hard disk, floppy disk, diskette.

magnetic domain small magnets on the surface of disks and tapes. The orientation of the magnets represents either a binary "1" or "0".

magnetic drum a persistent storage device in which data is stored on a magnetic layer on the surface of a cylinder. Now obsolete.

magnetic head *See* read/write head.

magnetic recording air gap term referring to two aspects of a magnetic recording system.

1. The gap between the poles of a read/write head is often referred to as the "air gap". Even though filled with a solid, it is magnetically equivalent to an air gap. With a recorded wavelength of λ and a head gap of d, the read signal varies as $sin(\pi x)/x$ where $x = \lambda/d$. The gap must then have $d < \lambda/2$ for reliable reading.

2. The space between the head and the recording surface is also referred to as an "air gap". The signal loss for a head at height h from the surface is approximately $55h/\lambda$ dB; if $d = \lambda/5$, the loss is 11 dB.

High recording densities therefore require heads "flying" very close to the recording surface and at a constant height – 0.1μm or less in modern disks. Such separations are achieved by shaping the disk head so that its aerodynamics force it to fly at the correct separation.

magnetic recording code method used to record data on a magnetic surface such as disk or tape. Codes that have been used include: return-to-zero (RZ), in which two signal pulses are used for every bit: a change from negative to positive pulse (i.e., magnetization in the "negative" direction) is used for a stored 0, a change from a positive to a negative pulse (i.e., magnetization in the "positive" direction) is used for a stored 1, and a return to the demagnetized state is made between bits; non-return-to-zero (NRZ), in which a signal pulse occurs only for a change from 1 to 0 or from 0 to 1; non-return-to-zero-inverted (NRZI), in which a positive or negative pulse is used for a — a sign change occurs for two consecutive 1s — and no pulse is used for a 0; double frequency (DF) (also known as frequency modulation (FM)), which is similar to NRZI but which includes an interleaved clock signal on each bit cell; phase encoding (PE), in which a positive pulse is used for a 1 and a negative pulse is used for a 0; return-to-bias (RB), in which magnetic transitions are made for 1s but not for 0s, as in NRZI, but with two transitions for each 1; and modified-return-to-bias (MRB), which is similar to RB, except that a return to the demagnetized state is made for each 0 and between two consecutive 1s.

RZ is self-clocking (i.e., a clock signal is not required during readout to determine where the bits lie), but the two-signals-per-bit results in low recording density; also, a 0 cannot be distinguished from a "dropout" (i.e., absence of recorded data). NRZ is not self-clocking and therefore requires a clocking system for readout; it also requires some means to detect the beginning of a record, does not distinguish between a dropout and a stored bit with no signal, and if one bit is in error, then all succeeding bits will also be in error up to the next signal pulse. NRZI has similar properties to NRZ, except that an error in one bit does not affect succeeding bits. PE is self-clocking.

magnetic resonance imaging (MRI) (**1**) a form of medical imaging with tomographic display that represents the density and bonding of protons (primarily in water) in the tissues of the body, based upon the ability of certain atomic nuclei in a magnetic field to absorb and reemit electromagnetic radiation at specific frequencies. Also called nuclear magnetic resonance.

(**2**) an imaging modality that uses a pulsed radio frequency magnetic field to selectively change the orientation of the magnetization vectors of protons within the object under study. The change in net magnetic moment as the protons relax back to their original orientation is detected and used to form an image.

magnetic tape a persistent, sequential-access storage medium in which data is stored on a magnetic layer on the surface of a tape. Often used for backups.

magnetic-tape track each bit position across the width of a magnetic tape read/write head, running the entire length of the tape.

magneto-resistive head *See* read/write head.

magnitude (**1**) the absolute value of a scalar.

(**2**) the norm of a vector, i.e., the square root of the sum of the squares of the vector components.

mail electronic messages sent from one user to another.

mailbox an operating system abstraction containing buffers to hold messages. Messages are sent to and received from the mailbox by processes.

mainframe a large centralized machine that supports hundreds of users simultaneously. Mainframes were often characterized by multiboard CPUs and relatively high cost.

main memory the memory directly accessible by the CPU. It may be addressed directly, or via a virtual memory system. Referred to in British usage as the store, and formerly referred to in the U.S. as core until magnetic core memory was replaced with solid state memory.

main program a software component that is called by the operating system.

maintainability (**1**) generally, the probability that an inoperable system will be restored to an operational state within the time t.

(**2**) for software, the ease with which it can be understood, corrected, adapted, and/or enhanced. Modification may include corrections, improvements, or adaptation of software to changes in the environment, and in requirements and functional specifications. *Maintainability* is defined as the effort to perform maintenance tasks, the impact domain of the maintenance actions, and the error rate caused by those actions.

maintenance assignment scope the amount of maintenance that one person or group can do on a certain system. When the maintenance assignment scope is high, the system is more maintainable.

maintenance department the corporate department that performs modification in software systems. These modifications may include adaptation, improvement, correction, porting, and so on.

maintenance factory a software tool that performs maintenance tasks on software in an automated fashion.

maintenance metric metric that tries to give a quantifying answer on how good a certain program is to maintain.

maintenance specialist a person that manages the maintenance process of a software system.

major cycle the largest sequence of repeating processes in cyclic or periodic systems.

major defect the relevant defect resulting in the failure of a software system or component that leads it to a non-operative condition. However, there are acceptable processing alternatives that could yield the desired result.

majority-logic decoding a simple, and in general suboptimal, decoding method for block- and convolutional codes based on the orthogonality of the parity-check sums.

majority protocol a concurrency control technique for distributed databases based on voting; if the transaction seeking a lock on an item receives a grant from the majority of the sites containing a copy, the lock is granted and the transaction may access the item.

make a Unix tool that allows the programmer to define a declarative model of the relationships between source files and generated files and that uses this model to build the system by recompiling only those files that have changed.

male connector a connector presenting pins to be inserted into a corresponding female connector that presents receptacles.

M-algorithm reduced-complexity breadth-first tree search algorithm, in which at most M tree nodes are extended at each stage of the tree.

Malhotra-Kumar-Maheshwari blocking flow given a flow function and its corresponding residual graph, the algorithm selects a vertex with the least throughput and greedily pushes the maximum flow from it to the sink. This is repeated until all vertices are deleted.

Malign failure *See* mishap.

malleability ease of modification of a software product. The product can be modified without modifying its design.

malleable job a parallel job that is able to handle changes in processor allocation during its execution even when the execution behavior of the job has not changed.

MAN *See* metropolitan-area network.

manager an entity that has to control other entities (of the same type) to accomplish a task.

Manchester adder an adder in which a Manchester carry-chain is used to propagate carries.

Manchester carry-chain a carry-chain that relies on the use of technology (pass transistors nowadays) to speed up the logical functions of the carry-chain.

mandatory access control (MAC) access controls used to enforce multiple levels of security by specifying classes of users and associating a set of privileges with each class, which cannot be modified by ordinary subjects. In military security, *mandatory access controls* are intended to prevent a user (or a Trojan horse executing on that user's behalf) from moving secret data to an unclassified file, where uncleared users might read it. Special trusted subjects used by system administrators may be permitted to violate these constraints in limited ways.

mandatory maintenance maintenance actions that need to be addressed in order to guarantee the system functionalities.

mandatory work first a static two-pass process that first traverses the minimal game tree and uses the provisional value found to improve the pruning during the second pass over the remaining tree.

Mandelbrot set the set of complex constants c_i for which the orbits of the function

$$f(z) = g(z) + c_i \ ,$$

evaluated at the initial condition of $z_0 = 0$, do not escape. The Mandelbrot set is usually found for the function $g(z) = z^2$.

Manhattan distance the distance between two points measured along axes at right angles. In a plane with p_1 at (x_1, y_1) and p_2 at (x_2, y_2), it is $|x_1 - x_2| + |y_1 - y_2|$. Also called city-block distance. *See also* Euclidean distance, l_m distance.

Manhattan routing a popular rectilinear channel routing model in which paths for disjoint nets can cross (a vertical segment crosses a horizontal segment) but cannot contain segments that overlap in the same direction at even a point.

manifestation of a latent fault the event in which a latent fault gives rise to a failure.

manifest constant a constant that has a symbolic name, but whose value is fixed and determinable at compile time. A special subclass of named constant.

manipulator workspace a manipulator workspace defines all existing manipulator positions and orientations which can be obtained from the inverse kinematics problem. The lack of a solution means that the manipulator cannot attain the desired position and orientation because it lies outside of the manipulator's workspace.

mantissa (**1**) the portion of a floating-point number that represents the digits. For example, if a floating-point number is represented as a computation $m \times b^e$, the value m is the mantissa. *See also* exponent, floating-point representation, significand.

(**2**) in the textual representation of a floating point number, the numeric characters (and optional sign) preceding the (optional) exponent indicator. For example, in $-3.4E^{-5}$, the value -3.4 is the mantissa.

many-one reduction a reduction that maps an instance of one problem into an equivalent instance of another problem.

MAP *See* maximum *a posteriori* estimator.

MAP estimator *See* maximum *a posteriori* estimator.

mapping the assignment of one location to a value from a set of possible locations. Often used in the context of memory hierarchies, when distinct addresses in a level of the hierarchy map a subset of the addresses from the level below.

MAR *See* memory address register.

mark-and-scan a means of garbage collection by which objects are located by starting at a set of root objects and following, from each object, its pointers to other objects. If, in this traversal, an object is found to be already marked, that traversal path is terminated. When there are no more traversals possible, the list of all allocated objects is examined (the scan) and any object which is not marked is determined to be unreachable, and the storage it occupies is returned to the heap (also known as the free storage pool). *See also* incremental garbage collection.

Markov algorithm a representation of a computation as a set of transformation rules. A Markov algorithm is Turing equivalent.

Markov chain a particular case of a Markov process, where the samples take on values from a discrete and countable set. Markov chains are useful signal generation models for digital communication systems with intersymbol interference or convolutional coding, and Markov chain theory is useful in the analysis of error propogation in equalizers, in the calculation of power spectra of line codes, and in the analysis of framing circuits.

Markovian routing a routing policy where the path from the current node to each destination depends only on the destination, and does not depend on any other factors such as the source node or the arriving switch port number.

Markov model a modeling technique where the states of the model correspond to states of the system and transitions between the states in the model correspond to system processes.

Markov process a discrete-time random process, $\{\Psi_k\}$, that satisfies $p(\psi_{k+1}|\psi_k,\psi_{k-1},\ldots) = p(\psi_{k+1}|\psi_k)$. In other words, the future sample ψ_{k+1} is independent of past samples $\psi_{k-1}, \psi_{k-2}, \ldots$ if the present sample $\Psi_k = \psi_k$ is known.

Markov random field an extension of the definition of Markov processes to two dimensions. Consider any closed contour Γ, and denote by Γ_i and Γ_o the points interior and exterior to Γ, respectively. Then a process ψ is a Markov random field (MRF) if, conditioned on $\psi(\Gamma)$, the sets $\psi(\Gamma_i)$ and $\psi(\Gamma_o)$ are independent. That is,

$$p\left(\psi\left(\Gamma_i\right), \psi\left(\Gamma_o\right) | \psi(\Gamma)\right) = p\left(\psi\left(\Gamma_i\right) | \psi(\Gamma)\right) \cdot p\left(\psi\left(\Gamma_o\right) | \psi(\Gamma)\right) .$$

See also Markov process, conditional statistic.

markup language one of any languages for annotation of source code to simply improve the source code's appearance with the means of bold-faced key words, slanted comments, etc. In computerized document preparation, a method of adding information to the text indicating the logical components of a document, or instructions for layout of the text on the page or other information which can be interpreted by some automatic system.

Marr-Hildreth operator (**1**) edge-detection operator, also called Laplacian-of-Gaussian or Gaussian-smoothed-Laplacian, defined by

$$\nabla^2 G = -\frac{1}{\sqrt{2\pi}\sigma^3}\left(1 - \frac{x^2+y^2}{\sigma^2}\right) e^{\frac{-(x^2+y^2)}{2\sigma^2}} .$$

It generates a smoothed isotropic second derivative. Zero crossings of the output correspond to extrema of first derivative and thus include edge points.

(**2**) the complete edge detection scheme proposed by Marr and Hildreth, including use of the $\nabla^2 G$ operator at several scales (i.e., Gaussian variances), and aggregation of their outputs.

m-ary hypothesis testing the assessment of the relative likelihoods of M hypotheses $H_1, H_2, \ldots, H_M$. Normally we are given prior statistics $P(H_1), \ldots, P(H_M)$ and observations $\mathbf{y}$ whose dependence $p(\mathbf{y}|H_1), \ldots, p(\mathbf{y}|H_M)$ on the hypotheses are known. The solution to the hypothesis testing problem depends upon the stipulated criterion; possible criteria include maximizing the posterior probability (MAP) or minimizing the expected "cost" of the decision (a cost C_{ij} is assigned to the selection of hypothesis j when i is true). *See also* binary hypothesis testing, conditional statistic, prior statistics, posterior statistics, maximum *a posteriori* estimator.

mask (**1**) in digital computing, to specify a number of values that allow some entities in a set, and disallow the others in the set, from being active or valid. For example, masking an interrupt.

(**2**) in image processing, a small set of pixels, such as a 3×3 square, that is used to transform an image. Conceptually, the mask is centered above every input pixel, each pixel in the mask is multiplied by the corresponding input pixel under it and the output (transformed) pixel is the sum of these products. If the mask is rotated 180° before the arithmetic is performed, the result is a 2D convolution and the mask represents the impulse response function of a linear, space-invariant system. Also called a kernel. *See also* convolution.

(**3**) for semiconductor manufacturing, a device used to selectively block photolithographic exposure of sensitized coating used for preventing a subsequent etching process from removing material. A mask is analogous to a negative in conventional photography.

maskable interrupt interrupt that can be postponed to permit a higher-priority interrupt by setting mask bits in a control register. *See also* non-maskable interrupt.

masking a phenomenon in human vision in which two patterns P_1 and $P_1 + P_2$ cannot be discriminated even though P_2 is visible when seen alone. P_1 is said to mask P_2.

mask programming programming a semiconductor read-only-memory (ROM) by modifying one or more of the masks used in the semiconductor manufacturing process.

mask register (1) in associative memory a device that is used to block out certain bits in the comparand that are not to be checked in the retrieval process.

(2) in interrupts it contains a bit map either enabling or disabling specific interrupts.

massive concurrency situation in database systems where there are a very large number of concurrent transactions.

massively parallel architecture a computer system architecture characterized by the presence of large numbers of CPUs that can execute instructions in parallel. The largest examples can process thousands of instructions in parallel, and provide efficient pathways to pass data from one CPU to another.

massively parallel processor a system that employs a large number, typically 1000 or more, of processors operating in parallel.

mass storage a storage for large amounts of data.

master the system component responsible for controlling a number of others (called slaves).

master control relay (MCR) used in programmable logic controllers to secure entire programs, or just certain rungs of a program. An MCR will override any timer condition, whether it be time-on or time-off, and place all contacts in the program to a safe position whenever conditions warrant.

master-slave flip-flop a two-stage flip-flop in which the first stage buffers an input signal, and on a specific clock transition the second stage captures and outputs the state of the input.

master theorem a theorem giving a solution in asymptotic terms for recurrence relations of the form $T(n) = aT(n/b) + f(n)$ where $a \geq 1$ and $b > 1$ are constants and n/b means either $\lfloor n/b \rfloor$ or $\lceil n/b \rceil$.

MAT medial axis transform. *See* medial axis.

match the result of comparing two instances of the same symbol.

matched edge an edge that is in a matching. *See also* free edge, matched vertex.

matched filter a operation employing an image that gives back high values when a mask is located on top of the object of interest and lower values elsewhere.

matched vertex a vertex on a matched edge in a matching, or one which has been matched. *See also* free vertex.

matching (1) a subgraph in which every vertex has a degree of at most one. In other words, no two edges share a common vertex.

(2) the problem of finding such a subgraph. *See also* perfect matching, free edge, matched edge, free vertex, matched vertex.

material the collective set of properties of the surface of an object that determines how it will reflect or transmit light.

math coprocessor a separate chip, additional to the CPU, that offloads many of the computation-intensive tasks from the CPU.

mathematical modeling a mathematical description of the interrelations between different quantities of a given process. In particular, a mathematical description of a relation between the input and output variables of the process. *See also* truth model and design model.

mathematical morphology an algebraic theory of non-linear image transformations based on set-theoretical (or lattice-theoretical) operations, which is generally considered a counterpart to the classical linear filtering approach of signal processing. This methodology in image processing arose in 1964 through the works of Matheron and Serra at Fontainebleau (France), was developed in the 1970s by Sternberg at Ann Arbor, Michigan, and became internationally known in the 1980s. It has been successfully applied in various fields requiring an analysis of the structure of materials from their images, for example biomedical microscopy, stereology, mineralogy, and petrography. *See* closing, dilation, erosion, morphological operator, opening.

matrix a two-dimensional array. The first index is the row; the second index is the column.

An n by m matrix is an array of $n \times m$ numbers of height n and width m representing a linear map from an m-dimensional space into an n-dimensional space. An example of a 2×2 real matrix is $A = \begin{bmatrix} 2 & 3 \\ 5 & 6 \end{bmatrix}$. *See also* circulant matrix, Hermitian matrix, orthogonal matrix, positive definite matrix, positive semi-definite matrix, singular matrix, Toeplitz matrix, square matrix, rectangular matrix, sparse matrix, lower triangular matrix, upper triangular matrix, ragged matrix, uniform matrix.

matrix-chain multiplication problem given a sequence of matrices such that any matrix may be multiplied by the result of multiplying all the matrices before it, finding the best parenthesizing such that the result is obtained with the minimum number of arithmetic operations. Usually dynamic programming yields the best parenthesizing.

matrix of configuration of information system in Pawlak's information system $S = (U, A)$ whose universe U has n members x_i and the set A consists of attributes $\mathbf{a}_j$, the elements of the universe are linked with each other. Connections between the elements of the universe U may be given, for example, by a function

$$\varphi : U \times U \to \{-1, 0, 1\} .$$

The pair (U, φ) is called the configuration of the information system S. The matrix of the information system S is an $n \times n$ matrix whose elements are

$$c_{ij} = \varphi\left(x_i, x_j\right) .$$

Matte shading *See* Lambert's law.

maturity the capability of the software product to avoid failure as a result of faults in the software. Attributes of software that bear on the frequency of failure by faults in the software.

maturity level name for the 5 levels of maturity in the Capability Maturity Model. *See also* capability maturity model.

Mauchley, John Graham (1907–1980) Born: Cincinnati, Ohio.

Mauchley is best known as one of the designers of ENIAC, an early electronic computer. It was Mauchley, in 1942, who wrote a proposal to the Army for the design of a calculating machine to calculate trajectory tables for their new artillery. Mauchley was a lecturer at the Moore School of Electrical Engineering at the time. The school was awarded the contract and he and J. Presper Eckert were the principal designers of ENIAC. They later went on to form their own company. Mauchley was the software engineer behind the development of one of the first successful commercial computers, UNIVAC I.

max-flow min-cut in a network with weighted edges, a cut-set is a set of edges the removal of which will disconnect a given source node from a given sink node. A minimal cut-set is a cut-set with the minimum possible sum of the weights of the edges. The maximum possible flow through the network from source to sink is equal to the sum of the weights of the edges of a minimal cut set.

maximal area density (MAD) for a magnetic disk, the maximum number of bits that can be stored per square inch. Computed by multiplying the bits per inch in a disk track times the number of tracks per inch of media.

maximal independent set a set of vertices in a graph such that for any pair of vertices, there is no edge between them and such that no more vertices can be added and it still will be an independent set.

maximum accumulated matching a defuzzification scheme for a classification problem in which a pattern is assigned to the class output by the rules that accumulate the maximum firing degree.

maximum *a posteriori* (MAP) estimator to estimate a random $\mathbf{x}$ by maximizing its posterior probability; that is,

$$\hat{x}_{MAP} = \arg_x \max p(\mathbf{x}|\mathbf{y}) ,$$

where **y** represents an observation. **x** is explicitly modeled as a random quantity with known prior statistics. *See also* maximum likelihood estimation, Bayesian estimator, Bayes' rule, prior statistics, posterior statistics.

maximum *a posteriori* probability the probability of a certain outcome of a random variable given certain observations related to the random variable; i.e., x was transmitted and y was received, then $P(x|y)$ is the *a posteriori* probability. The maximum *a posteriori* probability is found by considering all valid realizations of x.

For example, between two strings A and B is defined as

$$D_{MPP} = Pr\{B|A\} ,$$

where $Pr(B|A)$ is the probability that A is changed into B. Sometimes called maximum posterior probability (MPP) even though, strictly speaking it is a similarity rather than distance measure.

maximum bipartite matching *See* bipartite matching.

maximum distance separable code an (n, k) linear binary or nonbinary block code, with minimum distance $d_{\min} = n - k + 1$. Except for the trivial repetition codes, there are no binary maximum distance separable (MDS) codes. Nonbinary MDS codes like Reed-Solomon codes do exist.

maximum entropy a procedure that maximizes the entropy of a signal process.

maximum entropy restoration an iterative method of image restoration. At each iteration, an image is chosen whose Fourier transform agrees with that of the principal solution, i.e., the model of degradation, while maximizing the entropy.

maximum-likelihood the maximum of the likelihood function, $p(y|x)$ or equivalently, the log-likelihood function. Maximum-likelihood is an optimality criterion that is used for both detection and estimation.

maximum-likelihood decoding a scheme that computes the conditional probability for all the code words given the received sequence and identifies the code word with maximum conditional probability as the transmitted word. Viterbi algorithm is the simplest way to realize *maximum-likelihood decoding.*

maximum-likelihood estimation to estimate an unknown **x** by maximizing the conditional probability of the observations; that is,

$$\hat{x}_{ML} = \arg_x \max p(\mathbf{y}|\mathbf{x}) ,$$

where **y** represents an observation. **x** is considered unknown; it is not a statistical quantity. *See also* maximum a *posteriori* estimator, Bayesian estimator, Bayes' rule, prior statistics.

maximum matching a defuzzification scheme for a classification problem in which a pattern is signed to the class output by the rule that achieves the maximum firing degree.

maximum posterior probability distance *See* maximum *a posteriori* probability.

Max-Lloyd scalar quantization *See* Lloyd-Max scalar quantization.

Max-Lloyd SQ *See* Lloyd-Max scalar quantization.

max-min composition a frequently used method of composition of two fuzzy relations which, as the name implies, makes use of the min operation followed by the max operation.

max operation an operation on two or more variables where the resultant value is formed by taking the largest value, or maximum, among these variables.

MAX-SNP a complexity class consisting of problems that have constant-factor approximation algorithms, but no approximation schemes unless $P = NP$.

maxterms a Boolean sum term in which each variable is represented in either true or complement form only once. For example, $x + y' + w + z'$ is a maxterm for a four variable function.

MBCS *See* multibyte character set.

McCabe's metric a widely used code metric based on a measure of flow-of-control complexity called cyclomatic complexity.

The cyclomatic complexity is computed from the program flow graph and provides a hard measure of system complexity and reliability. For a given program flow graph, let e be the number of edges, n be the number of nodes, and p be the number of independent processes or disjoint flow graphs. Then the cyclomatic complexity is

$$C = 2e - n + 2p .$$

In theory, the higher the cyclomatic complexity, the lower the program reliability and program effort. For example, a piece of spaghetti code would tend to yield a very high cyclomatic complexity. However, a meaningful range can be developed based on historical data for a given application area.

McCabe's metric can be computed automatically and commercial programs exist to do this for a wide variety of programming languages. *See also* Halstead's metrics.

McCulloch-Pitts neuron originally a linear threshold unit that responded with a binary output at time $t + 1$ to an input applied at time t. In current usage, usually a linear threshold unit.

MC many-one reducibility a property of some formal languages. A language $\mathcal{L}$ is many-one reducible or NC reducible to $\mathcal{L}'$, written $\mathcal{L} \leq_m^{NC} \mathcal{L}'$ if there is a function f in FNC such that $x \in \mathcal{L}$ if and only if $f(x) \in \mathcal{L}'$.

MCR *See* master control relay.

M/D/1 queueing notation for a queue with a Poisson arrival (Markovian or "M" arrivals) process, a deterministic (D), or a fixed service and a single server. *See also* Kendall's classification.

MDA *See* monochrome display adapter.

MDR *See* memory data register.

MDS code *See* maximum distance separable code.

Mealy machine the model of a finite state machine in which the output actions are associated with the transition for changing state. *See* Moore machine.

mean *See* mean value.

mean ergodic theorem a mathematical theorem that gives the necessary and/or sufficient conditions for a random process to be ergodic in mean. Let $x(n)$ be a wide-sense stationary random process with autocorrelation sequence $c_x(k)$. A necessary and sufficient condition for $x(n)$ to be ergodic in the mean is:

$$\lim_{N->\infty} \sum_{k=0}^{N-1} c_x(k) = 0 .$$

The sufficient conditions for $x(n)$ to be ergodic in the mean are that

$$c_x(0) < \infty \quad \text{and} \quad \lim_{k->\infty} c_x(k) = 0 .$$

mean filter a filter that takes the mean of the various input signal components, or in the case of an image, the mean of all the pixel intensity values within the neighborhood of the current pixel.

mean of a random variable *See* expected value of a random variable.

mean of a stochastic process the expected value of a stochastic process at some point in time.

mean opinion score a subjective method of quantitatively assessing the quality of signals. Subjects rate the quality of presented signals (e.g., images) on a quality or impairment scale. The mean of these ratings is calculated as the mean opinion score.

mean queue length *See* mean queue size.

mean queue size the mean number of requests waiting in a queue. The mean queue length can be defined based on arbitrary observations (time-based mean) or as the value observed by arrivals or departures. Note that these three mean queue lengths generally are not equal to each other.

mean/residual vector quantization (MRVQ) a prediction is made of the original image based on a limited set of data, and then a residual image is formed by taking the difference between the prediction and the original image. Prediction data are encoded using a scalar quantizer and the residual image is encoded using a vector quantizer.

means-ends analysis an artificial intelligence technique that tries to reduce the difference between a current state and a goal state.

mean squared error (MSE) measure of the difference between a discrete time signal x_i, defined over $[1 \dots n]$, and a degraded, restored, or otherwised processed version of the signal $\hat{x}_i$, defined as $MSE = \frac{1}{n}\sum_{i=1}^{n}(x_i - \hat{x}_i)^2$. MSE is sometimes normalized by dividing by $\sum_{i=1}^{n}(x_i)^2$.

mean-square estimation an estimation scheme in which the cost function is the mean-square error.

mean time between failures (MTBF) the average time (usually expressed in hours) that a component works without failure. It is calculated by dividing the total number of failures by the total number of operating hours observed. Thus, the term is related to the time a user may reasonably expect a device or system to work before an incapacitating fault occurs.

mean time to failure (MTTF) the mean time elapsed until a failure occurs.

mean time to interrupt (MTTI) the mean time elapsed until a failure is detected and isolated.

mean time to repair (MTTR) the mean time that a device will take to recover from a non-terminal failure. The MTTR is usually part of a maintenance contract, where the user would pay more for a system whose MTTR was 24 hours, than for one of, say, 7 days. This means the supplier is guaranteeing to have the system up and running again within 24 hours (or 7 days). Fault tolerant devices have an MTTR of zero.

mean value the expected value of a random variable or function. The mean value of a function f is defined as

$$m_f = E(f) = \int_{-\infty}^{\infty} f p(f)\, df \,,$$

where $p(f)$ is the probability density function of f.

mean value analysis (MVA) an iterative technique for calculating the mean values of device utilization, throughput, queue length, and response time of a product-form queueing network. MVA iterates on the number of customers in the system, whereas convolution iterates on the number of devices in the system.

measure the number or category assigned to an attribute of an entity by making a measurement. Synonymous with metric.

measurement the process of empirical, objective assignment of numbers (or symbols) to properties of entities (objects and events) in the real world in such a way as to describe them. The use of a metric to assign a value from a scale (which may be a number or category) to an attribute of that entity. Quantitative evaluation (of a process or artifact).

measurement specialist a person who is engaged full time in measuring software. Typical job in a very large organization. Counting function points, defect removal, complexity, predicted quality, etc.

measurement system the sum of all stimulus and response instrumentation, device under test, interconnect, environmental variables, and the interaction among all the elements.

measurement tool a tool calculating metrics that are used to assign a value or a category to a predefined characteristic of a product.

mechanical part feeder mechanical device for feeding parts to a robot with a specified frequency and orientation. They are classified as vibratory bowl feeders, vibratory belt feeders, and programmable belt feeders.

mechanism an operating system function that can be used to implement many different policies without commitment to any specific policy.

medial axis a concept used in the development of a morphological skeleton.

Let X be a non-empty bounded set in a Euclidean space; assume that X is topologically closed, in other words it contains its border ∂X. For every point x in X, the Euclidean distance $d(x, X^c)$ of x to the complement X^c of X is equal to the distance $d(x, \partial X)$ of x to the border ∂X; let $B(x)$ be the closed ball of radius $d(x, \partial X)$ centered about x; it is the greatest closed ball centered about x and included in X. The medial axis transfer is the operation transforming X into its medial axis or distance skeleton $S(X)$ which can be defined in several ways:

1. $S(X)$ is the set of points x such that $B(x)$ is not included in $B(y)$ for any other y in X; in other words $B(x)$ is, among all balls included in X, maximal for the inclusion.

2. $S(X)$ is the set of points x such that $B(x)$ intersects ∂X in at least two points.

3. $S(X)$ is the set of points x at which the distance function $f(x) = d(x, \partial X)$ is not differentiable.

These three definitions coincide up to closure, in the sense that the three skeletons that they define may be different, but have the same topological closure. When X is connected and has a connected interior, its skeleton $S(X)$ is also connected and has the same number of holes as X. For digital figures and a digital distance d, one uses only the first definition: the skeleton $S(X)$ is the set of centers of maximal "balls" (in the sense of distance d) included in X; but then the skeleton of a connected set is no longer guaranteed to be connected. *See* distance, morphological skeleton.

medial axis transform (MAT) *See* medial axis.

median filter a filter that takes the median of the various input signal components, or in the case of an image, the median of all the pixel intensity values within the neighborhood of the current pixel, the median being defined as the center value of the ordered signal components.

medical imaging a multi-disciplinary field that uses imaging scanners to reveal the internal anatomic structure and physiologic processes of the body to facilitate clinical diagnoses. *See also* magnetic resonance imaging, positron emission tomography, and radiography.

medium access control (MAC) the layer defined by the IEEE 802 committee that deals with specifics of each type of LAN (for instance, token passing protocols on token passing LANs).

medium scale integration (MSI) an early level of integration circuit fabrication that allowed approximately between 12 and 100 gates on one chip.

megaflops (MFLOPS) millions of floating point operations per second. Usually applied as a measurement of the speed of a computer when executing scientific problems and describes how many floating point operations were executed in the program.

member a named component of a composite object. *See also* field, record, structure.

member function in object-oriented programming, a member function is the C++ name for a method defined inside the class that can operate freely on the data members. The main characteristic of member functions is the hiding of internal class architecture and design, providing interface to data and functionalities provided by the class.

member of a class in the intentional meaning, members of a class are its attributes. Extensionally, they are its instances, objects.

member record the record on the N-side of a relationship in the network model.

membership function a possibility function, with values ranging from 0 to 1, which describes the degree of compatibility, or degree of truth, that an element or object belongs to a fuzzy set. A membership function value of 0 implies that the corresponding element is definitely not an element of the fuzzy set while a value of 1 implies definite membership. Values between 0 and 1 imply a fuzzy (non-crisp) degree of membership.

Rigorously, let $\mu(.)$ be a membership function defining the membership value of an element x of an element of discourse X to a fuzzy set. If A is a fuzzy set, and x is an element of A, then the membership function takes values in the interval [a,b] of the real line ($\mu_A(x) : X \Rightarrow [a, b]\ ,\ a, b \in \Re$).

Usually fuzzy sets are modeled with a normalized membership function ($\mu_A(x) : X \Rightarrow [0, 1]$). Some examples of membership functions are shown in the figure. If A is a crisp (nonfuzzy) set, $\mu_A(x)$ is 1 if x belongs to A and 0 otherwise ($\mu_A(x) : U \Rightarrow \{0, 1\}$). *See also* crisp set, fuzzy set.

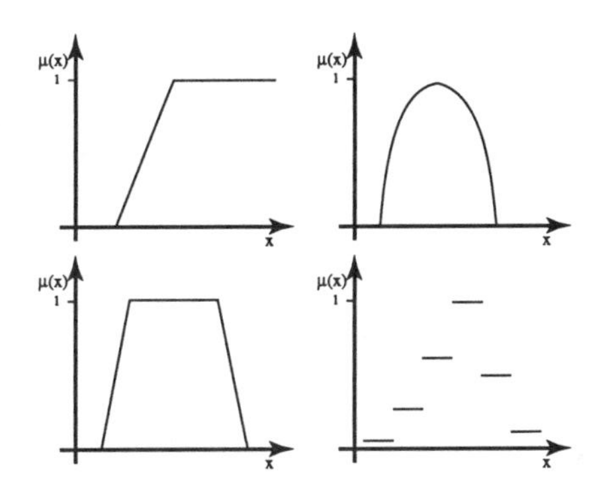

Membership function: some examples of normalized fuzzy sets.

membership grade the degree to which an element or object belongs to a fuzzy set. It is also referred to as degree of membership.

membership relationship a non-configurational whole-part relationship identified initially in the cognitive science literature and applied more recently to object modeling.

memory (1) area for storing computer instructions and data for either short-term or long-term purposes.

(2) the property of a display pixel that allows it to remain stable in an initially established state of luminance. Memory gives a display high luminance and absence of flicker.

memory access time the time from when a read (i.e., load) request is submitted to memory to the time when the corresponding data becomes available. Usually smaller than the memory cycle time.

memory address computation the computation required to produce an effective memory address; may include indexing and translation from a virtual to a physical address.

memory address register (MAR) a register inside the CPU that holds the address of the memory location being accessed while the access is taking place.

memory alignment matching data to the physical characteristics of the computer memory. Computer memory is generally addressed in bytes, while memories handle data in units of 4, 8, or 16 bytes. If the "memory width" is 64 bits, then reading or writing an 8 byte (64 bit) quantity is more efficient if data words are aligned to the 64-bit words of the physical memory. Data which is not aligned may require more memory accesses and more-or-less complex masking and shifting, all of which slow the operations.

Some computers insist that operands be properly aligned, often raising an exception or interrupt on unaligned addresses. Others allow unaligned data, but at the cost of lower performance.

memory allocation (1) the act of reserving memory for a particular process.

(2) the amount of random access memory (RAM) which is allocated by the operating system to a particular application during run time to support the application's specific activities.

memory bandwidth the maximum amount of data per unit time that can be transferred between a processor and memory.

memory bank a subdivision of memory that can be accessed independently of (and often in parallel with) other memory banks.

memory bank conflict conflict when multiple memory accesses are issued to the same memory bank, leading to additional buffer delay for such accesses that reach the memory bank while it is busy serving a previous access. *See also* interleaved memory.

memory block contiguous unit of data that is transferred between two adjacent levels of a memory hierarchy. The size of a block will vary according to the distance from the CPU, increasing as levels get farther from the CPU, in order to make transfers efficient.

memory bounds register register used to ensure that references to memory fall within the space assigned to the process issuing the references; typically, one register holds a lower bound, another holds the corresponding upper bound, and accesses are resricted to the addresses delimited by the two.

memory bus definition can be architecture-dependent. Typically, the bus is the high-speed interface connecting processor and secondary (or L2) cache to the memory board. All data from memory flow between the memory board and the secondary cache via this bus. Since processors can typically execute instructions much more quickly than the instructions and data can be fetched from the memory, the speed of this bus is a limiting factor on system performance.

memory caching a technique in which frequently used segments of main memory are stored in a faster bank of memory that is local to the CPU (called a cache). To avoid slowing down the processor with accesses on the memory bus, instructions and data fetched from main memory are placed in the cache, either connected to the processor by a special bus or on-chip with the processor. The higher the cache hit ratio, the better the processor throughput.

memory cell a part in a semiconductor memory holding one bit (a zero or a one) of information. A memory is typically organized as a two-dimensional matrix of cells, with "word lines" running horizontally through the rows, and "bit lines" running vertically connecting all cells in that column together. *See also* bit line.

memory compaction the shuffling of data in fragmented memory in order to obtain sufficiently large holes. *See also* memory fragmentation.

memory consistancy the programmer-visible mechanism that guarantees that multiple processor elements in a computer system receive the same value on a request to the same shared-memory address.

memory cycle the sequence of states of a memory bus or a memory (sub-)system during a read or write. A *memory cycle* is usually uninterruptible.

memory cycle time the time that must elapse between two successive memory operations. Usually larger than the memory access time.

memory data register (MDR) a register inside the CPU that holds data being transferred to or from memory while the access is taking place.

memory density the amount of storage per unit; specifically, the amount of storage per unit surface or per chip.

memory element a bistable device or element that provides data storage for a logic 1 or a logic 0.

memory fragmentation *See* internal fragmentation, external fragmentation.

memory hierarchy a system consisting of multiple levels with different types of memory at each level. Typically the read/write speed of the device increases and its storage capacity decreases as we move up the hierarchy and hence closer to the CPU. *See* hierarchical memory.

memory interleaving *See* interleaved memory.

memory latency the time between the issue of a memory operation and the completion of the operation. May be less than the time for a memory cycle.

memory leakage the loss of free memory space due to programming errors. These may be caused by dynamically allocating memory without de-allocating it after the exploitation of the reserved memory. The memory is no longer referenced and users cannot return automatically available without implementing recovery techniques such as a garbage collection.

memoryless property a nonnegative random variable X where

$$P[X > x + z | X > z] = P[X > x] ,$$

i.e., if X is taken to be the duration of a time interval, the distribution of the duration of the rest of the interval does not depend on the time that has already passed since the start of the interval, provided that the interval has not yet ended. The only distributions having the memoryless property are the exponential distribution for continuous random variables and the geometric distribution for discrete random variables.

memoryless system a system whose outputs only depend on present values, not past or future values, i.e., $y[n_0]$ depends only on $x[n_0]\ \forall\ n_0$.

memory management allocating memory to programs and routines.

memory management unit (MMU) a hardware device that interfaces between the central processing unit (CPU) and memory, and may perform memory protection, translation of virtual to physical addresses, and other functions. An MMU will translate virtual addresses from the processor into real addresses for the memory.

memory map a diagram that shows where programs and data are stored in a computer's memory.

memory-mapped I/O an input/output scheme where reading or writing involves executing a load or store instruction on a pseudo-memory address mapped to the device. *Contrast with* DMA, programmed I/O.

memory mapped I/O I/O scheme in which I/O control and data "registers" and buffers are locations in main memory and are manipulated through the use of ordinary instructions. Computers with this architecture do not have specific I/O commands. Instead, devices are treated as memory locations. This simplifies the structure of the computer's instruction set.

memory mapping **(1)** the extension of a processor-generated address into a longer address, in order to create a large virtual address or to extend a virtual address that is too short to address all of real memory when translated.

(2) the mapping between the logical memory space and the physical memory space, i.e., the mapping of virtual addresses to real addresses.

memory module a physical component used in the implementation of a memory. *See also* memory bank, interleaved memory.

memory partitioning when multiple processes share the physical main memory, the memory is partitioned between the processes. If the partitioning is static, and the amount of main memory for each process is not changed during the execution, the memory partitioning is fixed; otherwise it is said to be dynamic.

memory port an access path to a memory unit.

memory prefetching speculatively fetching a single or a collection of memory words in advance before an access to it is made with the hope of reducing the memory access delay.

memory protection a method for ensuring that multiple users or programs do not interfere with each other's memory space. For example, a process may have no access, or read-only access, or read/write access to a given part of memory. The control is typically provided by a combination of hardware and software.

memory reference a read of one item of data (usually a word) from memory or a write of one item of data to memory (same as memory reference).

memory reference instruction an instruction that communicates with virtual memory, writing to it (store) or reading from it (load).

memory refresh the process of recharging the capacitive storage cells used in dynamic RAMs. DRAMs must have every row accessed within a certain time window or the contents will be lost. This is done as a process more or less transparent to the normal functionality of the memory, and affects the timing of the DRAM.

memory select line a control line used to determine whether a unit of memory will participate in a given memory access.

memory stride the difference between two successive addresses presented to memory. An interleaved memory with a simple assignment of addresses performs best when reference strides are 1, as the addresses then fall in distinct banks.

memory swapping the transfer of memory blocks from one level of the memory hierarchy to the next lower level and their replacement with blocks from the latter level. Usually used to refer to pages being moved between main memory and disk.

memory utilization the amount of RAM consumed by an application normalized by the total RAM available within a processing platform. Value scales between 0 and 1, inclusive, with 1 implying consumption of all available RAM on the processing platform.

memory width the number of bits stored in a word of memory. The same as the width.

memory word the total number of bits that may be stored in each addressable memory location.

MEMS *See* microelectromechanical system.

mental simulation simulating the execution of all or part of a program or algorithm as a means of testing its correctness.

mercury delay-line memory *See* ultrasonic memory.

merge sort a sort algorithm which splits the items to be sorted into two groups, recursively sorts each group, and merges them into a final, sorted sequence. *See also* divide-and-conquer, external sorting.

meromorphic function a function that is analytic, except for a finite number of poles (at which the function ceases to be defined).

mesh processor a processor configuration that is similar to the linear array processor except that each processor element also communicates data north and south.

MESI protocol a cache coherence protocol for a single bus multiprocessor. Each cache line exists in one of four states: modified (M), exclusive (E), shared (S), or invalid (I).

mesopic formally, a description of luminances under which both human rod and cone cells are active. The mesopic luminance range lies between the photopic and scotopic ranges. Informally, describing twilight luminances. *See also* photopic, scotopic.

message a block of information that may be exchanged by concurrent program units as a means of communication. A codeword that identifies the receiver of the message and may also contain arguments. It can be viewed as a request for a service to be provided by the server object to the client object. A message can contain commands or data as well as a their combination.

message frequency the frequency with which inter-process messages are sent by one application to another.

message passing **(1)** in object orientation, the exchange of messages between objects. Objects communicating by sending messages is an underpinning tenet of the object-oriented paradigm.

(2) method of interprocess communication that relies on passing messages between processes via message queues.

message-passing system a multiprocessor system that uses messages passed among the processors to coordinate and synchronize the activities in the processors, as opposed to reading and writing common memory areas.

message size the size (in appropriate units, such as Kilobytes) of a message sent via simple messaging.

message switching a service-oriented class of communication in which messages are exchanged among terminating equipment by traversing a set of switching nodes in a store-and-forward manner. This is analogous to an ordinary postal system. The destination terminal need not be active at the same time as the originator in order for the message exchange take place.

metaball modeling a type of modeling based on the production of objects using spheres that attract and cling to each other according to their proximity to one another and their field of influence (the size of their attractive field). This form of modeling may also use cubes and other shapes, depending upon the modeler. *Metaball modeling* is particularly useful for creating organic objects and animation effects such as a group of mercury balls moving together and combining to form an object like a soda can.

meta-character a special character used in file name patterns and regular expressions.

metaclass (1) in object-orientation, the representation of a concept at the metalevel, i.e., a class which exists within the metamodel. For instance, whereas there are many classes in a typical model of an application, in the metamodel there is a single metaclass called Class. In fully interpretate object-oriented systems, the instances of a metaclass are the classes for the system.

(2) in database systems, a class whose members and instances are themselves classes rather than objects. Used to query the schema of an object-oriented database.

metadata data that describes other data. *See also* schema. The term "metadata" is often used in reference to information which describes data of a nontextual nature, such as images, sound clips, and video clips.

meta-EP a variation of evolutionary programming in which, instead of specifying the variances associated with the mutation process as a transformation of the fitness value, a process of self-adaptation is adopted to modify the variance according to the local topology of the fitness landscape. *See also* evolutionary programming.

metalanguage a language used to describe a language. Backus-Naur Form (BNF) is a classic metalanguage, as are regular expressions. A *metalanguage* expresses production rules of a grammar.

metal-oxide semiconductor memory memory in which a storage cell is constructed from metal-oxide semiconductor. Usually called MOS memory.

metameric match two colors that appear to be the same to a human. They may not have identical spectral distributions, but, because humans measure light using only three cone types, the differences are indistinguishable.

metamodel a model of a model. The model is described using a modeling language such as UML or OML and represents concepts in the system to be developed. The metamodel describes the rules of the modeling language itself.

metanotion in a W-grammar, a nonempty sequence of upper-case syntactic marks defined in the metaproduction rules of the grammar. *See also* protonotion, hypernotion.

metaphone an algorithm to code English words phonetically by reducing them to 16 consonant sounds. This reduces matching problems from incorrect spelling.

metaphor originally, indicating a rhetoric figure. The concept is used in the design of user interfaces for providing an analogy between some domain familiar to the user and the (typically more abstract) functionality of the system. The best-known example is a visual interface metaphor as in the desktop metaphor commonly adopted for personal computers (featuring cursors shaped like pencils, hard disk storage areas represented by pictures).

metaproduction rules one of the two context-free grammars that comprise a W-grammar. A metaproduction consists of a metanotion name, followed by "::", followed by a sequence of hypernotions separated by ";" and ending in ".". *See also* hyper rules, metanotion.

metasymbol a symbol used to denote a class of symbols or values.

metasyntax the syntax of a metalanguage.

method (**1**) a detailed approach to solving an engineering problem. A synonym for methodology, i.e., a set of rules and guidelines underpinning software development. A methodology is a way of doing things and is thus similar to the idea of a process.

(**2**) the implementation of an operation, i.e., the code describing how an operation is effected. The method describes the algorithm or procedure used in full detail, ready for execution. In C++, a method is called a member function.

methodology a general approach to solving an engineering problem. A synonym for method.

metric the defined measurement method with its measurement scale. Metrics can be internal or external, and direct or indirect. Metrics include methods for categorizing qualitative data. Synonymous with measure. A quantitative measure of the degree to which a system, component, or process possesses a given attribute.

metricator a tool for measuring metrics.

metric diagram a quantitative representation of shape and space used for spatial reasoning, the computer analog to or model of the combination of diagram/visual apparatus used in human spatial reasoning.

metric validation the verification process that has to be applied to a newly defined metric in order to assess if the metric is suitable for evaluating the characteristic under measurement. The validation may produce with an evaluation of metric suitability in measuring a targeted feature also a set of constant parameters.

metropolitan-area network (MAN) a computer communication network spanning a limited geographic area, such as city; sometimes features interconnection of LANs.

Mexican-hat function a function that resembles the profile of a Mexican hat. According to anatomy and physiology, the lateral interaction between cells in mammalian brains has a Mexican-hat form, that is, excitatory between nearby cells and inhibitory at longer range with strength falling off with distance. According to this phenomenon, Kohonen proposes a training algorithm for self-organizing system to update not only the weights of the winner but also the weights of its neighbors in competitive learning. *See also* self-organizing algorithm.

MFLOPs millions of floating point operations per second. Typically referred to as the number of floating point unit instructions processed in a second.

M/G/1 queueing notation for a queue with a Poisson arrival (Markovian or "M" arrivals) process, a general service (i.e., distribution of service time is general), and a single server. *See also* Kendall's classification.

microcode set of instructions that programs a computer to execute its nominal instruction set. *Microcode* is usually used as a replacement for "hard-wired" logic that interprets the instruction set because the execution unit of the computer can be made smaller. *See also* nanocode. The size of the program that can comprise the microcode is usually very limited. Microcode is usually extremely fast. *See* microinstruction.

microcommand an n-bit field specifying some action within the control structure of a CPU, such as a gate open or closed, function enabled or not, control path active or not, etc.

microcomputer a computer whose CPU is a microprocessor chip, and its memory and I/O interface are LSI or VLSI chips.

microcontroller a type of von Neumann architecture where there is no decoding of macroinstructions.

Microcontrollers are typically integrated circuit chips that are designed primarily for control systems and products. In addition to a CPU, a microcontroller typically includes memory, timing circuits, and I/O circuitry. The reason for this is to permit the realization of a controller with a minimal quantity of chips, thus achieving maximal possible minituarization. This, in turn, will reduce the volume and the cost of the controller. The microcontroller is normally not used for general-purpose computation as is a microprocessor.

micro-cycle one cycle in the evolutionary software development life-cycle process. According to this theory, development is performed by dividing the whole process into repetitive cycles. In each of these cycles, an evolution may be performed in an evolutionary manner proceeding for micro-cycles. The *micro-cycle* contains less phases of the whole development and is typically structured on the basis of the phase in which it is applied.

micro development the process that starts once the system architecture is well defined and stable. It consists in the analysis of requirement at subsystem and module level in order to define class interfaces. Once these are finalized, the development process can start.

microelectromechanical system (MEMS) micrometer-scale devices fabricated as discrete devices or in large arrays using integrated circuit fabrication techniques. Movable compact micromechanical or optomechanical structures and microactivators made using batch processing techniques.

micro estimating a bottom-up estimation approach, where the small parts are measured and so the total estimate is calculated.

microfacet an approximation used in developing an improved specular reflection component to surface shading. The Torrance-Cook (or Cook-Torrance) physical surface shading model assumes that a surface is composed of a set of tiny planar patches, each placed according to a distribution that depends on the surface. The microfacet model leads to a reflection function that gives more realistic values for the direction and intensity of the specular component of surface reflection.

microinstruction an instruction in a microcode program.

microkernel a small privileged operating system core that provides simple process scheduling, memory management, and communcation services.

micro-kernel a nano-kernel that also provides for task scheduling.

micromemory *See* control memory.

microphone a device that converts acoustical signals into electrical signals.

microprocessor a CPU realized on an LSI or VLSI chip.

microprogram a set of microcode associated with the execution of a program. If it supports macroinstructions, then each macroinstruction corresponds to some sequence of microinstructions.

microprogramming the practice of writing microcode for a set of microinstructions.

middle CASE support for the tasks to be performed in the middle of the software development effort, e.g., detailed design. *See* development process.

middle-out the software development is typically top-down or bottom-up. Approaches that try to combine both methods are typically called middle-out methods. This means that during the system design the system is both composed and decomposed according to methodology guidelines.

middleware software that mediates between an application program and a network. It manages the interaction between disparate applications across the heterogeneous computing platforms. The Object Request Broker (ORB), software that manages communication between objects, is an example of a middleware program.

MIDI *See* musical instrument digital interface.

migratable preemption a type of preemption for which a preempted thread of a parallel job can resume execution on any processor on the system.

migration the transfer of sufficient amount of state information from one machine to another to allow a program unit to execute on a target machine.

migration model one of the models in parallel genetic algorithm where multiple processors are adopted and the total number of individuals is divided into different sub-populations associated with each processor. Also known as the coarse-grain model, this approach restricts the processes of crossover and selection within each sub-population to allow exploration of different regions of the search space. Occasionally an individual is allowed to migrate from one sub-population to another for information exchanged between the sub-populations. The associated migration frequency is an important parameter which balances the possibility of diversity reduction among different sub-populations with the possibility of stagnation for an individual sub-population. Other important parameters include the number of migrants in each exchange, the selection strategy for these migrants and the determination of the set of original individuals to be replaced. *See also* diffusion model, parallel genetic algorithms.

milestone a major checkpoint in a software project. Completion of requirements, design, coding, etc. *Milestones* are typically located at constant time intervals (6 to 8 months) and are needed for controlling the project evolution. At the milestone a formal check of the availability of the planned results (deliverables, prototypes, progress report, cost statements) is checked.

millennium bug *See* millennium problem.

millennium problem the problem due to the definition of the date in software system by means of a couple of characters for storing the last two digits of the years. This causes problems when dates varying for the hundreds of years are manipulated. It has been called the *millennium problem* since it has been mainly highlighted in the vicinity of the year 2000. It is also known as the Y2K bug.

MIMD pronounced "mim-dee". *See* multiple instruction, multiple data.

MIMO system *See* multi-input-multi-output system. *See also* single-input-single output system.

minimal antichain decomposition a decomposition of a poset into the minimum possible number of antichains.

minimal cover set a cover set of functional dependencies FD where the removal of any of the functional dependencies causes FD to no longer be a cover set.

minimal representation for a positive Boolean function an equivalent representation where no product whose variable set does not contain the variable set of a distinct product can be deleted without changing the function.

minimax approximation an approximation in which the criteria for quality is the minimization of the maximum errors (at certain points) between the approximating and the approximated functions.

minimax estimate the optimum estimate for the least favorable prior distribution.

minimum distance in a forward error control block code, the smallest Hamming distance between any two code words. In a convolutional code, the column distance at the number of encoding intervals equal to the constraint length.

minimum free distance for any convolutional code, it is the minimum Hamming distance between the all-zero path and all the paths that diverge from and merge with the all-zero path at a given node of the trellis diagram.

minimum mean square error (MMSE) a common estimation criterion which seeks to minimize the mean (or expected) squared error,

$$\mathcal{E} = E\left[\mathbf{e}^T \mathbf{e}\right] ,$$

where **e** represents the error.

minimum mean square estimator (MMSE) a broad class of estimators based on minimizing the expected squared error criterion. Both the linear least squares estimator and the Bayesian least squares estimator are special cases. *See also* minimum variance unbiased estimator, linear least squares estimator, Bayesian estimator, maximum *a posteriori* estimator, maximum likelihood estimation.

minimum variance unbiased estimator an estimator $^{\theta}$ of a parameter θ is said to have minimum variance and to be unbiased if

$$E^{\theta} = \theta ,$$

and

$$E\left(^{\theta} - \theta\right)^2 \leq E(\theta - \theta)^2 ,$$

where θ is any other estimator of θ.

MINIX a microcomputer operating system produced for the purpose of educating students about the design and development of operating systems. The features of MINIX are compatible with those of Unix System VII, one of System V's predecessors. *See also* Linux.

Minkowski distance between two real valued vectors $(x_1, x_2, \ldots, x_n)$ and $(y_1, y_2, \ldots, y_n)$ a difference measure given by

$$D_{Minkowski} = \left(\sum_{i=1}^{n} |x_i - y_i|^{\lambda} \right)^{\frac{1}{\lambda}}$$

min-max control a class of control algorithms based on worst-case design methodology in which control law is chosen in such a way that it optimizes the performance under the most unfavorable possible effect of parameter variations and/or disturbances. The design procedure can be viewed as a zero-sum game with control action and uncertainty as the antagonistic players. In the case of linear models and quadratic indices, the min-max control could be found by solving zero-sum linear-quadratic games. Minimax operations may be performed on cost functionals, sensitivity functions, reachability sets, stability regions, or chosen norms of model variables.

min operation an operation on two or more variables where the resultant value is formed by taking the smallest value, or minimum, among these variables.

minor cycle a sequence of repeating processes in cyclic or periodic systems.

minor defect a defect that does not cause a failure, does not impair usability, and the desired processing results are easily obtained by working around the defect.

minterm a Boolean product term in which each variable is represented in either true or complement form. For example, $u \cdot v' \cdot w \cdot z'$ is a minterm for a four-variable function.

MIP mapping a technique of precomputing anti-aliased texture bitmaps at different scales, where each image in the map is one quarter of the size of the previous one. When the texture is viewed from different distances, the correct scale texture is selected by the renderer so that fewer rendering artifacts are experienced, such as Moiré patterns.

MIP is apparently an acronym relating to the latin "multum in parvo", meaning many things in a small place, since the texture contains the same content at different scales. A MIP mapped texture requires 4/3 times the storage of the original $(1 + 1/4 + 1/16 + \cdots)$.

MIPS millions of instructions per second, a measure of the speed of a computer. Typically the sum total of instruction cache, branch control, fixed point, and floating point instructions processed in a given second.

mirror to maintain a duplicate of, for example, a database.

MISD *See* multiple instruction stream, single data stream.

mishap an undesirable consequence of system failure, whose cost, whether economic, ecological, or in terms of injury or loss of life, is unacceptable. In some cases, it is a synonymous malign failure or accident.

MISO *See* multi-input-single-output system.

miss the event when a reference is made to an address in a level of the memory hierarchy that is not mapped in that level, and the address must be accessed from a lower level of the memory hierarchy.

miss penalty the time from when a miss occurs to when it is completely processed by the main memory or next level of cache.

miss probability the probability of falsely announcing the absence of a signal.

miss rate the percentage of references to a cache which do not find the data word requested in the cache, given by $1 - h$ where h is the hit rate. Also called miss ratio.

miss ratio *See* miss rate.

miss time the time to process a cache access that results in a miss.

mistake *See* human error.

mixed file a file that contains records of different record types.

mixed fragmentation *See* hybrid fragmentation.

mixed integer linear program a linear program with the added constraints that some of the decision variables are integer valued.

mixed language project a software project where more than one language is involved. For example, COBOL/CICS, C/Make, C/Assembly, etc. Mixed languages are common: about 30% of all projects use more than one language.

mixture model a probability model that consists of a linear combination of simpler component probability models.

ML (1) (archaic) machine language.

(2) a specific functional language.

See also maximum likelihood estimation.

M/M/1 queueing notation for a queue with a Poisson arrival (Markovian or "M" arrivals) process, exponential service, and a single server. *See also* Kendall's classification.

M/M/n queueing notation for a queue with a Poisson arrival (Markovian or "M" arrivals) process, exponential service, and "n" servers. *See also* Kendall's classification.

MMSE *See* minimum mean square estimator or minimum mean square error.

MMU *See* memory management unit.

MMX register a register that is designed to hold as many as eight separate pieces of integer data for parallel processing by a special set of MMX instructions. An MMX register can hold a single 64-bit value, two 32-bit values, four 16-bit values, or eight byte integer values, either signed or unsigned. In implementation, each MMX register is aliased to a corresponding floating-point register. MMX technology is a recent addition to the Intel Pentium architecture.

mobile robot a wheeled *mobile robot* is a wheeled vehicle that is capable of an autonomous motion because it is equipped for its motion, with actuators that are driven by an on-board computer. Therefore, mobile robot does not have an external human driver and is completely autonomous.

mock-up prototype an early version of a prototype typically produced for the feasibility study of a system.

modality a specific medical imaging technique, such as X-ray CT, magnetic resonance imaging, or ultrasound, that is used to acquire an image data set.

modal logic an extension of propositional logic with operators that express various "modes" of truth. Examples of modes are: necessarily A, possibly A, it has always been true that A, it is permissible that A. *See* temporal logic.

mode in a statistical sample, the value that occurs most often. *See also* mean.

mode filter a filter that takes the mode of the distribution containing the various input signal components, or in the case of an image, the mode of the distribution of all the pixel intensity values within the neighborhood of the current pixel. Complications can arise from the sparsity of the local pixel intensity distribution, and the mode should be that of the underlying, rather than the actual, intensity distribution.

model (**1**) a representation of reality of an artifact or activity intended to explain the behavior of some aspects of it. In creating a model, an abstraction technique is used. Thus, the model is typically less complex or complete than the reality modeled and can be regarded as an abstract description. This technique identifies commonalties and, in doing so, loses detail.

(**2**) a mathematical or schematic description of a computer or network system. Modeling usually involves an act of abstraction; i.e., the model only includes the most important properties of the original system.

model based image coding (**1**) image compression using stored models of known objects at both encoder and decoder, the encoder using computer vision techniques to analyze incoming images in terms of the models, and the decoder using computer graphics techniques to re-synthesize images from the models. The transmitted data are model parameters that describe at a relatively high level, the motion of the model parts. For example, if the stored model is of a human head, the transmitted data may be muscle flexions.

(**2**) the appreciation that all image coding relies on some model of the source, and the categorization of methods according to the type of model. In this scheme, (**1**) is termed "known-object coding" or "semantic coding".

model-based recognition approaches to recognizing objects in images that are based on comparing sets of stored model features against features extracted from an unknown image (generally geometric features).

model checking (1) a technique used to algorithmically check whether a model satisfies a specification. The model has to be operational (state machines, Petri Nets, etc.) and the specification language is usually a temporal logic.

(**2**) efficiently deciding whether a temporal logic formula is satisfied in a finite state machine model. *See also* Kripke structure.

modeling the process of building a model for an artifact or activity.

modeling assumption a proposition expressing control knowledge about modeling such as when a model fragment is relevant.

modeling time the time spent in producing the model.

model input an input parameter to a model; typical inputs include the rate of service at each device, the number of devices, and the number of customers in the system.

model of computation a formal, abstract definition of a computer. Using a model one can more easily analyze the intrinsic execution time or memory space of an algorithm while ignoring many implementation issues. There are many models of computation which differ in computing power (that is, some models can perform computations impossible for other models) and the cost of various operations. *See also* cell probe model, random access machine, big-O notation, alternation, alternating Turing machine, nondeterministic Turing machine, oracle Turing machine, probabilistic Turing machine, universal Turing machine, multiprocessor model, work depth model, parallel random access machine.

model output a value that is the result of solving a model; typical model outputs include average device queue length, average device or system throughput, or mean time to failure.

model-theoretic semantics the semantics of a deductive database DB with default negation usually is determined by a special subset $\mathcal{M}(\mathsf{DB})$ of the set of its Herbrand models, the so-called intended models of DB. For a database with definite rules and without default negation, there exists a unique intended model, the minimal model, which can also be derived procedurally by forward chaining or backward chaining.

modem abbreviation for modulator-demodulator. A device containing a modulator and a demodulator. The modulator converts a binary stream into a form suitable for transmission over an analog medium such as telephone wires or (in the case of a wireless modem) air. The demodulator performs the reverse operation, so two modems connected via an analog channel can be used to transfer binary data over the (analog) channel.

modem-FEC coding error control coding (ECC), applied to a digital signal such that feedforward error correction (FEC) can be used in the modem, thus detecting and often correcting transmission errors.

modification change to the design or the implementation of an item. The term change is used synonymously. Changes may be made for reasons of corrective, adaptive, or perfective maintenance.

modification anomaly *See* update anomaly.

modification state a specification of which modifications have been carried out on a particular example of an item in use on a given installation. The item may be a piece of hardware or an instance of a software component. The modification state must specify both the baseline version of the item, and all minor modifications that have been carried out.

modified-return-to-bias recording *See* magnetic recording code.

modified signed-digit computing a computing scheme in which a number is represented by modified signed-digit. This number system offers carry-free addition and subtraction. Instead of 0 and 1, numbers are represented by -1,0, and 1 for the same radix 2. If a number is represented by 0 and 1, we may need carry in the addition. However, since the number can be represented by three possibilities -1, 0, and 1, the addition and subtraction can be directly performed without carry following a specific trinary logic truth table for this number system.

modifier an operation that modifies the membership of a fuzzy set. Examples of modifiers are:

1. *very* $A = \mu_{con(A)}(u) = (\mu_A(u))^2$ (concentration);

2. *more or less*$A = \mu_{dil(A)}(u) = (\mu_A(u))^{.5}$ (dilatation).

See also fuzzy set, linguistic variable.

Modula a language designed by Nicklaus Wirth as a modern language for system implementation. Two dialects, Modula-2, from Wirth, and Modula-3, from a research consortium, are published.

modular decomposition *See* decomposition.

modular design modular is pertaining to the design concept, in which interchangeable units are used to create a functional end product. The modular design is referring to the process of composition/decomposition of a system.

modularity an attribute of a design that leads to the creation of highly reusable program components.

modularization *See* decomposition.

modular network a network whose overall computation is carried out by subnetworks whose outputs are combined in some appropriate way. The term is most commonly applied to networks that partition the input space so that the subnetworks operate on "local" data, but is also applied to the case where a problem can be decomposed into successive tasks, each being implemented by a suitable subnetwork.

modular programming a programming discipline where programs are divided into parts that can be shared, reused, and recompiled without affecting other parts of the system as long as the interfaces to modules are unchanged.

module (**1**) a self-contained software item with a specified function and a defined interface to the rest of the system. The concept is hierarchical: big modules may consist of composition of little modules, and a complete program may be thought of as a module of a larger system.

(**2**) applied to a self-contained hardware component or to one consisting of hardware and software.

(**3**) in programming languages, a piece of code that can be separately compilable.

module strength *See* cohesion.

module testing *See* component testing.

modulo an operation that represents the remainder of an integer division. Some languages support both a mod operator and a rem operator, one of which returns a signed remainder and the other of which returns an absolute value of the remainder. Other languages will implement the modulo (remainder) operation as one or the other of these choices. *See also* remainder.

modulus of a complex number z, is equal to the square root of the sum of the squares of its real and imaginary parts. *See also* residue number system.

modus ponens a rule of reasoning which states that given that two propositions, A and $A \Rightarrow B$ (implication), are true, then it can be inferred that B is also true.

modus tollens a rule of reasoning which states that if a proposition B is not true and given that $A \Rightarrow B$, then it can be inferred that A is also not true.

Moiré pattern a watered appearance usually provided as texture on the surface of objects. It arises from the interference between two overlapping patterns with a similar spatial frequency when two conditions are met. First, when the sampling frequency is close to the Nyquist Frequency for the signal (i.e., two times the highest frequency in the image signal). Second, the cutoff frequency of the reconstruction filter is located beyond one half the sampling frequency (e.g., a first order filter). These two problems result in mirror images of frequency components around the sampling frequency causing banding in the image signal.

The Moiré effect is seen in practical applications due to real world problems of finite filter lengths and errors in sampling rates.

moldable job a type of parallel job that has an architecture which can run with a different number of processors.

moment a statistic of a random variable. For example, the first moment is called the mean. In general, the nth moment is given by

$$m_n = E\left(X^T n\right) = \int_{-\infty}^{\infty} x^n X(x)\, dx \,,$$

where $f_X(x)$ is the probability density function of X. Moments are often used to aid the recognition of shapes. *See also* probability density function.

monitor (**1**) the main display device of a personal computer. Usually uses cathode ray tube (CRT) or liquid-crystal display (LCD) technology.

(**2**) a program that allows the control of the execution of other programs in an embedded system. It can be used for downloading code from a development computer into the embedded system (via communication channels, e.g., RS232, serial) and for debugging the loaded code during test and development. It is typically removed in the final delivered version, since the monitor is not an operating system.

(3) an abstract data type for which only one process may be executing procedures of this data type at any given time.

monitor display *See* monitor.

monochromatic light (or other source of electromagnetic radiation) having only one wavelength.

monochrome single color, as in a monochrome monitor.

monochrome display adapter (MDA) a monocrome video adapter with 25 lines and supporting 80 columns, proposed by IBM in 1981.

monocular vision a vision model in which points in a scene are projected onto a single image plane.

monotonically decreasing a function from a partially order domain to a partially ordered range such that $x \geq y$ implies $f(x) \leq f(y)$. *See also* monotonically increasing, strictly decreasing.

monotonically increasing a function from a partially order domain to a partially ordered range such that $x \geq y$ implies $f(x) \geq f(y)$. *See also* monotonically decreasing, strictly increasing.

monotonic reasoning a reasoning system based on the assumption that once a fact is determined it cannot be altered during the course of the reasoning process.

Monte Carlo *See* Monte Carlo method.

Monte Carlo algorithm a randomized algorithm that may produce incorrect results, but with bounded error probability.

Monte Carlo method a numerical technique that replaces a deterministic description of a problem with a set of random descriptions that have been chosen based on distributions that match the underlying physical description of the problem. This technique is widely used to investigate transport and terminal characteristics in small semiconductor structures.

Monte Carlo simulation an approach to estimating the mean value of a vector of random variables, often used in cases where analytic computation is not feasible. The steps of Monte Carlo simulation include generating a sequence of random variables from a uniform distribution on (0, 1), then calculating the mean of the corresponding values obtained for $f(x_1, \ldots, x_n)$, where f is the joint density function applied to n of the input values.

Moore machine the model of a state machine in which the output actions are associated with the state. *See* Mealy machine.

Moore's law the observation by Gordon E. Moore, in 1964, that the number of transistors on a chip was increasing by a factor of two per year. Since the late 1970s, the doubling time has been about 18 months.

MOPS millions of operations per second. A measure of performance of a processor to compare against other machines. For example, 200 MFLOPS and 100 MOPS correspond to two floating-point operations and one integer operation per clock cycle for a computer with a clock rate of 100 MHz.

morphing a continuous deformation from one keyframe or 3D model to another. In 3D this is often achieved by approximating a surface with a triangular mesh that can then be continuously deformed. In 2D, it is generally performed by either distortion or deformation.

morphological duality a morphological property. There are three standard meanings of duality for morphological operators:

1. Order-theoretic duality: To any property or concept corresponds the dual property or concept, where the relations $\subseteq, \leq$ and operations $\cup, \cap$, sup, inf, etc. are replaced by their duals $\supseteq, \geq$ and $\cap, \cup$, inf, sup, etc.

2. Duality under complementation (or gray-level inversion): Let ψ be a morphological op-

erator, and let N be the operation transforming an image I into its negative $N(I)$ (when I is a set, $N(I)$ is its complement, while when I is a gray-level numerical function, $N(I)$ has the gray-level inverted at each point). Then the dual of ψ is the operator ψ^* arising from applying ψ to the negative of an image; in other words ψ^* transforms I into $N(\psi(N(I)))$.

3. Adjunction duality: The dilation and erosion by B, namely $\delta_B : X \mapsto X \oplus B$ and $\varepsilon_B : X \mapsto X \ominus B$, form an adjunction, which means that for every X, Y we have

$$\delta_B(X) \leq Y \iff X \leq \varepsilon_B(Y) ,$$

"the dilation of X is below Y if and only if X is below the erosion of Y". This relation constitutes a bijection between the family of dilations and that of erosions, and it can thus be considered as a duality between them.

In the latter two cases, duality inverts the ordering relation $\leq$ between operators. *See* dilation, erosion, morphological operator.

morphological filter a morphological operator ψ that is both:

1. idempotent: applying it twice gives the same result as applying it only once; mathematically speaking, given an object X, we have $\psi(\psi(X)) = \psi(X)$;

2. increasing: it preserves the ordering relation $\leq$ between objects; given two objects X and Y, $X \leq Y$ implies $\psi(X) \leq \psi(Y)$. Openings and closings are morphological filters. *See* closing, morphological operator, opening.

morphological gradient the nonlinear analog to a linear gradient.

morphological operator an operation for transforming images, which does not arise from the traditional signal processing methodology using linear filters and Fourier analysis, but which is rather based on set-theoretical operations. Such an operator for sets (binary images) combines union, intersection, translation, and sometimes complementation. For numerical functions (gray-level images), union and intersection are replaced by supremum and infimum (upper and lower envelope), while gray-level inversion partially takes the role of set complementation. However, morphological operators for gray-level images can be visualized by applying the corresponding morphological operator for binary images to the umbra of a gray-level image. *See* closing, dilation, erosion, mathematical morphology, opening, umbra.

morphological pattern spectrum the normalization of a granulometric size distribution.

morphological processing low-level processing technique for binarized images involving the shrinking or growing of local image regions to remove noise and reduce clutter.

morphological skeleton an archetypal stick figure that internally locates the central axis of an image.

For sets in digital space, the approach is simplified as follows: For $r = 0, 1, 2, \ldots,$ one takes a bounded digital structuring element B_r, with B_0 being restricted to the origin, and B_r increasing with r; the *morphological skeleton* is made of all pixels p such that there is some integer $r \geq 0$ for which

$$p \in X \ominus B_r \quad \text{and} \quad p \notin (X \ominus B_r) \circ B_1 .$$

See erosion, medial axis transform, opening, structuring element.

morphological system a non-linear counterpart of linear shift invariant systems, based on morphological operators which are invariant under both spatial and gray-level shifts. *See* morphological operator.

morphology *See* mathematical morphology.

MOS memory *See* metal-oxide semiconductor memory.

motherboard in a computer, the main printed circuit board which contains the basic circuits and to which the other components of the system are attached. *See also* daughter board.

mother wavelet the wavelet from which wavelets in different scales are obtained by the translation and dilation operations. *See also* father wavelet.

motion analysis determination of the motion of objects from sequences of images. It plays important roles in both computer vision and image sequence processing. Examples for the latter include image sequence data compression, image segmentation, interpolation, image matching, and tracking.

motion blur blur which is due to the motion of an object: it frequently arises when images containing moving objects are grabbed by an insufficiently rapid digitizer.

motion compensation generation of a prediction image, an interpolated or extrapolated frame, from pixels, blocks, or regions from known frames, displaced according to estimates of the motion between the known frames and the target frame.

motion estimation strictly the process of estimating the displacement of moving objects in a scene within a video sequence, including camera movement and the independent motion of objects in the scene. In practice, refers to the determination of optical flow (the spatiotemporal variation of intensity) which may not correspond to real movement (when, for example, an object moves against a background of the same color). Gradient-based methods are based on the expansion of a Taylor series for displaced frame difference yielding an "optical flow constraint equation". This must be combined with other constraints to yield a variational problem whose solution requires the iterative use of spatial and temporal derivatives. Block-matching methods rely on finding minimum error matches between blocks of samples centered on the point of interest. Frequency-based techniques measure phase differences to estimate motion.

motion measurement the measurement of velocities. Usual motion measurement techniques such as optical flow give incomplete and unstable results which must be regularized by comparing velocities over a whole moving object (the latter must first be segmented).

motion segmentation the decomposition of a scene into different objects according to the variation of their velocities. *Motion segmentation* requires the measurement of velocities across the scene.

motion stereo a specific case in motion estimation in which the 3D scene is stationary and the only camera is in rigid motion.

motion vector a vector displacement representing the translation of a pixel, block, or region between two frames of video, usually determined by optical flow calculation or block matching (*See* motion estimation). For instance, in the case of dense motion field, say, an optical flow field, a motion vector is assigned to each pixel on the image plane. In the case of block matching for image coding, a motion vector is assigned to each block. In the case of computer vision, a motion vector sometimes refers to the velocity of a point or an object in the 3D space. Sometimes motion vectors are referred to as displacement vectors.

move instruction a computer instruction which transfers data from one location to another within a computing system. The transfer may be between CPU registers, between CPU registers and memory, or I/O interface in either direction. Some systems (such as Motorola M68000) permit transfer by a *move instruction* between memory locations (memory-to-memory transfer).

moving average (**1**) an Nth order moving average relationship between an input x and output y takes the form

$$y[n] = \sum_{i=0}^{N-1} \alpha_i x[n-i] \; .$$

A *moving average* process is any process y which can be expressed as the moving average of white noise x. *See also* autoregressive.

(**2**) the average value of an image over the pixels in a window when it is centered at a certain pixel.

moving maximum an image operation found by translating the image to a pixel and then taking the maximum of all values in the translate. *Contrast with* moving minimum.

moving median a type of binary image filter used to suppress salt-and-pepper noise and preserve edges.

moving minimum an image operation found by translating the image to a pixel and then taking the minimum of all values in the translate. *Contrast with* moving maximum.

Moving Picture Experts Group (MPEG) Group that standardizes methods for moving picture compression. MPEG-1 and MPEG-2 are very flexible generic standards for video coding, incorporating audio and system-level multiplex information. They use block-based motion-compensated prediction on 16×16 "macroblocks", and residue coding with the 8×8 discrete cosine transform. Frame types are Intraframes (coded without reference to other frames), P frames (where the prediction is generated from preceding I and P frames), and B frames (where the prediction is generated bidirectionally from surrounding I and P frames).

MPEG *See* Moving Picture Experts Group.

MPP distance *See* maximum *a posteriori* probability.

MRI *See* magnetic resonance imaging.

MRVQ *See* mean/residual vector quantization.

MSB in a binary word, a bit with the highest weight associated with it.

MS-DOS Microsoft's disk operating system.

MSE *See* mean squared error.

m-sequence maximal length sequence. A binary sequence generated by a shift register with a given number of stages (storage elements) and a set of feedback connections, such that the length of the sequence period is the maximum possible for shift registers with that number of stages over all possible sets of feedback connections. For a shift register with n stages, the maximum sequence period is equal to $2^n - 1$.

MSI *See* medium scale integration.

MTBF *See* mean time between failures.

MTTF *See* mean time to failure.

MTTI *See* mean time to interrupt.

MTTR *See* mean time to repair.

multiaccess link a link on which more than two nodes may reside.

multibus a standard system bus originally developed for use in Intel's Microcomputer Development System (MDS). This standard gives a full functional, electrical, and mechanical specification for a backplane bus through which a number of circuit boards may be interconnected. A full range of devices may be involved, including computers, memory boards, I/O devices, and other peripherals.

multibyte character set a representation which allows more than one character to represent a single glyph in a language. Originally designed to support ideographic languages such as Japanese and Chinese whose glyph set exceeds 256, the maximum that can be represented in a single byte. A multibyte character set typically intermixes sequences of one, two, three, or more characters, with transitions embedded in the character sequence to indicate transitions between the representations. In some cases, the use of a multibyte character set can be supplanted with Unicode. Many programming languages support the use of *multibyte character set* values either in the source file or in string constants. *See also* ASCII, ANSI, ISO.

multicast the ability to transmit a single message or packet that is received by multiple recipients.

multiclass queuing network a queuing network in which not all customers at a particular device are statistically identical. Rather, customers fall into categories, called classes, and all customers within the same class at a device are statistically identical. Different classes may have different mean service times. Customers can be routed from a class at one device to another class at another device.

multi-commodity flow a network flow problem involving multiple commodities, in which each commodity has an associated demand and source-sink pairs.

multiconstraint-based motion detection a gradient motion detection scheme where the optical flow estimation is transformed from an ill-posed to a well-posed problem using constraint equations.

multifunction card a card that provides multiple functions, such as serial and/or parallel connections, clock/calendar, and additional memory.

multigrid an efficient numerical algorithm for solving large sets of linear equations $A\mathbf{x} = \mathbf{b}$, particularly for "stiff" (nearly singular) A. The algorithm defines a hierarchy of grids, with interpolation and decimation operations defined between successive grids in the hierarchy. A system of equations $A_g\mathbf{x}_g = \mathbf{b}_g$ is defined on each grid g: the systems on finer grids yield higher resolution solutions; however, coarser grids yield much faster covergence times. Various empirical strategies have been devised which dictate the order in which different grids are used to contribute to the final solution. *See also* interpolation, decimation, multiscale.

multigrid block matching a block matching technique that is carried out with a hierarchical structure, known as a multigrid structure. In the highest hierarchical level an image may be decomposed into large blocks each with equal size. Each block in a higher level is further equally decomposed into sub-blocks, forming the next level. The block matching is first applied to the highest level. The matching results obtained are then propagated to the next hierarchical level. That is, the matching results obtained in the highest level are utilized as initial estimates and refined in the second highest level. This process continues until the last hierarchical level. It is noted that the term multigrid block matching is sometimes intermixed with multiresolution block matching in the literature. In the strict sense they are different. In the former, different levels in the hierarchical structure have the same resolution, while in the latter, different levels have different resolutions.

multilayer perceptron an artificial neural network consisting of an input layer, possibly one or more hidden layers of neurons (perceptrons), and an output layer of neurons. Each layer receives input from the previous layer and the outputs of the neurons feed into the next layer.

multilevel cache a cache consisting of two or more levels, (typically) of different speeds and capacities. *See also* hierarchical memory.

multilevel code in this scheme each bit (level) of a signal is encoded with different error correction codes of the same block length. It has inherent unequal error protection capability.

multilevel index an index with many levels. The first index built on a file can be viewed as an ordered file. A *multilevel index* comprises a primary index built on this first level and a primary index built on this second layer and so on.

multilevel memory *See* hierarchical memory.

multilevel optimization decision mechanism (or operation of such mechanism) whereby the decisions are made by solving a large-scale optimization problem, partitioned (decomposed) into several smaller problems; in a two-level optimization case local decisions are influenced by the coordinator — in the process of iterative coordination until overall satisfactory decisions are worked out; in a three-level or in a multilevel case the coordinating unit of the higher level coordinates the decisions of several subordinate units, which themselves can be coordinating units for the lower levels.

multilinear regression a mathematical method where an empirical linear function is derived from a set of experimental data that are distributed in an n-dimensional space (i.e., the estimation of the suitable weights of a linear combination of variables related to an explanatory variable by using a specific criteria of finding).

multimedia computing computing that involves computer systems with high-resolution graphics, CD-ROM drives, mice, high-performance sound cards, and multitasking operating systems that support these devices.

multimedia database a database used to store multimedia data such as video and audio.

multimedia PC a personal computer that includes a high performance processor, high resolution graphics, high quality sound generation, CD-ROM, and mouse. Special multimedia applications are designed to run on such configurations.

multi-member set a set such that member records may be of more than one type of record.

multimode a device that is compatible with many other types of devices.

multimode code a line code where each source word is represented by a code word selected from a set of alternatives. Code words are selected according to a predefined criterion which may depend in part on the statistics of the encoded sequence, thereby causing the same source word to be represented by a number of different code words throughout the encoded sequence.

multipass compilation (**1**) historically, in the days when source programs were larger than the machine could hold, refers to the fact that the source code would be read multiple times, each time constructing some additional partial information which was often stored on an intermediate storage device (usually magnetic tape).

(**2**) in modern compilers, a technique that divides the compilation process into two or more phases, each one of which may traverse the intermediate representation one or more times. More formally termed multiphase compilation.

multiphase compilation a technique that divides the compilation process into two or more phases, each of which performs some specific subtask of the compilation process. Also called multipass compilation.

multi-phase transaction a transaction that retrieves and processes two or more separate sets of data, at separate points in time.

multiple hashing a collision resolution policy whereby, upon collision, the hashed record is rehashed according to another hashing function.

multiple inheritance inheritance relationship in which there are two or more supertypes/superclasses. Not all object-oriented languages support multiple inheritance. *See also* inheritance, single inheritance.

multiple-input/multiple-output system a system that transforms two or more input signals to two or more output signals. Also known as SISO system and single variable (SV) system. *See also* system, single-input/single-output system.

multiple-input-multiple-output system (MIMO) also known as multivariable (MV) systems. A system that can transform two or more input signals to two or more output signals.

multiple-input-single-output system (MISO) a system that can transform two or more input signals to one output signal. *See also* single-input-single-output system, multiple-input-multiple-output system.

multiple instruction, multiple data (MIMD) a parallel processing system architecture where there is more than one processor and where each processor performs different instructions on different data values. Pronounced "mim-dee".

In an optical computer implementation, this can be done with a lenslet array, a hologram array, or a set of beamsplitters.

Typical MIMD computers include dataflow architectures and transputers. *See* Flynn's taxonomy.

multiple instruction stream, single data stream (MISD) a parallel processing architecture where more than one processor performs different operations on a single stream of data, passing from one processor to another. MISD computers include pipelined and very long instruction word architectures. *See* Flynn's taxonomy.

multiple-operand instruction a computer instruction that contains two or more data elements that are used while executing the instruction; i.e., ADD Z, X, Y is a multiple operand instruction that specifies the values of X and Y are added together and stored into Z.

multiple-valued logic any logic whose set of truth values is not restricted to true and false.

multiplex (**1**) to use a single unit for multiple purposes, usually by time sharing or frequency sharing.

(**2**) a unit used for multiple purposes. *See* multiplexer, multiplexing.

multiplexer a combinational logic device with many input channels and usually one output, connecting one and only one input channel at a time to the output.

multiplexer channel I/O channel that can handle several (time-multiplexed) I/O transactions at a time. *See also* byte multiplexer channel, block multiplexer channel, selector channel.

multiplexing of or being a communication system that can simultaneously transmit two or more messages on the same circuit or radio channel.

multiplexing gain a measure of the savings in resources that come from dynamically sharing the transmissions from several sources over a single communication channel. Its value indicates that in order to reach the same performance, the bandwidth per connection in a multiple user system is less than the bandwidth needed for a single connection system. For a given line bandwidth, traffic description of one connection and a number of connections, the bandwidth needed for each connection is known as the "effective bandwidth".

multiplexor channel I/O channel that can handle several (time-multiplexed) I/O transactions at a time. *See also* byte multiplexer channel, block multiplexer channel, selector channel.

multiplication pattern for arrays of similar elements, it is the product of the pattern of a single element and the array factor of the array.

multiplicative normalization a procedure used in the computation of mathematical functions, such as sine, cosine, etc. To compute $f(x)$, two variables, say X and Y, are maintained such that as X is reduced to some known constant (usually 0 or 1), Y tends to $f(x)$. The reduction of X is by multiplication with a sequence of constants.

multiplier an electronic system or computer software which performs the multiplication calculation.

multiplier recoding *See* Booth's algorithm.

multiplying D/A a D/A conversion process where the output signal is the product of a digital code times an analog input reference signal. This allows the analog reference signal to be scaled by a digital code.

multipoint bus a bus to which multiple components may be attached. The PCI bus is an example of a multipoint bus. *See also* point-to-point bus.

multiport a circuit presenting multiple access ports.

multiport card card that has multiple ports of a single type; for example, several parallel ports.

multiport memory one memory module can be accessed by two devices simultaneously. One example is a video memory that is interfaced to a graphics co-processor as well as the video interface of the monitor.

multiprefix a generalization of the scan (prefix sums) operation in which the partial sums are grouped by keys.

multiprocessing a modality by which two or more processes are concurrently executed by separate processors that typically have access to a common memory. *Compare with* multiprogramming. *See* multitasking, time sharing.

multiprocessing operating system an operating system where more than one processor is available to provide for simultaneity. *Compare with* multitasking operating system.

multiprocessor a computer system that has more than one internal processor capable of operating collectively on a computation. Normally associated with those systems where the processors can access a common main memory.

There are some additional stipulations for a "genuine multiprocessor":

1. It must contain two or more processors of approximately comparable capabilities,
2. all processors share access to a common, shared memory,
3. all processors share access to common I/O facilities, and
4. the entire system is controlled by a single operating system (OS).

See also mulitple-instruction multiple-data architecture, operating system, shared memory architecture.

multiprocessor model a model of parallel computation based on a set of communicating sequential processors.

multiprogramming a modality by which two or more tasks are executed in time sharing by the same processor. *Compare with* multiprocessing.

multirate signal processing a system in which there is at least one change of sampling rate. Typically an input signal is split into two or more sub-signals each with a sampling rate lower than the input signal.

multi-resolution analysis *See* multi-scale analysis.

multiresolution analysis analysis method that decomposes a signal into components at different resolution levels; the fine to coarse features are revealed in the fine to coarse resolution components.

multiresolution coding coding schemes that involve a multiresolution structure.

multiscale characterized by a parameterization or decomposition in scale, as opposed to (for example) time or frequency; in particular, an algorithm that computes or analyzes a function or an image at multiple resolutions. *See also* wavelet, multigrid, Laplacian pyramid, quadtree, resolution.

multi-scale analysis the analysis or transformation of a signal using analysis basis functions or analysis filters with differing time resolutions or spatial resolutions. Equivalently the basis functions have differing frequency resolutions. This is in contrast to the continuous or discrete Fourier transform whose analysis basis functions all have (roughly) the same frequency and time resolution. For discrete multiresolution analysis, a multi-level filter bank is often used: the discrete wavelet transform is a classic example. *See also* multiscale.

multiserver, multiserver queue a system that can process more than one task at the same time. The processing time for a single task remains the same as for a single server system, but some maximum number of additional tasks can be served at the same time without interference.

multispectral image an image that contains information from more than one range of frequencies. In this sense, an RGB color image is multispectral in that it contains red, green, and blue information. More usually, multispectral refers to frequencies of such number and type that it presents difficulties in interpretation and display. Adding infrared, ultraviolet, and radar information to an RGB image creates a problem, since the data cannot be simply drawn on a display. The use of many frequencies also provides a wealth of information sometimes needed to make fine distinctions between regions in an image, especially in satellite imagery. *See also* data fusion.

multi-stage subset decoding in this scheme decoding is done, first on the lowest partition level of the signal set and then gradually on the higher level with decoded information flow from the lower to the higher level.

multi-stage vector quantization a method for constrained vector quantization where several quantizers are cascaded in order to produce a successively finer approximation of the input vector. Gives loss in performance, but lower complexity than optimal (single-stage) vector quantization.

multi-stage VQ *See* multi-stage vector quantization.

multitasking a mode of operation that provides for concurrent execution of two or more tasks. *See also* multithreading.

multitasking operating system an operating system that provides sufficient functionality to allow multiple programs to run on a single processor so that the illusion of simultaneity is created. *Compare with* multiprocessing operating system.

multithreaded several instruction streams or threads executed simultaneously.

multithreaded execution a form of execution in which two or more threads of execution can be operational concurrently, or pseudo-concurrently. A multiprocessor computer can actually execute more than one thread of control concurrently; a uniprocessor computer can execute only one thread at a time, but by using preemptive scheduling, will give the illusion that multiple threads are actually running. *See also* fork, join, synchronization mechanism, cactus stack.

multithreading sharing a single CPU among multiple threads. This is accomplished by sharing as much as possible of the program execution environment between different threads so that the minimal information is saved and retrieved during the passage from a thread to the next (*See* context switch). Multithreading differs from multitasking in the sense that threads share their environment with each other. Threads may be distinguished only by the value of their program counters and stack pointers while sharing a unique address space and set of global variables. *Multithreading* can thus be used for fine-grain multitasking. Specific policies are provided for the selection of threads that have to be successively executed.

multi-valued attribute an attribute that may take on numerous values for any entity.

multi-valued dependency a property of a database relationship. Given a relational schema R, a relation r, and sets of attributes A and B ($A \subset R$ and $B \subset R$) there exists a multi-valued dependency between A and B, denoted $A \twoheadrightarrow B$, if when $t_1[A] = t_2[A]$ for $t_1, t_2 \in r$, then there also exists t_3 and t_4 in r such that $t_1[A] = t_2[A] = t_3[A] = t_4[A]$ and $t_1[A] = t_3[B]$ and $t_2[A] = t_4[B]$ and $t_1[A] \neq t_3[R - B]$ and $t_2[A] \neq t_4[R - B]$.

multi-variable system (MV) *See* multi-input-multi-output system.

multi-version data *See* multi-version technique.

multi-version technique a technique to increase concurrency, where a new version of a database object is written, rather than updating an existing version.

multiway merge the mechanism by which ω sorted runs are merged into a single run. The input runs are usually organized in pairs and merged using the the standard method for merging two sorted sequences. The results are paired again and merged, until just one run is produced. The parameter ω is called the order of the merge.

multiway tree a tree with any number of children for each node. *See also* binary tree.

Munsell color system a way of precisely specifying colors and showing the relationships between colors. In this system, the color space has three parameters: hue, value, and chroma (saturation). Munsell uses scales with visually uniform steps for each of these parameters. A

Munsell Book of Color displays a collection of colored chips arranged according to these scales. The parameters are written in (H, S, V) form, known as the Munsell notation. *See* HSV.

μ-recursive function a function that is a basic function (zero, successor, or projection) or one that can be obtained from other μ-recursive functions using composition and μ-recursion.

musical instrument digital interface (MIDI) a hardware and software standard for interfacing compatible musical instruments to computers and other electronic devices.

mutation random changes in genetic information due to replication errors in reproduction. In evolutionary computation, mutation is implemented in several different ways to allow the introduction of new features to the original population. In genetic algorithm, this is implemented as the random flipping of bit values in a chromosome with a certain mutation probability. On the other hand, in ES and EP, mutation is modeled as the random perturbation of each real-valued component of an individual using a Gaussian random variable. *See also* evolutionary programming, evolutionary strategy, genetic algorithm, program mutation.

mutation testing a technique for which two or more program mutations are executed and tested in order to detect if the used test cases are capable of detecting mutations.

mutex a synchronization construct that uses two operations, lock and unlock, and the notion of ownership. A special case of a semaphore in which the count value can be only 0 or 1. Also known as a binary semaphore. Some languages provide a *mutex* as a primitive data type. Some languages or systems permit a mutex to be acquired multiple times by a single thread of control while blocking access by multiple threads.

mutual exclusion a principle requiring that concurrent program units access the same critical section only one at a time.

mutual illumination *See* interreflection.

mutually exclusive events that cannot occur together.

mutual recursion a situation in which two functions call each other, for example, function A calls function B, and function B calls function A. This usually requires, in most languages, a forward declaration of one of the functions.

MV *See* multi-variable system.

MVA *See* mean value analysis.

N

name a sequence of characters representing a fundamental lexical entity in a programming language. Typically a sequence of letters, digits, and some punctuation symbols which define entities that represent values (for example, variables). The rules for forming a name are language-specific. *See also* symbol.

named COMMON a construct in the FORTRAN language which allows blocks of storage to be globally shared among subprograms. Unlike unnamed COMMON, the rules of modern FORTRAN do not require that the storage used for Named COMMON have extent commensurate with the program execution. A named COMMON block is required to exist only as long as some set of functions is actively using it.

name decoration transformations applied to exported function names to encode additional information and make the name unique. Particularly important when overloaded functions can be specified. The name actually referenced by the object code, and defined by the object code, is often unreadable except to those who know the decoration algorithm.

named parameter a parameter whose actual value is bound to its formal parameter by specifying its name, allowing parameters to be specified at the call site in any order convenient to the programmer. Often a named parameter mechanism is coupled to a default parameter mechanism by which the parameters that are not named are assigned specified (default) values. This mechanism is not in common use in most programming languages, but is often used in interactive languages, particularly command languages. *See also* command language, shell language. *Compare with* positional parameter.

named pipe a pipe that has at least one directory entry and is generally accessible.

name equivalence a form of type equivalence in which two types are considered equivalent if the names used to designate them are equivalent. The names must be textually equivalent and also represent the same instance of the name (for example, a type name declared in one scope is disjoint from the same textual name declared in a different scope). *See* structural equivalence, textual equivalence.

name mangling *See* name decoration.

namespace (**1**) a specification of how a name is interpreted, in particular, how lexically identical names may be disambiguated by contextual information.

(**2**) in many languages, a method of syntactically qualifying a name to disambiguate it from other names.

(**3**) in C++ and many other languages, a specification encompassing how names are to be interpreted.

NaN *See* not-a-number.

nand negated conjunction: 0 AND 0 = 1, 0 AND 1 = 1, 1 AND 0 = 1, 1 AND 1 = 0. *See also* AND, NOT, OR.

NAND gate a logic circuit that performs the operation equivalent to the AND gate followed by the inverter. The output of a NAND gate is Low only if all inputs are High.

nano prefix for metric unit which indicates division by one billion.

nanocode a very low level form of code that is used to write a program that interprets microcode.

nano-kernel code that provides simple thread-of-execution (same as "flow-of-control") management. It essentially provides only one of the three services provided by a kernel; that is, it provides for task dispatching.

N-ary association an association among three or more classes. Each instance of the association is an n-tuple of values from the respective classes.

0-8493-2691-5/01/$0.00+$.50

N-ary function (**1**) a function with n arguments.

(**2**) a function which takes any number of arguments, or a variable number of arguments. *See also* constant function, unary function, binary function, trinary function.

NAS parallel benchmarks the Numerical Aerodynamic Simulation (NAS) Parallel Benchmarks (NPB). These were developed in 1991 to evaluate the performance of parallel computing systems for workloads typical in NASA and the aeronautics community.

Nassi-Shneiderman chart *See* box diagram.

NASTRAN a widely used computer code for mechanical and structural analysis, such as for opto-mechanical and thermal analyses.

natural image an image as it comes directly from the camera.

natural join an equi-join where the repeated join attributes are removed.

natural language (**1**) a language with rules which depends on the usage rather than on strictly formalized rules. For example, English, Italian. *See* formal language.

(**2**) the branch of AI research that studies techniques that allow computer systems to accept inputs and produce outputs in a conventional language like English.

natural metric a metric that has a physical appearance: 5 lines of code or 3 dollars. As opposed to synthetic/indirect metric: function points, or the Dow-Jones index. Synonymous with direct metric.

N-best interface an interface between a speech recognition system and a natural language system in which the recognizer proposes N whole-sentence hypotheses, and then the natural language system selects the most plausible alternative form among the N theories. In an alternative tightly coupled mode, the natural language system is allowed to influence partial theories during the initial recognizer search.

NBS National Bureau of Standards, renamed in 1988 to the National Institute of Standards and Technology (NIST). *See* NIST.

NC the set of all languages $\mathcal{L}$ that are decidable in parallel time $(\log n)^{O(1)}$ and processors $n^{O(1)}$.

NCSS *See* non-commented source statements.

NDFSA *See* nondeterministic finite state automaton.

NDPDA *See* nondeterministic push-down automaton.

nearest neighbor algorithm a method of classifying samples in which a sample is assigned to the class of the nearest training set pattern in feature space; a special case of the K-nearest neighbor algorithm.

nearness query a form of spatial query which involves selecting objects that are located within a certain range from a specified location.

near pointer in some languages, running on machines which have more than one representation of a machine address, designates those addresses which can access a memory location within a limited range of some base location. Although commonly used in the context of Microsoft Windows running on segmented-address Intel processors, the concept predates that usage by at least 20 years. *Compare with* far pointer.

negation normal form (NNF) a form for logical formulas in which conjunction and disfunction are the only binary connectives and in which all negations are at the atomic level.

neighborhood in self-organizing system an area surrounding the winner in self-organizing competitive learning. According to so-called Mexican-hat interaction between the brain cells, Kohonen proposes that weight updating should be conducted not only for the winner but also for its neighbors. *See* self-organizing algorithm. *See also* Mexican-hat function.

neighborhood operation an operation, such as averaging or median filtering, that is dependent upon the locality of samples in a signal, not the signal as a whole. Also called a window operation. *See* windowing.

neocognitron a biologically inspired hierarchical network developed primarily for the recognition of spatial images. The network has up to nine layers. The lower layers respond to simple features and the higher layers respond to more complex features. Learning can be unsupervised or supervised.

nested loop join an implementation of the join operator. To execute $R \bowtie S$, for each tuple r in R, all records s in S are retrieved and compared to r for satisfaction of the join condition.

nested relation a relation that allows another relation as an attribute value.

nested scope a scope that is lexically or dynamically contained within another scope. In languages that support nested scope, it is generally possible to create names within the scope whose definition will supersede any definition of an enclosing scope. *See also* lexical scope, dynamic scope.

nested structure an information structure in which each player has access to the information acquired by all his precedents, i.e., players situated closer to the beginning of the decision process. If the difference between the information available to a player and his closer precedent involves only his actions, then the structure is ladder nested. This structure enables decomposition of the decision process onto static games that in turn results in recursive procedures for its solution. For dynamic infinite discrete-time decision processes, the ladder-nested structure results in the classical information pattern.

nested subroutine a subroutine called by another subroutine. The programming technique of a subroutine calling another subroutine is called nesting.

nested transaction a transaction initiated and completed while another transaction is pending. In general, it is expected that a nested transaction that is committed will still be subject to rollback if the transaction within which it occurs is terminated with a rollback.

net a set of terminals to be connected together.

netnews a distributed bulletin board system supported mainly by Unix machines and the people who post and read articles. Originally implemented in the 1970s at Duke University, it has swiftly grown to become international in scope and is now probably the largest distributed information utility in the world.

network (**1**) a set of nodes connected by links, as in a computer network, a queueing network model, or a neural network model.

(**2**) a conduit and software and hardware controls for connecting multiple computers.

(**3**) a directed graph with no cycles.

network architecture (**1**) a taxonomy for the layers in a network hierarchy. According to the OSI standard, network architecture is split among seven layers

1. physical layer
2. data link layer
3. network layer
4. transport layer
5. session layer
6. presentation layer
7. application layer

This architectural view covers all the aspects from physical connection to application development.

(**2**) the set of relationships among the networks and subnetworks of a large area or factory.

network caching to cut down on congestion in a computer network, as well as to provide faster response time, files may be kept in intermediate nodes and redistributed to others besides the original requesting node. For example, on the World Wide Web, Internet Service Providers may keep caches of frequently accessed files from remote locations to be accessed by their users. Difficulties include providing enough storage, efficient look-up, and managing stale data.

networked operating system an operating system that uses a network for sharing files and other resources.

network file system protocol (NFS) a network file-sharing strategy, originally developed by Sun Microsystems.

network interface card (NIC) the physical device or circuit used to interface the network with a local work-station or device.

network latency the difference between the time a message leaves its source and the time it arrives at its destination. *Network latency* is the sum of the queuing delay and the transmission time at each node on the path from source to destination. In general, network latency increases as the load on the network increases — the higher the throughput, as a percentage of the capacity, the greater the queuing delay and thus the greater the latency. *See also* end-to-end delay, transfer delay.

network layer the layer that forms a path by concatenation of several links.

network model a data model, in which the data is represented as collections of records and the relationships are represented by links. The records and links are organized as collections of graphs or networks. *See* network architecture.

network of queues an n-node network, in which each node consists of a single queue, and, when a customer has completed service at node i, it has a probability p_{ij} of leaving node i and proceeding to a particular node j. If it is an *open network*, there will also be some probability γ_i of a customer arriving at i from outside, and a probability $1 - \sum_1^n p_{ij}$ that a customer will leave the system at node i. Then the defining equation for the arrival rates at each node i is $\lambda_i = \gamma_i + \sum_1^n p_{ji}$. Both open and closed networks of queues have *product form solutions* for Markovian and certain other distributions.

network operating system an operating system that uses a network for sharing files and other resources.

network performance the ability of a network to deliver a requested or agreed service. *See also* quality of service, grade of service.

network prefetching fetching files across a network because traffic patterns indicate they are likely to be used soon. For example, on the World Wide Web, fetching the links within a file when the file is fetched, anticipating that the user will follow a link soon. In this example, the improvement in user response time is at the expense of a great increase in network load.

network pruning the process of cutting out nodes or unwanted connections in a neural network, to save computation or computational hardware. *See also* artificial neural network.

network service a service available through the network such as mail and file transfer.

network throughput sum of all output bit rates at one point of time.

neural network an information-processing device modeled after biological networks of neurons. Typically, it takes the form of a parallel distributed information processing structure consisting of processing elements, called neurons, interconnected via unidirectional signal channels called connections. Neurons can possess a local memory and can carry out localized information processing operations. Each neuron has a single output connection. The neuron output signal can be of any mathematical type desired. The information processing that goes on within each processing element can be defined arbitrarily with the restriction that it must be completely local (it depends only on the current values of the input signals arriving at the processing element and on values stored in the processing element's local memory, e.g., weights).

neural network model modeled on biological neurons that receive impulses from neighboring neurons and fire when a critical potential is reached, a *neural network model* contains artifical neurons that sum the inputs they receive via weighted links from other neurons and output a value that is a non-linear function of the input activation. Neural network models can be trained on a set of input data and used to predict time series; they have found wide applicability in modeling complex systems.

neural tree a tree-structured neural network. Such networks arise in the application of certain kinds of constructive algorithm.

neuro-fuzzy control system control system involving neural networks and fuzzy systems, or fuzzy neural networks. *See also* adaptive fuzzy system.

neuron a nerve cell. Sensory neurons carry information from sensory receptors in the peripheral nervous system to the brain; motor neurons carry information from the brain to the muscles.

neutrosophic logic a generalization of fuzzy logic which is the study of neutrosophic logical values of the propositions.

neutrosophic probability a generalization of the classical probability that studies the chance that a particular event will occur, where that chance is represented by three coordinates (variables): $t\%$ true, $i\%$ indeterminate, and $f\%$ false, with $t + i + f = 100$ and f, i, t belonging to $[0, 100]$.

neutrosophic set a generalization of the fuzzy set such that an element belongs to the set with a neutrosophic probability; i.e., $t\%$ is true that the element is in the set, $f\%$ false, and $i\%$ indeterminate.

neutrosophy a branch of philosophy which studies the origin, nature, and scope of neutralities, as well as their interactions with different ideational spectra.

new an operator that creates new instances of objects of a specific type. In object-oriented languages, the creation of such an instance will invoke the constructor for the object type. In other languages, there is no specific operator for creating new instances, but instead a general operation for allocating storage (for example, malloc in the C language). *See also* delete.

newline character an invisible line terminator/separator.

news *See* netnews.

Newton-Raphson approximation an iterative method for approximating the root of a nonlinear function. If x_i is an approximation to the root of f, then $x_{i+1} = x_i - a_i f(x_i)/f'(x_i)$, where f' is the first derivative of f and a_i is the step-size parameter. For use in the division of n by d, a good approximation to $1/d$ is found and then multiplied by N. In square-root evaluation, of $\sqrt{x}$, either the procedure is used to find a good approximation to $\sqrt{x}$, or it is used to find a good approximation to $1/\sqrt{x}$ and the result then multiplied by x.

Newton's method a class of numerical root-finding methods; that is, to solve $f(x) = 0$. The methods are based on iteratively approximating f using a low order polynomial and finding the roots of the polynomial. The most commonly used first-order method iterates the following equation:

$$x_{n+1} = x_n - \frac{f(x_n)}{f'(x_n)},$$

where f' represents the derivative of f.

NFS *See* network file system protocol.

***n*-gram language model** *See* statistical language model.

nibble four bits of information.

nibble-mode DRAM an arrangement where a dynamic RAM can return an extra three bits, for a total of four bits (i.e., a nibble), with every row access.

The typical organization of a DRAM includes a buffer to store a row of bits inside the DRAM for column access. Additional timing signals allow repeated accesses to the buffer without a row–access time. *See also* two-dimensional memory organization.

NIC *See* network interface card.

niche a set of environmental conditions to which a species has been adapted and which allows its continual survival. In evolutionary computation, the term also refers to stable sub-populations of individuals clustering around individual optima of a multi-modal fitness landscape. These sub-population clusters are usually formed using specialized techniques such as fitness sharing to maintain diversity among the individuals.

NIST National Institute of Standards and Technology, the national standards-setting body for the U.S., part of the Department of Commerce. Although it is concerned primarily with measurement standards, it does have some impact on computing technology. *See also* ANSI, IEEE, ISO.

nit an equivalent name for the unit of luminance: candelas per square meter.

NMI *See* non-maskable interrupt.

NNF *See* negation normal form.

node a queue and its associated servers in a queuing network.

node name a symbolic name or address by which a Unix system is known to other systems on a network.

node state a node can be excited into firing signals at different levels of activity. The state of the node describes how active the node is in firing.

noise (**1**) any undesired disturbance, whether originating from the transmission medium or the electronics of the receiver itself, that gets superimposed onto the original transmitted signal by the time it reaches the receiver. These disturbances tend to interfere with the information content of the original signal and will usually define the minimum detectable signal level of the receiver.

(**2**) any undesired disturbance superimposed onto the original input signal of an electronic device; noise is generally categorized as being either external (disturbances superimposed onto the signal before it reaches the device) or internal (disturbances added to the signal by the receiving device itself). Some common examples of external noise are crosstalk and impulse noise as a result of atmospheric disturbances or manmade electrical devices. Some examples of internal noise include thermal noise, shot noise, 1/f noise, and intermodulation distortion.

noiseless source coding theorem states that any source can be losslessly encoded with a code whose average number of bits per source symbol is arbitrarily close to, but not less than, the source entropy in bits.

noise smoothing any process by which noise is suppressed, following a comparison of potential noise points with neighboring intensity values, as for mean filtering or median filtering.

noisy channel vector quantization a general term for methods in vector quantizer transmission for noisy channels. *See* channel robust vector quantization, channel optimized vector quantization, channel matched vector quantization, redundancy-free channel coding.

noisy channel VQ *See* noisy channel vector quantization.

noisy source coding *See* lossy source coding.

non-binary code a code in which the fundamental information units or symbols assume more than two values. This is in contrast to binary codes for which the fundamental information symbols are two-valued or binary only. *See also* binary code, block code, convolutional code.

non-blocking cache a cache that can handle access by the processor even though a previous cache miss is still unfinished.

non-blocking call a call that does not require the recipient to be present for the data to be passed.

nonclustering index a dense index that puts no constraints on the table organization. Also known as secondary index. *See also* clustering index.

non-commented source statements (NCSS) this measurement refers to the data declarations and procedural code (executable statements) of a software system.

non-commutative algebra an algebraic system where the commutative laws do not hold. For example,

$$x \cdot y = y \cdot x ,$$

does not hold for quaternions.

non-conformance incorrect, incomplete, or superfluous implementation of an input product by an output product at some phase of system development. Synonymous with defect.

nonconformity nonfulfillment of a specified requirement(s).

non-dense index an index is non-dense if it does not contain an index entry for every record in the underlying file.

non-destructive readout when data are read from a particular address in a memory device, the contents of the memory at that address remain unaltered after the read operation.

non-determinism a freedom of choice in selecting the next action.

non-deterministic a property of a computation which may have more than one result. One way to implement a non-deterministic algorithm is using backtracking; another is to explore (all) possible solutions in parallel.

nondeterministic permitting more than one choice of next move at some step in a computation.

nondeterministic finite state automaton (NDFSA) a finite state automaton in which for each state and input, one or more next states may be possible.

nondeterministic finite tree automaton accepts finitary trees rather than just strings. The tree nodes are marked with the letters of the alphabet of the automaton, and the transition function encodes the next states for each branch of the tree. The acceptance condition is modified accordingly. *See also* deterministic finite tree automaton, nondeterministic tree automaton.

nondeterministic push-down automaton (NDPDA) a pushdown automaton that has the property that a mapping from input symbol, state, and stack symbol can result in more than one next state and/or stack state. A NDPDA can be used to parse a context-sensitive grammar, and is consequently more powerful than a deterministic PDA. *See also* Chomsky hierarchy, context-sensitive grammar.

nondeterministic tree automaton a nondeterministic finite automaton that accepts infinite trees rather than just strings. The tree nodes are marked with the letters of the alphabet of the automaton, and the transition function encodes the next states for each branch of the tree. The expressive power of such automata varies depending on the acceptance conditions of the trees. *See also* nondeterministic finite tree automaton.

nondeterministic Turing machine a Turing machine in which for each state and input one or more next states may be possible.

nonexecution-based testing performing reviews, inspections, analysis, or other forms of testing that do not involve the execution of code.

non-existent a state of a task or process, immediately before its creation or after deletion.

non-fuzzy output a function of the firing degrees and the fuzzy outputs of fired fuzzy rules regardless of what defuzzification method is used.

non-interleaved data ordering data ordering in which processing is performed component by component from left-to-right and top-to-bottom.

non-maskable interrupt (NMI) an external interrupt to a CPU that cannot be masked (disabled) by an instruction. *See* maskable interrupt.

nonmonotonic logic a logic in which the addition of new knowledge may invalidate previously inferrable conclusions.

nonmonotonic reasoning reasoning that can be revised if some values change during a session. It can deal with problems that involve rapid changes in values in short periods of time.

non-performing division division where the partial dividend is modified only if the result of the division is positive (assuming positive operands). (The main operation in division is the subtraction of a multiple of the divisor from the partial dividend). *See also* restoring division, non-restoring division.

non-preemptive priority the next customer to be serviced will be the one that is waiting and has highest priority number, but this customer cannot start until the customer currently being serviced has completed.

non-preemptive scheduling a scheduling strategy whereby a process does not release the processor until it has completed its work.

nonprivileged state an execution context that does not allow certain hardware instructions to execute.

nonprocedural language any specification that specifies a desired goal without specifying an algorithm to achieve it. Specific instances include query languages (for example, SQL), report generation languages (for example, RPG), or spreadsheets.

nonrecursive equation *See* recursive equation.

nonredundant number system the system where for each bit string there is one and only one corresponding numerical value.

non-redundant replication each fragment is stored exactly at each site and there is no overlap between fragments.

non-removable disk *See* removable disk.

non-repeatable read *See* unrepeatable read.

non-restoring division a division algorithm that is similar to the restoring-division algorithm, except that there is no restoration of negative partial dividends. Instead of a restoration, the next operation is an addition instead of a subtraction.

non-return-to-zero-inverted recording *See* magnetic recording code.

non-return-to-zero recording *See* magnetic recording code.

nonrigid body motion *See* rigid body motion.

nonseparable data *See* separable data.

non-serial schedule a schedule S of n transactions is said to be non-serial, if the operations of the transactions are interleaved.

nonterminal a name for a structure defined by a context-free grammar rule. Interior nodes of parse trees are labeled by nonterminals.

nonterminal symbol one of the set of symbols defined by a grammar. The set of symbols used to designate grammatical constructs in the language. In any grammar, the set of nonterminal symbols and the set of terminal symbols are disjoint sets. Also referred to as the nonterminal vocabulary. Sometimes designated by the metasymbol V_n.

nonterminal vocabularly the complete set of nonterminal symbols defined by a grammar.

nonuniform sampling *See* uniform sampling.

nonvolatile *See* nonvolatile memory.

non-volatile memory memory that retains its contents even when the power supply is removed. Examples are secondary memory and read-only memory.

nonvolatile memory the class of computer memory that retains its stored information when the power supply is cut off. It includes magnetic tape, magnetic disks, flash memory, and most types of ROM.

non-volatile random-access memory (NVRAM) SRAM or DRAM with non-volatile storage cells. Essentially each storage cell acts as a normal RAM cell when power is supplied, but when power is removed, an EEPROM cell is used to capture the last state of the RAM cell and this state is restored when power is returned.

nonzero-sum game one of a class of games in which the sum of the cost functions of the players is not constant. A number of players is not limited to two as in zero-sum games and in this sense one may distinguish between two-person and multi-person nonzero-sum games. Since the objectives of the players are not fully antagonistic, cooperation between two or more decision makers may lead to their mutual advantage. However, if the cooperation between the players is not admissible because of information constraints, lack of faith, or impossibility of negotiation, the game is noncooperative and an equilibrium may be defined in a variety of ways. The problem of solving the nonzero-sum game differs from the one for zero-sum games. The most natural is Nash equilibrium which is relevant to the saddle point equilibrium in zero-sum games. Its main feature is that there is no incentive for any uniteral deviation by any one of the players and their roles are symmetric. If one of the players has the ability to enforce his strategy on the other ones, then a hierarchical equilibrium called von Stackelberg equilibrium is rational. Yet another solution for the player is to protect himself against any irrational behavior of the other players and adopt min-max strategy by solving a zero-sum game although the original is nonzero-sum.

no-op a computer instruction that performs no operation. It can be used to reserve a location in memory or to put a delay between other instruction executions.

no-overwrite policy a policy in which the file system never rewrites existing data in a file. New data is always written into a new location on the disk.

nor negated disjunction: 0 NOR 0 = 1, 0 NOR 1 = 0, 1 NOR 0 = 0, 1 NOR 1 = 0.

NOR gate a logic circuit that performs the operation equivalent to the OR gate followed by the inverter. The output of a NOR gate is low when any or all inputs are high.

norm for a vector space V, a real-valued function N defined on V, satisfying the following requirements for every $v, w \in V$, and scalar λ:

1. $N(0) = 0$ and for $v \neq 0$, $N(v) > 0$.
2. $N(\lambda v) = |\lambda| N(v)$.
3. $N(v + w) \leq N(v) + N(w)$.

The most usual norms are the L^1, L^2, and L^∞ norms; for a vector x with n coordinates $x_1, \dots, x_n$ we have

$$\|x\|_1 = \sum_{k=1}^{n} |x_k| \ ,$$

$$\|x\|_2 = \left(\sum_{k=1}^{n} |x_k|^2 \right)^{1/2} ,$$

and

$$\|x\|_\infty = \max_{k=1}^{n} |x_k| \ ;$$

for a function f defined on a set E, these norms become

$$\|f\|_1 = \int_E |f(x)|\, dx \ ,$$

$$\|f\|_2 = \left(\int_E |f(x)|^2\, dx \right)^{1/2} ,$$

and

$$\|f\|_\infty = \min\{r \geq 0 \mid |f(x)| \leq r \text{ almost everywhere}\} \ .$$

Given a norm N, the function d defined by $d(x, y) = N(x - y)$ is a distance function on V. *See* chessboard distance, Euclidean distance, Manhattan distance.

normal distribution *See* Gaussian distribution.

normal form a specified criteria for a relation. We say a relation is in a given normal form if it satisfies certain criteria.

normalization the process of decomposing relations such that the resulting relations satisfy successive normal forms.

normalization algorithm *See* additive normalization, multiplicative normalization.

normalization shift a left shift to produce a normalized number (representation).

normalized a relation, R, where every functional dependency X functionally determines A, where A and the attributes in X are contained in R (but A does not belong to X), has the property that X is the key or a superset of the key of R. X functionally determines A if any two tuples with the same X values have the same A value. X is a key if no two records have the same values on all attributes of X.

normalized number a floating-point number in which the most significant digit of the significand (mantissa) is non-zero. In the IEEE 754 floating-point standard, the normalized significands are larger than or equal to 1, and smaller than 2.

normalized representation a representation in which the leading digit is not 0.

normal-order reduction a reduction strategy in which the leftmost redex is always reduced first. Also called call-by-name reduction.

normal tree a tree that contains all the independent voltage sources, the maximum number of capacitors, the minimum number of inductors, and none of the independent current sources.

normal vector a vector perpendicular to the tangent plane of a surface. Also, a vector that is geometrically perpendicular to another vector.

NOT a Boolean operation that returns the 1's complement of the data to which it is applied.

NOT 0 = 1, NOT 1 = 0. *See also* AND, OR.

not-a-number (NAN) in computer arithmetic, machine pattern that is used when the result of an arithmetic operation is not mathematically defined, e.g., the result of a magnitude subtraction of infinities of the square-root of a negative number.

no-write allocate part of a write policy that stipulates that if a copy of data being updated is not found in one level of the memory hierarchy, space for a copy of the updated data will not be allocated in that level. Most frequently used in conjunction with a write-through policy.

NP-complete the class of problems for which answers can be checked for correctness by an algorithm whose run time is polynomial in the size of the input (it is NP) and no other NP problem is more than a polynomial factor harder. Informally, a problem is NP-complete if answers can be verified quickly and a quick algorithm to solve this problem can be used to solve all other NP problems quickly. *See also* NP-hard.

NP-complete language a language in NP such that every language in NP can be reduced to it in polynomial time.

NP-hard a complexity class of problems that are intrinsically harder than those that can be solved by a Turing machine in nondeterministic polynomial time. When a decision version of a combinatorial optimization problem is proven to belong to the class of NP-complete problems, which includes well-known problems such as satisfiability, traveling salesman, etc., an optimization version is NP-hard. *See also* strongly NP-hard, AP.

N queens A problem in computational analysis. The probelm is stated as follows: place n chess queens on an $n \times n$ board such that no queen can attack another. Efficiently find all possible placements. *See also* eight queens.

NuBus an open bus specification developed at MIT and used by several companies. It is a general-purpose backplane bus, designed for interconnecting processors, memory, and I/O devices.

nucleus *See* kernel.

null (**1**) empty, as in "the null set".

(**2**) invalid, as in "null and void".

(**3**) rendered invalid, as in "nulls the effect of".

(**4**) in a language that has pointers or references, a designated value which indicates that the pointer or reference does not reference a value.

(**5**) in the C language, a designated value (traditionally zero, but not constrained to this value by the standard) representing a pointer which does not point to a valid value, and which by convention has a symbol NULL which is defined to represent that value.

(**6**) a type of device that (`/dev/null`) accepts input but has no output and no function.

nullary function a function with no arguments, also known as a constant function. *See also* unary function, binary function, N-ary function.

null modem cable a cable that allows two standard RS-232C devices to be connected directly without any other communications equipment such as a modem.

null pointer detection a form of execution semantics in which an attempt to access an object via a null pointer generates an error condition. What the error condition is, and how it is handled, is determined by factors such as the definition of the language, its implementation, and the machine on which it is running.

NULL value a special value used in relational databases. This value can belong to any domain.

number of machine instructions per executed line of code the quantity of low-level assembly commands required to execute a given line of high-level source code.

number system the representation of numbers as a sequence of digits with an interpretation rule which assigns a value to each sequence. The conventional number systems are fixed-radix and positional systems, where the digit in position i has a weight of r^i, where r is the radix.

NURBS Non-Uniform Rational B-Splines. A class of piecewise parametric curves or surfaces where each curve segment or surface patch is described by a ratio of Non-Uniform B-Spline polynomials. B-splines are a class of polynomials whose coefficients depend on a set of control points. For the Uniform B-Splines each curve segment or surface patch is defined by a parameter domain of fixed length or area, respectively, whereas in the Non-Uniform B-Splines the parameter domain does not have to be uniform. The Non-Uniform characteristic allows different levels of continuity between the curve segments and the surface patches, whereas it is restricted to p-1 levels for the uniform case, where p is the degree of the polynomial. Thus, Non-Uniform B-Splines can interpolate points more accurately. Furthermore, rational forms can represent conic curves and are invariant under rotation, scaling, translation, and perspective transformations. NURBS provide a superset of commonly used surfaces and have been adopted as the IGES (Initial Graphics Exchange Specification) for free-form surfaces.

NVRAM *See* non-volatile random-access memory.

N-well a region of n-type semiconductor located at the surface of a p-type substrate (or larger P-well) usually created in order to contain p-channel MOSFETs.

nybble *See* nibble.

Nyquist frequency *See* Nyquist rate.

Nyquist rate the lowest rate at which recovery of an original signal from its sampled signal is possible. If the highest frequency of the analog signal is f_H, the Nyquist frequency is the frequency of the samples f_S for proper recovery of the signal at the receiving end.

$$f_{\mathbf{Nyquist}} = f_S \text{ should be } > 2f_H$$

In digital transmission systems, the analog signal is transformed to a digital signal using an A/D converter.

Nyquist sampling theorem fundamental signal processing theorem that states that in order to unambiguously preserve the information in a continous-time signal when sampled, the sampling frequency, $f_s = 1/T$, must be at least twice the highest frequency present in the signal. *See also* aliasing.

O

O *See* big-O notation.

object (1) the basic functional data unit in an object-oriented system. An object is an encapsulation of state and behavior such that the internal information is hidden from external view.

(2) in object-orientation, an instance of a class definition.

object based a programming language (usually) that has many of the attributes of object-orientation but does not support inheritance.

object-based coding video compression based on the extraction, recognition, and parameterization of objects in the scene. Unknown-object coding finds geometrical structures representing 2D or 3D surfaces and codes these areas and their movement efficiently. Known-object coding relies on the detection of objects for which the system has a high-level structural model.

object cloning *See* cloning.

object code a representation of a source program in a lower-level representation more suited for execution. Object code is that code which is presented to a component that can execute the program. Object code may be machine code or byte codes. May also refer to an intermediate representation, relocatable code, which must be processed by a linker and perhaps a loader before it can be executed. *See* machine code.

object, composite *See* composite object.

object database management group (ODMG) a consortium of object-oriented database vendors responsible for defining standards for object-oriented database management systems.

object definition language (ODL) the data definition language proposed by the ODMG for designing the structure of object-oriented databases.

object detection the detection of objects within an image. This type of process is cognate to pattern recognition. Typically, it is used to locate products ready for inspection or to locate faults during inspection. An object usually denotes a larger characteristic in an image, such as an alignment of points, a square, a disk, or a convex portion of a region, etc.; its detection involves non-local image transforms, such as the *Hough transform* or the medial axis transform. *See* edge detection, Hough transform, key point detection, medial axis transform. *See also* feature detection.

object diagram (1) a synonym for class diagram.

(2) a synonym for collaboration diagram in which objects and their interactions/collaborations are shown. The object diagram may include relationships of is-part-of and those by means of method calls.

object identity the concept that each object is unique. This uniqueness is realized in terms of an object ID which is machine generated.

objective function *See* optimization problem.

object linking and embedding (OLE) a technology developed by Microsoft to represent composite objects in a persistent object store. It is implemented in terms of their underlying Component/Common Object Model technology. Languages are emerging which work directly with COM objects, for example, Visual Basic. *See also* COM.

object management group (OMG) an industry consortium aimed at setting standards in object-oriented development.

object manipulation language (OML) a language that allows objects in an object-oriented database to be created, updated, and destroyed. Normally implemented by adding extensions for persistence to an object-oriented programming language.

0-8493-2691-5/01/$0.00+$.50

object measurement the measurement of objects, with the aim of recognition or inspection to determine if products are within acceptable tolerances.

object model another name for an object diagram. Although it is really the model and not the diagram, the main way of visualizing the model and making it real is by drawing it on a diagram. Thus, the terms object model and object diagram have often come to be used synonymously.

object module a program unit that is the output of a translation of a source code.

object orientation (**1**) the set of concepts which constitute the object paradigm (object technology).

(**2**) measurement of the orientation of objects, either as part of the recognition process or as part of an inspection or image measurement process.

object-oriented (**1**) an adjective prefixed to any software term to indicate that it is operating within the constraints and definitions of the object-oriented paradigm rather than any other paradigm (such as the functional, procedural, or data abstraction paradigms).

(**2**) a model which assumes that data is represented as objects from one or more classes, and that methods of the classes are used to manipulate the objects.

object-oriented addressing a form of virtual addressing in which object numbers are mapped to memory regions and internal object references are mapped to offsets within an object's memory region.

object-oriented analysis (OOA) the activities in the life-cycle in which an object-oriented approach is being used and which are targeted towards understanding the problem. Perhaps better named "object-oriented modeling".

object-oriented database (OODB) (**1**) a database described in terms of classes of objects, which includes their attributes and behavior, and the relationships between those classes.

(**2**) a database which uses objects rather than relational tables to organize and manage its contents (objectbase).

object-oriented database management system the system which comprises all aspects of database management but focused towards the use of the object-oriented mechanisms.

object-oriented database management system (OODBMS) a database management system in which the logical database model is object-oriented. The OODBMS interface to the database is an object-oriented programming language.

object-oriented decomposition breaking a problem down into smaller pieces based on the rules and expectations of the object-oriented paradigm.

object-oriented design (OOD) the activities in the life-cycle in which an object-oriented approach is being used and which is targeted toward deriving a solution to a problem.

object-oriented language (OOL) a language for object-oriented programming. It has to be mainly conformant with the object-oriented paradigm, allowing the definition of classes and their relationships of is-part-of, is-a, is-referred-by, etc. It also has to provide an intrinsic mechanism for data hiding. Present examples include Java and C++.

object-oriented methodology an application development methodology that uses a top-down approach based on a decomposition of a system in a collection of objects communicating via messages.

object-oriented paradigm the set of concepts which define an approach to software development using object technology. Although not globally agreed, the underpinning concepts in the OO paradigm are encapsulation/information hiding; abstraction, especially classification; and modularity, identity, and polymorphism.

object-oriented process environment and notation (OPEN) full life-cycle, process-focused, third-generation OO "methodology". OPEN is underpinned by a process metamodel consisting of, inter alia, activities, tasks, techniques, and deliverables.

object-oriented programming (OOP) the activities in the life-cycle in which an object-oriented approach is being used and which are targeted towards translating the design into code.

objectory an outdated methodology which is the precursor to RUP.

object persistence the ability of an object to outlive the program that created it.

object program *See* object code.

object query language (OQL) a functional, statically typed query language used for object-oriented databases. All OQL query expressions consist of an operator applied to zero or more operands. The result of an OQL query can be any valid type defined in the OQL environment.

object recognition the process of locating objects and determining what types of objects they are, either directly or indirectly through the location of sub-features followed by suitable inference procedures. Typically, inference is carried out by application of Hough transforms or association graphs.

Object Relational Database Management System (ORDBMS) a DBMS that supports the Object Relational Model.

object/relational database management system an occurrence of a parent child relationship consists of one of the parent record types and zero or more of the child record types.

Object Relational Data Model a data model that allows extensions to the relational data model by allowing addition of object-oriented concepts.

object request broker (ORB) the means by which objects send and receive requests across a heterogeneous network. The ORB acts as a software bus to facilitate the interchange. This is now commonly known as middleware.

object sharing in object-oriented and based models, this is the sharing of an instance among two or more other objects. This mechanism can be implemented by means of is-referred-by relationship and may be the cause of severe errors when one of the objects sharing the other object deletes it without communicating the operation to the others.

object SQL (OSQL) a query language proposed by the Object Database Management Group for forming queries on object-oriented databases. Similar to SQL.

object table an index of (object identifier, object location) pairs used to dereference objects from their identifiers.

object type the type of an object determines the set of allowable operations that can be performed on the object. This information can be encoded in a "tag" associated with the object, can be found along an access path reaching to the object, or can be determined by the compiler which inserts "correct" instructions to manipulate the object in a manner consistent with its type.

object type library the collection of all object types and their inheritance hierarchies for an application.

occlusion visual obstruction. An *occlusion* occurs when an opaque surface prevents another surface from being seen. When rendering, it is necessary to determine which surfaces are not occluded, a problem known as the hidden surface problem.

occur-check a test performed during unification to ensure that a given variable is not defined in terms of itself (e.g., $X = f(x)$ is detected by an occur-check, and unification fails).

occurrence a string v occurs in a string u if v is a factor of u.

OCR *See* optical character recognition.

octal a numerical system of radix 8. Convenient for expressing computer value representations because three bits make up one octal digit. *See also* binary, hexadecimal.

octal number system a number system consisting of eight digits from 0 to 7; it is also referred to as base 8 system.

octree a space-occupancy representation used for representing 3D volumetric objects. It is a hierarchical representation, designed to use less memory than representing every voxel of the object explicitly. *Octrees* are based on subdividing the full voxel space containing the represented object into 8 octants by planes perpendicular to the three coordinate axes. Octants that completely contain a single object are denoted as being pure. Octants that contain multiple objects are recursively split into 8 new smaller octants. This splitting continues until all volumes are either pure, or some volume size limit is reached. A tree data structure can be used to represent the octree. Normally, the octree data structure will have only about as many nodes as there are voxels on the object surface, which can be much less than the total number of voxels in the object. Hence, an octree representation can save a lot of space when representing an object or scene.

The octree is the 3-dimensional generalization of the quadtree. *See* quadtree.

octtree *See* octree.

ODL *See* object definition language.

ODMG *See* object database management group.

ODMG-93 the standard for object-oriented databases defined by the ODMG.

ODMG data model an object model intended to be used by any object-oriented database that complies with the ODMG standard.

ODP *See* optimum distance profile code.

OFD code *See* optimum free distance code.

off-line (**1**) pertaining to the operation of a functional unit when not under the control of a computer.

(**2**) handled on a different computer.

off-line problem a decision problem in which an algorithm is given the entire sequence of inputs in advance.

offset distance, usually measured in addressable units, from a designated base address to another designated address. Often used within compilers to designate locations such as: (**1**) the location of a local variable as an offset from a stack pointer or frame pointer.

(**2**) the location of a parameter location as an offset from a stack pointer or frame pointer.

(**3**) the location of the field of a record relative to the base address of the record.

(**4**) in a string, the distance from the beginning to the end of a segment in that string.

offspring a new individual in the population which is derived, through the processes of recombination and mutation, from one or more original individuals in the previous generation.

off-the-shelf software product that is already developed and available, usable either "as is" or with modification, and provided by the supplier, the acquirer, or a third party. *See* commercial off-the-shelf.

OLAP *See* optical linear algebraic processor.

OLE *See* object linking and embedding.

Ω a theoretical measure of the execution of an algorithm, usually the time or memory needed, given the problem size n, which is usually the number of items. Informally saying some equation $f(n) = \Omega(g(n))$ means it is more than some constant multiple of $g(n)$. More formally, it means there are some c and k, such that $0 \leq cg(n) \leq f(n)$ for all $n \geq qk$. The values of c and k must be fixed for the function f and must not depend on n. *See also* big-O notation, asymptotic lower bound.

omega motion a code motion characterized by moving a computation to the tail of a branching construct such that it is executed only once, after all the branches are executed.

OMG *See* object management group.

omicron the Greek letter written as "o". (*See* little-o notation) or "O". (*See* big-O notation).

OML *See* object manipulation language.

OMZ *See* order-of-magnitude-zero.

one-copy equivalence a replica control policy which asserts that the values of all copies of a logical datum should be identical when the transaction that updates that item terminates.

one-dimensional cellular automaton a cellular automaton, where we trace the evolution of the system by observing a row of cells at time t followed by the row at time $t + 1$ and so on.

one-dimensional coding scheme a scheme in which a run of consecutive pixels of the same gray scale is combined together and represented by a single code word for transmission.

Scan lines are coded as "white" or "black" which alternate along the line. The scans are assumed to begin with white and are padded with white if this is not the case. Generally run-length coding is a one-dimensional coding scheme.

one-level memory an arrangement of different (in terms of speed, capacity, and medium) types of memory such that the programmer has a view of a single flat memory space. *See also* hierarchical memory, virtual memory.

one-level storage *See* one-level memory.

one-out-of-N coding a method of training neural networks in which the input vector and/or the output vector has only one non-zero element (usually equal to unity) for each training example.

one's complement (**1**) a notation used for the representation of negative numbers. In this representation a data word is organized such that negative numbers all contain a binary "one" in the leftmost bit, while positive numbers contain a "zero" in the leftmost bit. To obtain the one's complement of a given binary representation, every bit is inverted (which is equivalent to subtracting from 1 the value of the bit). *See also* diminished-radix complement.

(**2**) the operation of inverting a data word so that all ones become zeroes and vice versa.

one-sided surface a surface rendered in such a way that only one side is visible. That side is usually facing the same direction as the surface normal.

on-line pertaining to the operation of a functional unit when under the direct control of a computer.

on-line algorithm algorithm that processes the input sequentially.

on-line arithmetic *See* digit-online arithmetic.

online estimation a method to determine statistical characteristics of a time series while it is being observed. Due to the fact that long-term measurements will produce an unlimited number of observations, online estimators cannot rely on accessing every single observation in history. Instead, they must have a strategy of reducing older data to a number of variables that do not depend on the duration of observation.

on-line memory memory that is attached to a computer system.

on-line planning algorithm a planning algorithm in which planning computations and the execution of actions are carried out concurrently.

on-line problem a problem in which an algorithm receives a sequence of inputs, and must process each input in turn, without detailed knowledge of future inputs.

on-line support system digital form of assitance, such as tutorials, documentation, and help that aid computer-based learning and task-oriented activities.

on-line transaction processing the class of applications where the transactions are short, typically 10 disk I/Os or fewer per transaction; the queries are simple, typically point and multipoint queries; and the frequency of updates is high.

OOD *See* object-oriented design.

OODB *See* object-oriented database.

OOL *See* object-oriented language.

OOP *See* object-oriented programming.

op any one of the arithmetic operations, e.g., addition, subtraction, multiplication, or division; as in floating point operation (flop).

opaque impervious to light. An *opaque* surface will reflect light to some degree dependent on surface attributes. *See also* specular reflection, diffuse reflection.

opcode a part of an assembly language instruction that represents an operation to be performed by the processor. Opcode was formed from the contraction of "operational" and "code."

OPEN *See* object-oriented process environment and notation.

open an operation on a file or device that results in the initialization of this unit and its readiness for exchanging data.

open addressing a collision resolution policy whereby, upon collision, a search begins from the current free space until a free location is found.

opening a key morphological filter involving the union of translates.

For structuring element B, the opening is the composition of the erosion by B followed by the dilation by B; it transforms X into $X \circ B = (X \ominus B) \oplus B$. The opening by B is called an algebraic opening; this means that: (a) it is a morphological filter, (b) it is anti-extensive; in other words it can only decrease an object. *See* dilation, erosion, morphological filter, structuring element.

opening/closing filter *See* closing/opening filter.

open-loop planner a planning system that executes its plans with no feedback from the environment, relying exlusively on its ability to accurately predict the evolution of the underlying dynamical system.

open queueing network a queueing network in which there are arrivals to the system from outside (called external arrivals) and external departures from the system. In contrast to a closed queueing network, the number of customers in the network is not fixed.

open software foundation (OSF) a foundation created by nine computer vendors to promote open computing. It is planned that common operating systems and interfaces, based on developments of Unix and the X Window System will be forthcoming for a wide range of different hardware architectures. OSF announced the release of the industry's first open operating system.

open subroutine *See* inline function.

open system architecture a layered architectural design that allows subsystems and/or components to be readily replaced or modified; it is achieved by adherence to standardized interfaces between layers.

open systems interconnection model (OSI) a reference model for a network architecture and a suite of protocols (protocol stack) to implement it developed by ISO in 1978 as a framework for international standards in heterogeneous computer network architecture. The architecture is split between seven layers, from the lowest to the highest: physical layer, datalink layer, network layer, transport layer, session layer, presentation layer, application layer. Each layer uses the layer immediately below it and provides a service to the layer above.

operability attributes of software that bear on the user's effort for operation and operation control. The capability of the software product to enable the user to operate and control it. Aspects of suitability, changeability, adaptability, and installability may affect operability. Operability corresponds to controllability, error tolerance, and conformity with user expectations.

operand (**1**) an object or value on which an operation is performed. In machine code, the values manipulated by a single instruction.

(**2**) specification of a storage location that provides data to or receives data from the operation.

operand address the location of an element of data that will be processed by the computer.

operand address register the internal CPU register that points to the memory location that contains the data element that will be processed by the computer.

operating system a collection of software and firmware elements that control the execution of computer programs and provide such services as resource allocation, job control, input/output control, and file management.

operating system backup a database backup made using operating facilities.

operation specification of one or a set of computations on the specified source operands placing the results in the specified destination operands.

In object-oriented systems, an operation has a name and a list of parameters and is part of the external view of the class. The method which implements the operation is part of the internal view.

operationality boundary in a generalized explanation, any division between the root subtree and the peripheral subtrees such that the root subtree yields a useful concept. *See also* explanation.

operational model a model that can be specified in terms of state and transition between states. The operational models specify the systems by describing what the system has to do, instead of stating which are the properties of the systems such as the descriptive models. *See* executable specification.

operational profile a characterization of the conditions of use of a system, this is the set of operations that the software in a system can execute along with the probability with which each operation will occur, where an operation is defined as a group of runs that typically involve similar processing. The profile can sometimes be defined by a partitioning of the input space and estimate of the probability of encountering an input from each feature. A system trial must therefore be performed using a realistic operational profile. Definition of conditions of use is essential for measurement of reliability (and some other dependability attributes), since a system will generally exhibit different reliabilities under different conditions. Usage is synonymous.

operator (**1**) a symbol in a programming language which specifies a computation. Traditionally operators represent arithmetic computations (with some common representations shown in parentheses) such as add (+), subtract (-), multiply (*), divide (/), or negation (-), logical computations such as and (&, .AND.), or (|, .OR.), or not (, .NOT.), comparison (<, >, =, ==, <=, >=, .LT., .GT., .EQ., .LE., .GE.), and assignment (= or :=). Languages may define many other operators, for example, C++ defines new and delete operators among many others. Traditional notations use infix notation for binary operators and prefix notation for unary operators, and maintain the traditional algebraic operator precedence to determine operand association; other languages use prefix notation or postfix notation, or may redefine operator hierarchy or declare all operators to have the same hierarchy.

(**2**) in image processing any one of several teransformations. *See* Canny operator, Laplacian operator, Marr-Hildreth operator, morphological operator, Sobel operator.

operator function in object-oriented programming, a function that has been associated with one of the available operators of a class.

The concept of operator function is strictly related to operator overloading. This is usually performed by an operator function.

operator overloading (**1**) the characteristic where the meaning of an operator depends upon the context in which it is used. For example, in most languages, operators such as "+" and "-" may be interpreted as "integer add" or "floating point add" or to concatenate strings depending on the context. As a further example, "-" can be used as a monadic operator to negate an expression, or as a dyadic operator to return the difference between two expressions.

In the object-oriented paradigm, overloading is a good exploitation of polymorphism. User-defined operator overloading is provided by several object-oriented programming languages.

(**2**) the ability to define an operator to operate upon user-defined objects, not just the elementary types defined by the language. *See also* function overloading, polymorphism.

operator tree a logical representation of a query in which the inner nodes represent operators and the leaf nodes represent relations.

opportunistic design a term used to refer to a behavior of software design. In empirical observations of designers at work, the sequence of design steps that a designer actually takes appears to be strongly determined by specific events that have occurred earlier in the design activity rather than by any overall strategy. Often, this results in the designer working on some aspect of the problem as soon as it presents itself.

opportunistic planning planning episodes in which subproblems at different levels are analyzed simultaneously, with alternation among levels determined by emerging features of the solution.

optical adder an optical device capable of performing the function of arithmetic addition using binary signals. It can be constructed using a series of cascaded full adders where carry has to ripple through each full adder from least significant bit to most significant bit. It can also be constructed using cascaded layers. Each layer consists of a series of half adders. Carry also has to ripple through the cascaded layers. It is also known as digital adder.

optical addition adding operation using light. Two incoherent light beams have intensities A and B, respectively. When two beams are combined, i.e., illuminating the same area, the resultant intensity is $A + B$.

optical bus an optical channel used for transmitting a signal from a source to one or more detectors. A bus allows only the same interchange of information to take place at different detectors. A source is connected to many detectors.

optical character recognition (OCR) a process in which optically scanned characters are recognized automatically by machine. It is widely utilized in document storage, processing, and management.

optical computer (**1**) a general purpose digital computer that uses photons as an electronic computer uses electrons. The technology is still immature. The nonlinear operations that must be performed by a processor in a computer are difficult to implement optically.

(**2**) an analog optical processor capable of performing computations such as correlation, image subtraction, edge enhancement, and matrix-vector multiplication that are usually performed by an electronic digital computer. Such a computer is usually very specific and inflexible, but well adapted to its tasks.

optical computing the use of optics to aid in any type of mathematical computation. Categories of optical computing include:

1. analog processing such as using optical Fourier transforms for spectral analysis and correlation,

2. optical switching and interconnection using nonlinear, e.g., bistable, optical devices, and

3. the use of optical devices in digital computers.

optical demultiplexer a device which directs an input optical signal to an output port depending on its wavelength.

optical disk a disk on which data are stored optically. Data are written by altering the reflectivity of the surface, and read by measuring the surface reflection of a light source. Storage is organized in the same way as on a magnetic disk, but higher storage density can be achieved.

optical expert system an expert system that utilizes optical devices for performing logic operations. An expert system mostly requires logic gates to perform inferences. In an optical expert system, not only discrete optical logic gates are used, but the parallelism and capability of performing matrix-vector multiplication of optical processors are exploited as well. Sequential reasoning in an electronic expert system is replaced by parallel reasoning in an optical expert system, so its speed can be increased substantially. As with the optical computer, it is still immature.

optical flow the 2D field of apparent velocities of pixels on an image plane; i.e., the raw motion information arising from the displacement of points in the visual (optical) field. Let each image point p have an intensity I and a velocity $v = (v_x, v_y)$, which are both functions of p and time t, where the velocity represents the image plane projection of the point's three-dimensional velocity. Under the assumption that the point's intensity does not change along its trajectory, it follows that

$$(\partial I/\partial x)v_x + (\partial I/\partial y)v_y + (\partial I/\partial z)v_z + \partial I/\partial t = \nabla I \cdot v + \partial I/\partial t = 0\,.$$

See also aperture problem.

optical flow field the changes in light intensity values in the image plane. *See* optical flow.

optical flux *See* optical flow.

optical full adder an optical device forming part of an adder and able to receive three inputs (augend, addend, and carry) from the previous stage, and deliver two outputs (sum and carry). Several logic gates are required to provide the sum and the carry.

optical half adder an optical device forming part of an adder and able to receive two inputs (augend and addend), and deliver two outputs (sum and carry). The sum is the XOR function of two inputs, and the carry is the AND function of two inputs. An optical half adder consists of an optical XOR gate and an optical AND gate. It is also known as a one-digit adder.

optical interconnect an optical communication system in an electronic computer that consists of three primary parts:

1. sources,
2. optical paths with switches or spatial light modulators, and
3. detectors.

An optical signal from a source or an array of sources is transmitted to a detector or an array of detectors in *optical interconnects.* The transmitted optical signal from the source is converted from an electronic signal. The detector in turn reverts the optical signal into an electronic signal. The merits of optical interconnects include large bandwidth, high speed of light propagation (the velocity of electric signals propagating in a wire depends on the capacitance per unit length), no interference, high interconnection density and parallelism, and dynamic reconfiguration.

optical linear algebraic processor (OLAP) an optical processor that performs specific matrix algebraic operations as fundamental building blocks for optical computation and signal processing.

optical logic logic operations, usually binary, that are performed optically.

optical logic gate an optical device for performing Boolean logic operations. The basic idea is that since a computer is built by Boolean logic gates, if we can make optical logic gates, then we can eventually build a complete optical computer. Since light does not affect light, it cannot directly perform nonlinear operations represented by 16 Boolean logic gates. A thresholding device is required if intensities of two input beams are simply added in the logic gate. For example, to perform AND operation, 00-0, 01-0, 10-0, 11-1, the threshold is set 2. Only when both beams have high intensities is the output 1. Otherwise, the output is 0. To perform OR operation, 00-0, 01-1, 10-1, 11-1, the threshold is set at 1. The gate can be constructed by directing two input light beams onto the same detector. After passing through an electronic thresholding device, the resulting electric signal from the detector shows the logic output. To implement an optical logic gate without using an electronic circuit, optical thresholding devices such as MSLM or nonlinear optic materials can be used.

optical matrix-matrix multiplication an operation performing matrix-matrix multiplication using optical devices. For two-by-two matrices A and B, the matrix-matrix multiplication produces two-by-two matrix C, whose elements are

$$\begin{aligned} c_{11} &= a_{11}b_{11} + a_{12}b_{21} \\ c_{12} &= a_{11}b_{12} + a_{12}b_{22} \\ c_{21} &= a_{21}b_{11} + a_{22}b_{21} \\ c_{22} &= a_{21}b_{12} + a_{22}b_{22} \,. \end{aligned}$$

An optical implementation is as follows. Matrix elements b_{12}, b_{22}, b_{11}, and b_{21} are represented by four vertical lines on a spatial light modulator. Matrix A is represented by a two-by-two source array. Two images of matrix A are generated by an optical means, for example two lenslets. An image of A covers two lines of b_{12} and b_{22}, another image of A covers lines b_{11} and b_{21}. Using two cylindrical lenses, light passing through the spatial light modulator is focused into four points at the four corners of a square representing c_{11}, c_{12}, c_{21}, and c_{22}. Various optical arrangements for implementing matrix-matrix multiplication have been proposed. Optical matrix-matrix multiplication is needed for solving algebraic problems, which could be faster than an electronic computer because of parallel processing.

optical matrix-vector multiplication an operation performing matrix-vector multiplication using optical devices. For matrix W and vector X, the matrix-vector multiplication produces vector Y, whose elements are

$$\begin{aligned} y_1 &= w_{11}x_1 + w_{12}x_2 + w_{13}x_3 \\ y_2 &= w_{21}x_1 + w_{22}x_2 + w_{23}x_3 \\ y_3 &= w_{31}x_1 + w_{32}x_2 + w_{33}x_3 \,. \end{aligned}$$

A simple coherent optical processor to implement this operation is as follows. Vector elements x_1, x_2, and x_3 are represented by the transmittance of three vertical lines on a first spatial light modulator. A collimated coherent light beam passes through the first spatial light modulator. The modulated light then passes through a second spatial light modulator displaying three-by-three squares with transmittances of w_{11} to w_{33}. Two spatial light modulators are aligned such that the vertical line x_1 covers three squares of w_{11}, w_{21}, w_{31}, line x_2 covers w_{12}, w_{22}, w_{32}, and line x_3 covers w_{13}, w_{23}, w_{33}. The light passing the second spatial light modulator is focused by a cylindrical lens to integrate light in a horizontal direction. In other words, $w_{11}x_1$, $w_{12}x_2$, and $w_{12}x_3$ are summed up at a point on the focal line of a cylindrical lens. A detector placed at this point will produce y_1. Two other detectors will provide y_2 and y_3 in a similar way. Various optical arrangements are possible to implement matrix-vector multiplication, including correlators. Optical matrix-vector multiplication is important for crossbar switch and neural networks. It can also be used for finding eigenvalues and eigenvectors, solving linear equations, and computing the discrete Fourier transform.

optical multiplication multiplying operation using light. A light beam has intensity A. When the beam passes through a transparency with transmittance B, the intensity of the transmitting light beam is AB.

optical neural network optical processor implementing neural network models and algorithms. A neural network is an information processing system that mimics the structure of the human brain. Two features of neural networks are recognition capability and learning capability. In recognizing process, the neural net is formulated as

$$z_j = f\left(\sum_{ji} x_i + \theta_j\right) ,$$

where z_j is the output of the jth neuron, w_{ji} is the interconnection weight between the ith input neuron and the jth neuron, x_i is the input coming from the ith input neuron, θ_j is the bias in the jth neuron, and f is a nonlinear transfer function. Notice that z_j and x_i are binary. Nonlinear transfer function f could simply be a threshold function. The formation of an interconnection weight matrix is called the learning process. It appears that in the recognizing process, the neural network has to perform a matrix-vector multiplication and a nonlinear operation. The matrix-vector multiplication can be easily performed using an optical system. However, the nonlinear operation cannot be performed by optical means. It may be performed using electronic circuits. For two-dimensional neural processing, tensor-matrix multiplication is needed which can also be realized by optical means.

optical representation of binary numbers the representation of binary numbers 0 and 1 using light. Since an optical detector is sensitive to light intensity, it is a very logical choice to represent binary numbers 0 and 1 with dark and bright states or low and high intensity, respectively. However, some difficulty would occur when 1 has to be the output from 0, because no energy can be generated for 1 from 0 without pumping light. Binary numbers 0 and 1 can be represented by a coded pattern instead of the intensity of a single spot. They can also be represented by two orthogonal polarizations, although the final states should be converted to intensity, which is the only parameter that can be detected by a detector.

optical tensor-matrix multiplication an operation performing tensor-matrix multiplication using optical devices. For tensor W and matrix X, the tensor-matrix multiplication produces matrix Y, whose elements are

$$\begin{aligned} y_{11} &= w_{1111}x_{11} + w_{1112}x_{12} \\ &\quad + w_{1211}x_{21} + w_{1212}x_{22} \\ y_{12} &= w_{1121}x_{11} + w_{1122}x_{12} \\ &\quad + w_{1221}x_{21} + w_{1222}x_{22} \\ y_{21} &= w_{2111}x_{11} + w_{2112}x_{12} \\ &\quad + w_{2211}x_{21} + w_{2212}x_{22} \\ y_{22} &= w_{2121}x_{11} + w_{2122}x_{12} \\ &\quad + w_{2221}x_{12} + w_{2222}x_{22} . \end{aligned}$$

An incoherent optical processor can implement this operation as follows. The two-by-two matrix

$$\begin{pmatrix} x_{11} & x_{12} \\ x_{21} & x_{22} \end{pmatrix},$$

represented by four LEDs is imaged by a lenslet onto a part of the tensor that is a two-by-two matrix

$$\begin{pmatrix} w_{1111} & w_{1112} \\ w_{1211} & w_{1212} \end{pmatrix},$$

represented by a spatial light modulator. Light passing through the spatial light modulator is integrated by another lenslet to provide y_{11}. Thus, two sets of two-by-two lenslet arrays with four LEDs and a spatial light modulator displaying all tensor elements as four matrices are sufficient to perform this tensor-matrix multiplication. Various optical arrangements can implement tensor-matrix multiplication. Optical tensor-matrix multiplication is important for two-dimensional neural networks.

optimal (**1**) a solution to an optimization problem which has the minimum (or maximum) value of the objective function.

(**2**) the time, space, resource, etc. complexity of an algorithm which matches the best known lower bound of a problem.

optimal cost the minimum cost to process an input sequence.

optimal decision the best decision, from the point of view of given objectives and available information, which could be taken by the considered decision (control) unit; the term optimal decision is also used in a broader sense — to denote the best decision which can be worked out by the considered decision mechanism, although this decision mechanism may itself be sub-optimal; an example is model-based optimization as a decision mechanism at the upper layer of a two-layer controller for the steady-state process — the results of this optimization will be referred to as the optimal decision.

optimistic concurrency control *See* optimistic locking.

optimistic locking the acquisition of locks late in a transaction, with the assumption that no conflicting updates have been made by concurrent transactions, or with explicit tests to detect such updates.

optimization (**1**) a transformation by which a program is converted to an equivalent program which has better performance along one or more metrics (space, time, disk transactions, network transactions, etc.)

(**2**) a transformation as in (**1**) effected by a compiler.

(**3**) a transformation as in (**1**) effected by a programmer rewriting the program. Note that in general usage, "optimization" means achieving the lowest cost along the desired metrics, in all but a few well-defined cases it is actually impossible to prove that the transformed program is the best. It has been suggested that amelioration is a better designation for some of these transformations, particularly those performed by compilers.

(**4**) determining the values of the set of free parameters that minimizes or maximizes an objective function. The minimization or maximization may be subject to additional constraints. *See also* gradient descent, graph search, genetic algorithm, minimax estimate, simulated annealing.

optimization problem a computational problem in which the object is not to decide some yes/no property, as with a decision problem, but to find the best solution in those "yes" cases where a solution exists.

optimizer a program, or set of programs, used to choose an efficient execution strategy for a query from the possible strategies.

optimizing a performance criterion that requires maximizing or minimizing a specified measure of performance.

optimizing compiler a compiler which is able to perform optimizations during the compilation process to minimize the resulting code along one or more metrics such as code size, execution speed, data size, or other metrics established by the writer of the compiler. Since many of the computations are either NP-hard or involve undecidable problems, there is no guarantee that the resulting code is actually optimum.

optimum distance profile code (ODP) a convolutional code with superior distance profile given the code rate, memory, and alphabet size. ODP codes are suitable if, for example, sequential decoding is used.

optimum free distance code (OFD) a convolutional code with the largest possible free distance given the code rate, memory, and alphabet size. Often referred to as OFD code.

optoelectronics the interaction of light and electrons in which information in an electrical signal is transferred to an optical beam or vice versa, e.g., as occurs in optical fiber communications components.

OR the Boolean operator that implements the disjunction of two predicates, the true table for $\vee \equiv X$ OR Y is

X	Y	$X \vee Y$
F	F	F
F	T	T
T	F	T
T	T	T ,

n-ary ORs can be obtained as disjunction of binary ORs. *See also* AND, NOT, XOR, implies.

oracle Turing machine a Turing machine with an extra oracle tape and three extra states $q_?, q_y, q_n$. When the machine enters $q_?$, control goes to state q_y if the oracle tape content is in the oracleset; otherwise control goes to state q_n.

ordered set a set of values which has a total ordering relationship R (symmetric, transitive, and reflexive) on its elements, such that an iterator over the elements of the set always delivers the elements in an order such that for any two elements e_j and e_k, if $j < k$, $e_j R e_k$.

ordering among events the specification of the event ordering. It can be done by Since and Until operators in temporal logics. The ability of a temporal logic to express the order in which events are realized.

ordering field the field chosen on which the records of a file are physically ordered on disk.

ordering key a key field on which the records of a file are physically ordered on disk.

order-of-magnitude-zero an adjective describing a floating point number representation that has a significand of zero. Such numbers are evaluated to zero but actually represents many numbers other than zero because of the finite precision used.

order preserving a type of hash function which maintains the records in the hash file ordered according to the values of the hash field.

ordinary file a randomly addressable sequence of bytes organized and stored on some permanent media such as a magnetic disk.

OR gate a logic circuit that performs the OR operation. The output of the gate is high if one or all of its inputs are high.

orphan a process or computation for which no recipients exist.

orthogonal for two signals or functions, the condition that their inner product is zero. For example, two real continuous functions $s_1(t)$ and $s_2(t)$ are orthogonal if

$$\int s_1(t)s_2(t)\,dt = 0$$

orthogonal drawing an orthogonal drawing is a graph drawing in which each edge is represented by a polyuline, each segment of which is parallel to a coordinate axis.

orthogonally convex rectilinear polygon a rectilinear polygon P in which every horizontal or vertical segment connecting two points in P lies totally within P.

orthogonal matrix a matrix A whose inverse is A^T; that is, $A^T A = I$, where I is the identity matrix. More generally, A is a unitary matrix if its inverse is A^H (that is, the complex-conjugate transpose); consequently, a real unitary matrix is an *orthogonal matrix.* Although not recommended, the phrase "orthogonal matrix" is sometimes used to refer to a unitary matrix.

orthogonal persistence a form of persistence whereby any type of data can be made persistent without transforming or losing its type.

orthogonal set of functions for a real signal set $f_m(t)$ over $[t_1, t_2]$ such that

$$\int_{t_1}^{t_2} f_m(\tau) f_n(\tau)\,d\tau = \begin{cases} 0 & m \neq n \\ 1 & m = n\,, \end{cases}$$

orthogonal functions are necessary for transforms such as the FFT or the DCT.

orthogonal transform a transform whose basis functions are orthogonal. The transform matrix of a discrete orthogonal transform is an orthogonal matrix. Sometimes orthogonal transform is used to refer to a unitary transform. Orthogonal real transforms exhibit the property of energy conservation. *See* orthogonal.

orthogonal wavelet wavelet functions that form orthogonal basis by translation and dilation of a mother wavelet.

orthographic projection a type of parallel projection where the direction of projection is the same as the surface normal to the projection plane. Specialist types of orthographic projection include front-elevation, top-elevation, and side-elevation where the projection plane is perpendicular to the principle axis. Such projections, called isometric projection, are often used in engineering drawings as they preserve distances and angles.

orthonormal transform *See* orthogonal transform.

OSF *See* open software foundation.

OSI model *See* open systems interconnection model.

other the default user group consisting of all users who are not assigned to a specific group.

otherwise a designation of a control transfer point in a multiway selection statement (such as a case statement) which has the property that if no specifically designated control point is selected, a designated computation will be selected. In some languages this is designated the "otherwise" case, in C it is the "default" case.

OTP one-time programmable. *See* programmable read-only memory.

outer join a type of join between two relations which results in all the tuples from one of the participating relations being included together with the tuples from the second relation that satisfy the join condition.

outer union allows union of relations that are not union compatible. The union of the tuples of the two relations that are partially compatible are included in the results. Attributes that are not compatible are kept in the result. NULL values are inserted into the tuples that have no value for these attributes.

outlier a statistically unlikely event in which an observation is very far (by several standard deviations) removed from the mean. Also refers to points that are far removed from fitted lines and curves. In experimental circumstances, "outlier" frequently refers to a corrupt or invalid datum.

outline function a function whose code is executed through a function call. It is the opposite of inline functions, whose code is directly embedded in the calling function.

output assertion *See* post condition.

output block when the processor writes a large amount of data to its output, it writes one block, a certain amount of data, at a time. That amount is referred to as an *output block.*

output buffer when the processor writes to its output device, it must make sure that device will be able to accept the data. One common technique is that the processor writes to certain memory addresses, the output buffer, where the output device can access it.

output dependency the situation when two sequential instructions in a program write to the same location. To obtain the desired result, the second instruction must write to the location after the first instruction. Also known as write-after-write hazard.

output device a device that presents the results sent to the user, and typical output devices are the screen and the printer. (According to some interpretations, it also includes the stable storage where the processor saves data.)

output neuron layer a neuron (layer of neurons) that produces the network output (outputs). In feedforward networks, the set of weights connected directly to the output neurons is often also referred to as the output layer.

output product a product that is the result or the output of a process. A product that is generated by some phase of software development. An *output product* may be either an intermediate product or a final product.

output routine low-level software that handles communication with output devices. Handles the formatting of data as well as eventual protocols and timings with the output devices.

output vector a vector formed by the output variables of a network.

outsourcing contracting an outside company for delivering a software system.

outstar configuration a configuration in neural networks. An *outstar configuration* consists of neurons driving a set of outputs through synaptic weights. An outstar neuron produces a desired excitation pattern to other neurons whenever it fires. *See* outstar training. *See also* instar configuration.

outstar rule a learning rule that incorporates both Hebbian learning and weight decay. A weight is strengthened if its input signal and the activation of the neuron receiving the signal are both strong. If not, its value decays. A similar rule exists for instars, but has found less application.

outstar training neuron training where the weights are updated according to

$$w_i(t+1) = w_i(t) + \nu\,(y_i - w_i(t)) \;,$$

where ν is the training rate starting from about 1 and gradually reduced to 0 during the training. *See also* instar configuration.

overflow a condition in which the resulting value is too large to represent in the available machine representation. For example, adding 1 to the largest possible integer produces an overflow. In the case of floating point numbers, a value that exceeds the largest possible floating point number that can be represented. *See also* overflow detection, underflow, underflow detection.

overflow detection a form of execution semantics in which an arithmetic overflow produces an error condition. The exact nature of the condition and how it is handled are determined by factors such as the language, its implementation, and the platform on which it is running.

overhead (**1**) the set of operational costs additionally present to perform an activity. It is typically used in the definition of project costs and refers to a percentage on the personnel costs due to the cost of the structure. It is also used for identifying the time spent in scheduling and context switching in multitasking operating systems.

(**2**) the amount of time or work exceeding that needed to perform a required function.

overlap the process that the next phase in the software development process is entered before the current phase is entirely finished. This is typical of evolutionary models.

overlapped execution processing several instructions during the same clock pulses.

overlay (**1**) a mechanism by which sections of code are explicitly loaded into memory when required, replacing code which was already present. Although largely supplanted by virtual memory mechanisms, overlay techniques are still used on some embedded machines, real-time systems, and other specialized applications. Many languages for these machines have explicit support for overlay management.

(**2**) an image compositing method where an image is displayed over a background image.

overload a state in which a system has more work than it could process while still meeting all delay or loss requirements. A system can get into overload state if the offered load exceeds the throughput for which the system was dimensioned.

overloading principle according to which operations bearing the same name apply to arguments of different data type. In object-oriented systems, overloading is also known as ad-hoc polymorphism. *See* operator overloading, function overloading, polymorphism.

override a technique by which a subclass can change the behavior of a method of its superclass by defining a method with the same name and parameter types. In the absence of an override in the subclass, the method of the superclass will be invoked, providing the class declaration permits it. *See also* virtual method.

overriding a member of a class, both attribute and method, that is redefined in one of the derived classes, changing one or more of its characteristics without addressing polymorphism is considered overridden. For example, an attribute (an integer) of the base class can be overridden in the derived class by re-declaring it with the same name, but with a different type (a double).

oversampling sampling a continuous-time signal at more than the Nyquist frequency.

overshoot the amount by which an output value momentarily exceeds the ideal output value for an underdamped system.

owner with respect to a file, the user who is authorized to change a file's access mode and ownership; in general, a file's owner is the user who created it.

owner-coupled set the set-type in the network model. (The term co-set is used to distinguish it from the mathematical set).

P

P the set of all languages $\mathcal{L}$ that are decidable in sequential time $n^{O(1)}$.

package a grouping of several classes into a higher level chunk which has high internal cohesion and coupling and low inter-package coupling. It can be regarded as a subsystem. *See* software package.

packed decimal a data format for the efficient storage and manipulation of real numbers, similar to BCD, with digits stored in decimal form, two per byte.

packet usually refers to the smallest part of a message that can be routed independently through a network. The packet header contains the information necessary to route the packet through the network, as well as any information needed to reassemble the packet into a larger unit at the destination. Small packets are more likely to transmit correctly, but waste proportionately more transmission bandwidth by headers than larger packets.

packet delay the time elapsed from the moment when a packet is generated at its source node until it is received by its destination node. This delay can be expressed as the sum of the queueing delay, the access delay, the service delay, and the propagation delay. If these component delays are bounded, a simple sum of their maximum values gives the maximum packet delay. *See also* end-to-end delay, transfer delay.

packet filter a network security device that permits or drops packets based on the network layer address and often on the port numbers used in the transport layer.

packet-switched bus *See* split transaction.

packet switching a technique for transferring data through a computer network in which messages are divided into a sequence of short segments called "packets" and relayed from the source to the destination through a series of intermediate nodes. Each packet contains enough addressing information in its header so that the intermediate nodes can determine the next hop on a path to the destination. Packets from different sources may be interleaved on each link, the transmissions over each link are scheduled independently. Unlike a *circuit-switched* network, links are tied up only when they are being used, so throughput is better, but there is no guaranteed transmission time after a circuit is established. *See also* store-and-forward, virtual cut-through.

packing given a finite collection of subsets of a finite ground set, the process of finding an optimal subcollection that is pairwise disjoint. *See* covering.

packing and covering given a finite collection of subsets of a finite ground set, to find an optimal subcollection that are pairwise disjoint (packing) or whose union covers the ground set (covering).

PAC learning a supervised learning framework in which training examples x are randomly and independently drawn from a fixed, but unknown, probability distribution on the set of all examples. Each example is labeled with the value $f(x)$ of the target function to be learned. A PAC (probably, approximately correct) learning algorithm is one which, on the basis of a finite number of examples, is able, with high probability, to learn a close approximation to the target function.

padding argument a method for transferring results about one complexity bound to another complexity bound, by padding extra dummy characters onto the inputs of the machines involved.

page a fixed-length memory area that has a virtual address and that is transferred as a unit between main memory and auxiliary storage. *See* virtual memory.

0-8493-2691-5/01/\$0.00+\$.50

paged-segment a segment partitioned into an integral number of pages.

paged segmentation the combination of paging and segmentation in which segments are divided into equal sized pages. Allows individual pages of a segment rather than the whole segment to be transferred into and out of the main memory.

page fault event that occurs when the processor requests a page that is currently not in main memory. When the processor tries to access an instruction or data element that is on a page that is not currently in main memory, a *page fault* occurs. The system must retrieve the page from secondary storage before execution can continue.

page–fault–frequency replacement a replacement algorithm for pages in main memory. The is the reciprocal of the time between successive page faults. Replacement is according to whether or not the page-fault frequency is above or below some threshold.

page frame a block of main memory having the size of, and used to hold, a page. *See* virtual memory.

page lock a type of lock. Some database systems assign locks on entire pages containing a datum being read or updated by a transaction. This can cause unnecessary contention, since other data on the same page will then be unavailable for conflicting access by other transactions.

page miss penalty when a page miss occurs, the processor will manage the load of the requested page as well as the potential replacement of another page. This time involved, which is entirely devoted to the page miss, is referred to as the page miss penalty.

page–mode DRAM a technique that uses a buffer like a static RAM; by changing the column address, random bits can be accessed in the buffer until the next row access or refresh time occurs.

This organization is typically used in DRAM for column access. Additional timing signals allow repeated accesses to the buffer without a row–access time. *See also* two-dimensional memory organization.

page offset the index of a byte or a word within a page, calculated as the (physical as well as virtual) address modulus of the page size.

page printing a printing technique where the information to be printed on a page is electronically composed and stored before shipping to the printer. The printer then prints the full page nonstop. Printing speed is usually given in units of pages per minute (ppm).

page–printing printer a human–readable output device used for producing documents in a written form. The printer stores a whole page in memory before printing it (e.g., a laser printer).

pager a routine that initiates and controls the transfer of pages.

page replacement at a page miss, when a page will be loaded into the main memory, the main memory might have no space left for that page. To provide space for that new page, the processor will have to choose a page to replace.

page table a mechanism for the translation of addresses from logical to physical in a processor equipped with virtual memory capability. Each row of the page table contains a reference to a logical block of addresses and a reference to a corresponding block of physical storage. Every memory reference is translated within the CPU before storage is accessed. The page table itself may be stored in standard memory or may be stored within a special type of memory known as associative memory. A page table stored in associative memory is known as a translation lookaside buffer.

paging the process of transferring pages between main memory and secondary memory.

painter's algorithm a hidden-surface algorithm that sorts primitives in a back-to-front order, then an algorithm for hidden surface removal, where objects are assigned priorities based on proximity to the camera position. When the image is rendered to the buffer, the objects with higher priority overwrite those with lower priority (it "paints" them into the frame buffer in this order, overwriting previously "painted" primitives.) Although intuitive and simple to implement, this algorithm has been superseded by z-buffering.

PAL *See* programmable array logic.

palette the set of colors that may be used to compose an image.

pan camera rotation about an axis (vertically) perpendicular to the camera's view direction.

paper prototype a paper representation of an application. For example, story boards that describe the interaction at a human interface. Or using sticky notes to model a GUI.

paper tape strips of paper capable of storing or recording information, most often in the form of punched holes representing the values. Now obsolete.

paradigm refers to a process model, like the object-oriented paradigm refers to an object-oriented modality to analyze, design, and implement a software system.

parallax a monocular depth cue. *Parallax* is the difference in the apparent position of an object caused by a change in the point of observation.

parallel pertaining to the simultaneous transfer, occurrence, or processing of the individual parts of a whole, such as the bits of a character, using separate facilities for the various parts. *Contrast with* serial. *See* concurrent.

parallel adder a logic circuit which adds two binary numbers by adding pairs of digits starting with the least significant digits. Any carry generated is added with the next pair of digits. The term "parallel" is misleading since all the digits of each number are not added simultaneously.

parallel architecture a computer system architecture made up of multiple CPUs. When the number of parallel processors is small, the system is known as a multiprocessing system; when the number of CPUs is large, the system is known as a massively parallel system.

parallel bus a data communication path between parts of the system that has one line for each bit of data being transmitted.

parallel composition in a process algebra, the possibility to define a process as two subprocesses executing in parallel.

parallel computation thesis sequential space is polynomially related to a parallel time.

parallel computing computing performed on computers that have more then one CPU operating simultaneously.

parallel computing system a system whose parts are simultaneously running on different processors.

parallel database management system a database management system that is implemented on a tightly coupled multiprocessor.

parallel data transfer the data transfer proceeds simultaneously over a number of paths, or a bus with a width of multiple bits, so that multiple bits are transferred every cycle. A technique to increase the bandwidth over that of serial data transfer.

parallel discrete event simulation (PDES) the method of executing discrete event simulation programs on a parallel computer or on a distributed environment. PDES mechanisms broadly fall into two categories: conservative and optimistic. Conservative approaches process events only if it is safe to process them; they avoid the possibility of any causality error ever occurring. Optimistic methods, on the other hand, take risks and detect the causality errors if they occur. They recover from causality errors by invoking a rollback mechanism. One of the well-known optimistic approaches is the Time-Warp.

parallel divider a divider in which quotient bits are produced in the same cycle.

parallel genetic algorithms parallelized versions of a genetic algorithm where multiple processors are assigned to different subsets of the population for solution evaluation. In the migration model, the total number of individuals is divided into different sub-populations and the processes of crossover and selection are restricted to members within each sub-population. Occasionally an individual is allowed to migrate from one sub-population to another for information exchange. In the diffusion model, a local neighborhood structure is defined for each individual and the application of genetic operations is restricted within each neighborhood. In this way, information is allowed to propagate slowly throughout the population. *See also* diffusion model, migration model.

parallel input/output generic class of input/output (I/O) operations that use multiple lines to connect the controller and the peripheral. Multiple bits are transferred simultaneously at any time over the data bus.

parallel I/O *See* parallel input/output.

parallel I/O interface I/O interface consisting of multiple lines to allow for the simultaneous transfer of several bits. Commonly used for high-speed devices, e.g., disk, tape, etc. *See also* serial I/O interface.

parallelism concurrent execution of an algorithm. Many languages have explicit support for parallelism. In other cases, compilers for traditional languages (notably FORTRAN) attempt to discover those segments of the program which can be executed concurrently. *Parallelism* may be at the low-level computation level (multiple arithmetic units working concurrently), at the instruction set level (multiple instructions executing concurrently), or at the architecture level (more than one processor executing the same program). Parallelism often requires synchronization between concurrent paths of execution, either implicitly in the hardware, explicitly in the program, or implicitly in the implementation of the algorithmic transforms done by the compiler.

Although parallelism can be found in many electronic computing systems, parallelism is the inherent property of an optical system. For example, a lens, the simplest optical system, forms the whole image at once and not point-by-point or part-by-part. If the image consists of one million points, the lens processes one million data in parallel.

parallelism profile a description of processor usage by a parallel application. The parallelism profile is determined experimentally, and indicates the number of processors in use by the application as it executes as a function of time.

parallel manipulators manipulators that consist of a base platform, one moving platform, and various legs. Each leg is a kinematic chain of the serial type, whose end links are the two platforms. Parallel manipulators contain unactuated joints which makes their analysis more complex than that of serial type. A paradigm of parallel manipulators is the flight simulator consisting of six legs actuated by hydraulic pistons.

parallel multiplier a multiplier in which product bits are produced in the same cycle.

parallel port a data port in which a collection of data bits are transmitted simultaneously.

parallel processing **(1)** an environment in which a program is divided into multiple threads of control, each of which is capable of running simultaneously, at the same time instant.

(2) processing carried out by a number of processing elements working in parallel, thereby speeding up the rate at which operations on large data sets such as images can be achieved. Often used in the design of real-time systems.

parallel random access machine (PRAM) a theoretical shared-memory model, where typically the processors all execute the same instruction synchronously, and access to any memory location occurs in unit time.

parallel system a system with a low degree of autonomy.

parallel thinning a technique employing the hit-or-miss transform to restore noise-degraded images.

parallel–to–serial conversion a process whereby data whose bits are simultaneously transferred in parallel are translated to data whose bits are serially transferred one at a time. During the translation process, some timing information may be included (such as start and stop bits) or is implicitly assumed.

parallel-transfer disk a disk in which it is possible to simultaneously read from or write to multiple disk surfaces. Advantageous in providing high data transfer rates.

parallel transmission the transmission of multiple bits in parallel.

parallel window-aspiration search a method in which a multitude of processors search the same tree, but each with different (nonoverlapping) alpha-beta bounds.

parameter **(1)** a name declared as part of a function specification whose value is not defined until the function is called by the function call. Also known as a formal parameter.

(2) the value supplied to a subroutine at the time the subroutine is called. Also known as an actual parameter.

Argument is a synonym.

parameter binding *See* call-by-name, call-by-value, call-by-reference, call-by-value-result, call-by-result.

parameterized class a generic definition, not instantiated until the client provides the needed information. A parameterized class is also referred to as a template class. Based on the client supplied data types, the compiler may instantiate a run-time implementation for the class and its member functions.

parameterized element an element that presents one or more parameters that characterize its behavior or data information in general.

parameter space a domain formed by all possible values of the given parameters.

parametric an approach to shape representation in which a curve or a surface is defined by a set of equations expressed in terms of a set of independent variables (i.e., the parameters). This representation is convenient for curvature and bounds computation and the control of position and tangency.

parametric surface a surface defined explicitly by the range of values of a parametric function. For a parametric function $\mathbf{f}(\mathbf{a})$ that depends upon the parameter vector $\mathbf{a}$, the surface S can be defined formally as: $S = \{\mathbf{p} : \mathbf{p} = \mathbf{f}(\mathbf{a}), \forall \mathbf{a}\}$.

paramodulation in logic, any specialized inference rule for handling equality.

parent **(1)** in object-oriented computation, a parent class is a base class, superclass.

(2) in evolutionary computation, the original individual or set of original individuals from which a particular new individual is derived.

parent-child relationship type a 1:N relationship between two record types.

parent directory a directory that has at least one descendant directory, that is, a directory that is the predecessor of at least one other directory. *Compare with* child directory.

parent process a process that created a child process.

parent record type the type of a record acting as a parent in a hierarchical database.

parity property of a binary sequence that determines if the number of 1s in the sequence is either odd or even.

parity bit an extra bit included in a binary sequence to make the total number of 1s (including itself) either odd or even. For instance, for the following binary sequence 101 one would insert a parity bit $P(\text{odd}) = 1$ to make the total number of 1s odd; a parity bit $P(\text{even}) = 0$ would be inserted to make this number even. *See also* error detecting code.

parity-check code a binary linear block code.

parity check matrix a matrix whose rows are orthogonal to the rows in the generator matrix of a linear forward error control block code. A non-zero result of element-wise finite field multiplication of the demodulated word by this matrix indicates the presence of symbol errors in the demodulated word.

It is generated from the parity check polynomial of any linear (n, k) code and has dimension of $(n - k \times n)$. It is used by the decoder for error detection by checking the parity bits.

parity detection circuit a parity check logic incorporated within the processor to facilitate the detection of internal parity errors (reading data from caches, internal buffers, external data and address parity errors).

parking (**1**) on a bus, a priority scheme which allows a bus master to gain control of the bus without arbitration.

(**2**) the process of placing the heads of a disk on the landing zone before shutting off power.

Parnas partitioning a software design technique where modules are defined so as to isolate hardware-dependent or volatile sections of code.

parse (**1**) to analyze an input string according to the rules of a grammar.

(**2**) the results of such an analysis, for example, as represented by a parse tree.

parser generator a program that takes a specification of a grammar as input and produces as output a program (or data used by a standard program) which will parse input text. A parser of this type usually takes specific actions during the parse to construct an intermediate representation, and check for errors. Inputs not conforming to the grammar will be rejected.

parse table a representation of grammar rules that is processed by an interpreter. Generally constructed by parser generators, provides a general mechanism for constructing parsers for particular grammars without having to write much of the common code each time.

parse tree a graphical representation of the analysis of a sentence of a grammar, in which terminal symbols appear at the leaf nodes and nonterminal symbols appear in the internal nodes. The root node of a parse tree represents the designated root symbol of the grammar.

parsing the process by which an input string is analyzed using a grammar to determine if the input string satisfies the rules of the grammar. Also known as syntactic analysis.

partial borrow *See* assimilation of borrow.

partial carry *See* assimilation of carry.

partial correctness during the proving of the program correctness, it indicates that post-conditions follow directly from pre-conditions. *Compare with* total correctness, correctness.

partial difference *See* assimilation of borrow.

partial dividend at the start of a division, this is the dividend. Subsequently, it is the result of subtracting a multiple of the divisor. At the end it is the (final) remainder.

partial inheritance a mechanism by which a class may inherit only partially from the members of its superclass. Typically, this is formalized by specifying in the superclass the private members that are those that cannot be accessed by derived classes. It is the main cause of non-monotonic inheritance.

partial key a set of attributes that may be used in conjunction with the key of an identifying owner to uniquely identify instances of a weak entity type.

partially compatible two relations are said to be partially compatible if some of their attributes are union compatible.

partially decidable decision problem an important class of decision problems that can be transformed to a program that always halts and outputs 1 for every input expecting a positive answer and either halts and outputs 0 or loops forever for every input expecting a negative answer. *See* halting problem.

partially decidable problem one whose associated language is recursively enumerable. Equivalently, there exists a program that halts and outputs 1 for every instance having a yes answer, but is allowed not to halt or to halt and output 0 for every instance with a no answer.

partially dynamic graph problem problem where the update operations include either edge insertions (incremental) or edge deletions (decremental).

partial order an order defined for some, but not necessarily all, pairs of items. For instance, if two athletes lose to the same person, they both rank lower than the winner, but no rank between them is proved.

partial order relation in process scheduling, indicates that any process can call itself (reflexivity) if process A calls process B then the reverse is not possible (antisymmetry), and if process A calls process B and process B calls process C than process A can call process C (transitivity).

partial participation of an entity type E in a relation R specifies that each entity may exist without being involved in any relationship instance.

partial product those bits of a product that have been developed at a given stage in a multiplication. Sometimes confusingly used to refer to the multiplicand-multiples.

partial quotient those bits of a quotient that have been developed at a given stage in a division.

partial recursive function a partial function computed by a Turing machine that need not halt for all inputs.

partial remainder *See* partial dividend.

partial replication some fragments (and sections of which) may be replicated.

partial specification an incomplete specification of a system where some parts have been fully specified while other parts have only the interface specified.

partial sum *See* assimilation of carry.

participation constraints specifies whether an entity must be related to another entity via a relation.

participatory design (PD) any of a broad range of techniques designed to increase stakeholder participation in the design process.

particle system a technique for modeling irregular natural structures by a collection of independent objects, often represented as single points. Objects that have been represented using this technique include fire, smoke, clouds, fog, explosions, grass, etc. Each particle will have its own motion and property parameters, usually drawn randomly from a distribution (perhaps constrained by or linked to other particles, or other scene objects, such as grass being constrained to grow from a specified surface). Because natural effects based on *particle systems* need many particles for realistic appearance, rendering of particle systems often requires special-purpose methods that exploit the properties of the particular particle system.

partition (1) a portion of memory designated to hold a particular program or programs.

(2) division into parts. This is usually done by creating such a subdivision to the class and "overlaying" this division in such a way that each instance belongs to one or more partitions. Ideally, the partitions will not overlap so that an instance belongs to only one partition. When partitions overlap, care must be taken. In object technology, the use of discriminators can highlight potential overlapping problems and the use in UML of stereotypes can effectively create partition-overlap problems.

(3) a logical section or drive of the hard disk.

(4) a division of a set into nonempty disjoint sets which completely cover the set.

partitioned join a parallel join algorithm which involves partitioning the two input relations across the available processors, executing them locally and then merging the results. Suitable for equi-joins.

partitioned parallel hash join a parallel version of the hash join.

partitioned process controlled process which is considered as consisting of several subprocesses which can be interconnected; process partitioning is an essential step in control problem decomposition — for example, before decentralized or hierarchical control can be introduced; the term partitioned system is used when one is referring to the partitioned control system but is also often used to denote just the partitioned controlled process.

partitioned system controlled control system which is considered as consisting of several sub-systems which can be interconnected; process partitioning is an essential step in control problem decomposition — for example before decentralized or hierarchical control can be introduced; the term partitioned system is used when one is referring to the partitioned control system but is also often used to denote just the partitioned controlled process.

partitioning attribute an attribute chosen on which to partition a relation.

partitioning vector a range of values used to specify ranges in range partitioning. A partitioning vector ($< v_0, v_1, \ldots, v_N >$) (given N disks) is used as follows: all records with value in partitioning attribute less than v_0 are placed on disk 0, all records with values between v_i and V_{i+1} are placed on disk i.

partition table (1) a table that enables the grouping of states into equivalent sets.

(2) a disk file that contains information about how the disk is partitioned.

Pascal an Algol-class language defined by Niklaus Wirth in 1971, and named for the French mathematician Blaise Pascal.

passive object object that does not work autonomously and thus is only used by other objects in the systems. *See* active object.

password a unique string of characters that identifies a user to confirm his login identity beyond the username.

password aging a security precaution in which a system periodically requires users to change their passwords.

password file a system administration file that contains, among other things, the login name, home directory, and default shell for each system user. Some systems provide an additional measure of protection by hiding the location of the file that contains the encrypted user passwords.

password prompt a text string issued during the login sequence that indicates the system is ready to accept a user's password.

patch (1) traditionally, a modification made directly to an object module without reassembling or recompiling it from the source code.

(2) in modern systems with dynamic module linking, a patch is more often implemented by replacing a single module of a large set of modules.

path (1) in a file system, hierarchical sequence of directory names specifying the location of a file.

(2) the current location within the directory tree.

(3) in software engineering, a sequence of operations that are performed during the execution of a given program.

path condition a set of conditions that have to be satisfied in order to execute a program.

path dissolution an inference mechanism that operates on formulas in negation normal form. *See* negation normal form.

path expression a logical expression indicating the input conditions that must be met in order for a particular program path to be executed.

pathname a file reference that consists of at least a file name optionally preceded by a list of directory names. *See also* path, full path, relative path.

path planning the process of finding a continuous path from an initial robot configuration to a goal configuration without collision.

path system problem for a path system $P = (x, R, S, T)$, where $S \subseteq X, T \subseteq X$, and $R \subseteq X \times X \times X$, the problem of whether there is an admissible vertex in S. A vertex is admissible if and only if $x \in T$, or there exists admissible $y, z \in X$ such that $(x, y, z) \in R$.

path testing a testing action designed to execute the selected paths in a computer program.

path tracing an improvement on general ray-tracing techniques. Normal ray-tracing uses a constant factor to estimate the contribution of ambient light at a given surface point, but *path tracing* estimates the global illumination using, for example, Monte Carlo techniques. Images are thus generated using many paths through each pixel. Note that a degree of oversampling is always necessary, so this technique is computationally expensive.

Patricia tree a compact representation of a tree where all nodes with one child are merged with their parents.

pattern a rule that specifies equivalence classes of textual sequences. A pattern is often expressed in a notation such as a regular expression or some equivalent notation. Languages such as SNOBOL had elaborate pattern specifications which included rules for re-attempting a match after one attempt failed ("backtracking rules"). Modern languages which include string matching patterns as primitive operations include awk and Perl.

pattern element a positive (negative) pattern element is a "partial wildcard" presented as a subset of the alphabet Σ, with the symbols in the subset specifying which symbols of Σ are matched (mismatched) by the pattern element.

pattern matching (**1**) a process by which patterns may be recognized, potentially and sometimes in practice by template matching, but otherwise by inference following the location of features. Typically, inference is carried out by application of Hough transforms or association graphs.

(**2**) the detection, in an image (signal), of a subimage (time window) in which pixel (sample) values are structured according to a predefined schema.

pattern recognition ability to recognize a given subpattern within a much larger pattern. Alternatively, a machine capable of pattern recognition can be trained to extract certain features from a set of input patterns. There are statistical, syntactic, and structural methods for pattern recognition, and neural network methods are often considered a subset of pattern recognition.

pause operating system command to suspend the execution of a computer program.

pause instruction an assembly language instruction whose execution causes a momentary pause in program execution.

Pawlak's information system a system model denoted S can be viewed as a pair $S = (U, A)$, where $U = \{x_1, \ldots, x_n\}$ is a nonempty finite set of objects. The elements of U may be interpreted, for example, as concepts, events, goals, political parties, individuals, states, etc. The set U is called the universe. The elements of the set A, denoted $\mathbf{a}_j$, $j = 1, \ldots, m$, are called the attributes. The attributes are vector-valued functions. For example,

$$\mathbf{a}_j : U \to \{-1, 0, 1\} .$$

An example of a simple information system is shown in the following table.

	$\mathbf{a}_1$	$\mathbf{a}_2$	$\mathbf{a}_3$
x_1	-1	0	0
x_2	0	0	1
x_3	0	1	1
x_4	1	1	1

The first component of $\mathbf{a}_1$ being -1 may mean that x_1 is opposed to the issue $\mathbf{a}_1$, while

x_4 supports $\mathbf{a}_1$, etc. Such data can be collected from newspapers, surveys, or experts.

PC *See* program counter, personal computer.

PCB *See* printed circuit board.

p-Code (**1**) a specification of a virtual machine that executes programs compiled by the Portable Pascal Compiler.

(**2**) code generated by that compiler.

P-complete a concept from the theory of computation. A language $\mathcal{L}$ is P-hard under NC reducibility if $\mathcal{L}' \leq_m^{NC}$ for every $\mathcal{L}' \in P$. A language $\mathcal{L}$ is P-complete under NC reducibility if $\mathcal{L} \in \mathbf{P}$ and $\mathcal{L}$ is P-hard.

PC-relative addressing an addressing mechanism for machine instructions in which the address of the target location is given by the contents of the program counter and an offset held as a constant in the instruction added together. Allows the target location to be specified as a number of locations from the current (or next) instruction. Generally only used for branch instructions.

PCTE *See* portable common tool environment.

PD *See* participatory design.

PDA *See* pushdown automaton, personal digital assistant.

PDES *See* parallel discrete event simulation.

PDF *See* probability density function.

PDL *See* program description language.

PDU *See* protocol data unit.

PE *See* processing element.

peak rate (**1**) in processor and system performance, the rate guaranteed not to be exceeded, and rarely reached in practice. For example, peak rate processor execution is often quoted as the maximum possible instructions per second, assuming a 100% cache hit ratio.

(**2**) in communication networks, a traffic descriptor giving the bit rate of a source that may not be exceeded on a cell rate or packet rate basis.

peephole optimization an optimization applied to the generated instructions from a compiler, which detects redundant or unused computations, or discovers ways in which to combine computations, the goal of which is to reduce the number of instructions necessary to perform a computation. The optimizations are performed looking at a very limited adjacent-instruction context, hence the name: looking at the code stream through a small peephole.

peer process a process that is not related by the child/parent relationship with other processes.

penalty function *See* cost function.

pending event list a data structure where planned future changes to the system state are stored during the execution of a discrete-event simulation. Each entry contains a description of the event and the time at which it is scheduled to occur. Since pending events must be executed in chronological order, but additional events may be added to the schedule arbitrarily, the pending event list is usually implemented as a priority queue or heap data structure. *See also* event-driven simulation.

Pentium alternate name for Intel's 80586 series of microprocessors.

penumbra that part of a shadow due to a light source which receives partial illumination from the source. By definition the source will be an extended light source and the *penumbra* always surrounds the umbra.

perceptron one of the earliest neural algorithms demonstrating recognition and learning ability. Since the output of a neuron is binary, the neuron can classify an input into two classes, A and B. The perceptron model containing a single neuron is expressed as follows: $z = 1$ if $\sum w_i x_i > T$ and $z = 0$ if $\sum w_i x_i < T$ where z is the output, x_i is an input, w_i is an element of the interconnection weight matrix, and T is a generalized threshold. If the interconnection weight matrix is known, an input vector (x_i) can be classified into A or B according to the result of z. On the other hand, the interconnection weight matrix can be formed if a set of input vectors (x_i) and their desired outputs z are known.

The process of forming the interconnection weight matrix from known input-output pairs is called learning. The perceptron learning is as follows. First, the weight w_i and the threshold T are set to small random values. Then the output z is calculated using a set of known input. The calculated z is compared with the desired output t. The change of weight is then calculated as follows: $\delta w_i = \eta(t - z)x_i$ where η is the gain term that controls the learning rate, which is between 0 and 1. The learning rule determines that the corrected weight is

$$w_i = w_i(\text{old}) + \delta w_i \ .$$

The process is repeated until $(t - z) = 0$. The whole learning process is completed after all input-output pairs have been tested with the network. If the combination of inputs are linearly separable, after a finite number of steps, the iteration is completed with the correct interconnection weight matrix. Various optical systems have been proposed to implement the perceptron, including correlators.

The original perceptron was a feedforward network of linear threshold units with two layers of weights, only one layer of which (the output layer) was trainable, the other layer having fixed values.

perceptron convergence procedure a supervised learning technique developed for the original perceptron. If the output y_i of unit i in the output layer is in error, its input weights, w_{ij}, are adjusted according to $Dw_{ij} = h(t_i - y_i)x_j$, where t_i is the target output for unit i and h is a positive constant (often taken to be unity); x_j is the output of unit j in the previous layer which is multiplied by weight w_{ij} and then fed to unit i in the output layer.

perceptual coding involves the coding of the contextual information of the image features by observing the minimum perceptual levels of the human observer.

perceptualization a term perhaps more suitable than visualization in recognition of the efficacy of using auditory and tactile techniques for representing and communicating about scientific data.

perfect code a t-error correcting forward error control block code in which the number of non-zero syndromes exactly equals the number of error patterns of t or fewer errors. Hamming codes and Golay codes are the only linear nontrivial perfect codes.

perfect hash a method for constructing a hash table such that the expected number of operations required to locate an entry in the table, or verify that it is not present, approaches 1 as a lower bound. Generally the hash function $h(\text{value}) \rightarrow$ key is characterized by either algorithm selection or the selection of specific values for one or more additional parameters to the function $h(\text{value, constant,} \,|) \rightarrow$ key such that the desired lower bound is closely approximated. A true perfect hash guarantees that each equivalence class of hash keys under the selected hash function contains precisely one key.

perfective maintenance maintenance action that is usually taken to improve system characteristics without changing its functionalities.

perfect matching a matching, or subset of edges without common vertices, of a connected graph which touches all vertices exactly once. A graph with an odd number of vertices has one unmatched vertex.

perfect shuffle interconnects that connect sources $1, 2, 3, 4, 5, 6, 7, 8$ to detectors $1, 5, 2, 6, 3, 7, 4, 8$, respectively. The operation divides $1, 2, 3, 4, 5, 6, 7, 8$ into two equal parts $1, 2, 3, 4$ and $5, 6, 7, 8$ and then interleaves them. The size of the array must be 2^n. The array returns to its original order after n operations. When the option to exchange pairs of neighboring elements is added to a perfect shuffle network, any arbitrary permutation of the elements is achievable.

performability a measure that gives the probability that a system provides a certain level of performance, where the performance may be characterized in measures appropriate to the system under study.

performance the degree to which a system of components accomplishes its designated functions within given constraints, such as speed, accuracy, or memory usage.

performance analyst a person who is responsible for the computation, measurement, and/or analysis of the performance of a system.

performance animation measurement and recording of direct actions of a real person or animal for immediate or delayed analysis and playback.

performance bound an upper (lower) value for a performance measure of interest. System performance is not to exceed (to fall below) the performance bound.

performance evaluation the overall activities related to assessing performance.

performance requirement specification within a computer system architecture imposing specific constraints on the final performance of the system such as component, executable latency, throughput, and utilization. In designing, analyzing, and verifying a computer system architecture, measurements of these quantities must be made to verify that all specifications in system performance are met with acceptable margin. This is typical of real-time and critical systems.

performance specification a document that specifies the performance that the final system has to provide. This is typical of real-time and critical systems.

performance testing a test devoted to assess the performance of the system in critical conditions.

performance tuning the modification of the database schema or database management system to facilitate efficient performance.

periodic pertaining to a task or process executing repeatedly in cycles.

periodic convolution a type of convolution that involves two periodic sequences. Its calculation is slightly different from that of discrete linear convolution in that it only takes summations of the products within one period instead of taking summations for all possible products. Circular convolution is an operation applied to two finite-length sequences. In order to make its result equal to that of the linear convolution, the two sequences have to be zero-padded appropriately. These two zero-padded sequences can then serve as periods to formulate two periodic sequences. At this time, the result of the periodic convolution is equal to circular convolution.

periodic event an event that is produced/received with periodicity. The period can be constant or variable in a bounded interval. For example, if event e happens then the next occurrence of e is between 50 ms and 200 ms.

periodic task a task that is periodically started, generally because some information has to be periodically evaluated/updated. It is usually used to manage sampled continuous inputs that have to be evaluated before the next sample becomes available.

peripheral an ancillary device used to put information into and get information out of the computer. Also called input/output device.

peripheral adapter a device used to connect a peripheral device to the main computer; sometimes called an I/O card, I/O controller, or peripheral controller.

peripheral controller *See* peripheral control unit.

peripheral control unit a device used to connect a peripheral device to the main computer;

sometimes called an I/O card, I/O controller, or peripheral controller.

peripheral device a physical mechanism attached to a computer that can be used to store output from the computer, provide input to the computer, or do both. *See* I/O device.

peripheral processor a computer that controls I/O communications and data transfers to peripheral devices. It is capable of executing programs much like a main computer. *See* I/O channel.

peripheral transfer a data exchange between a peripheral device and the main computer.

peripheral unit a physical mechanism attached to a computer that can be used to store output from the computer, provide input to the computer, or do both. *See* I/O device.

Perl acronym for Practical Extraction Report Language. A language originally developed on Unix which is used for a variety of tasks, such as writing scripts for system administration. A scripting language.

permanent fault a fault that remains in existence indefinitely if no corrective actions are taken.

permission access right granted by an object's owner, usually represented as bits in the object's access code.

permutation a rearrangement of elements, where none are lost, added, or changed. A good algorithm to randomly permute N elements is to exchange each element e_i with a random element from i to N. Another algorithm is to tag each element with a random number, sort the array by the tags, then remove the tags. *See also* sort.

perplexity a measure associated with a statistical language model, characterizing the geometric mean of the number of alternative choices at each branching point. Roughly, it indicates the average number of words the recognizer must consider at each decision point.

persistence (**1**) systems support persistence to the degree that they provide a consistent long-term view of data objects. This support may include automatic translation from program data structures to long-term storage structures, and procedures for keeping snapshots of transactions that change persistent data, so that recovery is possible in the case of system error. Persistence can be implemented by using a database manager, a file system, or by creating an image of the whole application and reloading this information at the next program execution.

(**2**) the ability of data to outlive the program that created it.

persistence by reachability if an object is persistent, then any other object or value to which it refers must also be persistent. Persistence by reachability starts from a set of persistent root objects. Any object that can be followed by any path of references from a persistent root must also be persistent.

persistent class a class whose instances must be permanently stored in the database.

persistent data data that continues to exist after the process or transaction that created it has terminated. Persistent data may be used and possibly altered by many different transactions or processes. Files and databases are examples of persistent data in computer systems.

persistent data structure a data structure that preserves its old versions. Partially data structures allow updates to their latest version only, while all versions of the data structure may be queried. Fully persistent data structures allow all their existing versions to be queried or updated.

persistent object an object whose state must be made persistent. *See* persistence.

persistent object store a repository for the long-term storage of objects. Some form of persistent object store is at the heart of all object-oriented databases.

persistent programming language a programming language that provides support for persistence through orthogonal persistence.

persistent root used in systems that provide persistence by reachability, a persistent root is a value that is persistent. Furthermore, any value that is a component of the persistent root will also be persistent.

persistent store a long-term store of objects. Generally, objects in the persistent store have unique identifiers (UIDs) that allow a program to access them. Some programming languages incorporate the notion of a persistent store directly in the language, while others use mechanisms such as subroutine libraries to manage persistent stores.

persistent value a value that must be stored persistently at the end of the current process.

personal computer (PC) a generic term for a desktop, laptop, or hand-held microcomputer for general purpose use.

personal digital assistant (PDA) a small computer intended for tasks like maintaining phone lists, to do lists, and similar tasks.

perspective distortion a type of object distortion that results from projecting 3D shapes onto 2D image planes by convergence of rays toward a center of projection. *See also* perspective projection. This type of distortion is also called foreshortening.

perspective inversion a property of perspective projection in which planar (typically silhouetted) objects which are not perpendicular to the axis of projection will appear, in the absence of additional information, to have either of two possible orientations in space.

perspective projection the complete projection model of a scene onto an image plane via a pinhole camera model. The perspective projection of any set of parallel lines which are not parallel to the projection plane converge to a vanishing point. In 3D, the parallel lines meet only at infinity and there is an infinity of vanishing points, one for each of the infinity of directions in which a line can be oriented.

perspective transformation a matrix transformation which represents the perspective projection of objects in 3D scenes into 2D images, or the projection of one 2D image into another 2D image. Perspective transformations are conveniently carried out using 4×4 matrices representing homogeneous co-ordinate transformations.

pessimistic concurrency control *See* pessimistic locking.

pessimistic locking the acquisition of locks early in a transaction to prevent conflicting updates being made by concurrent transactions. Transaction isolation is a form of pessimistic locking.

PET *See* positron emission tomography.

Petri net a graphical language used in describing discrete parallel subsystems and allowing for the expression of concurrency concepts. Petri nets consist of a directed bipartite graph in which "places" (which can store tokens) are connected to "transitions". A transition can "fire" if each of its input places contain at least one token, removing one token from each of its input places and adding one token to each of its output place. Invented by Carl Adam Petri. *See also* stochastic Petri net, timed Petri net.

PGA *See* field-programmable gate array.

phantom (**1**) an artificial target, sometimes designed to mimic the size, shape, and attenuation characteristics of actual tissue, that is used to test and calibrate imaging hardware and software.

(**2**) an object in a selected set, retrieved a second or subsequent time the select is performed, which was not present in the set the first time the select was performed.

phantom read to retrieve a data set containing a phantom.

phantom record *See* phantom.

phase a computational component of a translator such as a compiler or assembler characterized as being specialized to some part of the analysis or code generation process.

phenotype observable physical and behavioral characteristics of an organism resulting from the external manifestation of its genotype and the interaction of the individual with the environment. *See also* genotype.

phoneme the smallest units in phonemics. *Phonemes* are produced by different manners of articulation (e.g., plosives, fricatives, vowels, liquids, nasals). The automatic speech recognition on large lexicons is often based on the recognition of units like phonemes in order to break down the complexity of the global recognition process, from speech frames to words and sentences.

phonemics the study of sound units in the framework of descriptive linguistics. Basically, unlike phonetics, the sounds are studied by taking into account the language and not only observable features in the signal.

phones the smallest units in phonetics, where the emphasis is placed on observable, measurable characteristics of the speech signal.

phonetic knowledge the knowledge on the acoustic structure of phones. It is of fundamental importance for the design of effective automatic speech recognition systems.

phonetics the study of the acoustic sounds where the emphasis is placed on observable, measurable characteristics of the speech signal.

Phong shading a shading model for surfaces based on the interpolation of local surface normals at the vertices of a triangular patch. The technique is used for more realistic rendering of glossy surfaces.

photometry making measurements from images. One example is creating a 3D scene description using stereo image analysis, and measuring the volume of an object in the model.

photopic formally, a description of luminances under which human cone cells are active. Informally, describing daylight luminances.

photo-realistic rendering the process of rendering images so that they closely resemble a photograph. Such renderings must take into account reflective properties, light sources, illumination, shadows, transparency, and mutual illumination.

phrase structure grammar a grammar which is classifiable as one of the members of the Chomsky hierarchy. A phrase structure grammar $\mathcal{G}$ is defined by a 4-tuple $\mathcal{G} =< V_t, V_n, P, S >$ where V_t is the set of terminal symbols, V_n is the set of nonterminal symbols, P is the set of productions, and S is the designated root symbol.

physical as seen by the human. *See* logical.

physical address *See* real address.

physical address space the total memory on board the system, which might be larger than the logical address space.

physical block the disk sector addresses associated with a logical block of a file. The file system finds the contents of a logical block in a file by using the logical block number as an index into an indirect block to find the disk sector address holding the requested data.

physical data independence the ability to change the internal schema without changing the conceptual schemas.

physical failure a failure that is solely due to physical causes, e.g., heat, chemical corrosion, mechanical stress, etc.

physical fault a fault that is the result of a physical failure.

physical page number the page-frame address (in main memory) of a page from virtual memory.

physical process a mechanism that can cause changes in the physical world, such as heat flow, motion, and boiling.

physical sensor an interface device at the input of an instrumentation system that quantitatively measures a physical quantity such as pressure or temperature.

pi-calculus a core calculus of message-based concurrency, in which everything is a process and all computation proceeds by communication on channels.

pickling a term used in the Modula language to describe the action of writing out an internal data structure to an external medium (often a disk file) in such a way that the data structure can be reconstructed to an equivalent structure when read back into the program. This involves converting pointers to some form of representation such as a relative pointer or relocatable pointer so that the pointer values may be reconstructed when the data is read back.

picture a collection of mutually disjoint subsets of an alphabet.

PICTURE a construct in the COBOL language which specifies external representation of a value, as well as its precision. *Compare with* USAGE.

picture description language a language in which parts of scenes are labeled and their relative positions are described in a special symbolic form. Typical relationships between objects are "inside", "adjacent to", "underneath", and so on. Such a language can be parsed and interpreted symbolically to build up a meaningful understanding of the picture.

PID *See* process ID.

piezo-electric effect the phenomenon in which a material possesses a non-centrosymmetric crystal structure that will generate a charge upon application of mechanical stress. As in the case of a pyroelectric material, this can be detected as either a potential difference or as a charge flowing in an external circuit.

pile file *See* heap file.

pin the electronic connection that allows connection between an integrated circuit or circuit board and some socket into which it is plugged.

pincushion distortion the geometric distortion present in an image in which both horizontal and vertical lines appear to collapse toward the display center. For a CRT image system, pincushion distortion is a result of the interaction of the long face plate radius of curvature and the short radius from the deflection center to the face plate. The interaction causes the top, bottom, left, and right raster center points to be closer to the center of the CRT face plate deflection than the top left, top right, bottom left, and bottom right raster corners. Consequently, a square grid has the appearance of a pin cushion.

pipe a one-way, first in/first out data channel to exchange data between two or more processes in a producer/consumer model.

pipe flush when a pipeline containing the stream of instructions or data must be emptied before execution can continue. *See* pipeline.

pipeline **(1)** a means by which computations are partially performed in a stepwise fashion, such that as the computation passes from one stage to the next, a new computation can enter the previous stage. The consequence is that the first result from the pipeline takes n time units where n is the number of stages of the pipe, but thereafter a new result will be produced on each subsequent time unit. However, the final result of a computation will not appear for $n - 1$ time units.

If the pipeline is kept full without stalls, an n-stage pipeline, in which each stage takes unit time, can operate on m instructions in time $m + n - 1$, rather than the mn time units that would be required if each instruction were executed sequentially without overlap. The term $n - 1$ in the total time is often referred to as the pipeline delay.

A pipeline has serious implications for a compiler writer, since the goal of the compilation process is to reorder the computations to maximize the utilization of the pipeline.

(**2**) in command line, a sequence of commands separated by pipe characters (|); in processes, an arrangement between two or more processes where the output of one is fed, as input, to a second; the output of the second is fed to a third, and so forth. A pipeline command line gives rise to a pipeline of processes.

Hardware-level parallelism can be achieved by arranging a computation into a pipeline. In the common case of a processor's instruction pipeline, different hardware units for instruction fetch, instruction decode, data fetch, and instruction execution operate simultaneously on different instructions.

pipeline chaining a design approach used in computers whereby the output stream of one arithmetic pipeline is fed directly into another arithmetic pipeline. Used in vector computers to improve their performance.

pipelined bus *See* split transaction.

pipelined cache a cache memory with a latency of several clock cycles that supports one new access every cycle. A new access can be started even before finishing a previous one. The access to the cache is divided into several stages whose operation can be overlapped. For instance, the cache can be pipelined to speed–up write accesses: tags and data are stored in independently addressable modules so that the next tag comparison can be overlapped with the current write access. Read accesses are performed in a single cycle (tag and data read at the same time).

pipelined divide-and-conquer a divide-and-conquer paradigm in which partial results from recursive calls can be used before the calls complete. The technique is often useful for reducing the depth of an algorithm.

pipelined parallelism parallelism achieved by pipelining. Pipelining refers to the using of the results from one operation as input to another operation before the original operation has fully completed. For example, given $R1 \bowtie_1 R2 \bowtie_2 R3$, the results from $R1 \bowtie_1 R2$ may be used as input to $\bowtie_2$ before the operation $\bowtie_1$ has fully completed.

pipeline interlock a hardware mechanism to prevent instructions from proceeding through a pipeline when a data dependency or other conflict exists.

pipeline processor a processor that executes more than one instruction at a time, in pipelined fashion.

pipelining a technique to increase throughput. A long task is divided into components, and each component is distributed to one processor. A new task can begin even though the former tasks have not been completed. In the pipelined operation, different components of different tasks are executed at the same time by different processors. Pipelining leads to an increase in the system latency, i.e., the time elapsed between the starting of a task and the completion of the task.

pixel (**1**) a single discrete sample point of an image. Image size and resolution are defined in terms of number of pixels.

(**2**) screen picture element capable of displaying one or more colors.

pixel adjacency the property of pixels being next to each other. The adjacency of pixels is ambiguous and is defined several ways. Pixels with four-adjacency or four-connected pixels share an edge. Pixels with eight-adjacency or eight-connected pixels share an edge or a corner. *See also* chain code, connectivity, pixel.

pixel depth the number of bits used to generate a color at each pixel. The number of different colors that can be displayed is equal to $2^{\text{pixel depth}}$. For instance, a pixel depth equal to one means that only black and white colors could be displayed; with a pixel depth equal to four, sixteen different colors could be displayed.

PJNF *See* project join normal form.

PL/1 acronym for Programming Language One, a language defined and implemented by IBM. PL/1 was a traditional procedure-oriented block-structured language that attempted to incorporate the features of all major programming languages of its era, including COBOL-like data specification features, FORTRAN-like computational features, dynamically allocated structures, representation specification mechanisms, and the like. As such, it was representative of a class of "universal programming languages" designed for business, scientific, and symbolic processing. Historically, no such language has been widely accepted or used.

PLA *See* programmable logic array.

placement argument in object-oriented programming languages, the additional arguments that can be used in the creation of new instances by means of the constructors.

plan a specification for acting that maps from what is known at the time of execution to the set of actions.

planar graph a graph that can be drawn in the plane with no crossing edges.

planarization informally, the process of transforming a graph into a planar graph. More precisely, the transformation involves either removing edges (planarization by edge removal), or replacing pairs of nonincident edges by 4-starts (planarization by adding crossing vertices). In both cases, the aim of planarization is to make the number of operations (either removing edges or replacing pairs of nonincident edges by 4-starts) as few as possible.

planar straight-line graph a graph that can be embedded in the plane without crossings in which every edge in the graph is a straight line segment. It is sometimes referred to as planar subdivision or map.

planning a process that involves reasoning about the consequences of acting in order to choose from among a set of possible courses of action.

platform specific computer hardware, as in the phrase "platform-independent". It may also refer to a specific combination of hardware and operating system and/or compiler, as in "this program has been ported to several platforms". It is also used to refer to support software for a particular activity, as in "This program provides a platform for research into routing protocols". Often synonymous with computer.

platter *See* disk platter.

playback the process during which a set of recorded actions on a computer program are reproposed for verifying the correct behavior of the system. *See* capture and playback.

PLC *See* programmable logic controller.

PLD *See* programmable logic device.

plenoptic function the 5-dimensional function representing the intensity or color of the light observed from every position and direction in three-dimensional space.

plug-and-play (PNP) hardware or software that, after being installed (plugged in), can immediately be used (played with), as opposed to hardware or software which first requires configuration. This feature depends on the structure of the operating system.

PNP *See* plug-and-play.

pointer Traditionally, a value that represents a machine address, and thus "points" to a location which represents a value or object in the language. Many languages replace the concept of "pointer" with a notion of a "reference" to a value or object. It is sometimes implied in such cases that a reference need not be implemented as a pointer.

pointer arithmetic a computation done on a pointer to derive a new pointer value. This may be explicitly available to the programmer (as in C and C++), implicit in the language (as in subscripting an array of objects), or hidden entirely (as in computing the address of a member of a record by the use of a base address and an offset).

pointer jumping in a linked structure, replacing a pointer with the pointer it points to. Used for various algorithms on lists and trees.

point feature a small feature which can be regarded as centered at a point, so that inter-feature distances can be measured, as part of a process of inspection, or prior to a process of inference of the presence of an object from its features. Commonly used point features include corners and small holes, or fiducial marks (e.g., on printed circuit boards). *See also* salient feature, key point detection.

point-in-time recovery recovery of a database to a specific point in time.

point light source a mathematically defined infinitely small point **l** from which light radiates. The point might be at infinity, in which all light rays are parallel, or it might be closer to the object, in which case light rays radiate outward in all directions. The amount of light radiated in different directions need not be uniform.

point operation an image processing operation in which individual pixels are mapped to new values irrespective of the values of any neighboring pixels.

point process a type of image processing in which the enhancement at any pixel depends only on the gray level at that pixel. This is in contrast to an operation whose value depends on the location of the pixel or on the values of neighboring pixels. *See also* gray level, neighborhood operation.

point sampling refers to algorithms that only solve for visibility at a finite number of discrete points. A typical example is ray tracing. They are generally used in simple renderers. There are four point sampling algorithms in common use today: z-buffering, painter's algorithm, ray tracing, and the scanline algorithm. Other point sampling algorithms are generally variations on these. They have advantages over continuous algorithms because they are easier to understand and implement, they are faster, and can generate a greater range of optical effects. It is difficult to generate photorealistic fully anti-aliased images using a point-sampling algorithm.

point-to-point bus a bus that connects and provides communication capability between two, and only two, components. It should be noted that buses support communications within the CPU, as well as external to the CPU. Most internal buses are point-to-point buses. *Contrast with* multipoint bus.

point-to-point motion the manipulator has to move from an initial to a final joint configuration in a given time t_f.

Poisson counting process *See* Poisson process.

Poisson distribution a probability distribution widely used in system modeling. A non-negative integer-valued random variable X is Poisson-distributed if $\text{Prob}(X = k) = e^{-a} a^k/k!$, where a is a positive parameter sometimes called the intensity of the distribution. For example, the number of "purely" random events occurring over a time interval of t often follows a Poisson distribution with parameter a proportional to t.

poissonization in a model, to replace a deterministic input with a Poisson process. *See also* depoissonization.

Poisson process a random point process denoting the occurrence of a sequence of events at discrete points in time. The time difference between the different events is a random variable. For a given time interval of length T, the number of events (points in the process) is a random variable with Poisson distribution given by the following probability law $P[N = k] = \frac{(\lambda T)^k e^{-\lambda T}}{k!}$, where N is the number of events that occur in the interval of length T, and λ is the expected number of occurrences of events per unit time. The time interval between any two events is a random variable with exponential distribution given by $F(\tau) = 1 - e^{-\lambda\tau}$, $\tau > 0$.

Poisson traffic traffic with negative exponentially distributed interarrival times.

policy a specific scheme for managing resources, independent of the means for implementing the scheme.

Polish notation (1) prefix expression in which the operator precedes its operand(s). Named after the Polish mathematician Jan Lukasewicz who defined it in 1951.

(2) postfix expression, properly called "reverse Polish notation", but shortened in common usage to "Polish notation".

polled interrupt a mechanism in which the CPU identifies an interrupting device by polling each device. *See also* vectored interrupt.

polling sequencing through a group of peripheral devices and checking the status of each. This is typically done to determine which device(s) is ready to transfer data.

polling system a server serving more than one queue. In addition to a normal queueing system, a polling is characterized by its service discipline(s) (the strategy determining how many customers to serve at one queue before switching to the next queue), the interqueue discipline (the strategy for selecting the next queue), and switchover times between one queue and the next.

polychotomy division into many distinct classifications.

polygon a plane figure which is a closed contour of straight lines. A basic primitive in the graphical representation of objects.

polygon fill a series of ordered planar vertices connected to form an enclosed area. This area is then completely rendered using a specified color or texture.

polygon mesh a common form of object representation using a set of planar facets or polygons that approximate the surface of an object.

polyhedron a 3D solid that is bounded by a set of polygons whose edges are each a member of an even number of polygons.

polyline a continuous line formed from one or more connected line segments. *Polylines* are specified by the endpoints of each segment.

polymorphism A characteristic of languages in which one can manipulate different derived objects (related data types) in a uniform and generic way. It is one of the most important mechanisms of the object oriented paradigm.

polynomial (1) any function that is the sum of constants times powers of the argument: $f(x) = \Sigma_{i=0}^{k} c_i x^{p_i}$.

(2) in complexity theory, the measure of computation, $m(n)$ (usually execution time or memory space), is bounded by a polynomial function of the problem size, n. More formally $m(n) = O(n^k)$. *See also* logarithmic, linear, exponential.

polynomial approximation the use of a polynomial to approximate a real function.

polynomial hierarchy the collection of classes of languages accepted by k-alternating Turing machines, over all $k \geq 0$ and with initial state existential or universal. The bottom level ($k = 0$) is the class **P**, and the next level ($k = 1$) comprises **NP** and co- **NP**.

polynomial time when the execution time of computation, $m(n)$, is bounded by a polynomial function of the problem size, n. More formally $m(n) = O(n^k)$. *See also* np complete, exponential, logarithmic.

polynomial time approximation scheme (PTAS) a meta-algorithm that for every $\epsilon > 0$ produces a polynomial time ϵ-approximation algorithm for a given optimization problem.

polynomial-time Church-Turing thesis an analog of the classical Church-Turing thesis, for which stating that the class **P** captures the true notion of feasible (polynomial time) sequential computation.

polynomial time reduction a reduction computable in polynomial time.

polynomial warp a type of commonly used image processing operation designed to modify an image geometrically. This type of operation is useful when an image is subject to some unknown physical spatial distortion, and then measured over a rectangular array. The objective is then to perform a spatial correction warp to produce a corrected image array.

pop an operation that removes the item from the top of a stack, revealing the item underneath. *See also* push.

pop instruction an instruction that retrieves contents from the top of the stack and places the contents in a specified register.

populating the process of adding chips to a card.

population in evolutionary computation, a set of individuals that represents potential solutions to an optimization problem. Instead of a single sequence of search points, the adoption of multiple potential solutions increases the efficiency of the search process by allowing simultaneous exploration of different regions of the search space, and decreases the probability of pre-mature convergence to local minima.

pop-up menu a menu that pulls down like a roller blind from a permanently visible title bar, usually at the top of the display or window.

port (**1**) a terminal pair.

(**2**) a place of connection between one electronic device and another.

(**3**) an input/output channel for the computer.

portability the ability for software to be compiled and successfully operate on a number of different environments. The environment may include organizational, hardware, or software features.

portable (**1**) a characteristic of a program which can run on more than one computer architecture, or combination of computer architecture and operating system.

(**2**) a characteristic of a program that is acceptable to more than one compiler.

(**3**) of a language, a subset of the language which is known to be acceptable (empirically, or by conformance to a standard) to more than one compiler.

portable application an application that can run on dissimilar computers and/or under dissimilar operating systems. This may be done by writing it in a language which compiles for all the desired environments, or in a language which compiles to an abstract machine that is then implemented via interpreters which run on the desired environments. *See* Java.

portable common tool environment (PCTE) a European Computer Manufacturers Association standard framework for software tools developed in the Esprit program. It is based on an entity-relationship object management system and defines the way in which tools access this.

portable compiler a compiler which is a portable application. A portable compiler can be used to bootstrap itself onto another machine by first using it as a cross-compiler executing on machine A to generate code for machine B, compiling the compiler itself via this mechanism, and thus producing an instance of the compiler which executes on machine B and which produces code for machine B.

portals a method for reducing framebuffer overdraw where visible areas of a 3D model are clipped before they are rendered. Small areas in the model are grouped as sectors and portals are the transition planes between them.

An initial view frustum is defined to be as large as the image plane. Then all visible polygons are clipped to this volume. A sector is rendered only if its portal is within the clipping frame. Then a new smaller frustum is defined at that position and polygons visible from it are clipped and so on, recursively.

port protection device device in line with a modem that intercepts computer communication attempts and requires further authentication.

pose recovery determining the position and orientation of an object in the world with respect to the camera coordinate system.

poset a set the elements of which are subject to a partial order.

positional number system a number system in which each digit in a sequence of digits has an associated weight that corresponds to its position.

positional parameter a parameter whose value is bound based on positional correspondence between the formal parameter list and the function call. This is the traditional parameter binding order in most languages. *See also* named parameter.

positive Boolean function a Boolean function that can be represented as a logical sum of products in which no variables are complemented. Also called an increasing Boolean function.

positive definite matrix a symmetric matrix A such that $x^T Ax > 0$ for any vector x not identically zero. The eigenvalues of a positive definite matrix are all strictly greater than zero.

positive semi-definite matrix a symmetric matrix A such that $x^T Ax >= 0$ for any vector x. The eigenvalues of a positive semi-definite matrix are all greater than or equal to zero.

positron emission tomography (PET) an imaging modality that uses injected positron-emitting isotopes as markers for physiological activity. The isotopes emit pairs of gamma photons which are detected using a gamma camera and coincidence detector.

POST *See* power on self test.

post condition a specification of the state following the execution of a segment of code, characterized by a predicate which evaluates to a "true" value if the state conditions are defined and satisfied. The lack of verification may cause the generation of error messages or the fault. *See also* precondition, weakest precondition.

posterior statistics the empirical statistics of a random quantity (scalar, vector, process, etc.), based on the *a priori* statistics supplemented with experimental or measured observations. *See also* prior statistics, Bayes' rule.

postfix *See* postfix notation.

postfix notation a notational or programming scheme in which both operands of a two-operand operation are written before the operator is specified. Example: $ab+$ is the postfix representation of a sum; this could be implemented in a programming model based upon an evaluation stack by the operation sequence PUSH a, PUSH b, ADD. Postfix notation is used in programming zero-address computers. Also known as "Reverse Polish Notation". *See also* Polish Notation.

postfix traversal a tree walk or traversal algorithm that first processes the subnodes in a specified order, traditionally left-to-right, then processes the node itself. *Compare with* prefix traversal.

postincrementation an assembly language addressing mode in which the address is incremented after accessing the memory value. Used to access elements of arrays in memory.

postmortem a dump that is produced upon abnormal termination of a computer program.

post mortem an evaluation of the good and less good aspects of a completed software project to improve future projects.

Post production a notation developed by Emil Post to describe computations. *Post productions* are a form of rewrite rules. Post productions are computationally equivalent to Turing machines.

Post's correspondence problem a problem in computability theory. The problem is stated as L given two sets of strings, find a sequence of words from the first set and a sequence, of the same length, of words from the second set, such that the concatenation of words in the first sequence is the same string as the concatenation of words in the second sequence. This is an undecidable problem.

PostScript a stack-oriented language with polymorphic operators. Unlike FORTH, PostScript has separate data and control stacks.

Originally developed by Adobe Systems to support page layout operations.

potential difference the difference between the electron energy level on one side of a circuit and another (measured in volts). Also called a voltage.

potpourri module a module that provides more than one service to a program. This form of module violates the idea of a module being considered a responsibility assignment. The existence of this form of module increases considerably the effort that a programmer has to expend on a maintenance operation, and increases the likelihood of an error being introduced to a program as a result of maintenance work.

power product of current and voltage, measured in watts.

power on self test (POST) a series of diagnostic tests performed by a machine (such as the personal computer) when it powers on.

power supply an electronic module that converts power from some power source to a form which is needed by the equipment to which power is being supplied.

power supply unit (PSU) *See* power supply.

pragma a construct in a programming language which allows the programmer to communicate information about the program to the compiler. *See* pragmatics.

pragmatics those aspects of a programming language which must be expressed outside the formal definitions of the language, that is, beyond the syntax and semantics.

PRAM *See* parallel random access machine.

PRAM model a multiprocessor model in which all the processors can access a shared memory for reading or writing with uniform cost.

preattentive processing visual stimuli that are processed at an early state in the visual system and in parallel. This processing is done prior to processing by the mechanisms of visual attention.

precedence graph a topological representation of a schedule.

precise interrupt an implementation of the interrupt mechanism such that the processor can restart after the interrupt at exactly where it was interrupted. All instructions that have started prior to the interrupt should appear to have completed before the interrupt takes place and all instructions after the interrupt should not appear to start until after the interrupt routine has finished. *Compare with* imprecise interrupt.

precision (**1**) for floating point numbers, or scaled integer numbers, a specification of the accuracy of the representation. Typically, a programmer has a choice of two precisions: single precision floating point and double precision floating point. Languages may extend or limit these choices depending on the computer on which the programs will execute. Some languages allow the programmer to specify the precision and the compiler is responsible for choosing a suitable underlying representation that will meet the programmer-specified precision constraints. *See also* single precision, double precision.

(**2**) the degree of exactness with which a certain measure has been performed.

(**3**) a feature of a set of documents. Given a document set of N documents comprising N_r relevant documents and N_n irrelevant documents, let R be the set of returned documents with R_r being the relevant returned documents, R_n being the irrelevant documents. The precision is defined to be $|Rr|/|R|$.

precondition an assertion that pertains to a point immediately preceding, in the execution sequence, a specified portion of a program. Characterized by a predicate that evaluates to "true" if all the conditions are well-defined and meet the constraints for correct execution. The lack of verification may cause the generation of error messages, the fault, and/or the non-execution of the associated operation. *See also* weakest precondition, post condition.

predecessor a function which produces the previous element of a set. In principle, any set which has a successor function is a totally ordered set, and a corresponding predecessor function is usually defined. Note that a totally ordered set of values may have a designated element for which the predecessor function is not defined. Generally, for programming languages, such sets include enumeration sets, character sets, and lists. Singly linked lists generally do not have a predecessor function. *See also* parent directory.

predeveloped software software that has been produced prior to the issuing of a contract or purchase order, or to satisfy a general market need. Predeveloped software includes purchased off-the-shelf software or reusable software that is already developed and available.

predicate a Boolean expression which is evaluated and produces one of two values, "true" or "false".

predicate calculus a calculus for expressing logic statements. Its formulas involve:

1. atoms: $P(T_1, T_2, \dots)$ where P is a predicate symbol and T_i are terms.

2. Boolean connectives: conjunction ($\wedge$), disjunction ($\vee$), implication ($\rightarrow$), and negation ($\neg$).

3. literals: atoms or their negations.

4. quantifiers: for all ($\forall$) and there exists ($\exists$).

5. terms (also called trees): constructed from constants, variables, and function symbols.

predicate lock a lock resulting from locking all data identified by a predicate, such as an SQL where condition.

predictability the property characterized by an upper bound on the overall reaction time of each task in the system to external stimuli.

predicted quality software quality that is estimated and predicted for the end software product quality at each stage of development, and which is based on fundamental design quality and intermediate product quality.

prediction (**1**) an estimation procedure in which a future value of the state (see the definition) is estimated based on the data available up to the present time.

(**2**) in branching, the act of guessing the likely outcome of a conditional branch decision. Prediction is an important technique for speeding execution in overlapped processor designs. Increasing the depth of the prediction (the number of branch predictions that can be unresolved at any time) increases both the complexity and speed.

prediction system a system consisting of a mathematical model together with a set of prediction procedures for determining unknown parameters, and interpreting results.

predictive coding compression of a signal by coding differences between samples and predictions from previously coded values. For example, in still image coding, a predictive encoder may predict a pixel by taking the average of the pixel's left neighbor and its above neighbor. With raster-order coding, these values are already available in the decoder, which can form the same prediction. The difference or prediction error values may be quantized. Their probability density function is approximately Laplacian, and further compression can therefore be achieved with entropy coding. Also called differential pulse code modulation.

predictive vector quantization the generalization of scalar predictive coding to vector coding.

predictive VQ *See* differential pulse-code modulation.

preempt a concept from the scheduling theory. A higher-priority task is said to preempt a lower-priority task if it interrupts the lower-priority task.

preemption the action of forceful removal of a currently executing program from the processor to replace it with a program unit with higher priority.

preemptive priority a priority scheme in which a higher-priority request interrupts an already ongoing service of a lower-priority request. After the interruption ends, the preempted request can either continue with the same service time from the point where it was interrupted (preemptive-resume), start over again from the beginning with the same service time (preemptive-restart), or select a new service time (preemptive-resample).

pre-emptive scheduling a scheduling strategy in which a running program unit is removed from the processor whenever there is a reason for another process to execute.

preemptive scheduling a scheduling where the scheduler can interrupt and suspend the currently running task in order to start or continue to run other tasks. The scheduler must ensure that when swapping tasks, sufficient state is saved and restored in order to guarantee that tasks do not interfere. *See* time sharing.

prefetch *See* fetch policy.

prefetching in the CPU, the act of fetching instructions prior to being needed by the CPU. *See* fetch policy.

prefetch queue in the CPU, a queue of instructions which has been prefetched prior to being needed by the CPU.

prefix a mathematical notation in which an n-ary operator precedes its n operands. The unary "negate" operator, Boolean "NOT" operator, and similar operators are traditional (unary) prefix operators, but in general the number of operands associated with the operator is not constrained to one. A string v is a prefix of a string u if $u = vu''$ for some string u''. *See* Polish Notation.

prefix code set of words such that no word of the set is a prefix of another word contained in the set. A prefix code is represented by a coding tree.

prefix sums a parallel operation in which each element in an array or linked-list receives the sum of the previous elements.

prefix traversal a tree walk or traversal algorithm which first processes a node, then processes its subnodes (if any) in a specified order, traditionally left-to-right. *See also* postfix traversal.

preformat information such as sector address, synchronization marks, servo marks, etc. embossed permanently on the optical disk substrate.

preincrementation an assembly language addressing mode in which the address is incremented prior to accessing the memory value. Used to access elements of arrays in memory.

preliminary design creates representation of the data and architecture.

premature commitment the extent to which a decision about a programming plan or implementation must be made before its consequences can be seen.

prenex normal form a form for first-order logical formulas in which all quantifiers appear in the front of the formula.

preprocessing a series of image enhancements and transformations performed to ease the subsequent image analysis process through, e.g., noise removal or feature extraction/enhancement.

preprocessor a computer program that carries out some preliminary processing steps prior to the primary processing.

For example, in a compiler, it processes the source program whose output is sent to the compiler. A preprocessor may be as simple as a macro processor (as in C), or a sophisticated language processor like the original implementations of C++, where the C++ "compiler" was actually a preprocessor that read C++ code and wrote C code that was eventually processed by the C compiler.

preventive maintenance the maintenance carried out at predetermined intervals or according to prescribed criteria and intended to reduce the probability of failure or the degradation of the functioning of an item.

Prewitt gradient masks a set of two images that when used in windowed convolution provides an edge filter.

primary area that part of the disk which holds the buckets of a hashing method accessible with one disk access, using the hashing function.

primary copy technique a technique for concurrency control in distributed databases where the distinguished copies are stored at different sites.

primary index an index built upon an ordering key field of a file.

primary key the candidate key chosen from those available.

primary port the first port of a particular type in the computer.

primary site the site chosen to host the distinguished copies in the primary site technique.

primary site technique a technique for concurrency control in distributed databases where all the distinguished copies are stored at one site, the primary site.

primary storage short-term storage consisting of the CPU's address space.

primary work job-related tasks, such as writing a newsletter, managing a budget, or learning a procedure, for which individuals rely on computer technologies for their successful completion.

prime attribute any attribute which is a member of any key of the relation.

prime-number interleaving *See* interleaved memory.

primitive-based coding any one of several schemes to detect edges, lines, and other local features of images then use them to code the image. For example, edges may be used to segment the image into regions which are then independently coded as simple surfaces, while the boundaries are compressed with a chain code.

primitive type a type directly available in a language or model. *See* atomic type.

Prim's algorithm an algorithm for computing a minimum spanning tree. It builds upon a single partial minimum spanning tree, at each step adding an edge connecting the vertex nearest to but not already in the current partial minimum spanning tree. *See also* Kruskal's algorithm.

Princeton architecture a computer architecture in which the same memory holds both data and instructions. This is contrasted with the Harvard architecture, in which the program and data are held in separate memories.

principal axis the optical axis of a lens or camera, usually normal to the image plane.

principal component notionally, the direction of greatest variability of a random vector or among a set of sample vectors. More specifically, the principal component is the direction of the eigenvector associated with the largest eigenvalue of the covariance matrix of the random vector (or the sample covariance of a sample set). More generally, the n principal components of a distribution are the eigenvectors corresponding to the n largest eigenvalues. Principal components are frequently used for data clustering, pattern analysis, and compression.

principal point the point at which the optical axis of the lens in a camera meets the image plane: also, the corresponding point in the image.

principal variation splitting (PV split) a static parallel search method that takes all of the processors down the first variation to some limiting depth, and then splits the subtrees among the processors as they back up to the root of the tree.

principle of locality *See* locality. *See also* sequential locality.

principle of optimality the observation, in some optimization problems, that components of a globally optimum solution must themselves be globally optimized.

printed circuit board (PCB) a substrate made from insulating material which has one or more sandwiched metallic conductor layers applied which are etched to form interconnecting traces useful for interconnecting components.

printer an output device for printing results on paper.

prioritization coding a coding scheme whereby the position of the symbol in the data stream indicates its weight.

priority the level of urgency assigned to a program unit. The *priority* associated with an activity (a job or operation for example) determines its relative importance with respect to other activities on the system. A higher priority activity is executed prior to a lower priority activity.

priority ceiling protocol a method used in interrupt driven systems to avoid priority inversion. It dictates that a task blocking a higher priority task inherits the higher priority for the duration of that task.

priority encoder an encoder with the additional property that if several inputs are asserted simultaneously, the output number indicates the numerically highest input that is asserted.

priority inheritance temporary reassignment of a higher priority to a task which accesses a resource potentially accessible by higher priority tasks.

priority interrupt an interrupt performed to permit execution of a program unit that has priority higher than the program unit currently executing.

priority inversion a situation in which a high priority process is blocked from execution by a potentially infinite number of lower priority processes due to a lower priority process holding a resource which the high priority process attempts to access.

priority mean value analysis a mean value analysis method for priority systems. Starting with the highest priority, the mean waiting time for each priority class is computed from a mean value analysis taking into account the mean waiting time due to services in the considered priority class and in all classes with higher priority.

priority queue an abstract data type which efficiently supports finding the node with the highest priority across a series of operations. The basic operations are: insert, find-minimum (or maximum), and delete-minimum (or maximum). Some implementations also efficiently support join two priority queues (meld), delete an arbitrary node, and increase the priority of a node (decrease-key). *See also* discrete interval encoding tree, best first search, heap, Fibonnaci heap.

prior statistics the statistics of a random quantity (scalar, vector, process, etc.) before any experimental or measured knowledge of the quantity is incorporated. *See also* posterior statistics.

prismatic joint a joint characterized by a translation which is the relative displacement between two successive links. This translation is sometimes called the joint offset.

private (**1**) a name which is not visible outside a designated context.

(**2**) in the C language, adding the qualifier "static" to a declaration which appears at the global file level makes the declaration private to the file.

(**3**) in C++, a keyword in a class declaration that indicates that the members which follow are visible only to the class and to friends of the class, but not to derived classes. *See* information hiding. *See also* protected, public.

private implementation that portion of an abstract data type, module, or object-oriented type/class that is hidden from the other portions of an application. Critical for achieving representation independence.

private key cryptography also known as secret key cryptography. In such a cryptographic system, the secret encryption key is known only to the transmitter and the receiver for whom the message is intended. The secret key is used both for the encryption of the plaintext and for the decryption of the ciphertext. *See also* public key cryptography.

private member in object-oriented, a class member having a visibility scope only for the class in which it is declared. Derived classes cannot access the private members and therefore they are subject to partial inheritance.

private type *See* private.

privileged instruction an instruction that can be executed only when the CPU is in privileged mode.

privileged mode a mode of execution of machine instructions in the CPU in which certain special instructions can be executed or data accessed which would otherwise be prohibited. *See also* user mode.

privileged state an execution context that allows all hardware instructions to be executed.

probabilistically checkable proof an interactive proof system in which provers follow a fixed strategy, one not affected by any messages from the verifier. The prover's strategy for a given instance x of a decision problem can be represented by a finite oracle language B_x, which constitutes a proof of the correct answer for x.

probabilistic models underlying probabilistic models that determine the input distribution (e.g., of generated strings).

probabilistic neural network a term applied loosely to networks that exhibit some form of probabilistic behavior but also applied specifically to a type of network developed for pattern classification based upon statistical techniques for the estimation of probability densities.

probabilistic splitting random splitting of a process. Consider a parent arrival stream that is split into two; arrivals from the parent stream are assigned to the child stream i, $i = 1, 2$, with probability p_i, $p_1 + p_2 = 1$. One consequence of random or probabilistic splitting of a process is that the individual (component) streams are independent.

probabilistic Turing machine a Turing machine in which some transitions are random choices among finitely many alternatives.

More formally, for a random $\mathbf{x}$ and any probabilistic event A, the probability density function $p_x(\mathbf{x})$ satisfies

$$\Pr(\mathbf{x} \in A) = \int_A \mathrm{d}p_x(\mathbf{x}) .$$

See also cumulative distribution function.

probability density function (PDF) a function describing the relative probability of outcomes of an experiment. For experiments with discrete outcomes, the PDF is analogous to a relative frequency histogram. For experiments with continuous outcomes, the PDF is analogous to a relative frequency histogram where the category bin widths are reduced to zero. The total area underneath a PDF must always be unity.

More formally, for a random variable $\mathbf{x}$ and any probabilistic event A, the probability density function $p_x(\mathbf{x})$ satisfies

$$\Pr(\mathbf{x} \in A) = \int_A \mathrm{d}p_x(\mathbf{x}) .$$

See also cumulative distribution function.

probability of delivery the proportion of messages that arrive at the destination node.

problem domain model *See* conceptual abstraction.

procedural abstraction separating out the details of an execution unit in such a way that it may be invoked in a program statement or expression.

procedural animation the creation of a motion by a procedure describing the motion specifically. Procedural animation may be specified using a programming language or an interactive system.

procedural cohesion cohesion among the several procedures/methods of a module/class. It is higher when most of the procedures collaborate for the production of a unique task. *See* cohesion, temporal cohesion, logical cohesion.

procedural design creates representations of algorithmic details within a module.

procedural language a term used to describe a language where the programmer specifies an explicit sequence of steps to follow to produce a result in contrast to a declarative language. Common *procedural languages* include Basic, Pascal, C, and Modula-2. Procedural languages are also referred to as imperative languages. A language focused on procedures instead of data types. *See* object-oriented, nonprocedural language.

procedural surface uses external parameters supplied to a model that determines how the surface will be generated. For example, a *procedural surface* that generates a polygonal representation of a sphere at a specified detail is procedural; the actual surface is generated by the specified sphere diameter and the number of polygons that will make up the surface. An advantage of using this approach is efficient storage and replication since individual polygons need not be explicitly specified.

procedural texture a texture generated by a model controlled by external parameters.

procedure a subroutine which does not return a value. A procedure is a self-contained code sequence designed to be re-executed from different places in a main program or another procedure. *See also* return instruction. Synonym for subroutine, function.

procedure call in program execution, the execution of a machine-language routine, after which execution of the program continues at the location following the location of the procedure call.

process (**1**) an executable program unit managed by an operating system.

(**2**) the context, consisting of allocated memory, open files, network connections, etc. in which an operating system places a running program.

(**3**) a set of tasks or activities that transform an input product into an output product by consuming resources (personnel, finance, facilities, equipment, tools, techniques, and methods). Resources are consumed in the course of a process. A process can be divided into smaller processes. Examples of processes are: "software development", "specification of requirements", "testing", and "maintenance". Each process at each level of abstraction consists of a technical process and a management process. The technical process receives products from a preceding process. A management process receives information about a technical process and/or its output product, and outputs management information to the succeeding process.

process algebra algebra on processes, that views a process as a type. Operators for parallel composition, sequential composition, and synchronization are usually defined. The most common process algebras are CSP and CCS. *See also* formal verification, formal method.

process assessment a disciplined examination of the processes used by an organization against a set of criteria to determine the capability of those processes to perform within quality, cost, and schedule goals by characterizing current practice and identifying strengths and weaknesses.

process control block an area of memory containing information about the context of an executing program. Although the PCB is primarily a software mechanism used by the operating system for the control of system resources, some computers use a fixed set of process control blocks as a mechanism to hold the context of an interrupted process.

process creation the action that creates a new process.

process decomposition definition of a process behavior with a set of communicating processes.

process descriptor a data structure in the kernel that represents a process.

process diagram a diagram that represents the set of communicating processes that model the system or a part of it. May also be depicted as the communication connections between processes if inputs/outputs communication model is used.

process environment part of the control scene which is outside the controlled process; within this environment are formed the uncontrolled, free inputs to the process; specified quantities related to the environment may be observed and used by the controller, for example when performing free input forecasting.

process ID an identifier which uniquely identifies a process.

processing element (PE) a processing module, comprising at least a control section, registers, and arithmetic logic, in a multiprocessor system. A processing element may be capable of operating as a stand-alone processor.

processing narrative a natural language description of a model or a program component.

process interface the set of service calls available to processes.

process management the direct control of the project evolution coordinating all the activities and controlling the progress and results.

process measure a measure defined for some attribute of a process.

process model a model for a set of partially ordered steps required to reach a goal.

process number an identifier that represents a process by acting as an index into the array of process descriptors.

processor the central processing unit (CPU) of a computer.

processor-centered view a view of computing that emphasizs the work of a processor.

processor element *See* processing element.

processor farm a collection or ensemble of processing elements to which parallel processing tasks are assigned and distributed for concurrent execution. In this model, tasks are distributed, or "farmed out", by one "farmer" processor to several "worker" processors and results are sent back to the farmer. This arrangement is suitable for applications that can be partitioned into many separate, independent tasks. The tasks are large and the communications overhead is small.

process oriented a characteristic of a specification model that uses processes to represent active parts of a system.

processor sharing (PS) a mathematical abstraction of round robin scheduling in which the quantum size approaches zero. When PS is used in a queuing network model, the server appears to divide service equally among the customers present. That is, each customer's service rate is divided by the number of customers present at the server at any point in time.

processor state privileged or nonprivileged state.

processor status word a register in the CPU that stores a collection of bits that, taken as a group, indicate the status of the machine at a given period of time.

process state the set of information required to resume execution of a process without interfering with the results of the computation.

process status the operating system record of the current details of a process's execution.

process support a foundation on which to build solutions supporting the execution and management of business tasks.

process swap the act of changing the execution point from one process to another.

process switch the action of directing the hardware to run a different process from the one that was previously running.

process tools tools that assist in managing the software development process, for example, controlling how programmers interact or what testing must be done before a file can be incorporated into a system.

producer a program unit that provides data to be used by other program units.

product a broader than a system. It incorporates all components that the producer uses, and all the items delivered to customers. Documentation and confidential source code, for instance, are part of the product but not part of the system. Result of activities or processes. Product may include service, hardware, processed materials, software, or a combination thereof. A product can be tangible (e.g., assemblies or processed materials) or intangible (e.g., information or concepts), or a combination thereof. Product can be either intended (e.g., offering to customers) or unintended (e.g., pollutant or unwanted effects). Examples of software products are also: specifications, test reports, executable object code, user manuals, and support documentation. Something that has value and is generated by a manufacturing or development process.

product analysis *See* product assessment.

product assessment a disciplined examination of the product against a set of criteria and metrics to determine the features of the product, for instance, considering quality, costs, etc. The *product assessment* can be performed by considering all the product components: code, documentation, assistance, service, etc.

product baseline in configuration management, the initial documentation defining the documentation details about the configuration items, the production, and maintenance.

product code two-dimensional burst and random error correcting code in the form of a matrix within each row and column of which are code words of two different linear codes.

product family a product family is a collection of all the components used to produce concrete systems and any other items delivered to customers. It is generic in the sense that many distinct product instances can be produced from one generic product.

product-form network a class of queuing networks studied by Jackson and by Gordon and Newell in which the calculation of the underlying state probabilities has the form of a product of the service times and routing probabilities; a Jackson network.

product-form queueing a product-form network.

product instance a product instance is what a customer buys. A product instance consists of a copy of executable implementation code and any other items sold with it, typically documentation.

production in a grammar, a rule which maps a sequence of terminal symbols and nonterminal symbols to a new sequence of terminal and nonterminal symbols. The rules which constrain what can be written in the productions determine the type of the grammar in the Chomsky hierarchy and consequently the type of automaton required to recognize sentences that are in the language defined by the grammar.

production library the entire set of applications available to an enterprise. This is also called a software portfolio.

production rate the amount of work that can be produced by a single worker in a certain amount of time.

production rule a rule in a grammar. Using production rules, one can generate sentences in the language, or recognize if a sequence of input symbols defines a valid sentence in the language.

productive work the effective time that an actual software engineer is working during a day. This figure is geographically dependent: in the U.S. it is about 5 hours, but in other countries it can be more or less.

productivity goods and services produced per unit of labor and expense. Measurable in function points.

productivity analysis the analysis of the productivity of each product and unit in a company.

productivity improvement plan plans to improve the productivity of software development, for instance, by adopting CMM, ISO 9001, etc.

productivity paradox a speculation on the basis of the fact that the estimated return on investment of the current information society is negative: each dollar invested returns about 80 cents, so it is a loss.

product management the definition and the coordination of all the activities for maintaining a product on the market. It includes the configuration management.

product measure a measure defined for some attribute of a product.

product of sums (PS) the AND combination of terms, which are OR combinations of Boolean variables.

product specification *See* specification.

profiling the activity of collecting and presenting timing data about a program.

program (**1**) a specification of an algorithm to be executed by a computer. A *program* may be represented as source code, an executable file, an applet, a byte code, or many other representations. The language in which a program is written may be a procedural language, functional language, or nonprocedural language, although traditional usage generally does not refer to nonprocedural specifications as "programs".

(**2**) any file which is executable by the operating system.

program analysis tools that are designed to aid the task of understanding existing source code by providing a large amount of detailed information about the program. Analysis tools help focus on the structure and attributes of the system. The relevant information can be extracted from a program by either analyzing the program text (static analysis), or by observing its behavior (dynamic analysis).

program concepts *See* program plan.

program counter (PC) the register in a CPU which is a pointer to the current instruction (while an instruction is being fetched) or the next instruction (while an instruction is being executed). A branch, jump, or skip instruction may change the value in the program counter to change instruction flow. In some machines with certain kinds of parallelism (particularly RISC machines), the modification to the program counter may not be effected until after the next instruction is executed.

program description language (PDL) any of a large class of formal and profoundly useless pseudo-languages in which management forces one to design programs. Too often, management expects PDL descriptions to be maintained in parallel with the code, imposing massive overhead of little or no benefit.

program design language (PDL) *See* program description language.

program generator an application that is capable of generating a program. *See* application generator.

program heuristic a general rule of programming concept that captures the conventions in programming and governs the composition of the program plans into programs.

programmable array logic (PAL) a programmable logic array with a fixed set of OR gates into which are fed sets of product terms.

programmable gate array (PGA) *See* field-programmable gate array.

programmable logic array (PLA) a programmable logic device which consists of an AND array forming logical products of the input literals and an OR array which sums these products to form a set of output functions.

programmable logic controller (PLC) a microprocessor-based system comprised of a set of modules for acquiring signals from the environment and others for producing effects on the environment. These effects are typically used for controlling electromechanical machines by means of actuators (motors, heating element, etc.). The rules for specifying the control law are Boolean expressions. These are sequentially and cyclically executed. More complex PLCs allow the description of the control law by means of a ladder diagram.

programmable logic device (PLD) an integrated circuit which is able to implement combinational and/or sequential digital functions defined by the designer and programmed into this circuit.

programmable radio system radios based on digital waveform synthesis and digital signal processing to allow simultaneous multiband, multiwaveform performance.

programmable read-only memory (PROM) a semiconductor memory device that has a primary function of storing data in a non-volatile fashion which can be programmed to contain predetermined data by means other than photomasking. PROMs may be one time programmable (OTP) or they may be either UV or electrically eraseable, depending on the particular semiconductor process technology used for manufacturing.

programmed I/O transferring data to or from a peripheral device by running a program that executes individual computer instruction or commands to control the transfer. An alternative is to transfer data using DMA or memory-mapped I/O.

programming (1) the action of writing code in a language consisting of both syntax and semantics. The resulting code is usually compiled or interpreted in order to obtain an executable form of the code allowing the performing of the task addressed with the code.

(2) human task during human-robot interaction: storing a behavioral repertoire via symbols, including words.

programming language a formalism for specifying the instructions that have to be executed by the computer. Several different programming languages are available. The selection of a specific programming language may influence the process of development.

programming system a set of programming languages and the support software necessary for using these languages with a given computer system.

program mutation (1) a program that has been intentionally changed in order to verify if test cases are capable of detecting the mutations.

(2) the process of creating mutated programs.

program plan an abstract representation of source code fragments. Comparison methods are used to help recognize instances of programming plans in a subject system, the focus being to identify similar code fragments with pattern matching at the programming language semantic level.

program plan recognition the use of program plans to identify similar code fragments. Existing source code is often reused within a system via "cut and paste" text operations. Detection of cloned code fragments must be done using program heuristics since the decision of whether two arbitrary programs perform the same function is undecidable.

program slice a fragment of a program in which some statements are omitted that are not necessary to understand a certain property of the program. For example, if someone is interested in how the value for a certain returned value of a function is arrived at, then only code that has a bearing, direct or indirect, on that value is relevant. *See* program slicing.

program slicing the process by which the slices of a program are identified and performed. This technique is frequently performed during maintenance and reengineering of programs for identifying the pieces of code that have to do with some variables. The several slices, if independent, can be reorganized. *See* program slice.

program state the collection of values of all program variables.

program status word (PSW) a combination program counter and status-flag register provided in IBM mainframe computers.

program structure diagram *See* structure chart.

program transformation a reengineering process. Transformations are the changes from unstructured code to structured code, updating design documents, or correcting specifications. It is assumed that the transformation improves the subject system according to some measurable criterion.

program translation transformation of source code from one language to another or from one version of a language to another version of the same language. For example, converting from COBOL-74 to COBOL-85.

program trouble report (PTR) *See* error report.

program understanding a term related to reverse engineering. *Program understanding* always implies that understanding begins with the source code while reverse engineering can start at a binary and executable form of the system or at high level descriptions of the design. The science of program understanding includes the cognitive science of human mental processes in program understanding.

Program understanding can be achieved in an ad hoc manner and no external representation has to arise. While reverse engineering is the systematic approach to develop an external representation of the subject system, program understanding is comparable with design recovery because both of them start at the source code level. Program comprehension is the process of acquiring knowledge about a computer program. Increased knowledge enables such activities as bug correction, enhancement, reuse, and documentation. While efforts are underway to automate the understanding process, such significant amounts of knowledge and analytical power are required that today program understanding is largely a manual task.

program visualization a mapping from programs to graphical representations. For visual programming languages, the visualization is only a way for showing the visual program.

progressive coding ordering of coded values such that the original signal can be recovered progressively. For example, in transform coding of a picture, transmission of the zero sequences coefficients for all blocks first, rather than transmission of all coefficients for the first block first. This allows the receiver to generate an approximate reconstruction early in the reception of data.

progressive DCT-based encoding image encoding in which the image is encoded in multiple scans, in order to produce a quick, rough decoded image when the transmission time is long.

progressive refinement an elaboration of the original radiosity method that enables a visualization of the solution to emerge as the equations are being solved. The solution, originally approximated with an ambient term, is gradually made more accurate.

progressive spectral selection algorithm a selection algorithm in which DCT coefficients are grouped into several spectral bands.

progressive successive approximation algorithm an approximation algorithm where all

DCT coefficients are sent first with lower precision, and then refined in later scans.

progressive transmission partial information of an image is transmitted. At each stage, an approximation of the original image is reconstructed at the receiver. The quality of the reconstructed image improves as more information is received.

project (1) the planning of activities, resources, deliverables, deadlines, and milestones for reaching a specific set of objectives.

(2) a relational operator for extracting specified attributes from a relation.

project control the supervision activity that guarantees the correct development, the respect of timelines, and the planned profile of a project.

project coordinator *See* project manager.

project database a repository in which documentation, code, fault-reports and quality manual of a project are stored.

projection (1) a mapping from a set to a subset of the set.

(2) in signal and image processing, the conversion of an n-dimensional signal into an $n-1$ dimensional version through some integration in the continuous case, or some summation in the discrete case. For instance, a 2D image can be viewed as a (perspective) projection of 3D scene via a camera. Another example exists in computed tomography. There a projection is a line integral along a straight ray.

projection of a fuzzy set for fuzzy set Q in a Cartesian product X^n, $X^n = X_1 \times \cdots \times X_n$, the, fuzzy set subspace R in X^i of X^n, $(i < n)$. The projection is usually denoted $R = \text{proj}(Q; X^i)$, in X^m with membership function obtained as the supremum of the membership function for the dimensions to be eliminated. *See also* cylindrical extension of a fuzzy set, fuzzy set, membership function.

projective invariant a measure that is independent of the distance and direction from which a particular class of object is viewed, under perspective projection. The cross-ratio is an important type of projective invariant, which is constant for four collinear points viewed under perspective projection from anywhere in space. Projective invariants are important in helping with egomotion (e.g., automatic guidance of a vehicle) and for initiating the process of object recognition in 3D scenes.

projective rule a database relationship. Given a set of attributes A, B, and C, if $A \rightarrow BC \models A \rightarrow C$.

project join normal form (PJNF) *See* fifth normal form.

project management system a software tool that aids the process of planning, organizing, staffing, directing, and controlling the production of a system. These tools usually aid the manager with views, i.e., Gantt, PERT, Kiviat diagrams, and charts.

project manager the leader of the project. He/she has the responsibility to:

1. control by facilitating the information flow among teams;
2. ensure that the project program, milestones, and time scales are maintained and deviations resolved and recorded with change control;
3. manage related documentation;
4. maintain relationships between partners;
5. control the behavior of project participants against the planned activities;
6. communicate progress;
7. establish external relationships according to the dissemination activity;
8. control quality of all activities and deliverables via specific tools;
9. administrate, e.g., collect cost statements on a timely basis, collect information for EC management, and collect progress reports on a timely basis;
10. act as a link between this project and other related projects;
11. arrange for the preparation of necessary project review documentation, etc.

project plan the schema of the project evolution along its development life-cycle considering both the macro- and the micro-cycles and the decomposition of the project into sub-projects and thus into subsystems. The project plan has to consider resources and time according to the technology used and the deadlines imposed. Typical project plans are given in Gantt diagrams.

project planning the activity that creates a description of the management approach for a project (the project plan).

project risk one of a set of potential project problems or occurrences that may cause the project to fail.

project scope a statement of basic requirements of the software to be built.

project size the dimension of a project on the basis of its size. For example, the cost of the project, the number of person months, the number of lines of code of the system, the number of attributes or methods counted of all system classes, etc.

project start-up the phase in which the project details and constraints are defined before the actual starting of the development phase. In this phase, decisions about the feasibility, the resources, the tools, etc. that will be used are made.

project team the personnel involved in the project development, set-up, preparation, and dissemination. The leader of the *project team* is the project manager.

project tracking the activity that enables a manager to understand the status of a project.

Prolog a symbolic programming language used in artificial intelligence based on predicate calculus.

PROM *See* programmable read-only memory.

prompt a symbol or message displayed by a computer system, requesting input from the operator.

proof (**1**) a series of logical mathematical steps that are used to demonstrate the truth of a mathematical expression. The steps can be characterized as a finite sequence of well-formed formulas, $F_1, F_2, \ldots F_n$, where each F_i is either an axiom or follows by some rule of inference from some of the previous F's, and F_n is the statement being proved.

(**2**) a formal methodology for demonstrating the correctness of a segment of software, usually by treating the code as a mathematical expression.

proof of correctness in the development of a system or of a component using a formal method, the activity that proves that the specification or the implementation is correct with respect to certain constraints or properties that a correct system must satisfy.

propagated borrow a borrow that is produced from the subtraction of two one-bit numbers, with a borrow-in included, i.e., from the subtraction of 1 from 1 with a borrow-in of 1. *See also* generated borrow.

propagated carry a carry that is produced from the addition of two one-bit numbers, with a carry-in included, i.e., from the addition of 0 and 1 with a carry-in of 1. *See also* generated carry.

propagated error in computer arithmetic, the error that results because a machine operator propagates, possibly with some amplification, errors in the operands.

propagation delay (**1**) the delay between transmission and reception of a signal. Caused by the finite velocity of electromagnetic propagation.

(**2**) the delay time between the application of an input signal to a chain of circuit elements and the appearance of the resulting output signal.

(**3**) the time it takes for a transistor switch to respond to an input signal, symbolized t_{pd}. It is calculated between the 50% rise point to the 50% fall point or vice versa (see graph of typical inverter gate).

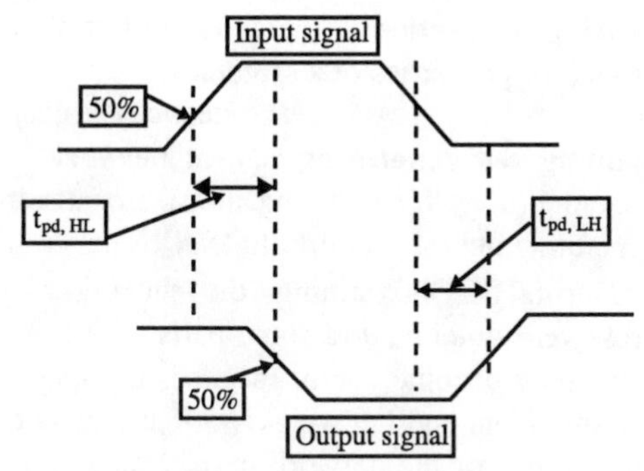

Propagation delay for a typical inverter gate.

The time from when the input logic level to a device is changed until the resultant output change is produced by that device.

propagation delay time *See* propagation delay.

proper qualities of a factor of a string that is not equal to the string itself.

proper subgraph a subgraph which does not contain all of the edges of the given graph.

property a statement that describes a characteristic of a system. It can be expressed in an informal way (with natural language) or with a formal language (logic, temporal logics, automata).

property-oriented a characteristic of a specification model that uses properties to describe the system. *See also* descriptive.

proportional selection a probabilistic selection mechanism adopted in a genetic algorithm where the fitness of each individual is interpreted as the probability of its survival into the next generation. More specifically, for each individual in the population, the associated probability of survival is evaluated using the following equation:

$$p(\mathbf{x}_i) = \frac{f(\mathbf{x}_i)}{\sum_{i=1}^{\mu} f(\mathbf{x}_i)},$$

where $p(\mathbf{x}_i)$ and $f(\mathbf{x}_i)$ are the probability of the survival and fitness value of the i-th individual, respectively. Roulette wheel selection is then adopted where the wheel is divided into sectors with angles proportional to the individual survival probabilities. An individual is selected according to whether a randomly generated angular value falls within the individual's associated sector in the wheel. A scaling function is usually required to transform the original objective function into the fitness function due to the positivity requirement for representing the survival probabilities. *See also* genetic algorithm.

propositional logic also known as predicate calculus. A formal language used to state logical constraints between propositions. The usual operators are: AND ($A \wedge B$), OR ($A \vee B$), NOT ($\neg A$), IMPLY ($A \rightarrow B$) and IFF ($A \leftrightarrow B$).

protected (**1**) referring to an entity whose access is limited by some rule of the language.

(**2**) in C++, a declaration that indicates the members of the class which follow are available only to the class itself, friends of the class, and derived classes. *See also* private, public.

protected member class member that can be accessed by the class in which it is declared or by the derived classes.

protection a hardware or software mechanism for protecting software/hardware/data from unauthorized use or copy. Diffuse protection mechanisms are encryption, watermarking, and fingerprint. The protection mechanism typically includes a code for identifying the product and the owner. Ring numbering was introduced in the Multics system as one basis for limiting access and protecting information. The term "security" is used when the constraints and policies are very restrictive.

protection exception an exception that occurs when a program tries to access a protected part of the memory of the system. The protected memory can be assigned to other processes as well as the operating system.

protection fault an error condition detected by the address mapper when the type of request is not permitted by the object's access code.

protocol a set of formal rules describing how to transmit data, especially across a network.

Low level protocols define the electrical and physical standards to be observed, bit- and byte-ordering, and the transmission and error detection and correction of the bit stream. High level protocols deal with the data formatting, including the syntax of messages, the terminal to computer dialogue, character sets, sequencing of messages, etc.

protocol data unit (PDU) the unit of exchange of protocol information between entities. Typically, a PDU is analogous to a structure in C or a record in Pascal; the protocol is executed by processing a sequence of PDUs.

protonotion in a W-grammar, a possibly empty sequence of syntactic marks designated by lower-case words. *See also* metanotion, hypernotion.

prototype (**1**) a working model of an application or system that is functionally equivalent to a subset of the product. Generally this model is inefficient and with many problems. It can be shown to the end-user/purchaser to check if the analysis is correct. Particularly important is the "User Interface Prototype" that shows how the user will interact with the system.

(**2**) synonym for signature, a specification of a function and its parameters, including parameter type and result type information.

prototyping the creation of a model and the simulation of all aspects of a product. CASE tools support different degrees of prototyping. Some offer the end-user the ability to review all aspects of the user interface and the structure of documentation and reports before code is generated. The software development process that follows a prototyping approach starts with an early prototype of the system and proceeds by means of a sequence of refinements adding new functionalities to the previous version. The versions are tested according to the requirement's needs, and the evolution should be planned according to a specific schedule.

provable a property of a system for which there exists a proof for it.

proving properties the activity of building proofs for properties of a system.

pruning self-generating neural network a methodology for reducing the size of a self-generating neural network (SGNN).

During SGGN training, the network may grow very quickly, and some parts of the network may become useful in either training or classification, and the weights of these parts of the neurons of the network never change after some training stage. We call these parts of the network dead subnets. It is obvious that the dead subnets of the network should be pruned away to reduce the network size and improve the network performance. One way to achieve this is to check the weights of each neuron in the network to see if they have been changed since the last training epoch (or during the last few epochs). If they are unchanged, the neuron evidently may be dead and should be removed from the network. If a neuron is removed from an SGNN, all of its offsprings should also be removed. *See* self-generating neural network. *See also* training algorithm of self-generating neural network.

PS *See* processor sharing, product of sums.

pseudo-associative cache similar to direct-mapped cache, but the processing of a miss involves first checking for a match with another cache entry. Also known as column-associative cache.

pseudo code *See* pseudo-code.

pseudo-code any form of informal algorithm description used to communicate an algorithm. Often a blend of notation from a programming language, other suitable notations such as mathematical expressions, and the writer's native language. Pseudo-code does not have to follow any syntax rules and can be read by anyone who understands programming logic.

pseudocolor *See* pseudocoloring.

pseudocoloring pseudocoloring recolors pixels with colored values as a function of the gray level value in the original monochrome image. Pseudocoloring is used because of the limitation of the human visual system to distinguish all the brightness range values.

The display in color of gray-level pixels in order to make certain gray-level pixels or gray-level patterns more visible. Typical pseudocoloring schemes are: 1. displaying all pixels in a range of gray levels in one color; 2. displaying each gray level at a fixed interval in a different color to produce colored contour lines; or 3. displaying the entire gray-level range as a rainbow. Pseudocoloring is usually carried out by look-up tables. *See also* false color, look-up table (LUT).

pseudo-exhaustive testing a testing technique which relies on various forms of circuit segmentation and application of exhaustive test patterns to these segments.

pseudofile an object that appears to be a file on the disk but is actually stored elsewhere.

pseudo-operation in assembly language, an operation code that is an instruction to the assembler rather than a machine-language instruction.

pseudo-random number generator a program that produces numbers from a particular distribution, often uniform or normal, that can be used instead of real data in simulations of systems. A huge variety of methods are known, and the choice of pseudo-random number generator is important to the validity of a simulation, since hidden dependencies can be quite subtle.

pseudo-random skewing *See* interleaved memory.

pseudo-random testing a testing technique (often used in BISTs) which is based on pseudo-randomly generated test patterns. The technique that uses a linear feedback shift register (LFSR) or similar structure to generate binary test patterns with statistical distribution of values (0 and 1), across the bits; these patterns are generated without considering the implementation structure of the circuit to which they will be applied. Test length is adapted to the required level of fault coverage.

pseudo-transitive rule a database relationship. Given sets of attributes A, B, C, and D, then if $A \rightarrow B$, and $BC \rightarrow D$, $\models AC \rightarrow D$.

PSU power supply unit. *See* power supply.

PSW *See* program status word.

psychovisual redundancy the tendency of certain kinds of information to be relatively unimportant to the human visual system. This information can be eliminated without significantly degrading image quality, and doing so is the basis for some types of image compression. *See also* image compression, Joint Photographic Experts Group (JPEG), quantization matrix.

PTAS *See* polynomial time approximation scheme.

PTR acronym for program trouble report. *See* error report.

p-tree a spatial access method that defines hyperplanes, in addition to the orthogonal dimensions, which node boundaries may parallel. Space is split by hierarchically nested polytopes (multidimensional boxes with nonrectangular sides). The R-tree is a special case which has no additional hyperplanes.

public (**1**) an entity, program, or system, is said to be public when it is visible and freely usable with limited restrictions. Public matter is currently made available on WWW and FTP sites.

(**2**) an entity, typically a name, which is exported from another entity in a compilation. The name may represent a variety of syntactic entities, such as variables, functions, or types, depending on the nature of the language and its implementation.

(**3**) in the C++ language, a declaration that indicates that subsequent names of a class are exported. *Compare with* private, protected.

public defect a defect found by someone other than the developer, as opposed to private defect, which is a defect found by the developer.

public key cryptography two different keys are used for the encryption of the plaintext and decryption of the ciphertext. Whenever a transmitter intends to send a receiver a sensitive message that requires encryption, the transmitter will encrypt the message using a key that the receiver makes available publically to anyone wanting to send them encrypted messages. On receiving the encrypted message, the receiver applies a secret key and recovers the original plaintext information sent by the transmitter. In contrast to secret key cryptography, *public key cryptography* does not suffer from the problem of having to ensure the secrecy of a key. *See also* private key cryptography.

public key cryptosystem system that uses a pair of keys, one public and one private, to simplify the key distribution problem.

public member a class member that can be accessed by external functions of classes not bounded by friendships or inheritance relationships.

pull-down menu a menu that pulls down like a roller blind from a permanently visible title bar, usually at the top of the display or window.

punched card a method, now obsolete, used to represent data and programs as cardboard cards where the values were represented by punched holes at appropriate places.

punctured code a code constructed from another code by deleting one or more coordinates from each codeword. Thus, an (n, k) original code becomes an $(n-1, k)$ code after the puncturing of one coordinate.

punctured convolutional code a code where certain symbols of a rate $1/n$ convolutional code are periodically punctured or deleted to obtain a code of higher rate. Because of the simplicity, it is used in many cases particularly in variable rate and high-speed application. *Compare with* punctured code.

puncturing periodic deletion of code symbols from the sequence generated by a forward error control convolutional encoder, for purposes of constructing a higher rate code. Also, deletion of parity bits in a forward error control block code.

purchaser customer in a contractual situation.

pure class a synonym that some languages use to denote an abstract class.

pure functional language a functional language that provides absolutely no mechanism for performing side effects, thus exhibiting referential transparency.

purely functional language a language that does not allow any destructive operation — one which overwrites data — such as the assignment operation. Purely functional languages are side effect-free; i.e., invoking a function has no effect other than computing the value returned by the function.

pure procedure a procedure that does not modify itself during its own execution. The instructions of a *pure procedure* can be stored in a read-only portion of the memory and can be accessed by many processes simultaneously.

pure virtual function in C++ language, abstract classes (for which an implementation cannot exist) define some functions as virtual, but when no implementation can be provided, they became pure virtual by the means of "=0" initialization.

purity the degree to which a color is saturated.

push operation that places data on the stack. *See also* pop.

pushdown automaton a restricted Turing machine where the tape acts as a pushdown store (or stack, where only the latest element can be read), with an extra one-way read-only input tape. *See also* Chomsky hierarchy, context-free grammar, nondeterministic push-down automaton.

pushdown stack (**1**) a data structure containing a list of elements that are restricted to insertions and deletions at one end of the list only. Insertion is called a push operation and deletion is called a pull operation.

(**2**) a pipeline that has the property that the intermediate results remain in the pipeline until forced out by subsequent additional computations. This is an important consideration for compiler writers since the optimization of such a pipeline requires complex analysis of the computations specified by the program.

PV split *See* principal variation splitting.

pyramid coding any compression scheme which repeatedly divides an image into two subbands, one a lowpass representation that is subsampled and used as input for the next level, and the other an error (difference) image. A small lowpass image plus a pyramid of difference images of increasing size is generated, allowing the lowpass information to be coded accurately with few codewords, and the highpass information to be coded with coarse quantization. *See also* Laplacian pyramid.

Q

QBE *See* query by example.

QoS *See* quality of service.

quadric surface a curved surface defined by the equation

$$ax^2 + by^2 + cz^2 + 2dxy + 2exz + 2fyz + 2gx + 2hy + 2iz + j = 0 .$$

Special cases of the surface include spheres, cones, cylinders, ellipsoids, hyperboloids, etc. The translation, rotation, and scaling of a quadric surface is easy, as is the calculation of its surface normal, intersection with a ray and calculation of the z-value (given the x and y values).

quadtree data structures that cater to the storage and retrieval of data in multi-dimensional data space, particularly an image space.

The approach is to recursively divide the data space into subdivisions. At each subdivision the subquadrants are assigned a "full", a "partially full", or an "empty" label depending on how much the quadrant intersects the region of the interest in the data space. The subdivision of partially full subquadrants continues recursively until all the subquadrants are homogeneous (full or empty) or a predetermined cutoff depth has been reached. The tree is such that each node is split along all d dimensions, leading to 2^d children.

Quadtrees can be used for facsimile coding, for representing contiguous regions in primitive-based coding, or for representing motion vector distributions. In facsimile, where the image consists of black and white values only, three codewords are required: S, B, W, meaning split-block, black-block, and white-block. Starting from large squares, each is analyzed and an appropriate codeword generated. If the codeword is S, then the same process is applied recursively to the four quarter-size blocks. *See also* octree.

quadtree complexity theorem the number of nodes in a quadtree region representation for a simple polygon (i.e., with nonintersecting edges and without holes) is $O(p+q)$ for a $2^q \times 2^q$ image with perimeter p measured in pixel widths. In most cases, q is negligible and, thus, the number of nodes is proportional to the perimeter. It also holds for three-dimensional data where the perimeter is replaced by surface area, and in general for d-dimensions where instead of perimeter we have the size of the $(d-1)$-dimensional interfaces between the d-dimensional objects.

quadtree compression compression of an image using a quadtree.

quadword a data unit formed from four words. On machines with a 16-bit word length, or in languages with a history of a 16-bit word length, a quadword represents 64 bits. Note that for some machines, 64 bits is the word length, so the usage depends upon the culture from which the language and machine evolved. *See also* byte, halfword, word, longword.

qualification a process for verifying if a system can be operatively used.

qualitative proportionality a qualitative relationship expressing partial information about a functional dependency between two parameters.

qualitative simulation the generation of predicted behaviors for a system based on qualitative information. Qualitative simulations typically include branching behaviors due to the low resolution of the information involved.

quality to be exchanged with the "degree of excellence" or "fitness for use" that only partially meets the definition. Software quality is defined in the ISO 9126 norm series. Software quality includes: functionality, reliability, usability, efficiency, maintainability, and portability. The quality of a system is the evaluation of the extent to which the system meets the abovementioned features. The response of the system to these features is called the estimated quality profile. Quality should not be used as a single term to express a degree of excellence in a com-

0-8493-2691-5/01/$0.00+$.50

parative sense nor should it be used in a quantitative sense for technical evaluations. To express these meanings, a qualifying adjective shall be used. For example, use can be made of the following terms: "relative quality" where entities are ranked on a relative basis in the "degree of excellence" or "comparative sense" (not to be confused with grade); "quality level" in a "quantitative sense" (as used in acceptance sampling) and "quality measure" where precise technical evaluations are carried out.

quality assessment *See* quality evaluation.

quality assurance all the planned and systematic activities implemented within the quality system, and demonstrated as needed, to provide adequate confidence that an entity will fulfill requirements for quality. There are both internal and external purposes for *quality assurance:* internal quality assurance: within an organization quality assurance provides confidence to management; external quality assurance: in contractual or other situations, quality assurance provides confidence to the customers or others.

quality attribute a feature or characteristic that affects an item's quality. In a hierarchy of quality attributes, higher level attributes may be called quality factors, lower level attributes may be called quality attributes. The features and characteristics of a software component that determine its ability to satisfy requirements.

quality circle a group of people who are professionally involved in quality assurance who meet on a regular basis to discuss quality assurance in order to improve the quality of software. Very popular in Japan.

quality control operational techniques and activities that are used to fulfill requirements for quality. *Quality control* involves operational techniques and activities aimed at both monitoring a process and at eliminating causes of unsatisfactory performance at all stages of the quality loop in order to result in economic effectiveness.

quality evaluation systematic examination of the extent to which an entity is capable of fulfilling specified requirements related to overall quality. A *quality evaluation* may be used to determine supplier quality capability. In this case, depending on specific circumstances, the result of quality evaluation may be used for qualification, approval, registration, certification, or accreditation purposes. An additional qualifier may be used with the term quality evaluation depending on the scope (e.g., process, personnel, system) and timing (e.g., pre-contract) of the quality evaluation such as "pre-contract process quality evaluation". An overall supplier quality evaluation may also include an appraisal of financial and technical resources. In English, quality evaluation is sometimes called "quality assessment", "quality appraisal", or "quality survey" in specific circumstances. The process of measurement, rating, and assessment of quality attributes for an evaluation item to determine if the specified software quality characteristics are fulfilled.

quality improvement the process on the basis of which the measures performed are analyzed for defining actions to improve the general quality of the production process or of a specific product.

quality indicator a measure that shows a trend in the relative quality of software or its products.

quality in use the quality that is estimated or predicted for the end software product at each stage of development for each quality in use characteristic, based on knowledge of the internal and external quality.

quality measure any assessment of software quality. For example, low quality may be defined as delivered software which does not work at all, or repeatedly fails in operation. A project where users report more than 0.5 bugs or defects per function point per calendar year are of low quality.

quality metric a quantitative measure of the degree to which an item possesses a given quality attribute.

quality model the set of characteristics and the relationships between them which provide the basis for specifying quality requirements and evaluation quality.

quality of service (QoS) performance of a communication network from the point of view of the user of a service. QoS measures can be, e.g., signal-to-noise ratio, packet loss rate, call blocking rate, achievable throughput, or cell delay variation.

quality profile a set of defined features that are considered relevant factors for the quality of a process or product. The *quality profile* also includes for each feature a value of reference that may depend on the company goals and product market profile. The quality profile is considered the reference profile for all the development processes and can be part of the contract.

quality system organizational structure, procedures, processes, and resources needed to implement quality management. The *quality system* should be as comprehensive as needed to meet the quality objectives. The quality system of an organization is designed primarily to satisfy the internal managerial needs of the organization. It is broader than the requirements of a particular customer, who evaluates only the relevant part of the quality system.

quantization (**1**) the subsetting of data or a resource to enable or speed up processing. An example of the former is where a device has no more than an 8-bit color capability thus requiring a 24-bit image to be requantized to 8-bit color for processing. Subsetting large data sets can also speed up processing. An example of resource quantization is where the processing of a screenfull of data in an image-based algorithm can be made much more efficient by subdividing the screen, perhaps on a binary basis, and applying the algorithm to smaller sections of the data.

(**2**) converting a continuous quantity into series of discrete values. For example, continuous images can be quantized into discrete pixels, color spaces can be quantized into a set of discrete colors, or continuous time can be quantized into discrete steps.

quantization matrix a matrix of numbers in the JPEG compression scheme that specifies the amount by which each discrete cosine transform coefficient should be reduced. The numbers are based on the human contrast sensitivity to sinusoids and are such that coefficients corresponding to frequencies which people are not very sensitive to are reduced or eliminated. *See also* contrast sensitivity, discrete cosine transform, Joint Photographic Experts Group, human visual system, psychovisual redundancy.

quantizer characteristic a measure of quantization performance. The performance of a scalar quantizer Q is typically characterized by its mean-squared quantization error

$$E\left[(Q(x)-x)^2\right] = \int (Q(x)-x)^2 f_X(x)\,dx\ ,$$

where $f_X(x)$ represents the probability distribution of the input X. In the special case where X is uniformly distributed, and where Q is a uniform quantizer (that is, the interval between successive quantization levels is Δ, and $Q(x)$ is the rounding of x to the nearest level), then the mean-squared quantization error is $\frac{\Delta^2}{12}$.

quantum mottle *See* speckle.

quaternion hyper-complex number (complex number pair) that transforms the magnitude and direction of a vector. Used in the generation of three-dimensional fractals.

QUEL the query language developed for the INGRES relational database management system. QUEL is a data definition and data manipulation language based on the relational calculus.

query a request to retrieve data from a data base. A *query* to a deductive database DB is usually formulated as a goal; e.g., the goal G {brother(cain, B)} asks for all brothers of cain in the database DB.

query by example (QBE) a query language developed by IBM in which the user formulates a query by filling in an on-screen template of relations rather than explicitly specifying a structured query. The names of attributes and relations are displayed as part of the template.

query compiler one of the components of a database management system. The *query compiler* is used to parse and analyze a query that is entered by the user. The query compiler generates calls to the run-time processor, which is responsible for executing the request.

query evaluation plan *See* execution strategy.

query expansion the process of adding new terms to a query automatically, for example, in response to relevance feedback.

query graph an internal representation of a query generated by the query compiler.

querying the process of retrieving data from a database using a query language.

query language a language used to access information stored in a database. In addition, a *query language* normally supports data definition and data manipulation. Examples include SQL and OQL. Query languages are usually nonprocedural languages.

query optimization a facility provided by database management systems to select the most reasonably efficient method of responding to a query from possibly many different strategies.

query processing the process that the database management system must follow in order to generate the result of a query on a database from a query expressed in a high-level query language. The process involves scanning, parsing, and validating a query in a high-level language, optimizing the query, generating the code to execute the query, and returning the result of the query to the user.

query tree *See* operator tree.

queue (1) a data structure maintaining a first-in-first-out discipline of insertion and removal. The earliest added item is at the head, and the item added most recently is at the tail. Basic operations are insert or enqueue and delete or dequeue. *Queues* are useful in many situations, particularly in process and event scheduling. *See also* FIFO memory, stack.

(2) a waiting area for tasks that arrive asynchronously to a serially reusable resource. *See also* buffer.

(3) a data structure characterized by having values added at one end and removed from the other.

queueing analysis the study of the behavior of queueing systems using mathematical techniques applied to a statistical model of the system.

queueing center a device at which a job joins a queue of waiting jobs if it finds the device busy upon arrival.

queueing discipline a scheduling discipline used for the selection of a job for processing from a set of jobs enqueued at a resource.

queueing network a particular approach for modeling computer and communication systems in which the system is modeled as a network of queues. The model consists of service centers that represent system resources and customers that represent units of work that need to be processed by the system. *See also* closed queueing network, open queueing network.

queueing network model a mathematically based software analysis/design technique for estimating and predicting performance for computing systems. Queueing models allow software engineers to identify bottlenecks and determine components that are I/O or computation bound at the earliest stages of the software design process.

queueing system a generic model for a system in which waiting and serving entities and sources of requests are found.

queueing time the delay experienced by a job before it starts receiving service at a resource.

queues with priorities a queueing system in which the scheduling discipline for selecting when each task will be executed depends on some parameters of the tasks. *See also* first-in-first-out, last-in-first-out, round-robin, processor sharing.

quicksort an in-place sort algorithm that uses the divide and conquer paradigm. It picks an element from the array (the pivot), partitions the remaining elements into those greater than and less than this pivot, and recursively sorts the partitions. There are many variants of the basic scheme above: to select the pivot, to partition the array, to stop the recursion on small partitions, etc. *See also* sort.

quincunx five points, four at the corners of a square and one in the center.

quincunx lattice a horizontally and vertically repeated pattern of quincunxes, which is identical to a square lattice oriented at 45°.

quincunx sampling the downsampling of a 2D or 3D signal on a quincunx lattice by removing every even sample on every odd line and every odd sample on every even line. In the frequency domain, the repeat spectrum centers also form a quincunx lattice. Quincunx sampling structures have been used for TV image sampling on the basis that they limit resolution on diagonal frequencies but not on horizontal and vertical frequencies.

quorum-based voting algorithm a replica control protocol where transactions collect votes to read and write copies of data. They are permitted to read or write data if they can collect a quorum of votes.

quoting a character *See* escaping a character.

R

R*-rounding *See* round-to-nearest-odd.

race a state in which two or more processes are continuously competing for resources and neither progresses.

race condition a situation where multiple processes access and manipulate shared data with the outcome dependent on the relative timing of these processes.

RAD *See* rapid application development.

radial basis function network a common network model consisting of a linear combination of basis functions, each of which is a function of the difference between the input vector and a center vector.

radial intensity histogram a histogram of average intensities for a round object in circular bands centered at the center of the object, with radial distance as the running index. Such histograms are easily constructed and, suitably normalized, form the basis for scrutinizing round objects for defects and for measuring radius and radial distances of cylindrically symmetrical features.

radiance a measure of the amount of electromagnetic radiation leaving a point on the surface. More precisely, it is the rate at which light energy is emitted in a particular direction per unit of projected surface area. The projected surface area is the projection of the surface onto the plane perpendicular to the direction of radiation. It is found by multiplying the surface area by $\cos\theta_r$, where θ_r is the angle of the radiated light to the surface normal.

radiography an imaging modality that uses an X-ray source and collimator to create a projection image. The image intensity is proportional to the transmitted X-ray intensity.

radiosity an image rendering algorithm that allows diffuse and mutual illumination effects by evaluating the radiation of light from light sources and reradiation among surfaces. Radiosity calculations determine the steady state in the radiative transport of light around a closed volume. Essentially, the illumination leaving a patch is a proportion of the light reaching the patch from all the other visible patches in the closed volume. Patch surface normals are typically distributed everywhere and some patches are occluded or partly obscured from each other. The accumulation of these radiation-attenuating effects is summed up as the form-factor between each pair of patches. The main and most time-consuming part of the radiosity calculation is the calculation of these form factors.

radix the base number in a number system. Decimal (radix 10) and binary (radix 2) are two examples of number systems.

radix complement in a system that uses binary (base 2) data, negative numbers can be represented as the two's complement of the positive number. This is also called a true complement.

radix sort a multiple pass sort algorithm that distributes each item to a bucket according to part of the item's key beginning with the least significant part of the key. After each pass, items are collected from the buckets, keeping the items in order, then redistributed according to the next most significant part of the key. Also called bucket sort.

Radon transform the Radon transform of a function $f(x, y), r(d, \phi)$ is its line integral along a line inclined at angle ϕ from the y axis and at a distance d from the origin.

ragged matrix a matrix having irregular numbers of items in each row. *See also* uniform matrix, sparse matrix.

RAID *See* redundant array of inexpensive disks.

RAM *See* random access memory.

0-8493-2691-5/01/$0.00+$.50

RAM neuron a random-access memory with n inputs and a single output. The inputs define $2n$ addresses and presentation of a particular input vector allows the contents of the 1-bit register at that address to be read, or to be written into. Training consists of writing 1s or 0s into the 1-bit registers, as required for the various input vectors in the training set.

random access (**1**) term describing a type of memory in which the access time to any cell is uniform.

(**2**) a method for allowing multiple users to access a shared channel in which transmissions are not coordinated (or perhaps are partially coordinated) in time or frequency.

(**3**) access to a file that is not sequential. This is possible with hash files, sorted files (of fixed length records), and relative files.

random access machine model of computation whose memory consists of an unbounded sequence of registers, each of which is capable of holding an integer. In this model, arithmetic operations are allowed to compute the address of a memory register.

random access memory (RAM) direct-access read/write storage in which each addressable unit has a unique hardwired addressing mechanism. The time to access a randomly selected location is constant and not dependent on its position or on any previous accesses. The RAM has a set of k address lines ($m = 2^k$), n bidirectional data lines, and a set of additional lines to control the direction of the access (read or write), operation, and timing of the device.

RAM is commonly used for the main memory of a computer and is said to be static if power has to be constantly maintained in order to store data and dynamic if periodic absences of power do not cause a loss of data. RAM is usually volatile. *See also* static random access memory, dynamic random access memory, non-volatile random-access memory.

random coding coding technique in which codewords are chosen at random according to some distribution on the codeword symbols. Commonly a tool used in the development of information theoretic expressions.

random failure failures that result from a variety of degradation mechanisms in the hardware. Unlike failures arising from systematic failures, system failure rates arising from random hardware failures can be quantified with reasonable accuracy.

randomized algorithm algorithm that makes some random (or pseudo-random) choices.

randomized complexity the expected running time of the best possible randomized algorithm over the worst input.

randomized decision rule a hypothesis decision/classification rule which is not deterministic (that is, the measurement or observation does not uniquely determine the decision). Although typically not useful given continuous observations, a randomized rule can be necessary given discrete observations.

randomized rounding a technique that uses the probalistic method to convert a solution to a relaxed problem into an approximate solution to the original problem.

random logic a digital system constructed with logic gates and flip-flips and other basic logic components interconnected in a nonspecific manner. *See also* microprogramming.

random orbit attracting point determined by iteration of random starting points by an appropriate geometric procedure.

random process a mathematical procedure for generating random numbers to a specific rule called a process, $\mathbf{x}$, which is defined on continuous $\mathbf{x}(\mathbf{t}), \mathbf{t} \in \mathcal{R}^n$, or discrete $\mathbf{x}(\mathbf{k}), \mathbf{k} \in \mathcal{Z}^n$ space/time. The value of the process at each point in space or time is a random vector. *See also* random variable, random vector, correlation, autocovariance.

random replacement algorithm in a cache or a paging system, an algorithm which chooses the line or page in a random manner. A pseudo-random number generator may be used to make the selection, or other approximate method. The algorithm is not very commonly used though it was used in the translation buffers of the VAX11/780 and the Intel i860 RISC processor.

random sampling using a randomly selected sample of the data to help solve a problem on the whole data.

random testing the strategy of selecting test cases at random according to the probability with which they are expected to be encountered in operation, in order to ensure that the operational profile used in test and trial is a reasonable approximation to reality.

random variable a continuous or discrete valued variable that maps the set of all outcomes of an experiment into the real line (or complex plane). Because the outcomes of an experiment are inherently random, the final value of the variable cannot be predetermined.

random vector a vector (typically a column vector) of random variables. *See also* random variable, random process.

range (1) for numeric data types, the limit of valid values for an instance of the data type. Some languages permit the programmer to specify the range of valid values for numeric data types. *See also* precision.

(2) the possible results of a function or relation. For instance, the range of cosine is $[-1, +1]$. *See also* domain.

range filter an edge detection filter that finds edges by taking the difference between the maximum and minimum values in a local region of the image. The range filter also accepts a weight mask the size of the local image region that controls pixel values before they enter the minimum and maximum calculations. The weight mask allows edges in certain directions to be searched for.

range partitioning a partitioning technique where relations are partitioned according to the range of the partitioning attribute. A partitioning vector is chosen which defines the set of intervals or ranges. All records within a certain range are placed on the disk associated with that range.

range query a query that contains a selection criteria specifying values in a range.

range relation specified by a condition on a relation.

rank for a given match, this is the number of matches in a longest chain terminating with that match, inclusive.

rank-based selection a selection mechanism in evolutionary computation where survival of an individual is determined by the ranking of its associated fitness value among all other individuals. In other words, the actual magnitude of the fitness value does not affect the selection process as long as its ranking remains the same within the list. Examples of rank-based selection strategy include the $(\mu+\lambda)$ and (μ, λ) strategy in ES, and the tournament selection strategy in EP. On the other hand, the proportional selection mechanism in GA is not a rank-based selection strategy due to the dependence of the selection probabilities on the magnitudes of the fitness values. *See also* evolutionary programming, evolutionary strategy.

rank filter an image transform used in mathematical morphology. Assume that to every pixel p one associates a window $W(p)$ containing it. Let k be an integer > 1 which is less than or equal to the size of each window $W(p)$. The rank filter with rank k and windows $W(p)$ transforms an image I into a filtered image I' whose gray-level $I'(p)$ at pixel p is defined as the k-th least value among all initial gray-levels $I(q)$ for q in the window $W(p)$. In a dual version, the k-th greatest value is selected. When each $W(p)$ is the translate by p of a structuring element W of size n, three particular cases are noteworthy: 1. $k = 1$: the rank filter is the erosion by W. 2. $k = n$: the rank filter is the dilation by $\widetilde{W}$, the symmetric of W. 3. n is odd and $k = (n+1)/2$: the rank filter is a median filter. *See* dilation, erosion, median filter, structuring element.

rapid application development (RAD) a technique in which users and developers interact within a workshop environment. RAD aims to facilitate, in a relatively short period of time, the development of the software system by identification of a readily agreed set of requirements in a collaborative manner.

rapid prototyping *See* prototyping.

raster an array of scanlines, painted across a CRT screen, which taken together form a rectangular 2D image. Often the term *raster* is used to refer to the 2D array of pixel values stored digitally in a frame buffer.

raster coordinates an artifact of the method of CRT image reconstruction where pixels are addressed and illuminated in a top-to-bottom, left-to-right fashion. Hence, *raster coordinates* are the 2D coordinates of the current drawing position either in the image window or the hardware frame buffer.

raster image *See* bitmapped image.

rate-compatible punctured convolutional code (RCPC code) one of a family of punctured convolutional codes derived from one low-rate convolutional parent code by successively increasing the number of punctured symbols, given that the previously punctured symbols should still be punctured (rate-compatibility). These codes have applications in, for example, variable error protection systems and in hybrid automatic repeat request schemes using additional transmitted redundancy to be able to correctly decode a packet.

rate-distortion theory Claude Shannon's theory for source coding with respect to a fidelity criterion, developed during the late 1940s and the 1950s. Can be viewed as a generalization of Shannon's earlier theory (late 1940s) for channel coding and information transmission. The theory applies to the important methods for vector quantization and predicts the theoretically achievable optimum performance.

rate distortion theory a theory aimed at quantifying the optimum performance of source coding systems. Using information theory, for several source models and distortion measures, rate distortion theory provides the optimum distortion function and the optimum rate function. The distortion function is optimum in that the distortion for a given rate is the theoretical minimum value of distortion for encoding the source at the given or lower rate. The rate function is optimum in that the rate at a given distortion is the minimum possible rate for coding the source at the given or lower distortion.

rate-monotonic scheduling a scheduling technique that assigns priorities to periodic tasks according to the principle: the shorter the task period the higher its priority.

rate-monotonic system a preemptive priority system where task priorities are assigned so that the higher the execution frequency, the higher the priority.

rating the action of mapping the measured value to the appropriate rating level. Used to determine the rating level associated with the software for a specific quality characteristic.

rating level a range of values on a scale to allow software to be classified (rated) in accordance with the stated or implied needs. Appropriate rating levels may be associated with the different views of quality, i.e., users, managers, or developers.

rational function a function that is the ratio of two polynomials. Rational functions often arise in the solution of differential equations by Laplace transforms.

raw data numbers, characters, images, or other methods of recording, in a form which can be assessed by a human or (especially) input into a computer, stored, and processed there, or transmitted on some digital channel. Computers nearly always represent data in binary. Same as data.

Rayleigh curve a roughly bell-shaped curve, representing the buildup and decline of manpower, effort, or cost, followed by a long tail representing manpower, effort, or cost devoted to enhancement or maintenance. According to this curve, allocation and de-allocation of human resources is performed. Also represents error and code rates. The Rayleigh equation is $f(x) = Cte^{-Dt^2}$ where C and D are constants and t is time.

ray tracing a rendering paradigm that aims to produce realistic images (rather than real-time) given a 3D model. The color of a pixel is determined by calculating the path of a ray of light passing through a point in the 3D model corresponding to the pixel. The path is traced back to a light source.

RBD *See* reliability block diagram.

r-command a set of commands (rsh, rlogin, rcp, rdist, etc.) that relies on address-based authentication.

RCPC code *See* rate-compatible punctured convolutional code.

reachability graph (**1**) a directed graph in which each node is a state from a stochastic process or finite-state machine, and an edge from state i to state j indicates that it is possible for the state of the system to change from i to j in one step.

(**2**) a graph used in formal verification to prove that every designed state of a process (e.g., a communication protocol) can be reached.

reaction time the time needed to react to an event and to produce a result.

reactive system a system that does not work in isolation but performs its task by interacting with others.

read to access contents of a storage device or data medium.

read-after-write hazard *See* true data dependency.

read ahead on a magnetic disk, reading more data than is nominally required, in the hope that the extra data will also be useful.

read authorization permission to read data from a relation.

read committed a transaction isolation level that permits reading of all committed changes. A read committed transaction prevents dirty reads, but not non-repeatable reads and phantom reads.

read instruction an assembly language instruction that reads data from memory or the input/output system.

read lock a database synchronization mechanism. If a transaction T holds a read lock on a data item X, then no other transaction can obtain a write lock on x.

read/modify/write an uninterruptible memory transaction in which information is obtained, modified, and replaced under the assurance that no other process could have accessed that information during the transaction. This type of transaction is important for efficient implementations of locking protocols.

read-modify-write cycle a type of memory device access that allows the contents at a single address to be read, modified, and written back without other accesses taking place between the read and the write.

read-mostly memory memory primarily designed for read operations, but whose contents can also be changed through procedures more complex and typically slower than the read operations. EPROM, EEPROM, and flash memory are examples.

read-once-write-all protocol the replica control protocol which maps each logical read operation to a read on one of the physical copies and maps a logical write operation to a write on all of the physical copies.

read-only memory (ROM) semiconductor memory unit that performs only the read operation; it does not have the write capability. The contents of each memory location are fixed during the hardware production of the device and cannot be altered. A ROM has a set of k input address lines (that determine the number of addressable positions 2^k) and a set of n output data lines (that determine the width in bits of the information stored in each position). An integrated circuit ROM may also have one or more enable lines for interconnecting several circuits and make a ROM with larger capacity. Plain ROM does not allow erasure, but programmable ROM (PROM) does. Static ROM does not require a clock for proper operation, whereas dynamic ROM does. *See also* random access memory, programmable read-only memory.

read past locking where a read omits locked data.

read permission grants or denies the associated class of users the right to read a file.

read phase the first portion of a transaction during which the executing process obtains information that will determine the outcome of the transaction. Any transaction can be structured so that all of the input information is obtained at the outset, all the computation is then performed, and finally all results are stored (pending functionality checks based on the locking protocols in use).

read/write head conducting coil that forms an electromagnet, used to record on and later retrieve data from a magnetic circular platter constructed of metal, plastic, or glass coated with a magnetizable material. During the read or write operation, the head is stationary while the platter rotates beneath it. The write mechanism is based on the magnetic field produced by electricity flowing through the coil. The read mechanism is based on the electric current in the coil produced by a magnetic field moving relative to it.

Less common are magneto-resistive heads, which employ non-inductive methods for reading. A system that uses such a head requires an additional (conventional) head for the writing. *See also* disk head, magnetic recording code.

ready a state of a task or process in which it is waiting to run on the processor.

real a value which nominally represents a real number. This is a misnomer, in that at best any given implementation on any specific machine can only represent an approximation to a real number, a floating-point number.

real address the actual address that refers to a location of main memory, as opposed to a virtual address which must first be translated. Also called a physical address. *See also* memory mapping, virtual memory.

real part in a complex number, the component that consists of a real number, that is, a number that can be found on the number line.

real time *See* real time.

real-time clock a hardware counter which records the passage of time.

real-time computing support for environments in which response time to an event must occur within a predetermined amount of time. Real-time systems may be categorized into hard, firm, and soft real-time.

real-time constraint *See* timing constraint.

real-time database a database that supports real-time applications. Queries submitted to a *real-time database* usually have an associated time constraint.

real-time logic logic used for the specification of real-time systems; generally this type of logic is an extension of a temporal logic with the ability to express timing constraints.

real-time operating system (RTOS) any operating system where interrupts are guaranteed to be handled within a certain specified maximum time, thereby making it suitable for control of hardware in embedded systems and other time-critical applications. RTOS is not a specific product but a class of operating systems.

real-time software software which must return an output within a certain real-time interval, in order to be able to affect some social or physical process. In general, real-time software presents a behavior that depends on the time and this behavior is defined in terms of temporal constraints that if unsatisfied may lead the system to critical and sometimes disastrous conditions. Examples are airline booking systems, chemical plant control, and avionics flight-control software.

real-time system a system whose correctness depends on the satisfaction of timing constraints. Typical real-time systems are process control systems. A class of real-time systems called hard real-time systems must have a fixed upper bound on their response times in order to prevent failure due to lack of response to a critical external process.

reantrant term describing a program that concurrently uses exactly the same executable code in memory for more than one invocation of the program (each with its own data), rather than separate copies of a program for each invocation. The read and write operations must be timed so that the correct results are always available and the results produced by an invocation are not overwritten by another one.

recall a feature of a set of documents. Given a document set of N documents comprising N_r relevant documents and N_n irrelevant documents, let R be the set of returned documents with R_r being the relevant returned documents, R_n being the irrelevant documents. The recall is defined to be $|Rr|/|Nr|$.

reception action about the receiving of a message.

receptive field the part of the visual field of a neuron within which a stimulus can influence the response (i.e., the firing rate) of the neuron.

receptive field function describes the response of a neuron to a small spot of light as a function of position.

recode changes to implementation characteristics. Language translation and control-flow restructuring are source code level changes. Other possible changes include conforming to coding standards, improving source code readability, renaming programming items, etc.

recoding of multiplier *See* Booth's algorithm.

recognize in the context of an automaton, the automaton is said to recognize a string if the last symbol in the string causes the automaton to enter an accepting state.

recombination an operation in evolutionary computational algorithms which generates a new individual from two or more original individuals in the population. There are several variants of this operation which can be summarized by the following equation:

$$x_i = \lambda x_i^{P_1} + (1 - \eta) x_i^{P_2} ,$$

where x_i denotes the ith component of the new individuals, $x_i^{P_1}$ and $x_i^{P_2}$ denote the corresponding components of the two parents, and η is a recombination factor which is defined differently for different types of recombination strategies. In discrete recombination, η is a binary random variable which is sampled anew for every i. In intermediate recombination, η is assigned the value of 0.5, which in effect performs an averaging operation on the two parents. In generalized intermediate recombination, η is a uniform random variable between 0 and 1. These operations can be generalized to more than two parents by replacing $x_i^{P_2}$ with the corresponding component of another randomly chosen individual in the population for every i. In genetic algorithms, a special form of recombination known as crossover is implemented which exchanges sub-strings between pairs of randomly chosen chromosomes. *See also* crossover.

reconstruction the process of forming a 3D image from a set of 2D projection images. Also applies to the formation of a 2D image from 1D projections. *See* image reconstruction, tomography and computed tomography.

reconstruction from marker in a binary image, this is the operation extracting all connected components having a non-empty intersection with a marker. This operation can be generalized to gray-level images by a morphological operator applying such a reconstruction on the gray-level slices of the image.

record (**1**) a collection of related fields.

(**2**) unit of data, corresponding to a block, sector, etc. on a magnetic disk, magnetic tape, or other similar I/O medium.

(**3**) in a database, a synonym for a row of a relational table.

(**4**) in many languages, a synonym for structure or class.

recording code a line code optimized for recording systems. *See* line code.

recording density number of bits stored per linear inch on a disk track. In general, the same number of bits are stored on each track, so that the density increases as one moves from the outermost to the innermost track.

record type a definition of the record format. Usually comprises a set of field names and their associated type.

recover to retrieve or reconstruct a database from a backup.

recoverability the capability of the software product to re-establish a specified level of performance and recover the data directly affected in the case of a failure. Following a failure, a software product will sometimes be down for a certain period of time, the length of which is assessed by its recoverability. *See* availability.

recovery The resumption of all or part of the required service by an item following a failure. Recovery restores the state of a process to an earlier configuration after it has been determined that the system has entered a state which does not correspond to functional behavior. For overall functional behavior, the states of all processes should be restored in a manner consistent with each other, and with the conditions within communication links or message channels.

recovery block scheme approach to fault tolerant software design in which diverse modules are called in sequence until the required output or transition in system state has been achieved successfully as judged by an acceptance test. The later modules in the sequence are only invoked if all earlier modules fail to satisfy the acceptance test.

rectangle detection the detection of rectangular shapes, often by searching for corner signals, or from straight edges present in an image. *Rectangle detection* is important when locating machined parts in images, e.g., prior to robot assembly tasks. *See also* square detection.

rectangular decomposition a space decomposition method that partitions the underlying space by recursively halving it across the various dimensions instead of permitting the partitioning lines to vary.

rectangular matrix an $n \times m$ matrix, or one whose size may not be the same in both dimensions. *See also* square matrix.

rectangular window in image processing, a rectangular area centered at a pixel under consideration. This area is known as a window, a mask, or a template. A square window of size 3×3 is used most often. *See also* neighborhood operation.

rectilinear distance, paths, lines, etc. which are always parallel to axes at right angles. For example, a path along the streets of Salt Lake City or the moves of a rook in chess. *See also* Euclidean distance, l_m distance.

rectilinear Steiner tree a minimum-length rectilinear tree connecting a set of points, called terminals, in the plane. This tree may include points other than the terminals, which are called Steiner points. *See also* Steiner tree, Euclidean Steiner tree.

recurrence relation the specification of a sequence of values in terms of earlier values in the sequence and base values. *See also* master theorem.

recurrent coding old name for convolutional coding.

recurrent network a neural network that contains at least one feedback loop.

recursion an algorithmic technique in which a function, in order to accomplish a task, calls itself with some part of the task. *See also* iteration, divide-and-conquer, tail recursion, collective recursion, divide and marriage before conquest, recursive procedure.

recursive (**1**) a type of data structure which is partially composed of other instances of the data structure. For instance, a tree is composed of smaller trees (subtrees) and leaf nodes, or a list may have other lists as elements.

(**2**) a type of algorithm which may call itself. *See also* recursion.

recursive closure a type of operation applied to a recursive relationship between tuples of the same type.

recursive decomposition an algorithm where space is divided into successively smaller pieces until a threshold is found. These algorithms can be used to draw curves by approximating them by a chain of line segments. This can also be used to render surfaces by subdivision algorithms, such as the hierarchical B-spline refinement algorithm.

recursive-descent a top-down parsing algorithm that translates context-free grammar rules into a set of mutually recursive procedures, with each procedure corresponding to a nonterminal. Recursive-descent parsing is usually the method of choice when writing a parser by hand.

recursive descent a parsing technique in which the state of the automaton is reflected in the control stack. There is one function for each nonterminal of the grammar. Well suited for parsing LL(k) grammars, particularly LL(1).

recursive-descent parser a parser that uses the recursive-descent algorithm. *See* recursive-descent.

recursive doubling the same as pointer jumping.

recursive equation a difference equation that is of the form

$$y(k) = \sum_{i=0}^{i=m} a_i x(k-i) - \sum_{i=1}^{i=n} b_i y(k-i) ,$$

where a_i and b_i are some proper real constants. When all the $b_i = 0$, it is called a nonrecursive equation.

recursive filter a digital filter that is recursively implemented. That is, the present output sample is a linear combination of the present and past input samples as well as the previously determined outputs. Traditionally, the term recursive filter is closely related to infinite impulse filter. In a nonrecursive filter, the present output sample is only a linear combination of the present and past input samples.

recursive function *See* recursive procedure.

recursive language a language accepted by a Turing machine that halts for all inputs.

recursively enumerable language a language which is defined by a Type 0 grammar of the Chomsky hierarchy. It is characterized by a set of productions of the form $\omega \rightarrow \mu$, where ω is a nonempty sequence of symbols of the vocabulary of the language and μ is a sequence of zero or more symbols of the vocabulary of the language.

recursive method method that estimates local displacements iteratively based on previous estimates. Iterations are performed at all levels, as in every pixel, each block of pixels, along scanning line, and from line to line or from frame to frame.

recursive procedure a procedure that can be called by itself or by another program that it has called; effectively, a single process can have several executions of the same program alive at the same time. Recursion provides one means of defining functions.

The recursive definition of the factorial function is the classic example:

```
if n = 0
    factorial(n) = 1
else
    factorial(n) = n * factorial (n-1)
```

recursive relationship a relationship in which the same entity type participates more than once.

recursive self-generating neural network (RSGNN) a recursive version of self-generating neural network (SGNN) that can discover recursive relations in the training data. It can be used in applications such as natural language learning/understanding, continuous spoken language understanding, and DNA clustering/classification, etc. *See also* self-generating neural network.

recursive set a set such that the owner and member record types are the same.

recursive subroutine a subroutine that can be called, either directly or indirectly, from within its own execution context.

redesign changes to design characteristics. Possible changes include restructuring a design architecture, altering a system's data model as incorporated in data structures or in a database, improvements to an algorithm, etc.

redex a subexpression in the form that is ready to be evaluated by a step of reduction.

redirecting standard error a technique whereby the output from standard error is redirected to a file or device.

redirection an operation of changing the flow of data from an input device or to an output device.

redo to re-apply a change made to a database from an after image held in a redo or after image log.

redocumentation the process of analyzing the system to produce new support documentation in various forms including user manuals and reformatting the system's source code listings. *Redocumentation* is the simplest and oldest form of reverse engineering. It is frequently performed for the requisition of a software system for its further maintenance. Thus, the inclusion of new views and prospects are typically planned in the new version of documentation.

redo log a log of after images of data changes used to re-apply the changes.

redo record a group of after images comprising a single, atomic change to the database.

reduced basis a basis for a lattice that is nearly orthogonal.

reduced instruction set computer (RISC) relatively simple control unit design with a reduced menu of instructions (selected to be simple), data and instructions formats, and addressing modes, and with a uniform streamlined handling of pipelines.

One of the particular features of a RISC processor is the restriction that all memory accesses should be by load and store instructions only (the so called load/store architecture). All operations in a RISC are register-to-register, meaning that both the sources and destinations of all operations are CPU registers. All this tends to significantly reduce CPU to memory data traffic, thus improving performance. In addition, RISCs usually have the following properties: most instructions execute within a single cycle, all instructions have the same standard size (32 bits), the control unit is hardwired (to increase speed of operations), and there is a CPU register file of considerable size (32 registers in most systems, with the exception of SPARC with 136 and AMD 29000 with 192 registers).

Historically, the earliest computers explicitly designed by these rules were designs by Seymour Cray at CDC in the 1960s. The earliest development of the RISC philosophy of design was given by John Cocke in the late 1970s at IBM. However, the term RISC was first coined by Patterson et al. at the University of California at Berkeley to describe a computer with an instruction set designed for maximum execution speed on a particular class of computer programs. Patterson and his team of researchers developed the first single-chip RISC processor.

Compare with complex instruction set computer.

reduce-reduce conflict in bottom-up parsers, a property of a state in which a parser has a choice of two productions which can be used to reduce the parsing stack, and both are legal for the amount of lookahead allowed. *Reduce-reduce conflicts* have no natural disambiguating rule.

reduction a computable transformation of one problem into another.

reduction in strength an optimization technique where the compiler is persuaded to replace arithmetic operations with equivalent, but faster operations.

redundancy (**1**) the use of parallel or series components in a system to reduce the possibility of failure. Similarly, referring to an increase in the number of components which can interchangeably perform the same function in a system. Sometimes it is referred to as hardware redundancy in the literature to differentiate from so-called analytical redundancy in the field of FDI (fault detection and isolation/identification). Redundancy can increase the system reliability.

(**2**) the duplication of data in a database.

(**3**) in robotics, the number n degrees of mobility of the mechanical structure, the number m of operational space variables, and the number r of the operational space variables necessary to specify a given task. Consider the differential kinematics mapping $v = J(q)\dot{q}$ in which v is $(r \times 1)$ vector of end-effector velocity of concern for the specific tasks and J is $r \times n$ Jacobian matrix. If $r < n$ the manipulator is kinematically redundant and has $(n - r)$ redundant degrees of mobility. Manipulator can be redundant with respect to a task and nonredundant with respect to another.

redundancy-free channel coding refers to methods for channel robust source coding where no "explicit" error protection is introduced. Instead, knowledge of the source and source code structure is utilized to counteract transmission errors (for example by means of an efficient index assignment).

redundancy statistics model refers to statistical similarities such as correlation and predictability of data. Statistical redundancy can be removed without destroying any information.

redundant array of inexpensive disks (RAID) standardized scheme for multiple-disk database systems viewed by the operating system as a single logical drive. Data is distributed across the physical drives allowing simultaneous access to data from multiple drives, thereby reducing the gap between processor speeds and relatively slow electromechanical disks. Redundant disk capacity can also be used to store additional information to guarantee data recoverability in case of disk failure (such as parity or data duplication). The RAID scheme consists of six levels (0 through 5), $RAID_0$ being the only one that does not include redundancy.

redundant number system the system in which the numerical value could be represented by more than one bit string.

reengineering the examination and modification of a system to reconstitute it in a new form and the subsequent re-implementation of the new form. Also known as renovation and reclamation. This process encompasses a combination of subprocesses such as reverse engineering, restructuring, redocumentation, forward engineering and retargeting. *Reengineering* also emphases the importance of a greater return on investment than could be achieved through a new development effort.

reentrancy the property of a software module such that it can be invoked while still executing one or more previous invocations.

reentrant a program that uses concurrently exactly the same executable code in memory for more than one invocation of the program (each with its own data), rather than separate copies of a program for each invocation. The read and write operations must be timed so that the correct results are always available and the results produced by an invocation are not overwritten by another one.

reentrant code code which does not maintain any local state in such a way that a concurrent execution would be affected. *Contrast with* serially reusable, recursive. *See also* reentrant.

reference in object-oriented languages *reference* identifies an alias for an object. In many languages and in common usage, reference is equivalent to pointer. However, there are languages in which a reference is implemented as a more complex construct than a simple pointer. In languages that support pointers, it is sometimes possible to perform pointer arithmetic and compute a new address based on an existing pointer address. Language designers, in order to distinguish a type of pointer on which address arithmetic cannot be performed (for example, as in the Java language), sometimes use the term reference to designate such a pointer.

reference count a means of managing heap storage in which each object contains a count of the number of references to it. Creating a new reference increases the *reference count;* deleting a reference decreases the reference count. When the reference count is zero, it means there is no outstanding reference to the object and the storage it occupies may be released for reuse. Often more efficient than garbage collection, but requires careful maintenance of the reference count. Requires special case analysis to determine reachability of objects when a cyclic graph of references can exist, since all the objects have references but the entire subgraph is unreachable from anywhere else in the program.

reference library the books, journals, and other material a company has for the software development staff. There is a direct correlation between quality/productivity and the size of the reference library.

reference monitor a system component that enforces system security policy by checking all accesses initiated by subjects to objects and permitting only those that are consistent with the policy. To be effective, the reference monitor must be tamperproof, enforced on every access and must not make mistakes.

reference semantics a semantic model in which values are always acted on by reference, so that changes always appear in the only instance of the value. *Compare with* copy semantics.

reference white in a color matching process, a white with known characteristics used as a reference. According to the trichromatic theory, it is possible to match an arbitrary color by applying appropriate amounts of three primary colors.

referential integrity states that any value of a foreign key in a relation occurs as a value of a primary key in a related relation or else must have the NULL value.

referential transparency a mathematical principle that states that if the value of an expression can be determined solely from the values of its subexpressions, and if any subexpression is replaced by an arbitrary expression with the same value, then the value of the expression remains unchanged. Many optimizations performed by compilers are based upon assumptions of referential transparency; many optimizations not performed by optimizing compilers are based upon incomplete knowledge of referential transparency.

refinement-specification process by which a specification is refined by adding details and verifying the consistency of the previous specification with the new one. This approach is typical of denotational/descriptive languages and models.

reflectance a measure of the ability of a surface to reflect electromagnetic radiation (e.g., light). It is equal to the ratio of the reflected flux to the incident flux. It is often important, especially in industrial inspection, to be able to determine the reflectance under varying illumination. *See also* illumination, industrial inspection.

reflexive association when a class instance is related with other instances of the same class, the resulting association is referred to as *reflexive association.* An example is a node of List that contains at least a reference to another node of the list.

reflexive rule a relationship in a database system. Given sets of attributes A and B, then if $B \subset A$, then $A \rightarrow B$.

reformatting the functional equivalent transformation of source code which changes only the structure to improve readability. Examples are pretty-printers and tools that replace GOTO loops with equivalent loops. The latter case in some cases is not only a simple process of reformatting but a real process of transformation.

refraction the phenomenon of a beam of light bending as the light's velocity changes. This occurs when the refractive index of the material through which the light is passing changes. Let **i** be the normalized incident ray vector (pointing towards the surface), which has unit surface normal **n**. If **t** is the transmitted (refracted) vector inside a transparent medium, then:

$$\mathbf{t} = \mu \mathbf{i} - \mathbf{n}(\mu(\mathbf{i} \cdot \mathbf{n}) + \left(1 - \mu^2\left(1 - \left(\mathbf{i} \cdot \mathbf{n}\right)^2\right)\right)^{\frac{1}{2}}\right) ,$$

where μ is the ratio $n_{\text{outside}}/n_{\text{inside}}$ of the refractive indices of the inside and outside media. *See* Snell's law.

refresh refers to the requirement that dynamic RAM chips must have their contents periodically refreshed or restored. Without a periodic refresh, the chip loses its contents. Typical refresh times are in the 5 to 10 millisecond range. *See also* memory refresh.

refresh cycle (**1**) a periodically repeated procedure that reads and then writes back the contents of a dynamic memory device. Without this procedure, the contents of dynamic memories will eventually vanish.

(**2**) the period of time taken to "refresh" a portion of a dynamic RAM chip's memory. *See also* refresh.

refresh period the time between the beginnings of two consecutive refresh cycles for dynamic random access memory devices.

region a portion of memory allocated dynamically to an application program.

region growing the grouping of pixels or small regions in an image into larger regions. Region growing is one approach to image segmentation. *See also* dilation, erosion, image segmentation, mathematical morphology.

region of interest (ROI) a restricted set of image pixels upon which image processing operations are performed. Such a set of pixels might be those representing an object that is to be analyzed or inspected.

region query a form of spatial query which requests objects that lie within a designated region.

register a circuit formed from identical flip-flops or latches and capable of storing several bits of data. Registers are usually limited in number, so that they can be designated by a small number of bits in each instruction (3 to 6 bits are typical). Most registers are disjoint from main memory, although this is not universally true (the DEC PDP-10, for example). Registers limited to use in address computations are designated index registers, those which can be involved in all arithmetic operations are designated accumulators, and most architectures that have registers which can serve any purpose are designated general registers. These distinctions become important for writing compilers.

register alias table *See* virtual register.

register allocation a method used by compilers for machines with limited register sets, which attempts to allocate intermediate computations and/or user-defined variables to the machine registers such that the number of times data has to be moved is reduced. Computationally, this problem is very difficult, and is often done by using a graph-coloring algorithm. Most commonly, a complete or optimal solution cannot be realized and a heuristic approximation is used.

register direct addressing (**1**) an instruction addressing method in which the memory address of the data to be accessed or stored is found in a general purpose register.

(**2**) a memory addressing scheme similar to direct addressing except that the operand is a CPU register and not an address.

register file a collection of CPU registers addressable by number.

register indirect addressing an instruction addressing method in which the register field contains a pointer to a memory location that contains the memory address of the data to be accessed or stored.

register renaming dynamically allocating a location in a special register file for an instance of a destination register appearing in an instruction prior to its execution. Used to remove antidependencies and output dependencies. *See* reorder buffer.

register set a designation of a storage class in a computer represented by one or more equivalent registers. The implementation of one register set may not necessarily be disjoint from a different register set; for example, the integer register set and the floating-point register set may be the same physical registers in some machines and disjoint physical registers in other machines. *Register sets* form the basis of one of the problems in most compilers, that of allocating intermediate computations to storage classes which are more efficiently accessed by the machine instructions.

register transfer notation a mathematical notation to show the movement of data from one register to another register by using a backward arrow. Notation used to describe elementary operations that take place during the execution of a machine instruction.

register window in the SPARC architecture, a set or window of registers selected out of a larger group.

registration the process of aligning multiple images obtained from different modalities, at different timepoints, or with different image acquisition parameters.

regression the methods which use backward prediction error as input to produce an estimation of a desired signal. Quantitatively, the regression of y on X, denoted by $r(y)$, is defined as the first conditional moment, i.e.,

$$r(y) = E(X|y) .$$

regression analysis a mathematical method where an empirical function is derived from a set of experimental data.

regression testing the test that is performed on a system when this has been produced in a new version/release. The *regression testing* has to verify: (i) all of the already tested functionalists that were correct and available in the previous version are also present in the new version without any problems, (ii) that the problems solved in the new version have been effectively solved without causing second level problems. The regression testing is very important in order to verify if the new release has solved the problems without adding new problems at the same time.

regular expression a recursive notation for a linguistic construct that can be analyzed by a finite state automaton. Equivalent to a Type 3 grammar in the Chomsky Hierarchy.

regularization a procedure to add a constraint term in the optimization process that has a stabilizing effect on the solution.

regularization-based motion detection a gradient motion detection approach which considers the optical flow estimation as an ill-posed problem according to the Hadamard theory.

regular language a language that can be described by some right-linear/regular grammar (or equivalently, by some regular expression).

regulator a controller that is designed to maintain the state of the controlled variable at a constant value, despite fluctuations of the load.

reinforcement learning learning on the basis of a signal that tells the learning system whether its actions in response to an input (or series of inputs) are good or bad. The signal is usually a scalar, indicating how good or bad the actions are, but may be binary.

rejection criteria criteria such as poor surface texture, existence of scratch marks, or out-of-tolerance distance measures, which constitute reasonable grounds for rejecting a product from a product line.

relation (1) a computation which takes some inputs and yields an output. Any particular input may yield different outputs at different times.

(2) formally, a mapping from each element in the domain to one or more elements in the range. *See also* function.

relational algebra algebra governing the relational operators.

relational calculus a declarative formal language for relational databases.

relational database a database in which a collection of tables contains database entries along with sets of attributes, such that the data can be accessed along complex dimensions using the structured query language (SQL). Such databases make it convenient to look up information based on specifications derived from a semantic frame.

relational operator an operator that returns a Boolean result based on a test between two values. Classically, arithmetic comparison operators such as less than, greater than, and equal are relational operators, but in languages that support them, operators such as set membership ("in"), subset relationships between sets ("$\subseteq$"), string relationships ("is an initial substring of", "is a substring of"), and pattern matching are all relational operators.

relational schema *See* relational type.

relational type comprises a relation name R and a list of attributes $A_1, \ldots, A_n$.

relationship connections between entities. Used in entity relationship and object modeling.

relationship instance an element of a relationship. Given the sets $E_1, E_2, \ldots E_N$, a relationship instance of the relation between $E_1, E_2, \ldots E_N$ is a member of the Cartesian product $E_1 \times E_2 \times \cdots \times E_N$.

relationship type a definition of associations between entities. A relationship between the entities $E_1, E_2, \ldots E_N$ is defined as the subset of the Cartesian product $E_1 \times E_2 \times \cdots \times E_N$.

relative-address coding in facsimile coding, represents the transition between levels on a particular scan line relative to transitions on the preceding scan line. A relative-address coding system has a pass mode codeword for indicating where a pair of transitions on the previous line does not have corresponding transitions on the current line, and a runlength coding mode applied when there is no nearby suitable transition on the previous line. CCITT Group IV facsimile uses a form of *relative-address coding*.

relative addressing an addressing mechanism for machine instructions in which the address of the target location is given by the contents of a specific register and an offset held as a constant in the instruction added together. *See also* PC-relative addressing, index register, base address.

relative error the ratio of the absolute error to the correct result, provided the latter is not zero; otherwise it is undefined.

relative file the address of a record is an integer value which reflects the position of the record relative to the first record.

relative form for time in temporal logics, when time is referenced to a general system clock and the value is expressed in time units. *See* absolute form for time.

relative input device an input device that reports its distance and direction of movement each time it is moved, but cannot report its absolute position. A mouse operates in this way. *See also* absolute input device.

relative path a pathname that starts with something other than a slash character and uses the current directory as a reference point. A given *relative path* may refer to different files, depending on the current directory.

relative pointer a pointer which is not itself a pointer to a memory location, but which when combined with a specific base value becomes a reference to a memory location. A relative pointer may be used to maintain location independence or because, given the nature of the information, a relative pointer may be encoded using fewer bits than a full pointer.

relaxation an enlargement of the feasible region of an optimization problem. Typically, the relaxation is considerably easier to solve than the original optimization problem.

relaxed two-phase locking two-phase locking, where locks can be released before the transaction terminates.

relay a device that opens or closes a contact when energized. Relays are most commonly used in power systems where their function is to detect defective lines or apparatus or other abnormal or dangerous occurrences and to initiate appropriate control action. When the voltage or current in a relay exceeds the specified "pickup" value, the relay contact changes its position and causes an action in the circuit breaker. A decision is made based on the information from the measuring instruments and relayed to the trip coil of the breaker, hence the name "relay". Other relays are used as switches to turn equipment on or off.

release a configuration management activity whereby a particular version of an item is made available for a specific purpose (e.g., released for test).

relevant failure a failure that should be included in interpreting test or operational results or in calculating the value of a reliability performance measure.

relevant incident an incident that should be included in interpreting test or operational results or in calculating the value of a dependability measure.

reliability the probability that a component or system will function without failure over a specified time period, under stated conditions. Limitations in reliability are due to faults in requirements, design, and implementation. Failures due to these faults depend on the way the software product is used and the program options selected rather than on elapsed time. The ability of a system to deliver its required service under given conditions for a given time. The reliability of a system is generally measured using operating time as the time domain, i.e., as the probability of non-failure in a given period of time during which the item is actually operating. For calculating the probability of failure over real time, operating time or execution time must be transformed into real time taking into account the frequency of use of the item, simultaneous operation of multiple copies, etc.

reliability block diagram (RBD) a parallel structure representing alternative paths for which a system is considered functional.

reliability model model that allows one to project consistently in accordance with specifications.

reliability performance the probability that an item can perform a required function under given conditions for a given time interval, (t_1, t_2).

relocatable (**1**) pertaining to a program that can be loaded into memory at any address, rather than at a specific address. *Compare with* absolute addressing.

(**2**) object code form that can be combined (linked) with other components such that any references within it may be modified depending on its relationship to the components with which it is linked. Note that a relocatable object file may become a component of a relocatable program.

relocatable pointer a pointer that does not reference a memory location but which has associated information, implicit or explicit, describing how to convert it to a pointer to a memory location. Often used as a means of storing complex data structures on disk or otherwise transferring them. *See also* pickling.

relocation the operation performed by a linker or loader which modifies relocatable values to bind them to actual memory addresses. Note that a linker may compute relative relocations which are not bound until the loader loads the program.

relocation register register used to facilitate the placement in varying locations of data and instructions. Actual addresses are calculated by adding program-given addresses to the contents of one or more relocation registers.

remainder an operation which returns as its result the remainder after integer division. Some languages support two operations, **rem** and **mod**, one of which returns an absolute value of the remainder and the other of which returns the signed value. Other languages will implement the remainder (modulo) operation as one or the other of these choices. *See also* modulo.

remodularization changing a module's structure in light of coupling analysis, in order to redefine the boundaries between modules and function of modules.

remote file system a file system accessible to one or more systems, other than its physical host, over a network; a file system residing on one host system that is attached to the root file system of a separate computer.

remote job entry the submission of jobs through a remote device.

remote procedure call (RPC) a call to a procedure executed on a remote machine.

remote sensing the use of radar, satellite imagery, or radiometry to gather data about a distant object. Usually, the term refers to the use of microwaves or millimeter waves to map features or characteristics of planetary surfaces, especially the Earth's. Applications include military, meteorological, botanical, and environmental investigations.

remote terminal unit (RTU) hardware that gathers system-wide real-time data from various locations within substations and generating plants for telemetry to the energy management system.

removable disk disk that can be removed from disk drive and replaced, in contrast with a non-removable disk, which is permanently mounted. *See also* exchangeable disk.

rename register *See* virtual register.

render to create an image from a description of a scene, its objects, light sources, and the viewer.

rendering the preparation of the representation of an image to include illumination, shading, depth cueing, coloring, texture, and reflection. Rendering often takes into account scene lighting and object texturing. Common rendering techniques include Phong and Gouraud shading; more complex rendering models such as raytracing and radiosity emphasize realistic physics models for calculating light interactions and texture interactions with objects.

rendezvous a synchronization mechanism in the Ada language in which both the sender and a receiver are blocked until a message is delivered.

reorder buffer a set of storage locations provided for register renaming for holding results of instructions. These results may not be generated in program order. At some stage, the results will be returned to the true destination registers.

repairability a property of a software system that allows for the correction of its defects with a limited amount of work.

repair time that part of active corrective maintenance time during which repair actions are performed on an item.

repeatable read a transaction isolation level that prevents dirty reads and non-repeatable reads but not phantom reads.

repelling point in an iterated function, a point that escapes.

repetition coding the simplest form of error control coding. The information symbol to be transmitted is merely repeated an uneven number of times. A decision regarding the true value of the symbol transmitted is then simply made by deciding which symbol occurred the greatest number of times.

replaceability the capability of the software product to be used in place of another specified software product for the same purpose in the same environment. For example, the replaceability of a new version of a software product is important to the user when upgrading. Replaceability may include attributes of both installability and adaptability.

replacement policy the paging policy that determines which page should be removed from primary memory if all page frames are full.

replay *See* reversible execution.

replication the storing of data at more than one site.

repository a place in which documents, source code, etc. are stored. For a tool, it represents the core, which is typically a database management system, where all generated documents are stored.

representation error the error that arises from the limitations of finitude imposed by the machine. For example, the decimal 1/3 cannot be represented exactly as a fixed-point number in decimal, nor can the decimal 1/10 be represented exactly as a fixed-point number in binary; both would require an infinite number of digits.

representation problem building models to understand software systems is an important part of reverse engineering. Formal and explicit model building is important because it focuses attention on modeling as an aid to understanding and results in artifacts that may be useful to others. The representation used to build models has great influence over the success and value of the result. Choosing the proper representation during reverse engineering is the *representation problem.*

representative level one of the discrete output values of a quantizer used to represent all input values in a range about the representative level. *See also* decision level.

reprocessing recovery of a database by manually or automatically reapplying all transactions since a backup to a database restored from the backup.

reproducible failure *See* systematic failure.

reproduction in evolutionary computation, the process of creating a new individual from one or more of the original individuals in the population.

required function a function of an item which is necessary to provide a given service.

required product quality the quality represented by the essential requirements stated in the Software Requirements Specification, which can be evaluated through validation.

required service the totality of functions required to be performed by an item on behalf of its users.

requirement an essential condition or feature that a system has to satisfy.

requirements acquisition the collection of requirements is the process in which all the features that have to be included in the systems are collected. Typically the collection of requirements is a document in natural language which may include technical details in some specific aspects. They can be incomplete and inconsistent. In some cases, in order to be capable of verifying their completeness and consistency they are formalized with formal languages.

requirements analysis a phase of software development life-cycle in which the business requirements for a software product are defined and documented. When the requirements are collected and formalized with a formal language or model, the analysis verifies their completeness and consistency.

requirements engineering a nexus between social science and computer technology. It includes the elicitation and analysis of the requirements of the user as well as ways of identifying sources. *Requirements engineering* leads to the production of a requirements specification which is understandable to the end user (client) and from which the software design can be easily created.

requirements for quality expression of the needs or their translation into a set of quantitatively or qualitatively stated requirements for the characteristics of an entity to enable its realization and examination. The requirements for quality should be expressed in functional terms and documented. It is crucial that the requirements for quality fully reflect the stated and implied needs of the customer. Quantitatively stated requirements for the characteristics include, for instance, nominal values, rated values, limiting deviations, and tolerances.

requirements model describes the functionality and behavior of a system. The *requirements model* is a functional design description of the system to be built.

requirements review the process in which the requirements are presented to project personnel: managers, developers, users, etc. for their discussion and approval. It may include hardware and software requirements.

requirements specification a document that describes the system/component requirements. It includes: functional requirements, interface requirements, design requirements, performance requirements, communication requirements, development standards and guidelines, and project profile for quality and costs. The requirements can be given in natural, semiformal, and formal models and languages.

requirements traceability the property that allows the tracing of the requirements along the development life-cycle up to the final product. The traceability of the requirements allow verification of how a required functionality has been implemented and which parts of the system have been influenced by its presence.

requirements validation the process that allows demonstration of the validity of requirements, showing that the requirements will lead to production of the right systems with respect to the needs. This is performed by verifying the requirements against test cases.

requirements verification the process that allows verification of the completeness and the consistency of the requirements. It has to demonstrate the system correctness.

rescalable an optimization where given any instance of the problem and integer $\lambda > 0$, there is an easily computed second instance that is the same except that the objective function for the second instance is (element-wise) λ times the objective function of the first instance. For such problems, the best one can hope for is a multiplicative performance guarantee, not an absolute one.

research and development a department in software organizations where new methods, systems, and software are explored. There is a correlation between long-term success of a company and efforts put in research and development.

reserved word a word (symbol) in a programming language whose meaning is predefined, and consequently which cannot be used by the programmer for any other purpose. *See also* keyword.

residence time the total time that includes both waiting time and service time which is experienced by a job at the system or at a system component.

resident pertaining to a computer program that remains in main memory at all times.

residual error the degree of misfit between an individual data point and some model of the data. Also called a residual.

residual vector quantization *See* interpolative vector quantization.

residue in computer arithmetic, the remainder from the division of one integer by another.

residue number system a system in which a number is represented by the set of its residues relative to a given set of numbers known as the moduli.

resolution (**1**) indicates the number of pixels per image. It is often represented in this format: $N \times M$ where N and M are the number of pixels per column and row, respectively.

(**2**) refers to the ability to resolve two point targets which are closely spaced in time or frequency. For a linear system, resolution can be measured in terms of the width of the output pulse produced by a point target.

(**3**) the minimum controllable displacement of a robotic manipulator.

resource (**1**) the consumable elements that are used by a process, i.e., decreased in value when used (e.g., personnel time — effort, finance, facilities, equipment, tools, energy, machinery, materials, facilities, transportation, paper, diskettes, ink, etc.). Software tools are not "consumed" in the normal sense of the word, but are still resources, e.g., computer hardware will be "amortized" during development, and software licenses will consume cash. Overhead costs (heating, lighting, rent of office space, etc.) also represent the consumption of resources.

(**2**) any abstract machine environment entity referenced by the program unit and explicitly allocated to this unit.

(**3**) in the world of Graphical User Interfaces, the arrangement of interface elements such as buttons, check boxes, and so on, within a window are often represented by storing their geometric information in an entity known as a resource. This entity is usually bundled into the executable file by creating a resource segment (or on the Macintosh, what is called a resource fork) to hold it. Many systems include other interface elements, including message strings, bitmaps, and the like as resource objects.

resource analysis the process in which the resources that are needed for a project are compared with those that are available.

resource authorization authorization to create new relations.

resource conflict the situation when a component such as a register or functional unit is required by more than one instruction simultaneously. Particularly applicable to pipelines.

resource deallocation the mechanism by which the resources assigned to a project are removed from it to return to the general resource pool of the factory, research group, etc.

resource editor an application which edits the values and objects that are placed in a resource segment.

resource fork the name Macintosh environment uses to refer to a resource segment.

resource management the process of identifying, estimating, allocating, and monitoring the resources of the project.

resource measure a measure defined for some attribute of a resource. Different types of measures have to be defined depending on the resource type.

resource segment a segment of the executable file which holds resources.

resource sharing the situation in which a resource can be accessed simultaneously by multiple processes.

resource utilization the capability of the software product to use appropriate amounts and types of resources when the software performs its function under stated conditions.

respecify changes to requirements characteristics. This type of change can refer to changing only the form of existing requirements. For example, taking informal requirements expressed in English and generating a formal specification expressed in a formal language such as Z. This type of change can also refer to changing system requirements. Requirement changes include the addition of new requirements or the deletion or alteration of existing requirements.

response store in associative memory the tag memory used to mark memory cells.

responsibility any purpose, obligation, or required capability of the instances of a class or package. A responsibility is typically implemented by a cohesive collection of one or more features. A responsibility can be a responsibility for doing, a responsibility for knowing, or a responsibility for enforcing.

responsibility-driven a suite of design techniques in which the focus is upon the identification of the responsibilities rather than the data or the functionality. One relevant and popular technique is CRC card modeling in which developers role-play objects. This is sometimes called anthropomorphizing.

restart the act of starting again a hardware or software process.

restore to recreate the current or earlier state of a database from a backup.

restoring division the main operation in division is the subtraction of a multiple of the divisor from the partial dividend. In *restoring division* the partial dividend is modified immediately after the subtraction but should the new value be negative (assuming positive operands), an addition is subsequently used to restore its original value before the next subtraction. *See also* non-performing division, non-restoring division.

restricted universe sorts algorithms that operate on the basis that the keys are members of a restricted set of values. They may not require comparisons of keys to perform the sorting.

restructuring the engineering process of transforming the system from one representational form to another at the same relative level of abstraction, while preserving the subject system's external functional behavior.

result equivalence where two or more schedules produce the same database state.

retargeting the re-engineering process of transforming and hosting or porting the existing system in a new configuration. This could be a new hardware platform, new operating system or a new CASE platform. *See* adaptive maintenance.

retention constraint a constraint in the network model which specifies whether a record of a member record type may exist in the database on its own or whether it must have a related owner.

retiming the technique of moving the delays around the system. *Retiming* does not alter the latency of the system.

retire unit in modern CPU implementations, the module used to assure that instructions are completed in program order, even though they may have been executed out of order.

retrace time amount of time that a blanked vertical retrace takes for a display device. Note that this time is less than the time the display is blanked.

retroreflector a type of surface with unusual reflectance characteristics, namely, that it reflects light mainly back in the direction from which it came. This makes retroreflecting surfaces appear much brighter than matte surfaces, if the light source is in the same direction as the viewer, and dark otherwise. Retroreflecting surfaces are often found on road markings and signs.

return a construct in a programming language which causes control to be transferred from a function back to the site that made the function call. In some languages, there may be no overt syntactic construct that performs this transfer, but it may be implicit in the structure of the program.

return address the address of an instruction following a call instruction, where the program returns after the execution of the call subroutine.

return instruction an instruction, when executed, gets the address from the top of the stack and returns the program execution to that address.

return-to-bias recording *See* magnetic recording code.

return-to-zero recording *See* magnetic recording code.

reusability the ability to use or easily adapt the hardware or software developed for a system to build other systems. *Reusability* can be applied at any scale, e.g., analysis diagrams, code, patterns, frameworks, or components.

reusable component a component whose interface and functionalities are designed to be reused in more than one project or system.

reuse using code developed for one application program in another application. Traditionally achieved by using program libraries.

Object-oriented programming offers reusability of code via mechanisms of inheritance and genericity. Class libraries with intelligent browsers and application generators are under development to help in this process. Polymorphic functional languages also support reusability while retaining the benefits of strong typing.

reuse analysis the process in which the possibilities for reusing software components are analyzed on the basis of the needs and the formal description of the available components.

reused code code that is being reused (included in a system different from that for which it has been developed) without being programmed from scratch.

reuse engineering the modification of software to make it more reusable, usually rebuilding parts to be put into a library.

reverse appraisal review of higher managers by their employees. Useful issue since many problems are caused by management and are not of a technical nature. Normally done via questionnaires on an annual basis.

reverse engineering the process of analyzing an existing system to identify its components and their interrelationships and create representations of the system in another form or at a higher level of abstraction. Reverse engineering is usually undertaken in order to redesign the system for better maintainability or to produce a copy of a system without access to the design from which it was originally produced.

reverse execution *See* reversible execution.

reverse Polish notation *See* postfix, Polish Notation.

reverse specification any technique intended to extract a description of what the examined system does. The description is made in terms of the application domain. Specification in this context means an abstract description of what the software does. In forward engineering, the specification tells us what the software has to do, but this information is not included in the source code. Only in rare cases can it be recovered from comments in the source code and from the people involved in the original forward engineering process. On one hand, this process must be bottom-up since the only reliable description of the behavior of software is its source code. On the other hand, reverse specification must be top-down also. Trivially, knowledge of the application domain is necessary to describe in terms of the application domain.

reversible execution a debugging approach in which the evolution of the execution is registered and then is reproposed under the user control. *See* capture and playback.

review a process of meeting in which the planned results and costs are compared with the obtained results and costs. This process is performed with the support of the project personnel: manager, sub-system managers, and users, customers, etc.

review cycle the cycle of approval that one goes through when working on large software projects. Can take time when it is on more than one location.

REVOKE an SQL command that may be used to revoke privileges previously granted.

revolute joint a joint characterized by a rotation angle which is the relative displacement between two successive links.

rewrite rule (1) in a formal grammar, a formal specification of a rule to transform a sequence of tokens into a different sequence of tokens.

(2) in processing a tree, a specification for transforming some subtree into a different subtree.

rewriting system a general mechanism which transforms input strings to output strings. All automata are rewriting systems, as are computationally equivalent mathematical systems such as Post productions or Turing machines.

REXX a scripting language developed on IBM mainframes and available on several other systems.

RGB color model the most widely used image representation, where color is represented by the combination of the three primary colors of the additive light spectrum.

If the R, G, and B components are defined as scalars constrained to a value between 0 (no intensity) and 1 (maximum intensity), all the definable colors will be bounded by a cube and it is typical to describe RGB combinations as coordinates on the cube (r, g, b). For example, pure red is (1, 0, 0) and the secondary color cyan is (0, 1, 1); darker colors have values closer to (0, 0, 0) (black) and lighter colors have values closer to (1, 1, 1) (pure white).

See also tristimulus value, color space.

RGB true color an RGB color system with 24 bits per pixel color resolution. This gives a choice of over 16 million colors per pixel. Such a system is generally known as a true color or full color system.

ρ-approximation algorithm an approximation algorithm that is guaranteed to find a solution whose value is at most (or at least, as appropriate) ρ times the optimum. The ratio ρ is the performance ratio of the algorithm.

Rice's model a method of complex asymptotics that can handle certain alternating sums arising in the analysis of algorithms.

ridge detection *See* edge detection.

right associative an interpretation of infix operators in which operations of equal operator precedence are interpreted as binding to the right, for example, $A * B * C$ is interpreted as $A * (B * C)$. A language may be right associative independent of its specified operator precedence, and some languages are right associative with no operator precedence. *See also* left-associative.

right outer join an outer join between two relations, whereby all the tuples from the rightmost operand (right relation) are included.

rigid body motion motion of bodies which are assumed not to change their shape at all, i.e., deformation is absent or is neglected. In contrast, non-rigid body motion takes deformation into consideration.

rigid job a parallel job for which the number of processors is determined outside the scheduler and the processor allocation remains unchanged until the job completes.

ringing in image processing, the occurrence of ripples near edges in an image processed by a lowpass filter with a steep transition band.

ripple adder *See* ripple-carry adder.

ripple-carry adder a basic n-bit adder that is characterized by the need for carries to propagate from lower- to higher-order stages.

RISC *See* reduced instruction set computer.

RISC processor *See* reduced instruction set computer.

RISC technology *See* reduced instruction set computer.

risk a measure of the cost of operating a system, derived by combining hazard probability, danger, and severity of mishap. Risk is defined as the possibility of loss or injury. Risk exposure is defined by the relationship

$$RE = P(UO) * L(UO) ,$$

where RE is the risk exposure, $P(UO)$ is the probability of an unsatisfactory outcome, and $P(LO)$ is the loss to the affected parties if the outcome is unsatisfactory. Examples of unsatisfactory outcome include schedule slips, budget overruns, wrong functionality, compromised non-functional requirements, user-interface shortfalls, and poor quality.

risk analysis the assessment of the loss probability and loss magnitude for each identified risk item. Prioritization involves producing a ranked and relative ordering of the risk items identified and analyzed.

risk identification the production of a list of project specific risk items that are likely to compromise a project's success. For example, *risk identification* is the generation of checklists of likely risk factors.

risk management the problem of identifying and controlling risks. *Risk management* is divided into the following tasks: risk assessment, risk identification, risk analysis and prioritization, risk control, risk management planning, risk resolution, and monitoring.

risk management and monitoring plan (RMMP) a plan for mitigating, monitoring, and managing risks.

risk management planning plans that lay out the activities necessary to bring the risk items under control. Activities include prototyping, simulation, modeling, tuning, etc. All management plans should be integrated to reuse parts of each where possible, and to be factored into the overall schedule.

risk prioritization the assessment of the loss probability and loss magnitude for each identified risk item. Prioritization involves producing a ranked and relative ordering of the risk items identified and analyzed.

risk resolution and monitoring production of a situation in which the risk items are eliminated or resolved. Risk monitoring involves tracking the project's progress towards resolving its risk items and taking corrective action where appropriate.

river routing a single-layer channel routing problem in which each net contains exactly two terminals, one at the top edge of the channel and one at the bottom edge of the channel. The nets have terminals in the same order along the top and bottom — a requirement if the problem is to be routable in one layer.

RLL *See* run-length limited code.

RMS *See* root-mean-squared error.

RMW memory cycle *See* read/modify/write.

Roberts gradient a morphological gradient calculated using a set of two images that when used in windowed convolution provides an edge filter.

robot-oriented programming using a structured programming language which incorporates high level statements and has the characteristics of an interpreted language, in order to obtain an interactive environment allowing the programmer to check the execution of each source program statement before preceding to the next one. Robot-oriented programming incorporates the teaching-by-doing method but allows an interaction of the environment with physical reality. *See* teaching-by-showing programming.

robot programming language a computer programming language which has special features that apply to the problems of programming manipulators. Robot programming is substantially different from traditional programming. One can identify several considerations which are typical to any robot programming method: The objects to be manipulated by a robot are three-dimensional objects, therefore a special type of data are needed to operate objects; robots operate in a spatially complex environment; the description and representation of three-dimensional objects in a computer are imprecise; and sensory information has to be monitored, manipulated, and properly utilized. *Robot programming languages* can be spliced into three categories:

1. Specialized manipulator languages which are built by developing a completely new language. An example is the VAL language developed by Unimation, Inc.

2. Robot library for an existing computer language. It is a popular computer language augmented by a library of robot-specific subroutines. An example is PASRO (Pascal for Robots) language.

3. Robot library for a new general-purpose language. These robot programming languages have been developed by first creating a new general purpose language, and then supplying a library of predefined robot-specific subroutines. An example is AML language developed by IBM.

robot vision a process of extracting, characterizing, and interpreting information from images of a three-dimensional world. This process is also called machine or computer vision.

robust box bound a set of upper and lower bounds for closed multiclass queueing networks that are robust in the sense that they are based on minimal assumptions about the stochastic nature of the system. They are called box bounds because they enclose a rectangular polytope of feasible throughput region in the multidimensional throughput space each dimension in which corresponds to the throughput of a particular class.

robust fuzzy controller a fuzzy controller with robustness enhancement or robust controller with fuzzy logic concepts.

robust fuzzy filter a fuzzy filter with robustness enhancement or robust filter with fuzzy logic concepts.

robustness the degree to which a system or component can function correctly in the presence of invalid inputs or stressful environmental conditions.

With image data, robust procedures are those that are able to detect objects without becoming confused by partial occlusions, noise, clutter, object breakages, and other distortions.

robust statistics the study of methods by which robust measures may be extracted from statistical or numerical data, thereby excluding measurements which are unlikely to be reliable and weighting other measurements appropriately, thereby increasing the accuracy of finally assessed values. Of specific interest is the systematic elimination of outliers from the input data. *See also* robustness, median filter.

ROI *See* region of interest.

role a class or an entity can play a role in a system a different role in another. In object-oriented programming, the relationship of specialization is sometimes defined by considering the specialization of the roles of the real entities which are internally modeled as classes.

role modeling identification and description of the roles played in a system. These are temporary and parallel reclassifications. Typically an object might have a permanent type, such as person, but then for a short period of time take on the classification of traveler or teacher.

rollback (1) an operation that undoes all of the computations performed after some previously stored state, typically through the use of recovery blocks. *See also* commit.

(2) SQL92 standard term for an abort.

roll forward to reinstate database changes by applying after images.

roll in to transfer data or computer program segments from auxiliary storage to main memory.

roll out (1) to transfer data or computer program segments from main memory to auxiliary storage for the purpose of freeing the main memory.

(2) to go live with a new software product.

ROM *See* read-only memory.

root (**1**) a user with special privileges, usually reserved for a system administrator.

(**2**) in overlaying memory management, the portion of memory containing the overlay manager and code common to all overlay segments, such as math libraries.

See also root node, root symbol.

root class in object-oriented modeling, it is the base/super class common to all the classes of a class tree. A system may have one or more *root classes.*

root directory the directory highest in a hierarchy.

root-mean-squared (RMS) error (**1**) the square root of the mean squared error.

(**2**) in robotics, the average square of deviations from a prescribed path; a measure of the quality of movements.

root node a node in a graph which has the property that it has no arcs leading into it. A graph in general may have multiple roots. A tree is a graph which has a single designated root, among its other properties. A special case of a tree which has only one node where the node is both a root node and a leaf node. If the graph is a lattice or semi-lattice, it also has a single root node, called the top node.

root symbol a designated nonterminal symbol in a grammar, the starting point for all sentences generated by the grammar.

root user *See* super user.

rotate a low-level operation available on many machines which computes a result by treating the representation as if it were represented in a circular fashion, with the low-order bit adjacent to the high-order bit, and thus computing the new result based on performing a shift operation where the bits "shifted in" to one side of the value are the bits that had been shifted out of the other side. Also known as a circular shift.

A common variant is to treat the machine value of n bits as consisting of $n+1$ bits for purposes of the shift, where the extra bit is a state value such as the carry flag. It is uncommon for a programming language to make the rotate operation visible to the programmer, but many compilers can take advantage of this operation to generate faster code for special cases.

rotation a geometric transformation that changes the orientation of an object, extended light source, or viewpoint. Specific rotations are often represented by a matrix R, which then transforms point p to the new position R_p. Rotation and many other simple transformations can be done simultaneously if positions and directions are represented in homogeneous coordinates.

rotational latency the time it takes for the desired sector to rotate under the head position before it can be read or written.

rotational position sensing mechanisms used in disks to recognize the different sectors in a track and synchronize the different bits in a sector.

rounding an operation which modifies a floating-point representation considered infinitely precise in order to fit the required final format. Usually an attempt is made to minimize the error involved. Common rounding modes include round to nearest, round toward zero, and round toward positive or negative infinity. *See also* chopping, round-to-nearest, jamming, truncation.

rounding bias the tendency of a rounding procedure to round in just one direction. A biased method is upward biased or downward biased, according to whether the expected value of the rounding error is positive or negative.

rounding digit after an alignment shift, the second most significant of the digits shifted out of the significand proper.

rounding error *See* round-off error.

rounding overflow overflow that occurs during rounding, usually as a result of upward rounding being down as an addition, in the least significant bit position, followed by truncation.

round-off error the representation error that occurs because of rounding.

round-robin arbitration a technique for choosing which of several devices connected to a bus will get control of the bus. After a device has had control of the bus, it is not given control again until all other devices on the bus have been given the opportunity to get control in a pre-determined order. The opportunity to get control of the bus circulates in a pre-determined order among all the devices.

round-robin partitioning a technique for partitioning a relation across a number of disks. Given N disks, the jth tuple is stored on the disk stored $j \bmod N$.

round-robin system (RR) (**1**) a scheduling discipline for a single processor and single ready queue that uses preemption based on a clock. A clock interrupt is generated at periodic intervals, called a quantum.When the interrupt occurs, the currently executing process is preempted and placed on the ready queue, and the next ready process is selected in a FCFS fashion to execute. RR is also known as time-slicing.

(**2**) a scheduling discipline in a system consisting of one server and multiple queues (polling system) where the different queues are served in a fixed sequential order. The service discipline (e.g., limited, exhaustive, gated) determines when the next queue is visited, and the queueing discipline (e.g., FCFS, LCFS, random) determines which job is the next to be served from the currently visited queue.

round-to-nearest a rounding procedure in which the outcome of the rounding is the representable value closest to the exact result. If the bits to be discarded represent a value less than $1/2$ ulp, then the rounding is downwards; if they represent a value greater than $1/2$ ulp, then the rounding is upwards; and if they represent a value of exactly $1/2$ ulp (a boundary case), then it is upwards or downwards equally often. This is one of the four rounding methods in IEEE-754; it is also known as unbiased rounding or symmetric rounding. *See also* round-to-nearest-even, round-to-nearest-odd.

round-to-nearest-even a version of round-to-nearest in which the boundary case is always settled by selecting from the two equally close representable values the one whose least significant bit is 0.

round-to-nearest-odd a version of round-to-nearest in which the boundary case is settled by selecting from the two equally close representable values the one whose least significant bit is 1. Also known as R*-rounding.

round-to-nearest-up a rounding method similar to round-to-nearest, except that the boundary case is always resolved in the upward direction. The method is also known as half-adjust rounding and is most commonly used in paper-and-pencil arithmetic.

round-towards-plus-infinity a rounding method in which the result is the value closest to and no less than the value rounded. One of the four methods in IEEE-754.

round-towards-zero *See* chopping.

round-trip acknowledgment on a computer network, when a message is sent from a sender to receiver, the *round-trip acknowledgment* is the message sent back to the sender from the receiver that confirms that the original message was received.

round-trip time the time it takes for a signal to be transmitted from a sender to a receiver and back again.

router a node, connected to multiple networks, which forwards packets from one network to another. It is much more complex than bridges that work between networks having compatible protocols. Also called gateway.

routine Synonym for function, subroutine, or procedure.

routing probability a concept from queueing theory. In a queueing network, the routing probability from node i to node j is the ratio of customers leaving i going next to node j over the total number of customers leaving node i.

routing protocol a mechanism by which network switches discover the current topology of the network.

row-access strobe *See* two-dimensional memory organization.

row decoder logic used in a direct-access memory (ROM or RAM) to select one of a number of rows from a given row address. *See also* two-dimensional memory organization.

row level lock a lock applied to a single row in a table.

RPC *See* remote procedure call.

RPG acronym for RePort Generator, a language developed by IBM to allow nonprocedural specification of reports from databases.

RPN acronym for reverse Polish notation. *See* Polish Notation.

RR *See* round-robin system.

RS flip-flop a single-bit, storage element, usually formed by connecting two NOR or NAND gates in series. RS stands for Reset-Set. For state variable Q and next state variable Q' the simplified truth table is given as

R	S	Q	Q'
0	0	0	0
0	0	1	1
0	1	0	0
0	1	1	0
1	0	0	1
1	0	1	1
1	1	0	X
1	1	1	X

the symbol "X" is used to denote an unknown state for the flip-flop. *See* JK flip-flop.

RSGNN *See* recursive self-generating neural network.

RTOS *See* real-time operating system.

r-tree an object hierarchy where associated with each element of the hierarchy is the minimum bounding rectangle of the union of the minimum bounding rectangles of the elements immediately below it. The elements at the deepest level of the hierarchy are groups of spatial objects. The result is usually a nondisjoint decomposition of the underlying space. The objects are aggregated on the basis of proximity and with the goal of minimizing coverage and overlap. *See also* b-tree, p-tree (2), p-tree (3).

RTTI acronym for run-time type information.

RTU *See* remote terminal unit.

rule a fundamental concept in database theory. A rule, "$\mathsf{A}\,\{B_1 \wedge B_2 \wedge \ldots \wedge B_n\}$" denotes a statement that an atomic formula A — called the head or conclusion — should be inferred, if the atomic formulas $\mathsf{B}_1, \mathsf{B}_2, \ldots, \mathsf{B}_n$ — their conjunction is called the body or precondition — have been inferred already. *See also* clausal form.

For example, the rule brother(P,B) $\{$father(P,F)$\wedge$ father(B,F)$\wedge$ different(P,B)$\wedge$ male(B)$\}$ states that if a person F is the father of two different persons P and B, and B is male, then B is called a brother of P.

rule modeling identification and description of business rules. They are "declarations of policy or conditions that must be satisfied" (OMG, 1988). Some authors discriminate between constraint rules (which specify policies or conditions) and derivation rules (for inferring or computing facts from other facts).

run a single, usually continuous, execution of a computer program.

run-length coding *See* run-length encoded.

run-length encoded a lossless compression technique where the image is stored as a sequence of triples specifying a pixel at which the image is black such that the pixel to the left is white for a specified run length.

run-length limited code (RLL) a line code which restricts the minimum and/or maximum number of consecutive like-valued symbols that can appear in the encoded symbol sequence.

runnable *See* ready.

running a state of a task or process, in which it is executing on a processor.

running digital sum the difference between cumulative totals of the number of logic ones and number of logic zeros in a binary sequence. It is a common measure in the performance of a line code.

run-time *See* runtime.

runtime (**1**) the execution time for a program or a system.

(**2**) adjective describing the support libraries needed to execute a program.

runtime type information information which is maintained as part of an object, generally not visible to the programmer except via special operations, which identifies the type of the object. This is the term C++ uses to describe the ability to query an object for its type during execution. In many other languages, this is implicit in the design.

rvalue A value that appears on the right-hand side of an assignment statement. Generally, an expression which yields a storable value. *See also* lvalue.

S

S-100 a 100-pin bus formerly used by computer hobbyists and experimenters.

safeness condition in the specification of a system, any condition expressing that something bad will not happen. For example, in a rail-crossing system a safeness condition can be "If the train is passing, the barriers are closed". Temporal logic is expressed with the formula

$$\square\,(\text{train } crossing \rightarrow closed)\ .$$

See liveness conditions.

safety (**1**) the probability that a system will either perform its functions correctly or will discontinue its functions in a well-defined, safe manner. For system safety, all causes of failures which lead to an unsafe state shall be included: hardware failures, software failures, failures due to electrical interference, and human interaction, and failures in the controlled object. The system safety also depends on many factors which cannot be quantified but can only be considered qualitatively.

(**2**) in a database, a rule where all variable symbols "X" occurring anywhere in the rule are safe. "X" is safe if it occurs in a regular body atom (i.e., an atom that does not have a built-in predicate symbol) or in an atom "$X = t$" or "$t = X$" with the built-in predicate symbol "=", such that all variable symbols in the term "t" are safe.

safety-critical a system whose failure may cause injury or death to human beings. For example, an aircraft or nuclear power station control system. Common tools used in the design of *safety-critical* systems are redundancy and formal methods.

safety-critical function a function that can cause or allow a hazard to exist.

safety-critical software software that performs (or controls a subsystem that performs) a safety-critical function.

safety-critical system a system that is intended to handle rare, unexpected, dangerous events.

safety integrity the likelihood of a safety critical system, function, or component achieving its required safety features under all the stated conditions within a stated measure of use.

safety property a program property asserting that the program will never move into a bad state during execution.

safety related software software which ensures that a system does not endanger human life, limb, and health, or the economics or environment of the capital equipment and control.

Sagittal projection a projection of a three-dimensional object onto a two-dimensional plane which intersects the object in a front to back direction dividing the objects into right and left halves. Typically with reference to an animal or human body.

salient feature a characteristic often local feature on an object which can be detected and used as part of the process of inferring the presence of an object from its features. Typical *salient features* include point features such as corners and small holes, or fiducial marks (e.g., on printed circuit boards), but may in addition include large-scale straightforwardly detected features such as large circular holes which can also aid the inference process.

salt and pepper noise *See* impulse noise.

SAM *See* standard additive model.

sample a single measurement that is taken to be representative of the measured property over a wider area, frequency range, or time period. When recording digital sound, a sample is a voltage measurement that reflects the intensity of the acoustic signal at a particular moment, and has a time period associated with it; that is, the

0-8493-2691-5/01/$0.00+$.50

sample represents the signal until the next measurement is made. In a digital image, a sample is a single measurement of light intensity at a particular point in the scene, and that measurement is used to represent the actual but unmeasured intensity at nearby points.

sample space the set of all possible samples of a signal, given the particular parameters of the sampling scheme.

sampling function a mathematical function used when sampling a signal. In particular, a sampling function $S(t)$ can be multiplied by the continuous function to be sampled, $F(t)$, to obtain the sampled version of F. S is most often a collection of equally spaced impulses.

sandbox a term used to describe a limited operating environment in which a program will be executed. The limited environment is intended to isolate the program so that it cannot read, modify, or damage the system in which it is executing. Languages such as JavaScript and VBScript usually execute in the context of a *sandbox.*

satellite imagery the acquisition of pictures of the earth from space. *Satellite imagery* can be used to enhance maps, collect resource inventories (e.g., forestry, water, land use), assess environmental impact, appraise damage following a disaster, and collect information on the activities of humans. Satellite imagery tends to be multi-spectral, including a wide range of optical frequencies and, more recently, infrared and radar. *See also* remote sensing.

satisfaction the degree to which the software system is appreciated by the end-users.

satisfiability relation a logic formula where there exist values of the variables in the formula for which the formula is true.

saturated logic logic gates whose output is fully on or fully off. The output voltage from such a switch or the current through such a switch is determined principally by the external circuit. *Compare with* active logic.

saturation (1) with respect to color, the amount or purity of the color seen. A pure color is said to be fully saturated, and the saturation decreases as white is added to the mix. The color pink, for example, is a less saturated version of red.

(2) the condition where the system load reaches its capacity. The server is busy 100% of the time.

save instruction an assembly language instruction that saves information about the currently executing process.

savepoint *See* checkpoint.

scalability the degree to which a system can be expanded by duplicating its components, without degradation of performance due to communication overhead and resource conflicts.

scalable video coding compression of video such that transmission at different data rates, or reception by decoders with differing performance, is possible merely by discarding or ignoring some of the compressed bitstream, i.e., without recoding the data. The compressed data are prioritized such that low-fidelity reconstruction is possible from the high-priority data alone; addition of lower-priority data improves the fidelity.

scalar (1) a quantity that has magnitude but no direction.

(2) a simple numeric value, such as integer or floating point.

(3) a value that represents a single object, rather than an aggregate of objects (such as an array).

scalar processor a CPU that dispatches at most one instruction at a time.

scalar quantization (SQ) quantization of a scalar entity (a number as opposed to vector quantization), obtained for example from sampling a speech signal at a particular time-instant. Each input value to the quantizer is assigned a reproduction value, chosen from a finite set of possible reproductions. A device performing *scalar quantization* is called a (scalar) quantizer.

scale (1) a set of values with defined properties. Examples of scales are: a nominal scale corresponding to a set of categories; an ordinal scale corresponding to an ordered set of scale points; an interval scale corresponding to an ordered scale with equidistant scale points; a ratio scale that not only has equidistant scale points but also possesses an absolute zero; an absolute scale corresponding to the counting of entities. Metrics using nominal or ordinal scales produce qualitative data, and metrics using other scales produce quantitative data. A numerical relation system into which a mapping from entities in the real world is defined (i.e., a measure) to characterize an attribute.

(2) a property of an image relating the size of a pixel in the image to the size of the corresponding sampled area in the scene. A large scale image shows object features in more detail than a small scale image. (*See also* resolution.)

(3) to change the size of (i.e., enlarge or shrink) an image or object while maintaining the overall proportions.

(4) one of two parameters of a wavelet, the other being translation. The scale specifies the duration of the wavelet.

scaled integer arithmetic a computational technique that represents fractional values by multiplying the desired value by a scaling factor so that the resulting value is represented as an integer, for example, representing distances in units of nanometers or time in microseconds. Often used in computations involving currencies to avoid the intrinsic errors of floating point representations. Also used on machines with no floating point unit or very slow floating point computations to improve performance.

scaled number an optimization technique where the least significant bit (LSB) of an integer variable is assigned a real number scale factor.

scaled processor architecture (SPARC) name for a proprietary class of CPUs.

scaling function the solution to the multiscale equation; it can be obtained by iterating a low pass filter in the two-channel filter bank an infinite number of times.

scan a parallel operation in which each element in an array receives the sum of all previous elements.

scan design a technique whereby storage elements (i.e., flip-flops) in an IC are connected in series to form a shift-register structure that can be entered into a test mode to load/unload data values to/from the individual flip-flops.

scan line in a digital image, a contiguous set of intensity samples reflecting one row or column of the image. A class of image processing algorithms, called scan line or scan conversion methods, looks at the image one or two scan lines at a time in order to achieve the goal.

scanline algorithm an algorithm that renders an image one row at a time, e.g., generates the image values for pixels left-to-right as it scans across the image. After one row is generated, the algorithm proceeds to the next row. One advantage of this algorithm is it can use less memory to generate the results for only single rows at a time. Another potential advantage is a reduction in computation as the set of object primitives that need to be rendered at each pixel along a scan line may not change very often, so some results calculated at one pixel can be used at the next.

scanner (1) a device used for scanning written documents or printed pictures by tracing light along a series of many closely spaced parallel lines.

(2) any device that deflects a light beam through a range of angles, using mechanisms such as diffraction from electro-optic or acousto-optic gratings or mechanical deflectors.

scanning process for converting attributes of a display at raster coordinate locations, such as color and intensity, into a fixed set of numerical attributes for manipulation, transmission, or storage of the display.

scan tape *See* helical scan tape.

scenario the description of a context of use of a system. A *scenario* is typically used for describing the system functionalities during description of system requirements. It may involve the description of the typical operations that the users of the system may need to perform on the system with the results that are obtained.

scene analysis the process of extracting specific features from a larger picture or scene. Typically, this process will involve object and feature detection, inference of the presence of objects from their features, projective invariance properties, analysis of the play of light on surfaces, including approaches such as texture analysis, and many other types of procedures.

schedulability analysis the compile time prediction of execution time performance.

schedulability the property of a set of tasks that ensures that the tasks all meet their deadlines.

schedule (**1**) to select units of concurrency that are to be dispatched.

(**2**) a group of transactions where each transaction's operations are (and must be) executed in order; however, operations from two or more transactions may be interleaved.

scheduler a part of the operating system for a computer that decides the order in which programs will run.

scheduling in an operating system, scheduling of CPU time among competing processes.

schema (**1**) a description of the structure of an object. Plural: schemata. *See also* metadata.

(**2**) in evolutionary computing, a symbol string defined over the alphabet {0, 1, #}, where # is a wildcard symbol denoting either 0 or 1 for the purpose of matching. The specified positions in a *schema,* i.e., those positions with definite bit values rather than the # symbol, uniquely characterize a set of binary symbols with identical bit values in the corresponding positions. For example, the schema 11## refers to any of the strings 1100, 1101, 1110, and 1111. The order of a schema is the number of specified positions, and the defining length refers to the maximum length between specified positions. For example, the schema 1###0 is of order 2 and defining length 3. *See also* schema theorem.

schema diagram (**1**) diagrammatic representation of the database schema.

(**2**) any object in the database schema.

schema evolution a problem in persistent object storage in that as a system evolves, the data representation of objects may be modified to reflect changes in the requirements or implementation. This requires that objects that already exist either have to be modified or newer versions of the application need to be able to read older versions of the objects.

schema theorem a statement which characterizes how the number of individuals associated with a particular schema varies from generation to generation in genetic algorithms. Formally, the theorem can be summarized by the following inequality:

$$N(H, t+1) \geq N(H, t)\left(\frac{f(H, t)}{\overline{f}(t)}\right)\left(1 - p_c\frac{d(H)}{L-1} - p_m O(H)\right),$$

where $N(H, t)$ and $N(H, t+1)$ are the number of individuals associated with schema H in the current and next generation, respectively, $f(H, t)$ is the average fitness of those individuals associated with H, $\overline{f}(H, t)$ is the average fitness of all individuals in the population, p_c is the crossover probability, and p_m is the mutation probability. In addition, $O(H)$ and $d(H)$ refer to the order and defining length of schema H, respectively. Essentially, the theorem states that for those schema H with above average fitness (i.e., $f(H, t) > \overline{f}(t)$), a low order $O(H)$, and a short defining length $d(H)$, its corresponding number of representatives $N(H, t)$ in the population will increase at an exponential rate. *See also* building block, schema.

scientific notation a notation in which a value is adjusted so that it is expressed as a numeric value in the range $-10.0 < n < 10.0$, followed by a power-of-ten multiplier. For example, 9999.7 is expressed as 9.9997×10^3. Scientific notation is used, with some syntactic changes, to enter and display floating point numbers into a computer. For example, most languages would allow the constant to be written as $9.9997E3$, where E stands for "10 to the". *See also* floating-point constant.

scientific visualization the use of computer graphics techniques to represent complex physical phenomena and multidimensional data in order to aid in its understanding and interpretation.

scope a context that controls the visibility of, and interpretation of, names in a programming language. *See also* dynamic scope, lexical scope, unbound name.

scoreboard term originally used for a centralized control unit in the CDC 6600 processor which enabled out-of-order issue of instructions. The scoreboard unit held various information to detect dependencies. Now sometimes used for the simpler mechanism of having a single valid bit associated with each operand register.

scotopic formally, a description of luminances under which human rod cells are active. Informally, describing dim or night-time luminances.

SCP *See* service control point.

scrambling randomization of a symbol sequence using reversible processes that do not introduce redundancy into the bit stream. *See also* self-synchronizing scrambling.

screen *See* monitor.

screen door transparency a technique for rendering the transparency of an object. The key idea is to render only some of the pixels associated with the object, depending on how transparent the object is.

script a program written in a command language.

scripting language a programming language designed for writing sequences of commands that would otherwise have to be typed repetitively. Many scripting languages are highly idiosyncratic, rarely have a consistent syntax or semantics, and are often a blend of several languages including, in some cases, the native command language of the underlying operating system shell. Also called command language.

SCSI pronounced "scuzzy". *See* small computer systems interface.

SCSI adapter card which provides an input/output port attachment from main processor memory to external storage devices and other peripherals using the SCSI interface.

SCSI throughput the speed with which the SCSI interface adapter can send and receive data to and from a device connected to it.

sculptured surface a highly flexible surface generated by the combination of surface patches which have both their boundary curves and interior blending functions defined by polynomials of, usually, at least order three. *See also* b-spline and Bézier curve.

SDL *See* specification and description language.

search to look for a value or item in a data structure. There are dozens of algorithms, data structures, and approaches. *See also* array, binary search, b-tree, dictionary, hash table, heap, inverted index, linked list, quadtree.

search space the set of all potential solutions to a problem, from which the correct solutions are to be chosen.

search tree a tree where every subtree of a node has values less than any other subtree of the node to its right. The values in a node are conceptually between subtrees and are greater than any values in subtrees to its left and less than any values in subtrees to its right.

Formally, each node in the search tree contains $p-1$ search values and p pointers to other nodes, i.e., $(P_1, K_1, P2, \ldots K_{q-1}, P_q)$. Two constraints are enforced:

1. $K_1 < K_2 < \ldots K_q$ for all nodes and

2. for all values Z in a subtree referenced by P_j, $K_{j-1} < Z < K_j$.

See also binary search tree, b-tree.

secondary cache a buffer element between slow speed peripheral devices, such as disks, and a high speed computer.

secondary hash a technique for implementing a hash table such that if a hash collision occurs, a new hash position is computed by applying the original or a different function to the original value to obtain a new hash key.

secondary index an index built on some non-ordering field of a file.

secondary key a non-ordering key field upon which a secondary index has been built.

secondary memory generic term used to refer to any memory device that provides backup storage besides the main memory. *Secondary memory* is lower-level, larger capacity and usually a set of disks.

Only data and programs currently used by the processor reside in main memory. All other information (not needed at a specific time) is stored in secondary memory and it is transferred to main memory on a demand basis. It is the highest (big but slow) level in the memory hierarchy of modern computer systems.

secondary port the second port of a particular type in the computer.

secondary storage in computer devices such as hard disks, floppy disks, tapes, and so forth, the storage that is not part of the physical address space of the CPU. *See* secondary memory.

second-line manager a manager to whom a layer of subordinate managers report. The latter are the first-line managers. *See* sub-system manager.

second normal form (2NF) a relation that is in first normal form and where no non-prime attribute is partially dependent on the key, i.e., every non-prime attribute is fully functionally dependent on the key.

second-order type system a language system that includes existential or universal quantification over type variables.

sector *See* disk sector.

sector mapping a cache organization in which the cache is divided into sectors where each sector is composed of a number of consecutive lines. A complete sector is not transferred into the cache from the memory, only the line requested. A valid bit is associated with each line to differentiate between lines of the sector that have been transferred and lines from a previous sector. Originally appeared in the IBM System/360 Model 85.

security **(1)** attributes of software that bear on its ability to prevent unauthorized access, whether accidental or deliberate, to programs and data. The ability of a system to deliver its required service and data under given conditions without unauthorized access.

(2) the extent to which an operating system provides protection against unauthorized use of computer resources.

security monitor an abstraction representing security policy, a system mechanism that can enforce it, or both. May be used to inform users how the system behaves, guide the system's design and implementation and reason about its properties.

seek time the time that it takes to position the read/write device over a desired track of information.

segment **(1)** a region in computer memory defined by a segment base, stored in a segment base register, and, usually, a segment limit, stored in a segment limit register. *See* virtual memory.

(2) a portion of a computer program that may be executed without the entire program being resident.

(3) in pipelining a disjoint processing circuit, also called a stage.

(4) the substring of a pattern delimited by two don't cares or one don't care and one pattern boundary.

segmentation (1) an approach to virtual memory when the mapped objects are variable-size memory regions rather than fixed-size pages.

(2) the process of dividing programs into independently addressed segments.

(3) the partitioning of an image in mutually exclusive elements in which visual features are homogeneous. Region-based segmentation relies on the analysis of uniformity of gray level or color; contour-based segmentation relies on the analysis of intensity discontinuities.

segmentation-based coding a coding scheme that is based on segmentation. *See* segmentation.

segmented architecture in computer architecture, a scheme whereby the computer's memory is divided into discontinuous segments.

segment mapping table a memory table within a computer that is used to translate logical segment addresses into physical memory addresses.

segment register a register that stores the base, or starting memory address, of a memory segment.

segment table a table that is used to store information (e.g., location, size, access permissions, status, etc.) on a segment of virtual memory.

SELECT (1) a relational operator for selecting from relation tuples which satisfy some selection criteria.

(2) SQL statement used for retrieving information from a database.

selection the process in evolutionary computation where individuals with high fitness in the original population are incorporated into the new population, while individuals with low fitness are displaced from the population. Both deterministic and probabilistic approaches have been adopted for this purpose: the $(\mu + \lambda)$ and (μ, λ) strategies in ES are deterministic, while the proportional selection strategy in GA and the tournament selection strategy in EP are examples of probabilistic approaches. *See also* proportional selection, rank-based selection.

selective testing testing only a selected set of program paths and data inputs.

selectivity of a condition, the ratio of the number of tuples that satisfy the condition to the number of tuples in the relation.

selector channel I/O channel that handles only one I/O transaction at a time. Normally used for high-speed devices such as disks and tapes. *See also* multiplexor channel.

self in object-oriented languages, a designation of the object being operated upon.

self-adaptation the adaptation of the strategy parameters, which specify the variances and covariances of the random variables for mutation, in addition to the object variables in ES and meta-EP. This is to allow adaptive specification of the search region for each individual according to the local structure of the fitness landscape. Provisions are also made in particular implementations of self-adaptation to ensure the validity of the updated strategy parameters. For example, in ES, the logarithmic normal random variable is adopted for updating the standard deviation of the mutation process to ensure its positivity. *See also* strategy parameter.

self-generating neural network (SGNN) networks of self-organizing networks each node network of which is an incomplete self-organizing network. For this kind of network of neural networks, not only the weights of the neurons but also the structure of the network of neural networks are learned from the training examples.

While SGNN can be as complex as an acyclic directed graph, the most frequently used SGNN takes a tree structure and is called a self-generating neural tree (SGNT) which is very similar to a self-organizing tree but with a much higher ratio of neuron utilization. Since less neurons participate in the competition during the training and classification, the speed of SGNT is faster. SGNN has found applications in diagnosis of communication networks, image/video coding, large scale Internet information services, and speech recognition. *See* self-generating neural tree. *See also* self-organizing neural tree.

self-generating neural tree (SGNT) a simplified version of self-generating neural network with a tree structure. SGNT is normally much faster in training and classification but with less descriptive power compared to the corresponding SGNN because of its simple topological structure. However, compared to the SONT, the SGNT has the same descriptive power if the number of network nodes is the same, a higher ratio of neuron utilization, higher speed, and may end up with higher accuracy since large-scale networks can be generated and trained quickly. *See* self-generating neural network, self-organizing neural tree.

self-modifying code a program using a machine instruction which changes the stored binary pattern of (usually) another machine instruction in order to create a different instruction which will be executed subsequently. Definitely not a recommended practice and not supported on all processors.

self-occlusion a surface satisfying the following conditions:

1. light cast from behind the surface does not illuminate it.

2. the light source is in front of the surface but some closer portion of the surface blocks the incoming light.

3. the light source is in front of the surface and the surface is illuminated, but some closer portion of the surface blocks the light coming from the surface.

self-organizing a neural network that is capable of changing its connections so as to produce useful responses for input patterns without the instruction of a smart teacher (i.e., the networks can train themselves to accomplish tasks).

self-organizing algorithm a training algorithm for a self-organizing system consisting of the following main steps:

1. Calculate the similarities of the training vector to all the neurons in the system and compare them to find the neuron closest to the training vector, i.e., the winner.

2. Update the weights of the winner and its neighborhood according to

$$w_i(t+1) = w_i(t) + \alpha\left(x_i(t) - w_i(t)\right) ,$$

where $w_i(t)$ is the i-th weight of the neuron at time t, $x_i(t)$ is the i-th component of the training vector at t, and α is a training rate. The neighborhood of the winner starts from a bigger area and reduces gradually during the training period.

self-organizing map (SOM) a neural network that transforms a multi-dimensional input to a 1D or 2D topologically ordered discrete map. Each neuron has a weight vector $\vec{w}_j$ and for a given input $\vec{x}$, the output is the index of the neuron with minimum Euclidean distance. During training, the weights are updated as

$$\vec{w}_j(n+1) = \begin{cases} \vec{w}_j(n) + \alpha(n)\left[\vec{x} - \vec{w}_j(n)\right], & C_j \in N(C_i, n) \\ \vec{w}_j(n), & C_i \notin N\left(C_j, n\right) \end{cases} ,$$

where α is a learning parameter in the range $0 < \alpha < 1$, and $N(C_i, t)$ is the set of classes that are in the neighborhood of the winning class C_i at time t. Initially, the neighborhood may be quite large during training, e.g., half the number of classes or more. As the training progresses, the size of the neighborhood shrinks until, eventually, it only includes the one class.

self-organizing neural tree (SONT) a tree-like network of self-organizing neural networks each node of which is a Kohonen network. Each of the neurons in the higher level networks has its child network in the lower level of the tree hierarchy. The training method is similar to that of Kohonen's method, but is conducted hierarchically. From the top (root) of the tree down, the winner of the current self-organizing network is found as the closest neuron to the training example. The weights of the winner and its neighbors are updated and then the child network of the winner will be selected as the current network for further examination until a leaf node network is encountered. This kind of network of neural networks can be useful for complex hierarchical clustering/classification. However, the utilization of the neurons may become poor as the network size grows if the uniform tree structure is adopted. The utilization may be improved if a carefully designed structure is used, but how to obtain an optimum structure remains an issue. Self-generating neural network (SGNN) may be a solution to this problem. *See* self-organizing system. *See also* self-generating neural network.

self-organizing strategies heuristic that reorders a list of elements according to how the elements are accessed.

self-organizing system a class of unsupervised learning systems which can discover for itself patterns, features, regularities, correlations, or categories contained in the training data, and organize itself so that the output reflects these discoveries.

self-similar in an image, when the structure of the whole is often reflected in every part.

self-similar data traffic Network traffic that consists of slow decaying "bursts" across all time scales. Unlike Poisson, or Markov-like distributions with short-range dependence and fast decaying correlations, studies have shown the computer network traffic, such as Ethernet, WWW, and variable bit-rate video, is characterized by long-range dependence and burstiness and can be more accurately modeled by a self-similar stochastic process.

self-similarity a fractal-like model of random error perturbations, introduced by Benoit Mandelbrot in a 1965 paper, for data that appears to come in bursts, as in the early hydrological data studied by Hurst. Such data shows self-similarity over different time scales. *See* self-similar stochastic process, Hurst parameter.

self-similar stochastic process a stochastic process X_t is self-similar if the m-aggregated series $X^{(m)}$ obtained by summing the original series X_t over non-overlapping blocks of size m has the same autocorrelation function $r(k) = E[(X_t - \mu)(X_{t+k}) - \mu)]$ as the original series for all m. The distribution of the aggregated series is thus the same, except for the changes in time scale, as the original.

self-synchronizing scrambling a technique that attempts to randomize a source bit stream by dividing it by a scrambling polynomial using arithmetic from the ring of polynomials over $GF(2)$. Descrambling is performed with only bit-level synchronization through continuous multiplication of the demodulated sequence by the same scrambling polynomial. The division and multiplication procedures can be implemented with simple shift registers, enabling this technique to be used in very high bit rate systems.

self-test a test that a module, either hardware or software, runs upon itself.

self-test and repair a fault tolerant technique based on functional unit active redundancy, spare switching, and reconfiguration.

semantic analysis *See* syntactic analysis, lexical analysis.

semantic description language a notation that incorporates both the concept of a formal syntax for a language and the concept of language semantics, such that sentences generated by this language are both syntactically correct and semantically correct.

semantic modeling a conceptual modeling technique based on the concept of a class and its properties.

semantic net a diagram using a graph structure to represent knowledge. Nodes represent things and arcs joining them represent relationships. There are many kinds of *semantic nets* including class diagrams, collaboration diagrams, and context diagrams.

semantics the meaning of a string or sequence of toke symbols in some language, as opposed to syntax which describes how symbols may be combined independent of their meaning. The semantics of a programming language are a transformation from programs to answers. A program is a closed term and, in practical languages, an answer is a member of the syntactic category of values. The two main kinds are denotational semantics and operational semantics.

See also static semantics, execution semantics, syntax, pragmatics.

semaphore a variable used to synchronize concurrent program units by two operations: lock and unlock. *Semaphores* were promulgated by Edsger Dijkstra in 1965 (although they had been invented in 1961). *See also* mutex, binary semaphore, critical section.

semidefinite programming a generalization of linear programming in which any subset of the variables may be constrained to form a semidefinite matrix. Used in recent results obtaining better approximation algorithms for cut, satisfiability, and coloring problems.

semi-join a technique used to join relations located at different sites. Given relations $R1$ and $R2$ stored at site $S1$ and $S2$, respectively, the semi-join of $R1$ and $R2$ (i.e., $R1 \ltimes_{A=B} R2$) is equivalent to performing a $project_B$ $(R2)$ at site $S1$, transferring the result to $S2$ and performing the join there.

semi-lattice a directed acyclic graph that has a single designated root node, called the top node. Unlike a lattice, a *semi-lattice* has no bottom node. A tree is a special case of a semi-lattice in that in a tree there is a unique path from the root node to any other node. The data types of many programming languages form a semi-lattice, particularly in object-oriented systems with a class structure. If multiple inheritance is not a capability of the type system, the types will form a tree; if multiple inheritance is possible, the types will form a semi-lattice.

semiorthogonal wavelets wavelets whose basis functions in the subspaces are not orthogonal but the wavelet and scaling subspaces spanned by these basis functions are orthogonal to each other.

sense amplifier in a memory system, circuitry to detect and amplify the signals from selected storage cells.

sensitive dependence on initial conditions, a system that is subject to great variance in later states due to only slight variance in the initial conditions.

sensor a transducer or other device whose input is a physical phenomenon and whose output is a quantitative measurement of that physical phenomenon. Physical phenomena that are typically measured by a sensor include temperature or pressure to an internal, measurable value such as voltage or current.

sensor alignment alignment of sensors so as to correct the time delay differences arising from spatial differences.

sentence **(1)** a sequence of symbols generated by applying the rules of a grammar.

(2) a sequence of symbols which cause an automaton to enter an accepting state.

separable data a 2D signal that can be written as a product of two 1D signals.

separable kernel for a 2D transform, a kernel that can be written as the product of two 1D kernels. For higher dimension transforms, a separable kernel can be written as the product of several 1D kernels. *See* separable transform.

separable queueing network a set of interconnected queues in which the joint distribution of the states of all queues factors into the product of marginal distributions. *See also* Jackson's theorem, product-form network.

separable transform a 2D transform that can be performed as a series of two 1D transforms. In this case the transform has a separable kernel. The 2D continuous and discrete Fourier transforms are separable transforms. In higher dimensions a separable transform is one that can be performed as a series of 1D transforms.

separation theorem a theorem showing that two complexity classes are distinct. Most known *separation theorems* have been proved by diagonalization.

separator a syntactic mark that is used between syntactic constructs. For example, a comma is traditionally a *separator* in the parameters of a function call. *See* terminator.

SEQUEL acronym for Structured English Query Language, developed by IBM for database queries. Became SQL.

sequence diagram a kind of interaction diagram in which time is shown linearly for each of the interacting objects. Messages passed between objects are also shown. Sequence diagrams are useful for focusing on temporal ordering of messages.

sequence number attack an attack based on predicting and acknowledging the byte sequence numbers used by the target computer without ever having seen them.

sequence point a point in a program, either explicit (such as a semicolon, or just before a function call after all parameters have been evaluated) or implicit (such as an end-of-line) that indicates that all computations preceding it must be completed before any computation following it is performed. Note that with an optimizing compiler the evaluation order implied by a sequence operator may be changed by the compiler if such a change does not violate the semantics of the program; that is, the resulting program performs the identical computation as one which is not optimized.

sequency coefficients the coefficients generated when multiplying an image by the Hadamard matrix.

sequential access (**1**) access to a file in a linear manner.

(**2**) retrieving data from sequential devices like cartridge tapes. Before the nth datum is retrieved, the first through $n-1$th data item must be bypassed. *Compare with* direct access, random-access.

sequential-access storage storage, such as magnetic tape, in which access to a given location must be preceded by access to all locations before the one sought. *See also* random-access device.

sequential consistency the situation when any arbitrary interleaving of the execution of instructions from different programs does not change the overall effect of the programs.

sequential DCT-based encoding image encoding in which each image component is encoded in a single left-to-right, top-to-bottom scan.

sequential decoding a sub-optimum decoding method for trellis codes. The decoder finds a path from the start state to the end state using a sparse search through the trellis. Two basic approaches exist: depth-first algorithms and breadth-first algorithms.

sequential divider a divider in which quotient bits are produced at a rate of one per addition/subtraction cycle.

sequential fault a fault that causes a combinational circuit to behave like a sequential one.

sequential file *See* sorted file.

sequential locality part of the principle of (spatial) locality, that refers to the situation when locations being referenced are next to each other in memory. *See also* principle of locality.

sequential multiplier a multiplier in which product bits are produced at a rate of one per addition cycle.

sequential-parallel divider a divider in which quotient bits are produced at a rate of a few bits (concurrently) per addition/subtraction cycle.

sequential-parallel multiplier a multiplier in which product bits are produced at a rate of a few bits (concurrently) per addition cycle.

serial executed sequentially, one operation at a time, until completion. Generally, there is an explicit or implicit interdependence of operations performed in a serial fashion such that any other sequence would produce a different result. In many cases, underlying machine architectures are obliged to maintain serial semantics even when the machine can perform one or more operations in parallel. Serial semantics may be based strictly on the original coding of the program, or a compiler may be permitted to change the order, subject to the restrictions of a sequence point. *See also* collateral execution, parallelism.

serial adder a one-digit adder in which n-digit addition is performed by feeding in operands one digit-pair at a time.

serial bus a data communication path between parts of the system that has a single line to transmit all data elements.

serial divider a one-digit divider in which n-digit addition is performed by feeding in operands one digit-pair at a time.

serial I/O interface I/O interface consisting of a single line over which data is transferred one bit at a time. Commonly used for low-speed devices, e.g., printer, keyboard, etc. *See also* parallel I/O interface.

serializable a transaction isolation level that guarantees concurrent serializable transactions produce the same result, when the same transactions are run serially. A serializable transaction prevents dirty reads, non-repeatable reads, and phantoms.

serialization (**1**) forcing operations to occur in a serial fashion, even when the operations could be performed in parallel. *Serialization* is usually accomplished by one or more synchronization mechanisms. Serialization is mandatory when the correctness of the computation depends upon, or might depend upon, the exact order of computation. Common examples of serialization are multiple processors that want to write into the same region of memory, or two packets, arriving on different input links to a node in a network, that both want to exit on the same output link.

(**2**) converting an internal data structure representation to an external one, generally with the implication that the external representation could be converted to an equivalent internal representation at a later time. *See also* pickling.

serialization graph a means of representing a transaction history, showing interactions between reads and updates of data.

serially reusable a description of a section of code which can be executed by at most one thread of control at a time. If other threads of control must be locked out, a synchronization mechanism may be necessary to avoid concurrent access. *Contrast with* reentrant code.

serial multiplier a one-digit multiplier in which n-digit addition is performed by feeding in operands one digit-pair at a time.

serial operation data bits on a single line are transferred sequentially under the control of a single signal.

serial port a communications interface that supports bit by bit data transmission.

serial printing printing is done one character at a time. The print head must move across the entire page to print a line of characters. The printer may pause or stop between characters. Printing speed is usually given in units of characters per second (cps).

serial schedule a schedule where all operations for each transaction in the schedule are executed together; i.e., operations from different transactions are not interleaved.

serial transmission a process of data transfer whereby one bit is transmitted at a time over a single line.

server (**1**) a computer that serves, through storage of data and execution of programs, a number of peripheral computers or terminals connected through a computer network.

(**2**) a program that provides services to other programs (named client). The connection among clients and server is normally managed by means of message passing, often over a network. Sometimes a special protocol to encode or encrypt the client's requests and the server's responses is adopted. In general, the server runs continuously (e.g., a daemon) waiting for the client requests.

(3) in a queueing model, a serially reusable resource for processing tasks.

server stub system software residing in a server machine that implements the called procedure.

server utilization the fraction of time within a given time period that a resource is busy working on requests. The fraction can vary between 0 and 1. *See also* utilization.

service result generated by activities at the interface between the supplier and the customer and by supplier internal activities, to meet the customer needs. A service may be linked with the manufacture and supply of tangible product.

service control point (SCP) an on-line, real-time, fault-tolerant, transaction-processing database which provides call-handling information in response to network queries.

service demand the amount of work to be performed by a particular device for a job. Typically the service demand of a job at a device is expressed in terms of the overall service time required per job from the device.

service management system (SMS) an operations support system which administers customer records for the service control point.

service primitive the name of a procedure that provides a service; similar to the name of a subroutine or procedure in a scientific subroutine library.

service time the time a request needs to complete service in a server. In queueing models, the service time distribution can be specific to classes of requests or to servers.

session (1) the time and work performed by a computer system between logging in and logging off by a specific user.

(2) the context created by a database login.

(3) an instance of one or more protocols which provides the logical endpoints through which data can be transferred.

set a collection of unique items.

SET *See* specification for exchange of text.

set-associative cache a cache that is divided into a number of sets, each set consisting of groups of lines and each line having its own stored tag (the most significant bits of the address). A set is accessed first using the index (the least significant bits of the address). Then all the tags in the set are compared with that of the required line to find the line in the cache and to access the line. *See also* direct mapped cache, associativity.

set associative cache a cache in which a line or block from main memory can only be placed in a restricted set of places in the cache. A set is a group of two or more blocks in the cache. A block is first mapped onto a set (direct mapping defined by some bits of the address), and then the block can be placed anywhere within the set (fully associative within a set). *See also* direct mapped cache, fully associative cache.

set-membership uncertainty a model of uncertainty in which all uncertain quantities are unknown except that they belong to given sets in appropriate vector spaces. The sets are bounded and usually compact and convex. The estimation problem in this case becomes one of characterizing the set of states consistent both with the observations received and the constraints on the uncertain variables. Control objectives are usually formulated in terms of worse case design tasks, target sets reachability, guaranteed cost control, robust stability, or practical stabilization of the uncertain systems. For linear systems with energy-type ellipsoidal constraints imposed on the uncertain variables representing initial conditions, the additive disturbance and observation noise solution of the state estimation problem is given by the estimator similar to the Kalman filter, and a control problem in the form of min-max optimization of a given quadratic criterion leads to the linear-quadratic game. In the case of instantaneous ellipsoidal constraints, the exact solution of estimation and control problems is difficult to obtain, nevertheless by bounding recursively the sets of possible state approximating ellipsoids leading to suboptimal filtering and control laws similar to the optimal ones for the energy-type constraints. Generally efficient results might be found only for bounding sets parameterized by a little number of parameters. Except for the ellipsoids, such property is endowed only by polyhedral sets bounding uncertain variables. In this case efficient results could be reached by the use of linear programming algorithms.

set partitioning rules for mapping coded sequences to points in the signal constellation that always result in a larger Euclidean distance for a trellis coded modulation system than for an uncoded system, given appropriate construction of the trellis. Used in coded modulation for optimizing the squared Euclidean distance.

set theory theory governing sets and the set operators.

set type a description of a 1:N (1 to many) relationship between two record types.

setup program special program provided with a computer to set or reset the system configuration parameters.

severity synonymous with criticality. *See* failure severity.

SGNN *See* self-generating neural network.

SGNT *See* self-generating neural tree.

SGVQ *See* shape-gain vector quantization.

shadow casting logic gate an optical logic gate originally using shadow casting technique. The principle of shadow casting logic gate can be explained as follows. First, NOT A and NOT B are generated from inputs A and B. Second, four products of AB, A (NOT B), (NOT A) B, and (NOT A) (NOT B) are produced by passing a light beam through two transparencies that could be spatial light modulators representing A and B, A and NOT B, NOT A and B, and NOT A and NOT B. Third, the four products are added optically. The 16 combination of four products are the 16 Boolean logic operations.

shadowing (**1**) excess propagation loss resulting from the blocking effect of obstacles such as buildings, trees, etc.

(**2**) the statistical variation of propagation loss in a mobile system between locations the same distance from a base station, usually described by a log normal distribution.

shadow map a pre-computed array used to test if points on object surfaces are in shadow. The array contains depth values from the viewpoint of a point light source giving the distance to the first object surface encountered. If a given pixel in the environment is not contained in the array, then it is in shadow. This method is useful for quickly re-rendering an image from several different viewpoints or when several light sources are used — each would then have its own shadow map.

shadow paging a database recovery technique that uses copies of physical database pages.

shadow RAM an area of RAM into which the BIOS is copied in order to speed operating system execution.

shallow binding the implementation of dynamic name binding in languages such as LISP wherein the values associated with a name are maintained in a stack associated with the name. Entry into a scope in which a new name is declared adds the new value for the name to the head of the stack of values, and exiting the scope removes the value. *See also* deep binding.

shallow copy a mirror image of one object created by copying the contents of the source object. The copy is performed also on pointers and, therefore, the new object shares the same external resources of the original. The de-allocation of an external resource on an object causes an unpredictable behavior in the latter. *See* cloning.

Shannon, Claude (1916–1989) considered to be the founding father of modern electronic communications theory. His contributions include the application of Boolean algebra to analyze and optimize switching circuits and, in his classic paper "The Mathematical Theory of

Communication", he established the field of information theory by developing the relationship between the information content of a message and its representation for transmission through electronic media.

Shannon information the information content of an event x with a probability of occurrence of $p(x)$ defined as

$$I(x) = -\log p(x) .$$

The unit of $I(x)$ depends on the base of the logarithm — "bits" for base 2, "nats" for the natural logarithm. *See also* entropy.

Shannon's law *See* Shannon's source coding theorem.

Shannon's sampling theorem states that when an analog signal is sampled, there is no loss of information and the analog signal can be reconstructed by low-pass filtering, if and only if the largest (absolute value) frequency present in that signal does not exceed the Nyquist frequency, this being half the sampling frequency.

Shannon's source coding theorem a major result of Claude Shannon's information theory. For lossy source coding it gives a bound to the optimal source coding performance at a particular rate ("rate" corresponds to "resolution"). The theorem also says that the bound can be met by using vector quantization of (infinitely) high dimension. For lossless source coding the theorem states that data can be represented (without loss of information) at a rate arbitrarily close to (but not lower than) the entropy of the data. *See also* entropy, rate-distortion theory.

shape analysis the analysis of shapes of objects in binary images, with a view to object or feature recognition. Typically, shape analysis is carried out by measurement of skeleton topology or by boundary tracking procedures including analysis of centroidal profiles.

shape from the recovery of the 3D shape of an object based on some feature (e.g., shading) of its (2D) image.

shape function a function that gives the possible dimensions of the layout of a component with a flexible (or not yet completely determined) layout. For a shape function $s : [w_{\min}, \infty] \rightarrow [h_{\min}, \infty]$ with $[w_{\min}, \infty]$ and $[h_{\min}, \infty]$ subsets of $\Re^+$, $s(w)$ is the minimum height of any rectangle of width w that contains a layout of the component.

shape-gain vector quantization (SGVQ) a method for vector quantization where the magnitude (the gain) and the direction (the shape) of the source vector are coded separately. Such an approach gives advantages for sources where the magnitude of the input vector varies in time.

shape measure a measure such as circularity measure (compactness measure), aspect ratio, or number of skeleton nodes, which may be used to help characterize shapes as a preliminary to, or as a quick procedure for, object recognition.

shaping a traffic policing process that controls the traffic generation process at the source to force a required traffic profile.

shared-disk architecture a parallel database management system architecture where any processor has access to any disk unit through the interconnect but exclusive (nonshared) access to its main memory.

shared file system files residing on one computer that can be accessed from other computers.

shared lock a lock that permits concurrent readers access to an object.

shared memory characteristic of a multiprocessor system in which all processors in the system share access to the main memory. In a physically shared memory system, any processor has access to any memory location through the interconnection network.

shared memory architecture a computer system having more than one processor in which each processor can access a common main memory. *See* shared memory.

shared variable a variable that can be accessed by two or more concurrent program units.

sharpening the enhancement of detail in an image. Processes that sharpen an image also tend to strengthen the noise in it. *See also* edge enhancement, gradient, image enhancement, Laplacian operator, noise, Sobel operator.

shell (**1**) program that accepts user input and performs the operations specified by such input. This may involve performing specific operations, executing programs, and modifying state. A synonym is command interpreter.

(**2**) the language used by a shell. Often a full programming language constitutes a shell language. *See* shell language.

(**3**) common name for a Unix command processor.

shell language a language used to write programs for a shell or command interpreter.

shell program *See* shell script.

shell script a program file consisting of a series of shell commands and shell programming constructs. *Shell scripts* are ASCII text files.

shell sort a diminishing increment sort where the first increment is the smallest power of 2 less than half the number of items to be sorted ($2^{\lfloor \log_2 n \rfloor - 1}$) and each succeeding increment is half.

shell variable a name-value pair created and maintained by the shell; often used to specify operational parameters to the shell.

shift a low-level operation on the bits of a value in which the result is computed by changing the positions of the bits by a linear motion towards the high-order bit ("shift left") or the low-order bit ("shift right"). If the high-order bit represents the sign, there may be variants such as signed shift and unsigned shift. *Contrast with* rotate. In some languages, these operations are directly visible to the programmer. Whether these operations are visible or not, compilers may take advantage of the existence of such operations to generate more efficient code.

shifter that component of the computer which performs a shift or rotate operation. The properties of a *shifter* are usually of great interest to a compiler writer, since the performance of the compiled code can be greatly improved if an efficient shifter is known to exist.

shift instruction a program instruction in which data in a register or memory location is shifted one or more bits to the left or right. Data shifted off the end of the register or memory location is either shifted into a flag register, used to set a condition flag, or dropped, depending on implementation of the instruction.

shift register a register whose contents can be shifted to the left or right.

Shockley, William (1910–1989) Born: London, England.

Shockley is best known as one of the developers of the transistor. In 1956 Shockley, along with John Bardeen and Walter Brattain, received the Nobel Prize for his work. Schockley led the group at Bell Labs responsible for the semiconductor research that led to the development of the "point-contact transistor". In later life Shockley became known for his controversial public pronouncements on various political and genetic issues.

short-circuit evaluation a term applied to the evaluation of Boolean expressions which implies that only as much of the expression needs to be evaluated as will uniquely determine the result. For example, in the expression (A and B), if A evaluates to false, B need not be evaluated at all, since its evaluation will not change the value of the expression. Generally used in programming languages so illegal conditions can be bypassed; for example, "if n is not zero and q divided by n is greater than 5" means that if n is zero, the division will not occur because the conjunction is false. Note that some languages will not short-circuit an evaluation if they detect that there are side effects from the evaluation of later terms. Other languages state that either the side effects are guaranteed to not occur, or that the meaning of the program is undefined because the side effects may or may not occur.

shortcutting *See* pointer jumping.

short duration transaction a transaction that completes in a short time period.

shortened code a code constructed from another code by deleting one or more message symbols in each message. Thus, an (n, k) original code becomes an $(n - 1, k - 1)$ code after the deletion of one message symbol.

shortest common superstring the shortest possible string that contains as substrings a number of given strings.

shortest job first a scheduling discipline that selects the job with the smallest execution time to run on the node.

shortest job next a scheduling policy that relies on running the program unit with the shortest job.

show stopper a problem that is so severe that it stops or cancels an entire project.

shrinking phase the phase in which locks are released by a transaction operating under the two-phase locking protocol.

shuffle *See* permutation.

sibling node a sibling is a node in a tree that has the same parent node as another node. The two nodes are siblings.

side effect an effect, intended or unintended, of performing a computation. Even intended *side effects* can change the meaning of a program if they violate referential transparency; for example, "$h(x) + g(x)$" appears to be an addition and, consequently, under the commutative rule should be equivalent to "$g(x) + h(x)$", but only if the computations of function g cannot affect the computation of function h and vice-versa. When side effects are present, the language's specification of evaluation order may be critical, or the compiler's ability to reorder evaluations may be limited by the language's specification of evaluation order. *See also* aliasing, short-circuit evaluation.

Sierpinski gasket a fractal created by repeatedly dividing a square into nine equal sized squares and removing the middle one.

Also known as a Sierpinski carpet.

Sierpinski triangle a fractal generated by repeatedly dividing a triangle into four self-similar ones and removing the inner fourth one.

sieve of Eratosthenes an algorithm to find all prime numbers up to a certain N. Begin with an (unmarked) array of integers from 2 to N. The first unmarked integer, 2, is the first prime. Mark every multiple of this prime. Repeatedly take the next unmarked integer as the next prime and mark every multiple of the prime.

sigmoid function a compressive function that maps inputs less than -1 to approximately zero, inputs greater than 1 to approximately 1, and maps values from -1 to 1 into the range 0 to 1. A common sigmoid function is

$$\frac{1}{1 + e^{-\frac{x}{T}}} .$$

Sigmoid functions are often used as activation functions in neural nets. *See also* activation function.

signal (**1**) in synchronization, an indication that a semaphore, mutex, or other synchronization primitive may permit one or more threads of control waiting on the synchronization primitive to proceed in execution.

(**2**) in some languages, a mechanism that causes a control transfer to a specified handler. This transfer may be asynchronous, for example, as an interrupt, or synchronous, performed at a point chosen by the programmer.

signal averaging an averaging process which is used to enhance signals and suppress noise, thereby improving the signal to noise ratio. *See also* averaging.

signal decimation *See* decimation.

signal detection detecting the presence of a signal in noise.

signaling channel a means of information flow inherent in the basic model, algorithm, or protocol and, therefore, implementation invariant.

signal processing a generic term which refers to any technique that manipulates the signal, including but not limited to signal averaging, signal conditioning, and signal recognition. When applied to images, it is normally referred to as image processing, the term signal processing usually being reserved for 1D signals.

signal recognition the recognition of signals by appropriate analysis, often with the help of filters such as matched filters or frequency domain filters.

signal variance *See* variance.

sign-and-magnitude notation the computer equivalent of paper-and-pencil representation of signed numbers. A number is represented by one bit or digit used for the sign and other digits used for the unsigned magnitude.

signature (**1**) a characteristic easily computed feature or function by which a particular object or signal may be at least tentatively identified. An example is the centroidal profile for an object having a well-defined boundary.

(**2**) a prototype of a function. Provides a definition, often as a forward reference, to a function to simplify the design of a compiler.

(**3**) a value that is used as a component of a data structure to identify the nature of the information contained therein.

signature analysis an analysis of the signature to extract the desired (signal) information.

sign bit the bit, typically the high-order bit of a numeric value, that indicates its sign. In most machines, 2's-complement representation is used, and the high-order bit is 0 for positive values and 1 for negative values. *See also* unsigned integer.

signed-digit representation a fixed-radix number system in which each digit has a sign (positive or negative). In a binary signed-digit representation, each digit can assume one of the values -1, 0, and 1.

signed shift a shift operation in which a shift toward the low-order bit creates new high-order values by copying the sign bit as often as needed to supply the new values. *Compare with* unsigned shift.

sign flag a bit in the condition code register which indicates whether the numeric result of the execution of an instruction is positive or negative (1 for negative, 0 for positive).

significand the mantissa portion of a floating-point number in the IEEE 754 floating-point standard. It consists of an implicit or explicit leading integer bit and a fraction.

significant digit an approximate value represented in radix r is said to be correct to n significant digits if the absolute error from the corresponding exact result is no greater than $r^{-n}/2$.

sign-magnitude representation a number representation that uses the most significant bit of a register for the sign and the remaining bits for the magnitude of a binary number.

signum function the function

$$\mathrm{sgn}(t) = \begin{cases} +1 & t > 0 \\ 0 & t = 0 \\ -1 & t < 0\,. \end{cases}$$

Used in modeling numerous types of system functions.

silver bullet syndrome the idea that a single tool or method can make large improvements solving all problems in productivity or other high level feature.

SIMD pronounced "sim-dee". *See* single instruction stream, multiple data stream.

similarity measure the reciprocal concept of distance measure. *See* distance measure.

SIMM *See* single-inline memory module.

simple attribute *See* atomic attribute.

simple cell refers to the possibility of dividing the receptive fields of visual cells into separate excitory and inhibitory zones.

simple tandem queue a queueing system where customers follow one routing chain from the outside world to the first queue, then to the second queue, and finally return to the outside world. *See also* tandem queues.

SIMULA a language originally designed as a discrete event simulation language. A later implementation, SIMULA-67, is among the earliest object-oriented languages. SIMULA-67 is a direct ancestor of C++.

simulated annealing an optimization technique that seeks to avoid local minima by allowing the search trajectory to follow paths that not only decrease the objective function but also sometimes increase it. The probability that an increase in the objective function is allowed by the technique is governed by a quantity that is analogous to temperature. The scheme commences with a high temperature, under which the probability of allowing increases in the objective function is high, and the temperature is gradually reduced to zero, and from then on no further increases in the function are allowed.

simulation (**1**) a replication of a real-world system, typically computerized; the replication may be of the form or functionality of the system, or of both, and it may be, but is not necessarily, interactive in real-time. *See* computer simulation.

(**2**) for executable specification models, permits to animate the system to find the behavior of the system in response to particular inputs. It is used to validate the system (or sub-system) specification comparing the desired results with the effective results. For complete validation it is infeasible and model checking is preferred.

simulation language a language designed to simulate some real-world phenomenon. *Simulation languages* are often domain-specific languages with particular features that make the type of simulation they perform easier to express. Discrete Event Simulation Languages simulate systems characterized by the arrival and processing of distinct events, such as queuing systems. Continuous Simulation Languages simulate systems characterized by complex but mathematically continuous problems such as fluid dynamics, stress analysis, and the like. Simulation languages are used when the problem does not have a closed form mathematical solution and the results must be computed at each individual step.

simulation model an abstraction of a real system allowing its main characteristics to be simulated.

simulation theorem any theorem showing that one kind of computation can be simulated by another kind within stated complexity bounds. Most known containment or equality relationships between complexity classes have been proved this way.

simulation time the time in a system that is being modeled. Usually referred to as the physical system, viewed as being composed of some number of physical processes that interact at various points in simulated or virtual time. The time in the simulator of the physical system is referred to as the real time or the wall-clock time.

simulator a program used to predict the behavior of a circuit.

simultaneous pertaining to the occurrence of two or more events at the same time.

simultaneous contrast the phenomenon in which the brightness (perceived luminance) of a region on a dark background is greater than the brightness of an identical region on a light background. Illustrates that brightness (perceived luminance) is different from lightness (actual luminance). *See also* brightness, human visual system, Mach band.

simultaneous resource possession a complicating feature of certain types of performance models (for example, central server models) where customers must have control of several different types of resources at the same time (such as main memory and the CPU) before service can begin.

single-address computer a computer based on single-address instructions. In such a machine each instruction specifies at most one operand, the second operand (when required) being implicitly a single designated register, the accumulator. *See also* two-address computer, zero-address computer.

single-address instruction a CPU instruction defining an operation and exactly one address of an operand or another instruction.

single-address machine *See* single-address computer.

single-chip microprocessor a microprocessor that has additional circuitry in it that allows it to be used without additional support chips.

single-destination shortest-path problem a problem in computability theory. The problem is to find the shortest path from each vertex in a weighted, directed graph to a specific destination vertex. Equivalent to the single-source shortest-path problem with all directions reversed. *See also* graph, all pairs shortest path.

single-extended precision *See* single precision.

single inheritance a subclass that has only one class as a parent class. The obtained class (named derived class or subclass) inherits the instance variables (attributes) and methods having predefined visibility scopes (i.e., in C++ public and protected) and may add new attributes and new methods or overwrite the existing methods, adding functionalities. *See* inheritance, multiple inheritance.

single-inline memory module (SIMM) a miniature circuit board that contains memory chips and can be plugged into a suitable slot in a computer motherboard in order to expand the physical memory.

single inline packaging (SIP) a method of packaging memory and logic devices on small PCBAs with a single row of pins for connection.

single-input–single-output (SISO) system a system that transforms one input signal to one output signal. Also known as single variable (SV) system. *See also* multiple-input–multiple-output system.

single instruction stream, multiple data stream (SIMD) a computer where each processing element is executing the same (and only) instruction, but on different data. SIMD computers include systolic, wavefront, and vector processors. Pronounced "sim-dee". *See* Flynn's taxonomy.

single instruction stream, single data stream (SISD) a computer architecture where the CPU processes a single instruction at a time and a single datum at a time. von Neumann processors are the most common form of SISD processor. *See* Flynn's taxonomy.

single layer perceptron an artificial neural network consisting of a single layer of neurons (perceptrons) with an input layer. *See also* multilayer perceptron, perceptron.

single point failure a system failure due to the failure of a single component.

single precision in a given computer system, the smallest format used for the representation of floating-point numbers; in IEEE-754, the single-precision format is 32 bits wide. Single-extended precision uses more bits but no more that twice single precision; in IEEE-754, single-extended precision is between 43 and 64 bits wide.

Historically, single precision computations could be performed more quickly than double precision computations, hence their reflection into programming languages.

single program, multiple data (SPMD) the dominant style of parallel programming, where all of the processors utilize the same program, though each has its own data.

single-server queue a model in which a queue is served by one server.

single-source shortest-path problem a problem in computability theory. The problem is to find the shortest paths from a specific source vertex to every other vertex in a weighted, directed graph. *See also* all pairs shortest path, single-destination shortest-path problem, Dijkstra's algorithm.

single-step to operate a processor in such a way that only a single instruction or machine memory access cycle is performed at a time and enables the user to examine the status of processor registers and the flags. A common debugging method for small machines.

single-system model the computational model in which all computation takes place on a single processor.

single-user benchmark a benchmark where only a single user makes requests.

single-valued attribute an attribute that may have only one value for any entity.

singly linked list a list in which each cell has a pointer only to its successor cell.

singularity analysis a complex asymptotic technique for determining the asymptotics of certain algebraic functions.

singular matrix a square matrix A is singular if its rows (or columns) are not linearly independent. Singular matrices cannot be inverted, have zero determinants, and have linearly dependent columns and rows.

singular set a set with no owner record type.

singular value decomposition (SVD) useful decomposition method for matrix inverse and pseudoinverse problems, including the least-squared solution of overdetermined systems. SVD represents the matrix A in the form $A = U\Lambda^{\frac{1}{2}}V$, where Λ is a diagonal matrix whose entries are the singular values of A, and U and V are the row and column eigenvector systems of A. Any matrix can be represented in this way. In image processing, SVD has been applied to coding, to image filtering, and to the approximation of non-separable 2D point spread functions by two orthogonal 1D impulse responses.

sink node **(1)** a node in a network model which receives traffic on its inputs, but does not generate any traffic back into the network.

(2) an output node.

sinusoidal coding parametric speech coding method based on a speech model where the signal is composed of sinusoidal components having time-varying amplitudes, frequencies, and phases. *Sinusoidal coding* is mostly used in low bit rate speech coding.

SIP *See* single inline packaging.

SISD *See* single instruction stream, single data stream.

SISO *See* single-input–single-output system.

six connected *See* voxel adjacency.

six-sigma quality level this refers to a quality improvement method using statistics called six sigma. The six sigma quality level is equivalent to saying that the defect removal efficiency is very high (about 99.9999%).

size distribution for a family of objects, a function measuring the number or volume of all objects in any size range. In mathematical morphology, this notion is developed by analogy with a family of sieves: each sieve retains objects larger than a given size, and lets smaller objects go through; when two sieves are put in succession, this is equivalent to using the finest sieve. This idea is formalized by taking a one-parameter family of morphological operators γ_r $(r > 0)$ such that for all objects X, Y, and $r, s > 0$, one has

1. $\gamma_r(X) \leq X$, that is, γ_r is anti-extensive;
2. $X \leq Y$ implies that $\gamma_r(X) \leq \gamma_r(Y)$, that is, γ_r is increasing;
3. $\gamma_r(\gamma_s(X)) = \gamma_{\max(r,s)}(X)$.

In particular, each γ_r is an algebraic opening, and for $s > r$ we have $\gamma_s(X) \leq \gamma_r(X)$.

See mathematical morphology, morphological operator, opening.

skeleton a framework capturing the structure of an object or shape constructed from a series of points connected by thin lines. In a similar way to a wireframe representation, a skeleton is used to increase the performance of the rendering system since it is not necessary to render solid surfaces. Objects represented using a skeleton can be given a skin by specifying a diameter from the skeleton used to render the surface. In cartoon animation, a skeleton is literally a line structure representing the position of the limbs of a figure and is not necessarily oriented along the medial axis.

A typical application relates to interpretation of hand-drawn characters and script.

skew *See also* tape skew.

skewed addressing *See* interleaved memory.

skip instruction an assembly language instruction that skips over the next instruction without executing it.

Skolem standard form a form for first-order logical formulas in which all existentially quantified variables are replaced by constants or by functions of constants and universally quantified variables.

SLDNF-resolution a concept in database theory. Extends SLD-resolution for deductive databases with default negation in rule bodies. The SLDNF-resolution of a default negated literal $\natural C$ succeeds (fails), iff the SLD-resolution for C fails (succeeds).

SLD-refutation a concept in database theory. If through a sequence of steps of SLD-resolution for a deductive database DB starting with a goal "$G_0 \{A_1 \wedge \ldots \wedge A_m\}$" the goals $G_0, G_1, \ldots, G_k$ have been constructed, where the final goal "$G_k \{\}$" has zero body atoms, i.e., G_k denotes the textitcontradiction, then this shows that $DB \cup \{G_0\}$ is inconsistent, which in turn gives a proof of the conjunction $A_1\theta \wedge \ldots \wedge A_m\theta$ from DB, where θ is the composition of the unifiers that have been used.

SLD-resolution a concept in database theory. Stands for linear resolution with selection function for definite clauses. Given a goal $G \{A_1 \wedge \ \ldots \wedge A_m\}$ and a selection function s such that $s(G) = A_i$ is the selected atom of G, a resolving definite clause $A \{B_1 \wedge \ \ldots \wedge B_n\}$ is searched for in the database DB, such that there exists a unifier θ for A and A_i. Then a new goal $G' \{(A_1 \wedge \ \ldots \wedge A_{i-1} \wedge B_1 \wedge \ldots \wedge B_n \wedge A_{i+1} \wedge \ldots \wedge \mathbf{A_m})\,\theta\}$ is constructed by replacing A_i in G.

sleep a command to suspend a process for a predefined time interval.

slice for an n-dimensional array, specifying no more than $n - 1$ subscripts defines a subset of the array called a slice.

slicer a device that estimates a transmitted symbol given an input which is corrupted by (residual) channel impairments. For example, a binary slicer outputs 0 or 1, depending on the current input.

slicing floor plan a floor plan which can be obtained by the recursive bipartitioning of a rectangular layout using vertical and horizontal line segments.

sliding window in an ARQ protocol, the (sliding) window represents the sequence numbers of transmitted packets whose acknowledgments have not been received. After an acknowledgement has been received for the packet whose sequence number is at the tail of the window, its sequence number is dropped from the window and a new packet whose sequence number is at the head of the window is transmitted, causing the window to slide one sequence number.

slow start in TCP a characteristic of the sliding window protocol that is used in TCP to guarantee reliability between the sender and the receiver. The sending window starts out small and gradually increases in size as a function of the measured round-trip time and receipt of successful acknowledgments. This characteristic causes the sender to initially pause and wait for acknowledgements on the first packets that are sent. If acknowledgments are received, then the sending window is increased so that the sender no longer has to wait.

small computer systems interface (SCSI) a high-speed parallel computer bus used to interface peripheral devices such as disk drives. Pronounced "scuzzy".

Smalltalk an object-oriented polymorphic language developed at Xerox Palo Alto Research Center.

smart card credit-card-sized device containing a microcomputer, used for security-intensive functions such as debit transactions.

smoothing any process by which noise is suppressed, following a comparison of potential noise points with neighboring intensity values, as for mean filtering and median filtering. Also, a process in which the signal is smoothed, e.g., by a low-pass filter, to suppress complexity and save on storage requirements.

smoothing filter an operation applied to an image to suppress random additive pixel noise.

smooth shading a method of polygon shading where calculations are performed for the vertices, and values for pixels inside the polygon are derived from linear interpolation of the vertex values.

SMP *See* symmetric multiprocessor.

SMS *See* service management system.

snake energy-minimizing contour that combines internal constraints on its shape, such as smoothness of the contour, and external constraints from the image, such as brightness or gradient magnitude. Also called an active contour.

snapshot a dynamic dump of the contents of one or more specified memory areas.

Snell's law a law defining how light is bent or refracted when it passes through a boundary between two dielectric media of different indices of refraction, such as air and glass or air and water. It is expressed by $n_1 \sin(\phi_1) = n_2 \sin(\phi_2)$ where n_1 and n_2 are the index of refraction of the two media. ϕ_1 and ϕ_2 are the angles which the boundary surface normal makes with the incident light ray and the refracted light ray, respectively.

SNOBOL a string-pattern-matching language. *See also* ICON, awk.

snooping bus a multiprocessor bus that is continually monitored by the cache controllers to maintain cache coherence.

snoopy cache a type of cache used in symmetric multiprocessing computers. In such a configuration each processor usually has a private cache as well as shared main memory. With a snoopy cache, when one processor performs a write to main memory across the shared system bus, the other processor will monitor the bus to detect this write, and update any values that are in the local cache at the time of the write. The next time a read is performed from the second processor, the value in the local cache of this processor will be valid.

snow noise noise composed of small, white marks randomly scattered throughout an image. Television pictures exhibit snow noise when the reception is poor.

Sobel masks a set of two images that when used in windowed convolution provides an edge filter. *See* Sobel operator.

Sobel operator a common digital approximation of the gradient ∇f, often used in edge detection. It is specified by the pair of convolution masks

$$\frac{\partial f}{\partial x} \approx \begin{matrix} -1 & 0 & 1 \\ -2 & 0 & 2 \\ -1 & 0 & 1 \end{matrix}, \quad \frac{\partial f}{\partial y} \approx \begin{matrix} -1 & -2 & -1 \\ 0 & 0 & 0 \\ 1 & 2 & 1 \end{matrix}.$$

The respective mask responses, g_x, g_y, are components of the local vector gradient, from which edge magnitude and direction can be calculated straightforward.

soft computing an association of computing methodologies centering on fuzzy logic, artificial neural networks, and evolutionary computing. Each of these methodologies provides us with complementary and synergistic reasoning and searching methods to solve complex, real-word problems. *See also* fuzzy logic, neural network.

soft deadline a deadline that can always be missed without severe consequences.

soft decision demodulation that outputs an estimate of the received symbol value along with an indication of the reliability of that value. It is usually implemented by quantizing the received signal to more levels than there are symbol values.

soft-decision decoding decoding of encoded information given an unquantized (or finely quantized) estimate of the individually coded message symbols (for example the output directly from the channel). *Compare with* hard-decision decoding.

soft failure *See* fail soft.

soft link a directory entry that refers to a file via another directory entry and provides indirect access to a file. *Compare with* hard link.

soft real-time system a real-time system in which failure to meet deadlines results in performance degradation but not necessarily failure. *Compare with* firm real-time system, hard real-time system.

software all or part of the programs, procedures, associated data, use, operation, rules, and associated documentation of an information processing system. Software is an intellectual creation that is independent of the medium on which it is recorded.

software bottleneck a saturated software process that limits the increase in system performance. Typically occurs in multi-layered synchronous client–server systems in which a server is busy almost 100% of the time either executing or waiting for response from lower level servers.

software characteristic *See* characteristic.

software complexity *See* complexity.

software component a component that consists solely of software.

software defect a perceived departure in a software product from its intended properties, which if not rectified, would under certain conditions contribute to a software system failure (departure from required system behavior during operational use).

software design a phase of software development life-cycle that maps what the system is supposed to do into how the system will do it in a particular hardware/software configuration.

software development cycle *See* life-cycle.

software development life-cycle a way to divide the work that takes place in the development of an application.

software engineering (**1**) a systematic approach to the analysis, design, assessment, implementation, test, maintenance, and reengineering of software; that is, the application of engineering to software. In the software engineering approach, several models for the software life cycle are defined, and many methodologies for the definition and assessment of the different phases of a life-cycle model are characterized and exploited. Software engineering methodologies and techniques often use CASE tools for aiding the process of development and assessment.

(**2**) the study of the techniques and approaches according to (**1**).

software evolution the accommodation of perfective, corrective, and adaptive maintenance, which may involve some reengineering activity. Those activities which are geared toward improving the software itself, rather than increasing one's understanding of the same. Examples include restructuring, redocumenting, and data reengineering. All are meant to evolve the subject system from its current form to one that better meets the new requirements.

software factory a form of software development first popularized in Japan. The idea is to develop software in the same manufactured manner as tangible items in a normal factory.

software failure system failure due to the activation of a design fault in a software component. All *software failures* are design failures because software consists solely of design and does not wear out or suffer from physical failure.

software fault a design fault located in a software component. *See* fault.

software house a company mainly devoted to produce software products.

software inspection a process whereby a team of software developers either follows sequences of state changes or possible execution paths resulting from a particular series of events or inputs or, using a checklist of potential errors, determines if similar errors are present in the code.

software interrupt a machine instruction that initiates an interrupt function. Software interrupts are often used for system calls because they can be executed from anywhere in memory and the processor provides the necessary return address handling. Also known as a Supervisor Call instruction (SVC) (IBM mainframes) or INT instruction (Intel X86).

software library a controlled collection of software modules and related documentation designed to aid in software development, use, and maintenance.

software life cycle the sequence of processes performed when developing and maintaining software. *See* life-cycle process.

software maintenance the set of modifications to be performed on a software system in order to address new functionality, correct errors, and improve system characteristics.

software methodology the study of how to navigate through each phase of the software process model (determining data, control, or use hierarchies, partitioning functions, and allocating requirements) and how to represent phase products (structure charts, stimulus-response threads, and state transition diagrams).

software metric a metric for assessing software through measures. *See* metric, measure.

software off-the-shelf *See* off-the-shelf software.

software package a complete and documented set of programs (synonymous with application software) supplied to several users for a generic application or function. Some software packages are alterable for a specific application. *See* package.

software problem report *See* error report.

software process improvement the mechanism of improvement of the software development process. The improvement is typically reached by establishing a quality and cost control and by taking decision on the basis of the experience made on the quantitative analysis of the assessment result. In this way, a virtuoso mechanism may be established for continuously improving the process of software development.

software product a set of computer programs, procedures, and possibly associated documentation and data. *See* product.

software psychology area of study that attempts to discover and describe human limitations in interacting with computers. These limitations can place restrictions on and form requirements for computing systems intended for human interaction.

software quality the totality of features and characteristics of a software product that bear on its ability to satisfy stated or implied needs, including internal and external aspects and contributing to functionality, reliability, usability, efficiency, maintainability, and portability of the system under development.

software quality assessment criteria the set of defined and documented rules and conditions which are used to decide whether the total quality of a specific software product is acceptable or not. The quality is represented by the set of rated levels associated with the software product.

software quality assurance (SQA) a series of activities that assist an organization in producing high quality software. *See* assessment, quality.

software quality characteristic one of a set of attributes of a software product by which its quality is described and evaluated. A *software quality characteristic* may be refined into multiple levels of sub-characteristics. These characteristics are functionality, reliability, usability, efficiency, maintainability, and portability.

software quality metric a quantitative scale and method that can be used to determine the value a feature takes for a specific software product.

software reengineering the reverse analysis of an old application to conform to a new methodology.

software renovation *See* software reengineering.

software requirements specification a deliverable that describes all data, functional and behavioral requirements, constraints, and validation requirements for software.

software reuse the process that describes the ability to reuse existing software in new applications. When software is reused in its entirety without changes, a gain in productivity is attained. Critical for object-oriented design.

software science a set of metrics also known as Halstead metrics.

software size a measure of the dimension of a software program. This measure is typically performed by counting:

1. the lines of code,
2. the number of tokens used (according to a specific language), or
3. the number of lines (carriage return) of code.

Different size measuring techniques may have different consistency and reliability on different languages and systems. Some of these may also depend on the style in which the program has been written. This is obviously an unsuitable feature.

software system architecture a definition of the software system in terms of computational components and interactions among the components. Components are things such as clients and servers, databases, filters, and layers in a hierarchical system. Interactions among components can be procedure call and shared variable access, or client-server protocols, database-accessing protocols, asynchronous event multicast, and piped streams.

software testing the process of examining the behavior of a program or part of a program by executing it on selected or exhaustive sets of input data. Testing is performed to discover defects in a program. Testing is not the process of confirming that a program is correct, nor is it a demonstration that defects are not present.

software tool a computer program used to help develop, test, analyze, configure, or maintain another computer program or its documentation; for example, automated design tools, compiler, test tools, maintenance tools.

software transformation *See* program transformation.

software translation *See* program translation.

software volume a measure of the system volume with specific metrics that take into account the number of operators and operands, an example is the program volume metric of Halstead. *See* Halstead's metrics.

solid state disk (SSD) very large-capacity, but slow, semiconductor memory that may be used as a logical disk, extended main memory, or as a logical cache between main memory and conventional disk. SSD is typically constructed from DRAM and equipped with a battery to make it non-volatile. First used in IBM 3090 and Cray-XMP computer systems.

solvable problem a computational problem that can be solved by a halting GOTO program. The problem may have a nonbinary output.

SOM *See* self-organizing map.

sonorant the class of phonemes with a formant-like spectrum. For example, vowels and nasals exhibit a spectrum that is based on formants.

SONT *See* self-organizing neural tree.

SOP *See* sum of products.

sort the problem of arranging items in a predetermined order. There are dozens of algorithms, the choice of which depends on factors such as the number of items relative to working memory, knowledge of the orderliness of the items or the range of the keys, the cost of comparing keys vs. the cost of moving items, etc. *See also* quicksort, insertion sort, radix sort, bucket sort, merge sort, shell sort, permutation.

sorted file a file where the records are maintained in the file in order of the ordering key.

sorting array a data structure where the data to be sorted is placed in an array and access to individual items can be done randomly. The goal of the sorting is that the ascending order matches the order of indices in the array.

sorting liked list a data structure where the data to be sorted is a sequence represented as a linked list. The goal is to rearrange the pointers of the linked list so that the linked list has the data in sorted order.

sort merge a sorting algorithm used to sort records stored on disk. Initially, N blocks of file are read into memory, sorted in memory, and then written to disk. The second stage involves merging these sorted sections.

sort-merge join an implementation of the join operator which is applicable if both relations are physically sorted on the join attribute. Both files are scanned in linear order with join attributes of tuples in the two files checked during the scan.

sound deductive system a deductive system that produces only true statements of a theory.

soundness in artificial intelligence, a property of inference rules. An inference (or rewrite rule) is sound if every inferred formula is a logical consequence of the original formula.

source traditionally, a textual representation of a program. Typically converted to a more suitable representation by another program, variously designated a compiler or assembler. Some source notations may be read directly as they are being executed, by a program often referred to as an interpreter. Many programming systems actually store the source code in a non-textual form and generate the textual representation on demand.

source code **(1)** computer instructions and data definitions expressed in a form suitable for input to an assembler, compiler, or other translator.

(2) a set of codewords used to represent messages, such that redundancy is removed, in order to require less storage space, or transmission time.

source code metric a metric that is evaluated directly on code and that evaluates internal characteristics of the system. *See* McCabe's metric, Halstead's metrics.

source coding the process of mapping signals onto a finite set of representative signal vectors referred to as codewords.

source compression *See* source coding.

source encoder a device that substantially reduces the data rate of linearly digitized audio signals by taking advantage of the psychoacoustic properties of human hearing, eliminating redundant and subjectively irrelevant information from the output signal. Transform source encoders work entirely within the frequency domain, while time-domain source encoders work primarily in the time domain. Source decoders reverse the process, using various masking techniques to simulate the properties of the original linear data.

source module a program unit that is the input of a translation.

source node a node in a network model which generates traffic, but does not receive any traffic from other nodes in the network; an input node.

source operand in ALU operations, one of the input values.

space-constructible function a function $s(n)$ that gives the actual space used by some Turing machine on all inputs of length n, for all n.

space sharing in multiple processor systems, if more than one application is to be run on the system at the same time, then space sharing occurs if some processors are allocated to the first application, some to the second application, and so on, at the same time. An alternative is time-sharing, where all processors are allocated to all executing applications at the same time, and each processor is time-shared by the processes in the applications.

space-time the product of the amount of memory and the amount of time used by a process.

space-time constraint method for creating automatic character motion by specifying what the character has to be, how the motion should be performed, what the character's physical structure is, and what physical resources are available to the character to accomplish the motion.

spaghetti code a programming style in which the control is so complex that it is likened to attempting to follow a thread of spaghetti through a plate of spaghetti. The term is generally attributed to Edsgar Dijkstra. Modern programming languages attempt to eliminate the temptation to write such complex code by providing well-structured control mechanisms that reduce the need for the use of the GOTO statement. Some languages (notably BLISS) eliminated the GOTO entirely. *See also* McCabe's metric.

spanned organization a file organization in which records may span more than one block. This is achieved by maintaining a pointer in one block referencing the remainder of a record in another block.

spanning tree a connected, acyclic subgraph containing all the vertices of a graph.

SPARC *See* scaled processor architecture.

sparse array an array that has very few values defined. A programming language or its implementation may choose to provide a more compact representation for a sparse array.

sparse graph a graph in which $|E| \ll |V|^2$.

sparse index *See* non-dense index.

sparse matrix a matrix that has relatively few non-zero (or "interesting") entries. It may be represented in much less than $n \times m$ space. *See also* ragged matrix.

sparse vector in computer instruction processing, a matrix in which most elements have such small values that they are treated as zeros. Special representation schemes can be used to save memory space, with a cost of increased execution time to access single elements of the matrix.

sparsification technique for designing dynamic graph algorithms, which when applicable transform a time bound of $T(n, m)$ onto $O(T(n, n))$, where m is the number of edges, and n is the number of vertices of the given graph.

sparsity used here to refer to LCS problem instances in which the number of matches is small compared to the product of the lengths of the input strings.

spatial database databases that store data and information pertaining to spatial locations. Support is provided for efficient querying and indexing of spatial locations.

spatial locality *See* locality.

spatial navigation the process of orienting and moving through a virtual environment.

spatial partitioning a technique used to divide a large task into a series of smaller ones. The basic approach is to devise a preprocessing stage which determines spatially coherent groups for processing. This strategy is particularly appropriate for parallel architectures where the groups can be sent to different processing units.

spatial query a query against a spatial database. Usually involves querying data referring to spatial locations.

spatial resolution (**1**) the ability to resolve two closely spaced points or a periodic pattern. Rayleigh proposed the criterion that two stars could be resolved when the maximum in the image pattern from one star coincides with the first minimum in the other. Units of spatial resolution are lines or line pairs per millimeter.

(**2**) a measure of the ability of a system to resolve spatial details in a signal. For a discrete image, spatial resolution generally refers to the number of pixels per unit length, giving possibly different horizontal and vertical spatial resolutions.

spatio-temporal motion detection a motion detection technique for three-dimensional images where filters are applied to the volume image in space-time. *Compare with* gradient and correspondence motion detection.

spawn to create a new concurrent program unit.

SPEC *See* System Performance and Evaluation Cooperative.

SPEC benchmarks a series of benchmarks put forward by the Standards Performance Evaluation Corporation (SPEC). *SPEC benchmarks* enable a common measure of performance among processors of different architectures and is source code independent.

special file a point of interface to one of the computer's hardware devices or a synchronized communications channel (pipe) between cooperating programs.

specialization (**1**) a technique by which more specific or specialized classes are created from more general classes.

(**2**) the relationship that links the specialized class to the more general parent. In object-oriented modeling, this is the specialization or specialization inheritance relationship. *See* specialization, is-a and is-a-kind-of, relationship.

special-purpose digital signal processor digital signal processor with special feature for handling a specific signal processing application, such as FFT

specification (**1**) a statement of the design or development requirements to be satisfied by a system or product.

(**2**) document stating requirements. A qualifier should be used to indicate the type of specification such as product specification or test specification. A specification should refer to or include drawings, patterns, or other relevant documents and indicate the means and the criteria whereby conformity can be checked.

specification and description language (SDL) a language standardized by the ITU-T well suited to functional design of reactive systems comprising concurrent processes with state-transition behavior. More recently, an object-oriented version of the language has been proposed.

specification fault a design fault of an item which results from its required function having been incorrectly or incompletely defined. *Specification faults* often give rise to usability problems in operation, but can lead to other types of incident also. They can only be detected by validation, not verification.

specification for exchange of text (SET) a data exchange standard for textual information.

specification inheritance a type of inheritance in which the subclass conforms to its parent class; i.e., subtyping. This use of *specification inheritance* supports polymorphism.

specification language a language for the specification of systems. It is typically a formal language with the support of techniques for the verification and validation of the specifications. In some cases, the specification can be used as the semantic level of CASE tool. The CASE tool results in the interface by means of which the specifications are given in a simplified and supported manner.

specification list in the C++ language, an optional part of function declaration that identifies the exceptions that can be thrown.

specification model a model for the specification of systems. It is typically a formal model supporting techniques for the verification and validation of the specifications. In some cases, the specification can be used as the semantic level of CASE tool. The CASE tool results in the interface by means of which the specifications are given in a simplified and supported manner.

speckle granular image noise due to fluctuations in the number of photons arriving at an image sensor. *Speckle* often occurs in night-vision equipment and X-ray images. Also called quantum mottle.

specular component a component of light reflected from a point in the mirror direction — as if the surface was a perfect mirror. In practice this component is "empirically spread" to simulate a practical glossy surface.

specular highlight a bright spot or highlight on an object caused by angular-dependent illumination. Specular illumination is dependent on the surface orientation, the observer location, and the light source location.

specular reflection one component of light reflection at a surface point (*see also* diffuse reflection). *Specular reflection* is observed on "shiny" surfaces and is characterized by highlights on the surface. The amount and direction of specular reflection depend on the directions of the incident light and the viewing direction with respect to the surface normal.

speculative evaluation (**1**) technique for performing a computation which may later be discarded. This may be performed either by the computer hardware or as a feature of a programming language.

(**2**) when implemented in a language, it is often handled by mechanisms for backtracking, for example, in performing artificial-intelligence-style searching. A speculative evaluation is handled as a transaction, with the backtracking mechanism providing the rollback.

(**3**) a technique for implementing a computer with an instruction pipeline in which the instructions that follow a conditional branch are decoded and executed, but the results are not committed until it is determined which way the branch instruction will transfer control. Depending on the sophistication of the computer, either one of the branch executions is selected for speculative execution, or both are. A compiler for such machines should generate code that takes advantage of the machine's speculative execution characteristics.

speculative execution synonym for speculative evaluation when implemented in computer hardware.

speech analysis process of extracting time-varying parameters from the speech signal which represent a model for speech production.

speech coding source coding of a speech signal. That is, the process of representing a speech signal in digital form using as low a rate (in terms of, e.g., bits per second) as possible.

speech compression the encoding of a speech signal into a digital signal such that the resulting bit rate is small and the original speech signal may be reproduced with as little distortion as possible. The transformation of a coded speech signal into another coded speech signal of lower bit rate in such a way that there is insignificant loss in speech quality of the decoded and play-back signal.

speech enhancement improvement of perceptual aspects of speech signals.

speech pre-processing the first step in all problems of speech processing. In preprocessing the objective is to condition the signal so as to come up with more compressed and informative representations. Within portions of about 10 ms (frames), in practice the speech signal turns out to be quasi-stationary. For each frame, all relevant speech pre-processing approaches return a vector of parameters that make it possible to reconstruct the signal. *Speech pre-processing* is mainly carried out by using frequential approaches (e.g., the short-time Fourier transform) or linear prediction.

speech recognition the process of recognizing speech portions carrying out linguistic information. The recognition can involve phonemes or single and connected words. Because of the crucial role of time, most successful approaches to automatic *speech recognition* are currently based on HMM (hidden Markov models) that very naturally incorporate the time dimension.

speech recognizer system for performing speech recognition.

speech synthesis the process of turning information into synthesized speech. When the synthesis involves restrictive linguistic domains (e.g., announcements in railway stations), the process often consists simply of playing back speech recorded in E-PROM memories using proper coding (e.g., ADPCM). However, if one makes no restrictions on the information to synthesize, then only artificial speech production is possible, which is commonly based on systems that predict phonetic units from linguistic information.

speed-up the improvement in performance that is obtained by executing a parallel application on N processors. Speedup is calculated as:

$$\texttt{speedup} = \frac{\texttt{execution time on one processor}}{\texttt{execution time on } N \texttt{ processors}}.$$

See also Amdahl's law.

spindle *See* disk spindle.

spin lock a mutual exclusion mechanism where a process spins in an infinite loop waiting for the value of a variable to indicate a resource availability.

spiral computer tomography (CT) an imaging modality that uses a rotating X-ray source and detector revolving around a continuously moving gantry. As viewed from the gantry, the X-ray source appears to travel in a spiral. A continuous set of projection images is gathered around the spiral and is interpolated to obtain traditional transverse cross-section images. Also known as helical CT.

spiral CT *See* spiral computer tomography.

spiral life-cycle a software development life-cycle in which the typical activities (analysis, development, coding, etc.) occur repetitively, notably risk evaluation. It introduced into life-cycle modeling the important issue of risk-driven considerations. It is an evolutionary approach for software development.

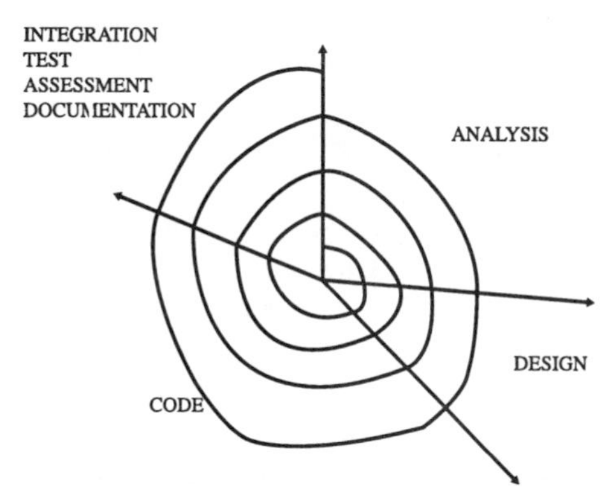

The spiral life-cycle.

spiral model *See* spiral life-cycle.

spline piecewise polynomial, with a smooth fit between the pieces. *See* spline curve.

spline curve a curve defined using a set of control points $(p_0, p_1, \ldots p_L)$. Every control point p_k has an associated blending function, $R_k(t)$, which is described within each span (t_i, t_{i+1}). The blending function is a continuous piecewise polynomial, continuous at each knot and weighted by the polynomials. This gives a curve $p(t)$, which is the union of the piecewise polynomials where all segments meet.

spline wavelet wavelet that is in the form of a spline.

split and merge procedure often used in image or signal segmentation. The procedure involves splitting, iteratively applied if needed, the inhomogeneous regions of an image or sections of a discrete signal and followed by merging similar regions or sections is a split and merge.

splitting criterion in artificial intelligence, the basis for selecting one of a set of possible tests.

split transaction a bus transaction (e.g., memory read or write) in which a request and the corresponding response are sent in two different bus transactions.

split-transaction bus a multiprocessor system where all the processors, memories, and I/O devices are equally accessible without a master-slave relationship.

SPMD *See* single program, multiple data.

spool acronym for "simultaneous peripheral operation on-line". Area managed by a process (called a spooler) where data from slow I/O operations are stored in order to allow their temporal overlapping with other operations.

spooler the program that initiates and controls spooling.

spooling the use of auxiliary storage as a buffer storage to reduce processing delays when transferring data between peripheral equipment and the processor.

sporadic pertaining to a process or task executing in a non-repeatable fashion, at some bounded rate.

sporadic system a system with interrupts occurring sporadically.

sporadic task *See* aperiodic task.

spreadsheet a nonprocedural notation for specifying computations based on individual cells, rows, and columns of a table. Values are computed by equations, rather than by specifying an algorithm for evaluation.

spurious interrupt unwanted, random interrupt.

SQ *See* scalar quantization.

SQA *See* software quality assurance.

SQL *See* structured query language.

square detection a special case of rectangle detection.

square matrix an $n \times n$ matrix, i.e., one whose size is the same in both dimensions. *See also* rectangular matrix.

square pixel *See* pixel.

SRAM *See* static random access memory.

SRT division a division algorithm in which a guess is made of the quotient digit (from a few bits of the partial dividend and a few bits of the divisor) before the main subtraction is complete. SRT is commonly used in the implementation of high-speed dividers. The name is derived from the first initials of its inventors: Sweeney, Robertson, and Toch.

SSD *See* solid state disk.

stability **(1)** generally, the ability of a system to achieve an equilibrium state.

(**2**) in a queueing or scheduling system, the property of a system that the service rate is greater than or equal to the arrival rate. A stable scheduling algorithm will not make a stable system unstable.

(**3**) attributes of software that bear on the risk of unexpected effects of modifications. The capability of the software product to avoid unexpected effects from modifications of the software.

stable database the portion of the database that is stored in secondary storage.

stable equilibrium a system which can not easily be moved to a chaotic state.

stack a storage area that is nominally accessed solely in a last-in-first-out manner. In most machines, however, there is at least a means of accessing stack locations relative to a stack pointer, and most of these additionally support a frame pointer that allows a compiler to more readily determine the offsets to locations representing formal parameters and local variables. Also called a pushdown stack, pushdown list, or LIFO.

stack algorithm a sequential decoding algorithm for the decoding of convolutional codes, proposed by Zigangirov in 1966.

stack architecture *See* zero-address computer.

Stackelberg equilibrium a hierarchical equilibrium solution in non-zero-sum games in which one of the players has the ability to enforce his strategy on the other players. The player who holds the powerful position is called the leader, while the other players who react to the leader's strategy are called the followers. In the case of multi-person games, there exists a variety of possible multi-level decision making structures with many leaders and followers. Thus, the definition of the von Stackelberg equilibrium is uniquely and clearly set only for two-person decision problems but it could be adopted for any given hierarchical structure. If J_1, J_2 denote cost functions of the leader and the follower, respectively, and d_1, d_2 their admissible strategies, then the set $R(d_1)$ defined as $\{d$ (admissible for the follower): $J_2(d_1, d) \leq J_2(d_1, d_2)$ for each admissible $d_2\}$ is called the optimal response or rational reaction set of the follower. Then a strategy d_1^* is a Stackelberg strategy for the leader if:

$$\begin{aligned} J_1^* &= \max_{d_2 \in R(d_1^*)} J_1\left(d_1^*, d_2\right) \\ &\leq \max_{d_2 \in R(d_1)} J_1\left(d_1, d_2\right) , \end{aligned}$$

for all admissible d_1. J_1^* is the Stackelberg cost of the leader and any $d_2^* \in R(d_1^*)$ is an optimal strategy for the follower that is in equilibrium with d_1^*. The pair (d_1^*, d_2^*) is a Stackelberg solution and corresponding values of the cost functions give the *Stackelberg equilibrium outcome.* The Stackelberg outcome of the leader may be lower than his Stackelberg cost. If the rational reaction set of the follower is a singleton, then they are equal and they are not worse than the outcome which could be achieved by the leader.

stack filter a positive Boolean function used as a filter in conjunction with threshold sets.

stack frame a segment of a stack which provides the necessary context for the execution of a function. A *stack frame* will include at least space for the parameters to the function (when present), the return address to the caller, and space for the local variables of the function. Depending on the target architecture, additional information such as locations to save register values, information for exception handling, and similar information may also be part of the stack frame. *See also* coroutine, frame pointer.

stack local variable a variable that is allocated on the stack when a context is entered and deallocated when the context is exited. Synonym for automatic variable. Sometimes referred to as a local variable although a name with local scope does not necessarily have local extent.

stack machine *See* zero-address computer.

stack overflow a condition in which the total available space for the stack has been consumed, due either to excessive recursion or to large blocks of stack-allocated storage being used.

stack pointer a register in a processor that holds the address of the top of the stack memory location. The address varies as information is stored on or retrieved from the stack; it always points to the top of the stack.

stage *See* segment.

staircasing image processing process in which lines are scan-converted to fixed pixel grid points. The illuminated pixels often do not lie on the true path of the line. The result is that displayed lines are normally jagged in appearance, an effect commonly known as the jaggies or *staircasing*. The effect can be reduced or eliminated by antialiasing.

stakeholder anyone who has an interest in a software development project. Typically, *stakeholders* include the end users of the product, the purchasers of the product (who may not be the end users), the managers of the users, and the developers and their managers.

stale read *See* dirty read.

stand-alone pertaining to hardware or software or a system that is capable of performing its function without being connected to other components.

standard additive model (SAM) a fuzzy system that stores IF-THEN rules that approximate a function $F : X \to Y$. In a simple SAM the rules may have the form "IF $x = A_j$ THEN $y = B_j$", where $x \in X$, $y \in Y$, and A_j, B_j are fuzzy sets. The SAM then computes the output $F(x)$ given the input x using a centroidal defuzzifier. An example of a centroidal defuzzifier is

$$F(x) = \frac{\sum a_j(x)c_j}{\sum a_j(x)} ,$$

where a_j is a membership function of the fuzzy set A_j and c_j is the centroid of the fuzzy set B_j. The term SAM was coined by B. Kosko. *See also* fuzzy system.

standard array decoding during decoding of a forward error correction code, the process of associating an error pattern with each syndrome by way of a look-up table.

standard error the I/O stream, usually associated with the user's terminal display, where (by convention) a program outputs error messages and diagnostic information.

standard input the I/O stream, usually associated with the user's keyboard, where (by convention) a program reads its input.

standard output the I/O stream, usually associated with the user's terminal display, where (by convention) a program writes its output, possibly excluding error and diagnostic messages.

Standard Template Library (STL) a library of classes and methods defined for the C++ language as a set of templates (generic classes and methods). These classes and methods simplify the use of C++ by providing a standardized and unified set of facilities such as iterators, enumerators, and more to C++ programmers.

star-shaped polygon a polygon P in which there exists an interior point p such that all the boundary points of P are visible from p. That is, for any point q on the boundary of P, the intersection of the line segment $\overline{p, q}$, which is the boundary of P is the point q itself.

start bit the first bit (low) transmitted in an asynchronous serial transmission to indicate the beginning of the transmission.

start symbol a designated nonterminal symbol in a grammar, from which all sentence generation starts, or which all parsing attempts to reach.

starvation a condition when a process is indefinitely denied access to a resource while other processes are granted such access. This situation can prevent the denied program unit from proceeding and may ultimately lead to system failure.

state (1) in an automaton, represents a computational stage of the process of processing an input stream.

(2) in any computation, the current set of values being operated upon.

state automaton *See* finite state machine.

state diagram (1) a form of diagram showing the conditions (states) that can exist in a logic system and what signals are required to go from one state to another state.

(2) a simple diagram representing the input-output relationship and all possible states of a convolutional encoder together with the possible transitions from one state to another. Distance properties and error rate performance can be derived from the *state diagram.*

state machine a software or hardware structure which can be in one of a finite collection of states. Used to control a process by stepping from state to state as a function of its inputs. *See also* finite state machine.

statement in most programming languages, the basic unit of sequential computation. The exact definition of what constitutes a *statement* is language-dependent, but in traditional programming languages an elementary computation that produces a result or modifies the flow of control of the program is a statement (as distinguished from an expression or a declaration). Statements are usually basic objects of program sequencing, and often are part of the specification of evaluation order ("all computations of one statement shall be completed before the next statement begins computations" is a typical specification).

statement coverage selecting and executing a series of test cases to ensure that every statement in the code is tested at least once.

statement function a construct in the FORTRAN language (which has equivalents in some other languages) which allows a programmer to declare a single-line ("statement" in FORTRAN) definition of a function. In this case, the word "function" is interpreted in the particular semantics of the FORTRAN language to mean "a computation returning a value which can be used in an expression, and which has no side effects". Often implemented by compilers as inline functions.

state-oriented model a model based on state machines or their extensions.

state space conditional code an approach where the number of codes is much less than with conditional coding. The previous $N - 1$ pixels are used to determine the state s_j. Then the j-th variable word-length is used to code the value.

state-space operator a representation for an individual action that maps each state into the state resulting from executing the action in the previous state.

state transition in a state machine, the event related to a change of state. Generally the transition is performed when a certain condition becomes true while staying in the departing state of the transition.

The transfer of control from one state to another (possibly the same) state is affected by a production rule, which is a mapping from the current input symbol, the current state, and the values in the store (usually a pushdown stack), to the new state. A state transition usually implies reading one symbol from the input string. *See also* lambda.

state transition diagram (STD) a diagram consisting of circles to represent states and directed arcs to represent transitions between the states. One or more actions (outputs) may be associated with each transition. The diagram represents a finite state machine.

state-translation function a function that maps each state and action deterministically to a resulting state. In the stochastic case, this function is replaced by a conditional probability distribution.

static (1) pertaining to an event or process that occurs before the execution of a computer program.

(**2**) term used to describe memory that does not need to be refreshed periodically due to gradual discharge.

(**3**) in many programming languages, a variable whose scope is limited but whose extent is permanent. Algol-60 is one of the earliest languages to have used this designation.

(**4**) in C++, a static component is a component shared by all instances of objects of the type. *See also* static member, static function.

(**5**) in C, a name which is not exported from the module in which it appears. (Synonym: private).

static allocation the mechanism by which the memory reserved for a variable is allocated at loading time and statically reserved at compiling time. The size of the allocate memory is determined on the basis of the variable declaration.

statically checked language a language where good behavior is determined before execution.

static analysis the analysis of a system or component directly performed on the code and documentation without compiling it. *See* assessment.

static assertion an assertion about the system that depends on elements that do not change their truth value with time.

static binding association of the call of a procedure with a specific procedure at compiling time. In object-oriented programming and in other dynamic typed systems, it implies the recognition at compile time of the type to which an instance belongs. It is also known as early binding. *See* dynamic binding.

static checking a collection of compile-time tests, mostly consisting of type checking.

static-column DRAM a DRAM that is organized in the same manner as a page-mode DRAM but in which it is not necessary to toggle the column access strobe on every change in column address.

static data member a data member, declared as static, that becomes common to and shared by all instances of a class.

static function (**1**) in C, a function whose name is not exported from the module in which it appears.

(**2**) in C++, a function that is not specific to any object instance, and which may be called without having any object instance available.

static Huffman encoding *See* Huffman encoding.

static member in C++, the designation of a value whose one instance is shared by all objects of the type. Synonym for class global member.

static model in object-oriented analysis, it is the definition of the system architecture based on classes definition. The model also involves the description of internal design of classes, such as attributes, member functions, cardinality, access level, responsibility, etc.

static object instantiation the memory reservation for an instance of a variable performed at compile-time.

static random access memory (SRAM) random access memory that, unlike dynamic RAM, retains its data without the need to be constantly refreshed.

static scheduler a scheduler that can manage the execution of processes that are known in advance. The scheduler establishes offline if the execution of the processes can be done with the timing constraints specified. *See* dynamic scheduling.

static scheduling (**1**) the scheduling process that can be performed offline because all the timing constraints of the involved tasks to be scheduled are known. *See* dynamic scheduling.

(**2**) a scheduling approach, usually for multiprocessor systems, in which the processor allocation to a job does not change after the job starts running on the system.

static semantics semantics that are based upon static analysis of the program. A program that conforms to the rules of the grammar may still be invalid because of rules that are not part of the formal grammar. An example of static semantics is type checking. *See also* execution semantics.

static storage (**1**) storage allocated at compile time.

(**2**) storage allocated when a program is loaded into memory, and whose addresses are therefore fixed (static) for each object in the storage. *Compare with* dynamic storage.

static type-checking the type-checking performed by the compiler in order to check if the constructs and expressions written in terms of tokens and variables are consistent with the types to which they refer.

static variable a variable whose scope is limited by where it syntactically appears, but whose extent is permanent.

statistical language model a powerful mechanism for providing linguistic constraints to a speech recognizer. The model specifies the set of follow words with associated probabilities, based on the preceding $n - 1$ words. *Statistical language models* depend on large corpora of training data within the domain to be effective. Also called n-gram language model

statistical pattern recognition methods for carrying out the recognition of patterns on the basis of statistical analysis. These methods are typically based on the learning of unknown pattern probability distributions from examples.

statistical quality assurance a technique for process improvement based on the statistical analysis of measurements of the product and the process.

status register a register in a processor that holds the status of flags; individual bits in the register represent flag status.

s-t cut a partitioning of the vertex set into S and T such that $s \in S$ and $t \in T$.

STD *See* state transition diagram.

steady state a state where characteristic system measures can be described as a stationary random process. The system measures can be, e.g., queue occupation, delay, or utilization. If there are no inherent instabilities, a system can reach *steady state* some time after a stationary input process has been switched on. Reaching the steady state is an important prerequisite for evaluating mean values in a simulation.

steal/no-force a buffer management policy that allows committed data values to be overwritten on nonvolatile storage and does not require committed values to be written to nonvolatile storage. This policy provides flexibility for the buffer manager at the cost of increased demands on the recovery subsystem.

steepest descent algorithm *See* gradient descent.

Steiner point (**1**) a point that is not part of the input set of points, for instance, a point computed to construct a Steiner tree.

(**2**) a point with a particular geometric relation to a triangle.

Steiner ratio for a given variant of the Steiner tree problem, the maximum possible ratio of the length of a minimum spanning tree of a set of terminals to the length of an optimal Steiner tree of the same set of terminals. Usually written ρ (rho).

Steiner tree a minimum-weight tree connecting a designated set of vertices, called *terminals,* in a weighted graph or points in a space. The tree may include non-terminals, which are called Steiner vertices or Steiner points. *See also* Euclidean Steiner tree, rectilinear Steiner tree.

Steiner vertex a point that is not part of the input set of points, for instance, a point computed to construct a Steiner tree.

stemming the reduction of morphological and derivational variations of words to a common root form.

step edge an idealized edge across which the luminance profile takes the form of a step function, i.e., a line separating two regions having different average gray-levels. *See* edge.

stepwise refinement a technique for accomplishing functional decomposition or procedural design (also called partitioning).

steradians the unit of solid angle. The solid angle corresponding to all of space being subtended is 4π *steradians.* Solid angle is defined as the surface area of a unit sphere which is subtended by a given object.

stereo the use of two images to generate a 3D description. For example, two slightly different images are displayed in each eye of a virtual reality head mounted display in order to induce an impression of 3D. Stereo matching is a process by which two images of the same scene are compared in order to deduce 3D information.

stereo imaging *See* binocular imaging.

stereopsis the recovery of depth (or relative depth) by finding corresponding points in two or more images of the same scene. *See* stereo vision.

stereotype a user defined submetatype in UML. It represents a second and coincident classification. Thus, objects of an Apple class might also be stereotyped as belonging to the class called GreenObject. This extra annotation would be given in UML as "GreenObject".

stereo vision a vision model in which imaged objects are projected onto two image planes, to extract depth information from the scene. *See* binocular vision.

sticky bit after an alignment shift, the logical OR of all the bits shifted out except for the guard digit and the rounding digit. The bit is used during rounding, to check for the boundary case in round-to-nearest rounding.

stiffness of a manipulator arm an attribute of robot arm. Assume that a force is applied to the end-effector of a manipulator arm. The end-effector will deflect by an amount which depends on the stiffness of the arm and the force applied. In another words, the stiffness of the arm's end-effector determines the strength of the manipulator arm. Usually the actuator itself has a limited stiffness determined by its feedback control system, which generates the drive torque based on the discrepancy between the reference position and the actual measured position. We model the stiffness by a spring contact that relates the small deformation at the joint to the force or torque transmitted through the joint itself. It is called the joint stiffness.

still image stationary image or single frame as opposed to moving image or video. Includes photographic images, natural images, medical images, and remote sensing images. Usually implies multilevel (grayscale or color) rather than bilevel.

still image coding compression of a still image. A coder consists of the four steps: (i) data representation (typically by transform, decomposition into subbands, or prediction), (ii) quantization (in which data is approximated or discarded according to some measure of its importance), (iii) clustering of nulls (in which runs or blocks of zero values are coded compactly), and (iv) entropy coding (in which the statistical properties of the data are exploited in lossless compression).

Stirling's approximation $n! \approx \surd(2n\pi)(n/e)^n$. *See also* factorial, gamma function.

Stirling's formula $\surd(2n\pi)(n/e)^n < n! < \surd(2n\pi)(n/e)^n(1 + \frac{1}{12n-1})$. *See also* Stirling's approximation.

STL *See* Standard Template Library.

stochastic neuron an artificial neuron whose activation determines the probability with which its output will enter one of its two possible states. The most commonly used expression for the probability that the neuron output y takes on the value $+1$ is $Pr\{y = +1\} = 1/(1 + e^{-2\mathrm{net}/T})$, where net represents the activation of the neuron and T is a quantity analogous to temperature which controls the uncertainty in the neuron output. When T is infinite, a positive activation leads to an output of $+1$ with probability 0.5, and when T is zero, a positive activation leads to an output of $+1$ with probability 1.0.

stochastic ordering a method to compare two random variables. We say that random variable X is stochastically larger than Y, written as $X \geq_{st} Y$ if $P[X > z] \geq P[Y > z]$, for all real z. A strong ordering between two random variables.

stochastic Petri net a Petri net in which the elapsed time from the moment that all of its input places are filled until a transition fires is determined by a random variable. Typically, these random variables are exponentially distributed, in which case the model is equivalent to a Markov process.

stochastic process a collection of vector random variables defined on a common probability space and indexed by either the integers (discrete stochastic process) or the real numbers (continuous stochastic process). A stochastic process $x = x(t)$ is a vector function of both time t and the sample path.

stochastic sampling a type of sampling that varies the time intervals between samples. *Stochastic sampling* allows for a signal to be sampled at a lower apparent sampling frequency achieving equal results to a signal sampled at a much higher sampling frequency. The apparent benefits of stochastic sampling are counterbalanced by the fact that the sampling interval, since it is changing, must be recorded in addition to the signal samples, in order to reconstruct the signal correctly.

In image processing, stochastic sampling mitigates the visual effects of aliasing by sampling in an irregular manner, rather than on a regular grid. Recognizable aliasing artifacts are replaced by noise, which viewers find less objectionable. *See also* jittering.

stochastic search algorithm a form of optimization searching based on random sampling (Monte Carlo) methods where points v from a set V are randomly chosen with probability $1/ \mid V \mid$. The minimum values of $f(v)$ are recorded as the random sampling proceeds, and the sampling does not terminate arbitrarily as might occur in a deterministic search. Simulated annealing is an example of a stochastic algorithm.

stochastic signal processing the branch of signal processing that models and manipulates signals as stochastic processes rather than as deterministic or unknown functions. *See also* signal processing, random process.

stop an operating system command to terminate the execution of a computer program.

stop bit the last bit (high) transmitted in an asynchronous serial transmission to indicate the end of a character. In some serial transmissions, one and half to two bits are used as *stop bits.*

stopping criterion in artificial intelligence, the conditions under which a set of instances is not further subdivided.

stopping time an event that can be recognized in real time without any knowledge of the future that is used to terminate a measurement period. For example, the end of a fixed time period, or the departure of a fixed number of customers, or the entry into a particular system state (e.g., server idle) could all be used as *stopping times,* while a time instant where the system will remain idle for at least 30 minutes would not qualify.

stop-word any word that contributes little to the semantic interpretation of a document. Usually includes prepositions, conjunctions, and articles. These are usually removed before comparisons in an information retrieval system.

storage class a designation of types of storage implemented on a machine. *Storage classes* may or may not be reflected in a programming language. Examples of storage classes are integer registers, floating point registers, main memory, and the stack, among others.

storage compaction a method by which allocated storage in a heap is moved so that all allocated storage is contiguous, leaving all the unallocated storage contiguous.

storage fragmentation a phenomenon experienced in languages which allows for storage allocation, in which the repeated allocation and deallocation of blocks of storage of variable size result in fragmenting the heap, eventually resulting in a collection of unusably small fragments which cannot be coalesced (usually after a compaction).

store (**1**) the act of placing a value into storage.

(**2**) in British usage, the main memory of the computer, once referred to in the U.S. as core.

(**3**) in an automaton, the auxiliary storage used to hold state, such as a pushdown stack.

store-and-forward a method for relaying packets through an intermediate switching node in a network. Packets that arrive on each incoming link are stored in a buffer. When the entire packet has been received, the node checks it for transmission errors and queues the packet for transmission over the next hop towards the destination. Packets may experience queueing delays at an intermediate node, or even be dropped if the node runs out of buffer space for storing packets. *See also* packet switching, virtual cut-through.

stored program computer a computer system that is controlled by machine instructions stored in memory. The instructions are executed one after the other unless otherwise directed.

store instruction a machine instruction which copies the contents of a register into a memory location. *Compare with* load instruction.

straight edge detection the location of straight edges in an image by computer. Often accomplished with the Hough transform.

straight-line drawing a graph drawing in which each edge is represented by a straight line segment.

straight through cable a cable that allows a standard RS-232C device to be connected to communications equipment such as a modem.

strange attractor when the attracting set of an iterated function or procedure is a fractal.

strategy parameter the set of miscellaneous parameters associated with each individual in evolutionary systems and evolutionary programming which specifies the variances and covariances of the random variables for the mutation process. In other words, these parameters determine the step size and direction of the search mechanism. They are typically modified by a self-adaptation process according to the local structure of the fitness landscape. *See also* self-adaptation.

stratified negation a concept in database theory. A database DB is locally stratified, if for all rules of DB none of the default negative literals C_i in the body depends on the atom A in the head. Stratification can be defined in terms of atoms (local stratification) or predicate symbols (global stratification, or just stratification).

A stratified database DB can be evaluated by partitioning it into layers that are evaluated subsequently based on the results of the previous layer.

streakline a line showing the path taken by all particles that pass through a given location in a vector field.

stream the sequence of data or instructions that flows into the CPU during program execution.

stream cipher an encryption system or cipher in which the information symbols comprising the plaintext are transformed into ciphertext individually. An important property of a stream cipher is that like-valued plaintext symbols are not necessarily transformed into the same ciphertext. A stream cipher normally acts in an additive sense and in the case of bits being encrypted, the information bits, X_n are added modulo-2 to the bits, Z_n, generated by the so-called running-key generator. The ciphertext, Y_n, is therefore given by $Y_n = X_n \oplus Z_n$, $n =$

$1, 2, \ldots, N$ where the plaintext consists of N bits and $\oplus$ denotes modulo-2 addition. Generally, the running key bits and the encryption key bits are not the same. The encryption key merely specifies the mechanism used to generate the running-key bits. Such a mechanism could be a number of linear feedback shift registers whose outputs are combined to form the running-key bits. *See also* block cipher, encryption.

streamline a line drawn in a vector field such that, at any instant, the tangent to the line at any point on the line is the direction of the flow. Often restricted to fields with steady flow, in which case the streamline shows the path of a tracer particle.

stream traffic traffic that needs to be delivered at a certain rate. *Stream traffic* often has real-time requirements.

strength reduction a global optimization that is characterized by discovering common partially specified computations and performing the common part once. For example, in accessing elements in multiple arrays, the offset into corresponding arrays is usually identical except for the value of the base address of the array. The access path becomes a partially specified computation (base, computed offset), and the computed offset is computed once for all arrays sharing the computation.

stress testing a type of testing where the system is subjected to a large disturbance in the inputs (for example, a large burst of interrupts), followed by smaller disturbances spread out over a longer period of time.

strict coherency the property of a distributed file or memory system in which each read or load retrieves the value produced by the most recent write or store to that location.

strict functional language a function language adopting applicative-order evaluation.

strictly decreasing a function from a partially ordered domain to a partially ordered range such that $x > y$ implies $f(x) < f(y)$. *See also* strictly increasing, monotonically decreasing.

strictly increasing a function from a partially ordered domain to a partially ordered range such that $x > y$ implies $f(x) > f(y)$. *See also* strictly decreasing, monotonically increasing.

strict schedule a schedule where a value is not read or written until all transactions that write the value have committed.

strict two-phase locking two-phase locking where locks are only released when the transaction terminates.

stride *See* memory stride.

string (**1**) value that is an ordered sequence of characters which can be represented on the machine on which the program runs.

(**2**) informal designation of a string constant.

(**3**) in the context of a formal grammar, a sequence of symbols of the grammar which can be generated by the rules of the grammar.

(**4**) in the context of a formal grammar, a sequence of symbols presented to an automaton which recognizes the string as either being one which can be generated by the grammar or one which cannot.

string constant a representation of an ordered sequence of characters in the source code of a program. The means of representing a *string constant* are defined by individual programming languages, although there are traditional representations such as the use of single quotes or double quotes to mark the start and end of a string constant.

string editing problem for input strings x and y, the problem of finding an edit script of minimum cost that transforms y into x.

string literal *See* string constant.

string operator operator that applies to string values. Typical string operators include single character extraction (subscripting), string concatenation, substring extraction, character membership, substring membership, and comparison operations.

string searching with errors searching for approximate (e.g., up to a predefined number of symbol mismatches, insertions, and deletions) occurrences of a pattern string in a text string.

string searching with mismatches the special case of string matching with errors where mismatches are the only type of error allowed.

striping the spreading of data across several disks in order to improve performance. *See* disk array.

strong bisimulation partitioning an algorithm to partition a set of states by strong bisimulation equivalence. This is used to minimize a state space by combining the states in the same partition into one. *See also* bisimulation equivalence, formal verification.

strong entity type an entity type that possesses its own key.

strong fairness the kind of fairness (with respect to a scheduling discipline) where when a process makes a request infinitely, often it eventually will be granted.

strongly checked language a language where no forbidden errors can occur at run time (depending on the definition of forbidden error).

strongly connected graph a directed graph in which there is a path from each vertex to every other vertex. *See also* connected graph, bridge.

strongly normalizing a lambda-expression for which every sequence of reductions terminates in a normal form.

strongly NP-hard a problem that is still NP-hard even when any numbers appearing in the input are bounded by some polynomial in the length of the input. *See also* NP-complete.

strongly typed a language that is based on a strictly rigorous type checking.

strong-neighbor mask an image mask used in the moving median filter.

strong typing a feature of programming languages and models that requires the definition of the type of each data. This approach allows the application of a strong type checking and the execution of algorithms only on the right data types.

structural aspect the aspect related to the definition of the composition/decomposition of a system.

structural decomposition the approach of system decomposition.

structural equivalence a means of type equivalence in which an object is considered to be of a specific type if its structure is identical to the structure of the specific type. For example, under structural equivalence, any record consisting of exactly two integers is equivalent to any other record consisting of exactly two integers. *See also* name equivalence, textual equivalence.

structural pattern recognition method for carrying out the recognition of pattern on the basis of a structured representation. For instance, in many interesting problems, the patterns can effectively be given linguistic descriptions based on grammars.

structural sharing mechanism by which an entity/class shares the structure with another entity/class. In object-oriented programming, this is implemented via is-a relationships. The subclass shares the inherited data structure with its superclasses.

structure a heterogeneous collection of information. Known also as a record, class, or composite object.

structure chart a diagram reporting modules, activities, or other entities of a system considering their structural decomposition, or breakdown into smaller components. This graph is typically different from the call graph since the structure graph is mainly oriented in decomposing data while the call graph describes the functional decomposition.

Synonymous with program structure chart. *Compare with* call graph.

structured design A process of system design that works in accordance to the system decomposition approach, top-down design. *See* top-down, structure chart.

structured language a programming language where the program may be broken down into blocks or procedures which can be written without detailed knowledge of the inner workings of other blocks, thus allowing a top-down design approach.

structured light patterns of light projected onto objects that are to be viewed by cameras and interpreted by computer. For example, a grid of parallel straight lines of light projected on to a curved object will appear from a separate viewpoint to be curved and will provide information on the 3D shape of the object.

structured noise noise that is not random but which is typically periodic, or contains elements of some unwanted signal. This category of noise includes clutter, crosstalk, easily recognized spikes, and so on.

structured program a program constructed of a basic set of control structures, each one having one entry point and one exit point.

structured programming a discipline of programming advocated initially by Edsgar Dijkstra, in which programs are developed in such a way that a high-level abstraction is used as a specification, then is decomposed into a set of lower-level operations, proceeding downward until actual code must be written. A fad in the 1970s, it has been largely supplanted by better techniques, such as modular decomposition.

structured query language (SQL) a query language used for relational databases. SQL is an ISO and ANSI standard. It is often embedded in other programming languages. SQL provides basic language constructs for defining and manipulating tables of data, language extensions for referential integrity and generalized integrity constraints, facilities for schema manipulation and data administration, and capabilities for data definition and data manipulation. Development is currently underway to enhance SQL into a computationally complete language for the definition and management of persistent, complex objects. This includes: generalization and specialization hierarchies, multiple inheritance, user-defined data types, triggers and assertions, support for knowledge-based systems, recursive query expressions, and additional data administration tools. It also includes the specification of abstract data types (ADTs), object identifiers, methods, inheritance, polymorphism, encapsulation, and all of the other facilities normally associated with object data management.

SQL is also a data management language and a data definition language since it provides facilities for data definition, query, and update. The result of an SQL query is a table of tuples.

structured software software implemented as a set of independent modules, each with a defined function and interface.

structure editor a text or graphical editor that allows a source program to be edited in terms of its syntactic constructs. Such editors typically allow the user to select and operate on whole syntactic units and to apply transformations to the underlying parse tree rather than to the program text.

structure estimation determination of the structure of objects, i.e., the 3D coordinates of surface points of objects, from sequences of images. It is a task sometimes closely related to motion estimation.

structuring element an image or shape that is used in a morphological operator as a probe interacting with the image to be analyzed, thus leading to a transformation of that image. It can be either a set of points (a colorless shape), or a gray-level image (a shape with a gray-level profile on it). In contrast with the natural image to be processed, the *structuring element* is chosen by the user and generally has a small support. *See* morphological operator.

stub object an instance of a stub class/type. It provides little or no functionalities allowing a test environment to narrow the focus of a test. A stub creates an emulation layer of other elements providing the same interface but containing only partial implementation or predetermined responses.

style guide a set of principles and rules for consistently designing user interfaces according to a particular interface style (often for a specific platform). General industry *style guides* are usually independent of a particular application domain; in-house or application style guides may provide additional rules for designing particular classes of applications.

subadditive ergodic theorem if a stationary and ergodic process satisfies the subadditive inequality, then it grows almost surely linearly in time.

sub-block a part of a cache line that can be transferred to or from the cache and memory in one transaction. This is applicable in the cases where the complete line cannot be transferred in one transaction. Each *sub-block* requires a valid bit.

subchannel I/O the portion of a channel subsystem that consists of a control unit module, the connections between the channel subsystem and the control unit module, and the connections between the control unit module and the devices under its control. In earlier versions of the IBM channel architecture, the subchannel was known as an I/O channel.

subclass *See* derived class.

subclassing mechanism of producing derived classes from a class. *See* subtyping, type, class, subtype, subclass.

sub-directory *See* child directory.

subgraph isomorphism the problem of deciding if there is a subgraph of one graph that is isomorphic to another graph.

subject an active entity within a system, such as a process or device. Forms a unit for assigning authorizations, usually associated with (acting on behalf of) some user. *Subjects* are also objects.

subprocess a process that is used in the composition/decomposition of another process. *See* process decomposition.

subquery an expression that identifies data from tables. *Subqueries* are used as expressions in the context of structured query language (SQL) statements.

subroutine a section of a program that defines a computation that can be requested from other locations of the program by using a subroutine call, also known as a function call. The computation of a *subroutine* may be modified by values supplied at the time of the subroutine call by specifying parameters. Synonyms: function, procedure, routine. In some programming languages, these terms are defined more specifically. For example, in FORTRAN, a "function" can return a value and be used as a part of an expression, while a "subroutine" is defined as the target of a CALL statement and cannot return a value or be used in an expression.

subroutine call a construct in a programming language that causes control to transfer to a subroutine. Control resumes following the *subroutine call* after the subroutine returns.

subroutine call and return a specialized JUMP or BRANCH instruction that provides a means to return to the instruction following the call instruction after the subroutine has been completed. A RETURN instruction is usually provided for this purpose.

subroutine library a collection of commonly used subroutines, usually in a format that allows them to be selected by the linker.

subroutine linkage the sequence of instructions used to implement a subroutine call. The optimum sequence depends upon the specification of the language and the available instructions on the machine on which the program will execute. A particular language or compiler may use more than one method of calling a subroutine. A *subroutine linkage* specification includes information as to how to pass information to the subroutine, how to call it, and how to obtain the results it computes.

sub-sampling pyramid a spatial domain hierarchy is generated by repeatedly subsampling the original image data. The reconstruction at any level simply uses the subsampled points from all previous levels in conjunction with the new points from the current level.

subscript a notation used to identify a value or set of values from a collection of values. The nature and interpretation of subscripts depends upon the programming language, as does the notation used to encode them. For example, a simple integer can select one member of an array, an ordered set of values. However, some languages permit collections to be subscripted with noninteger values for performing selections. *See also* associative array.

subsequence of a string, any string that can be obtained by deleting zero or more symbols from that string.

subspace method any method of reducing the dimensionality of image matching or recognition problems by representing the images in terms of their projection into lower-dimensional space.

substate a hidden state included in a more general state which shares with them most of the general state features. In state diagrams, states that include *substates* are differently drawn and evidenced.

subsumption a fundamental rule of subtyping, asserting that if a term has a type A, which is a subtype of a type B, then the term also has type B.

subsystem a component of a larger system which interacts with other components via a defined interface and which is sufficiently complex to be considered a system in its own right. In object-oriented programming, an independent group of classes that collaborates to fulfill a set of responsibilities; a separate branch of classes, a set of classes addressing a specific problem.

subsystem composition the process of constructing composite software components out of building blocks such as variables, procedures, modules, and subsystems.

subsystem manager the middle level managers that under the control of the project manager manage one or more project subsystems and related personnel.

subtask a portion of computation associated with a given application task. For current heterogeneous computing applications, *subtasks* are typically defined by the programmer. Ideally for automatic heterogeneous computing, subtasks may be determined through the process of task profiling. For this case, computational requirements of each subtask should be relatively homogeneous.

subtracter a circuit which subtracts two values.

subtractive adder an adder in which addition is realized as a subtraction, i.e., $A + B$ is realized as $A - (-B)$.

subtransaction *See* nested transaction.

subtype *See* derived type.

subtyping mechanism of producing derived types from a type. *See* subclassing, type, class, subtype, subclass.

successful termination termination of an algorithm without problems.

successor a function that produces the next element of a set. In principle, any set which has a successor function is a totally ordered set, and a corresponding predecessor function is usually defined. Note that a totally ordered set of values may have a designated element for which the successor function is not defined. Generally, for programming languages, such sets include enumeration sets, character sets, and lists. Singly linked lists generally do not have a predecessor function. *See also* child directory.

sudden underflow a method for dealing with underflow. When underflow occurs, the computed value is replaced with zero.

suffix a string v is a suffix of a string u if $u = u'v$ for some string u'.

suffix automation smallest automation accepting the suffixes of a string.

Sugeno fuzzy rule a special fuzzy rule in the form *if* x *is* A *and* y *is* B *then* $z = f(x, y)$ where "if x is A and y is B" is the antecedent, A and B are fuzzy sets, and the consequent is the crisp (nonfuzzy) function $z = f(x, y)$. *See also* fuzzy IF-THEN rule, fuzzy inference system.

suitability the capability of the software product to provide an appropriate set of functions for specified tasks and user objectives. Examples of appropriateness are task-oriented composition of functions from constituent subfunctions, and capacities of tables. *Suitability* also affects operability.

summative evaluation a type of evaluation occurring after online support systems are finished or in very late stages of development. In general, they measure the success of final design decisions.

sum of products (SOP) a standard form for writing a Boolean equation that contains product terms (input variables or signal names either complemented or uncomplemented ANDed together) that are logically summed (Ored together).

superacuity the ability to perceive visual effects with a resolution that is finer than can be predicted from the spacing of receptors in the human eye.

superclass in an object-oriented language, the class from which another class has been derived. A subclass inherits the data members and methods of its *superclass,* but not all of these may be available to it, depending on the nature of the language (for example, the private data members and methods of a C++ class). In some object-oriented languages the terms more frequently used are base class or parent.

supercomputer at any given time, the most powerful class of computer available.

superconic generalization of conic curve in which the trigonometric terms in the formula of the curve are raised to an arbitrary power to control the smoothness of the curve. It can be expressed by:

$$\begin{aligned} x &= a(\cos\theta)^\gamma \\ y &= b(\sin\theta)^\gamma \end{aligned}$$
$$a, b \in \mathcal{R}\,, -\pi/2 \leq \theta \leq \pi/2,\ \gamma \geq 0\,.$$

superinterleaving *See* interleaved memory.

superkey of a relation, any set of attributes of that relation such that any two instances of that relation will not have the same values for those attributes.

superpipelined processor a processor where more than one instruction is fetched during a cycle in a staggered manner. That is, in an n-issue *superpipelined processor,* an instruction is fetched every $1/n$ of a cycle. For example, in the MIPS R4000, which is two-issue superpipelined, a new instruction is fetched every half cycle. Thus, in effect, the instruction pipeline runs at a frequency double that of the system (in the R4000 the pipeline frequency is 100 MHz, while the external frequency is 50 MHz). Superpipelined processors usually have a relatively deep pipeline, of about 7 stages or more (8 stages on the R4000).

superpipelining a pipeline design technique in which the pipeline units are also pipelined internally so that multiple instructions are in various stages of processing within the units. The clock rate is increased accordingly.

superposition the combination of stochastic or random point processes, as for example, when two or more arrival streams merge into a single arrival stream. The *superposition* of two Poisson processes with parameter λ_i, $i = 1, 2$, respectively, is also a Poisson process with parameter $\lambda_1 + \lambda_2$.

superquadric a class of parametric surfaces, derived from the class of quadric surfaces, in which the trigonometric terms of the quadric equation, written in parametric form, are raised to a power. The exponents are known as the squareness parameters and are used to pinch or square off parts of the original quadric shape. A special case is the superconics.

super-resolution the process of combining data from multiple, similar images of the same object to form a single image with increased spatial resolution.

superscalar processor a processor where more than one instruction is fetched, decoded, and executed simultaneously. If n instructions are fetched and processed simultaneously, it is called an n-issue superscalar processor. For example, the Pentium is a two-issue, and the DEC 61164 is a four-issue superscalar processor. This feature was implemented both on CISC (Pentium) and RISC (61164) processors.

supertype a type from which another type (named "subtype") has been derived.

super user a user who is granted omnipotent access rights to a system; generally the system's administrator, who is treated as the owner of the entire system.

supervised learning the collection of techniques where analysis uses a well-defined (known) dependent variable. All regression and classification techniques are *supervised learning* techniques.

There are two ways for supervised learning in SGNN. The first is the same as that of supervised learning for a self-organizing system. The second is to make use of information gains of the attributes to the classification. That is, use the inner product of the training vector and the information gain vector corresponding to its attributes to train the network. Experiments show that this way of supervised learning for SGNN can significantly improve both the performance of the network and the training speed. *See* self-generating neural network, information gain.

supervised neural network neural network that requires input-output pairs to form the interconnection weight matrix of a network. The Hopfield model, perceptron, and backpropagation algorithm are *supervised neural networks.*

supervisor *See* executive.

supervisor call instruction (SVC) *See* software interrupt.

supervisor instruction any processor instruction that can be executed when the processor is running in supervisor mode. The separation of supervisor instructions is required to isolate the system's control information from tampering by user programs.

supervisor mode one of two CPU modes, the other being user mode. Sometimes called privileged mode, this mode allows access to privileged system resources such as special instructions, data, and registers.

supervisor state one of two CPU states, the other being user state. When the CPU is in *supervisor state,* it can execute privileged instructions.

supplier organization that provides a product to the customer. In a contractual situation, the "supplier" may be called the "contractor". The *supplier* may be, for example, the producer, distributor, importer, assembler, or service organization. The supplier can be either external or internal to the organization.

support code code that is used to test a product or code for building a tool that helps in the development of other software.

support family any of a collection of families used to compose support systems. The generic support is likely to be layer structured. Therefore, several general support families are likely, for example, operating system, I/O system, communication system, user interface, and database management system. It is part of the implementation design to determine which support families to use, their composition, and how instances are to be configured.

supporting technology an input of a software process that is not consumed by the process. Examples are principles, concepts, methodologies, software development environments, data, etc. A *supporting technology* may also be an output product of an external process.

support manual the document containing the information necessary to service and maintain a system or component throughout its life-cycle.

support of a fuzzy set the crisp set of all points x in X with membership positive ($\mu_A(x) > 0$), where A is a fuzzy set in the universe of discourse X. *See also* fuzzy set, membership function.

support software software that is used to develop and/or maintain other software. For example, a compiler.

support system system that contains the support needed to actually execute an application system, e.g., the operating system, the user-interface library. It will normally consist of several layers of support where the lower layers provide services to the higher layers.

supremal decision unit control agent or a part of the controller of the partitioned system, which perceives the objectives and the operation of this system as a whole and is concerned with following these overall objectives; in case of a large scale system with hierarchical multi-level (two-level) controller, the coordinator unit is often regarded as the *supremal decision unit.*

surface-based modeling refers to techniques for modeling the three-dimensional surfaces of objects.

surface normal any surface that is smooth enough for at least one derivative calculation at a given point has a *surface normal.* This is a unit vector **n** that is perpendicular to the plane tangent to the surface at the given point. It is usually taken to be pointing outward away from the surface. Smooth surfaces have surface normals at every point. Planar surfaces have the same surface normal at every point that is not at the edge of the surface. At crease or fold edges, the surface normal is undefined.

surface patch (**1**) a small piece of surface with arbitrary shape and size surrounding a surface point with a given surface normal.

(**2**) a primitive element of a geometric surface description, such as a spline or triangulation patch. Graphics techniques that use the different *surface patch* representations are mainly related to surface representation, visibility analysis, illumination, and reflectance.

surface rendering *See* rendering.

surface texture *See* texture, texture analysis, texture modeling.

surge protector a device used to protect the computer against sudden surges in power.

suspend to break the execution of a program unit temporarily.

suspension the state of a task or a process as a result of a suspend operation.

SVC *See* supervisor call instruction.

SVD *See* singular value decomposition.

SVID acronym for System V Interface Definition, pertaining to a version of the Unix operating system.

SVR3 System V Release 3, pertaining to a version of the Unix operating system.

SVR4 System V Release 4, pertaining to a version of the Unix operating system.

swap in assembly language, an instruction that swaps two values one for the other.

swapping the process of interchanging the contents of an area of main storage with the contents in an auxiliary storage.

sweeping the definition of a new object in a higher dimension produced by arbitrary movement of the originating object along a path in the space of the higher dimension. For example, one can create a cylindrical surface by *sweeping* a line about another line which is parallel.

switch (1) in the C language, the keyword for the multiselection control construct similar to the case statement of most languages.

(2) in ALGOL-60, a simple form of multiselect statement in which one of a set of target statements designated by labels would be selected. *See* computed goto.

(3) in some command languages, a modification or specification of an option, often preceded by a syntactic mark such as a hyphen or slash symbol.

(4) in some command languages, a specification of a keyword parameter to a command.

switch box a rectangular routing region containing terminals to be connected on all four sides of the rectangle boundary and for which the entire interior of the rectangle can be used by wires (contains no obstacles).

switching node a computer or computing equipment that provides access to networking services.

switchover time the time occurring in a polling system between service to one queue and service to the following queue. During a switchover interval, the server is idle and at the same time unavailable to requests.

symbol (1) in a programming language, this is typically a name that conforms to the lexical specifications for that language, but may also include other character sequences defined by the language.

(2) in a definition of an automaton, a set of entities that can be used to form an ordered sequence called an input string which is processed by the automaton, or alternatively what are called sentences in the language the automaton is processing.

(3) in the definition of a grammar, the set of entities that can comprise a sequence generated by the rules of the grammar, called a sentence.

(4) in the context of a formal grammar, an entity defined by the grammar that can form a sequence of input to the automaton. This is often referred to as an input string. All input strings are not sentences in the grammar.

symbol error rate a fundamental performance measure for digital communication systems. The *symbol error rate* is estimated as the number of errors divided by the total number of demodulated symbols. When the communication system is ergodic, this is equivalent to the probability of making a demodulation error on any symbol.

symbolic constant a value that is a constant, but which is represented by a name whose meaning and type are determined at the time the name is declared. Some languages do not permit *symbolic constants.* Some languages permit symbolic constants but whose values are fixed at compilation time, so the compiler knows the exact value for each symbolic constant. Some languages permit a name to be assigned a value at execution time, with the restriction that once the value is assigned it may not be changed while the name scope exists.

symbolic evaluation a form of evaluation in which an expression is transformed from one symbolic form to another. For example, the result of evaluating "$(a + b) * (c + d)$" is not a numeric result based on the values of the variables a, b, c, and d, but the symbolic computation "$(a*c+a*d+b*c+b*d)$". Many languages provide for such symbolic evaluation, applying mathematical rules to apply reductions to the resulting expressions. The intended result is a closed-form solution to a problem. Symbolic integration and symbolic differentiation apply transformations of calculus; query optimization applies transformations of relational calculus; proof methods apply transformations of predicate calculus.

symbolic language a programming language expressing operations, instructions, and addresses by using symbols that can be understood by humans rather than in machine language. Assembly as well as all the high level languages are *symbolic languages.*

symbolic link *See* soft link.

symbolic model checking extension of model checking for systems with infinite states (i.e., hybrid automata). States are merged in classes of states.

symbolic processing use of symbols, rather than numbers, combined with rules-of-thumb (or heuristics), in order to process information and solve problems.

symbolic substitution *See* symbolic substitution logic gate.

symbolic substitution logic gate an optical logic gate using a specific algorithm developed for optical computing called symbolic substitution. In symbolic substitution, one or more binary input data are together represented by an input pattern. In its original method, four identical patterns are duplicated from the input pattern. The four patterns are shifted to different directions. The shifted patterns are added. The added pattern is thresholded. The previous procedure is then repeated. The thresholded pattern is split into four identical patterns. The four identical patterns are shifted to different directions and then combined as output pattern. All these steps are equivalent to first, recognizing input pattern, and then substituting it with output pattern. In the improved method, input pattern can be substituted with output pattern using two correlators, the first correlator for recognition and the second for generating output pattern, or using a holographic associative memory. Output pattern can be any of 16 Boolean logic operations or their combinations.

symbolic trace a trace reporting the execution evolution in symbolic manner, thus reporting symbols instead of the actual values that can be typically obtained during the effective execution.

symbol synchronization a technique to determine delay offset or rate of symbol arrival from the received signal. Can be based on either closed or open loop methods.

symbol table the data structure in a compiler which holds information describing the names used by the program being compiled.

symmetric half plane field the class of image models which can be implemented recursively pixel by pixel. That is, if the pixels in an image are ordered lexicographically (either by rows or by columns), then a symmetric half plane model is one in which a pixel p is a function of only those pixels preceding p in the ordering. *See also* Markov random field.

symmetric multiprocessor (SMP) a class of multiprocessor consisting of two or more CPUs configured together via a common memory interface bus and with each process maintaining its own cache, but configured and attached to the same RAM. The operating system operates across all processors within the symmetric multiprocessor and, hence, from the user's perspective, is transparent as to which processor is supporting use functions. The term symmetric is used to refer to the fact that all processors within the multiprocessor have "equal rank"; that is, no one processor is responsible for maintaining cache coherency or for communication with external interface devices. Rather, the operating system itself holds the functionality — and the responsibility — for maintaining consistency among processor cache segments and for evenly distributing workloads across all processors (level-loading).

symmetry breaking a technique to break the symmetry in a structure such as a graph, which can locally look the same to all vertices. Usually implemented with randomization.

synchronization Controlling concurrent access to code and/or data.

synchronization mechanism an implementation of a means of synchronization. This may refer to constructs in a programming language, or low-level machine-specific means of implementing synchronization. *See also* semaphore, mutex, monitor, critical section, rendezvous.

synchronize to update a table or database to be an exact image of another table or database.

synchronized refinement a systematic approach to detecting design decisions in source code and relating the detected decisions to the functionality of the system.

synchronous pertaining to two or more processes whose interaction depends upon the occurrence of a specific global event such as a common timing signal.

synchronous bus a bus in which bus transactions are controlled by a common clock signal and a fixed number of clock periods is allocated for specific bus transactions. *Compare with* asynchronous.

synchronous circuit a sequential logic circuit that is synchronized with a system clock.

synchronous communication a type of communication that permits the communication only if all the participants (usually the sender and the receiver) assert that they want to communicate. It is also called rendezvous. *See* asynchronous communication.

synchronous event an event that occurs at predictable times such as the execution of a conditional branch instruction or a hardware trap.

synchronous message passing synchronization protocol where the message sending process requires both sender and receiver to synchronize at the moment of message transmission.

syntactic analysis analyzing a program for correctness of meaning. Typically, syntactic analysis does not check that the input string "makes sense", that is, is semantically correct (although in some cases semantic information can be embedded in a powerful grammar). In many compilers, the process of *syntactic analysis* and semantic analysis are combined, but that is often an issue of the compiler writer combining these two processes. *See also* parsing, lexical analysis, semantic analysis.

syntactic error *See* syntax error.

syntax the part of a formal definition of a language that specifies legal combinations of symbols that make up statements in the language.

syntax checked a modified compiler consisting of only the lexical analyzer and parser. Used to verify that a program is valid under the program language in question.

syntax-directed analysis a compilation technique characterized by the use of formal syntactic analysis techniques. Historically, this was a significant step up from ad hoc compilation and analysis techniques which had preceded it. Rarely heard in modern usage.

syntax error an error in which the source code for a program or function fails to adhere to the rules of the source language.

synthesized attribute *See* attribute grammar, inherited attribute.

synthetic data data generated probabilistically to have certain good statistical properties.

synthetic image a computational model of the natural image.

synthetic metric a metric that is derived from data instead of a directly measurable metric. As opposed to a natural metric. Function points, the Bang metric, and Mark II metric are all synthetic metrics.

synthetic traffic data arrivals and request sizes produced according to a stochastic model.

system a collection of components, items, or equipment organized or designed to accomplish a specific function or set of functions. It can be based on one or more processes, hardware, software, facilities, and people. *System* is a hierarchic concept: a system at one level may be a subsystem viewed from a higher level.

system administrator the person or persons responsible for the operation of a system.

system analysis the study of user needs using interviews, use cases, rapid prototypes, and the like.

system assessment the process of assessment of a system. In the case of software system, *see* product assessment.

systematic failure a failure related in a deterministic way to a certain cause, which can only be eliminated by a modification of the design or of the manufacturing process, operational procedures, documentation, or other relevant factors. A *systematic failure* can usually be reproduced at will by simulating the failure cause.

systematic fault a fault that manifests itself as a systematic failure.

system availability the instantaneous probability that the system is operational at time t. If the limit of this function exists as t goes to infinity, it expresses the expected proportion of time the system is available to perform useful computations.

system building the process of transforming descriptions using tools to create some less abstract description. This may involve converting designs to source programs to object code. However, the building process may include other transformations such as the construction of system documentation from document fragments.

system bus in digital systems, the main bus over which information flows.

system call a call to a software module executed by an operating system kernel.

system catalog in a relational database, the system catalog is a collection of relations containing information regarding the information to be stored in the database. Names and types of relations and attributes are stored in the system catalog together with information pertaining to security and indexes.

system clock clock internal to the CPU used for synchronization and timing.

system complexity the level of objective difficulty of the system. *See* complexity. For a system, the complexity may include the complexity of requirements, developing or maintaining the system measured by the time, number of steps or arithmetic operations, or memory space required.

system composition the process for system analysis/specification/implementation which creates components by using other components previously specified or taken from libraries. *See* structural decomposition, bottom-up.

system CPU that component of CPU time spent supporting system functions. This CPU component is different from user CPU in that user CPU is booked to a particular user application, whereas *system CPU* is that component of kernel processing required for the system to support any user.

system decomposition the process for system analysis/specification/implementation which creates components by decomposing their complexity in those of a set of smaller components (a divide and Impera approach) that are produced in the next step. This is performed with the aim of composing the development components for obtaining the complete system. *See* structural decomposition, top-down, structural decomposition.

system design *See* design.

system development *See* development.

system development cycle *See* life-cycle.

system engineering an interdisciplinary approach and means to enable the realization of successful systems. It focuses on defining customer needs and required functionality early in the development life-cycle, documenting requirements, then proceeding with design synthesis and system validation while considering the complete problem: operations, performance, test, manufacturing, cost & schedule, training & support, and disposal. *System engineering* integrates all the disciplines and specialty groups into a team effort forming a structured development process that proceeds from concept to production to operation. System engineering considers both the business and the technical needs of all customers with the goal of providing a quality product that meets the user's needs.

system failure a failure that is observed at the interface of a system with its environment. This is distinct from component failure, local failure, and internal failure. Such events within the system may or may not eventually lead to a system failure. *See* error recovery.

system flow chart *See* flow chart.

system generation the process of selecting optional parts of an operating system and creating its particular configuration.

system implementation a phase of software development life cycle during which a software product is integrated into its operational environment.

system life-cycle *See* life-cycle.

system load the volume of work to be processed by the system. It is usually expressed in normalized form as a fraction between 0 and 1 (i.e., its utilization) or as a percentage of its capacity.

system log *See* transaction log.

system maintainability the ability of a system to be repaired and to evolve over time.

system model the abstraction of a real system including stochastic models for arrivals and service processes. *System models* are used both for simulation and for mathematical analysis. Simulation models, however, can generally be more complex than analytically tractable models.

system modeling a technique to express, visualize, analyze, and transform the architecture of a system. Here a system may consist of software components, hardware components, or both, and the connections between these components. A system model is then a skeletal model of the system. It is intended to assist in developing and maintaining large systems with emphasis on the construction phase.

system owned set *See* singular set.

System Performance and Evaluation Cooperative (SPEC) a standards body for performance benchmarks.

system profile the set of features with the corresponding reference values that describe the values that have to be reached at the end of the development life-cycle or at the end of the reengineering process.

system program software used in the management of computer resources.

system reliability the conditional probability that the system has survived the interval $[0, t]$ given that it was operational at time $t = 0$.

systems engineering an approach to the overall life-cycle evolution of a product or system. Generally, the systems engineering process comprises a number of phases. There are three essential phases in any *systems engineering* life cycle: formulation of requirements and specifications, design and development of the system or product, and deployment of the system. Each of these three basic phases may be further expanded into a larger number. For example, deployment generally comprises operational test and evaluation, maintenance over an extended operational life of the system, and modification and retrofit (or replacement) to meet new and evolving user needs.

system size the dimension of a system evaluated by means of a size-based metric.

system software software designed to facilitate the operation and maintenance of a computer system and its associated application programs. *Compare with* application software.

system test *See* system testing.

system testing testing conducted on a complete, integrated system to evaluate the compliance with its specified requirements. *See* software testing.

system thread a collection of processing functionality which encapsulates the performance of a given system capability. For instance, in the case of an automatic teller machine (ATM), one such thread is the cashing of checks. This thread has associated with it functionality and processes which must be accomplished to satisfy the objective of the *system thread.*

systolic flow of data in a rhythmic fashion from a memory through many processors, returning to the memory just as blood flows from and to the heart.

systolic processor a SIMD computer consisting of a set of interconnected cells, each capable of performing a simple operation and synchronized by an external clock or "heartbeat".

T

table a collection of rows (or tuples) of data. Each relation may be viewed as a table with each column representing an attribute.

tableau method an inference mechanism that operates on formulas in negation normal form. *See* negation normal form.

table lock a lock acquired on an entire database table.

tactical view the process of design in which a local part of the system is analyzed in detail.

tag (**1**) that part of a memory address held in a direct mapped or set associative cache next to the corresponding line, generally the most significant bits of the address.

(**2**) a field attached to an object to denote the type of information stored in the object. The tag can flag control objects to prevent misuse. Tags can be used to identify the type of each object and thereby to simplify the instruction set, since, for example, only one ADD instruction would be necessary if each numeric object were tagged with its type (integer vs. real, for example).

tagged image file format (TIFF) a popular image-file format that is very flexible. TIFF can hold compressed or uncompressed images, or different types of pixels, and is usable on different operating systems. *See also* file format, image compression, Lempel-Ziv-Welch coding.

tagged value a pair (tag name, value) attached to a model element to document additional information, generally secondary to the element's semantics. Example: author = Sally.

tail recursion a special form of recursion where the last operation of a function is a recursive call. The recursion may be optimized away by executing the call in the current stack frame and returning its result rather than creating a new stack frame. *See also* collective recursion, iteration.

Takagi–Sugeno–Kang fuzzy model a fuzzy model that was studied by Takagi, Sugeno, and Kang. It is called the TSK or just TS fuzzy model, and it can be viewed as a special case of SAM. *See also* fuzzy system.

tandem queues a system where the output process from one queue becomes the input process to the next queue.

tape skew misalignment of magnetic tape during readout, leading to a difference between bit positions as written and recognized for reading. Generally not a serious problem for low recording density or low tape speed; otherwise, a correction is required, e.g., by the use of "deskewing buffers".

target architecture the architecture of a system that is being emulated on a different (host) architecture.

target entity target entities can be divided into five classes: input products, supporting technology, resources, process, and output products. Synonymous with evaluation item.

target language the language that is generated for its adoption on a target hardware and machine.

target machine the computer on which a prepared code has to be executed. The code can be produced with a CASE tool or with classical editors for programming languages.

target system a computer system on which the output of a compiler is intended to execute. Although compilers commonly produce code for the same system on which they operate, a class of compilers referred to as cross-compilers generate code for systems other than the one on which they run. *Contrast with* host system.

task (**1**) a sequence of instructions treated as a unit that can execute simultaneously with other instruction sequences.

0-8493-2691-5/01/$0.00+$.50

(**2**) In concurrent programming, a generic word used for both process and thread.

(**3**) the smallest unit of work that can be project managed.

(**4**) an activity of a set of actions undertaken by an agent assuming a specific role in order to change the current state of a system to another state (specified by the goal).

(**5**) from the Ada specification, a synonym for thread, with the inclusion of a particular synchronization mechanism, the rendezvous.

task bar a linear array of buttons which usually floats on top of application windows for providing quick access to frequently needed functions or applications.

task control block (TCB) in an operating system, a region of memory that defines status of individual tasks.

task coordinates variables in a frame most suited to describing the task to be performed by a manipulator. *Task coordinates* are generally taken as Cartesian coordinates relative to a base frame.

task domain the set of knowledge, skills, and goals possessed by users that is specific to a kind of job or task.

task interaction graph a graph that divides a program into maximal sequential regions connected by edges representing task interactions. Task interaction graphs can be used to generate concurrency graph representations, and both facilitate analysis of concurrent programs.

task-level programming *See* object-oriented programming.

task migration the act of moving a task from one node to another within the system.

task performance an activity carried out at a user interface and aimed at completing a task.

task point a decomposable unit of a task that can be counted. This count is called the task point metric. It has been proposed as an object-oriented equivalent of the function point in that it can be collected early in the lifecycle and provide the basis for cost and effort estimation.

task profiling the process of decomposing an application task into subtasks, each of which is homogeneous with respect to computational requirements, and characterizing those computational requirements. In general, a task and its subtasks consist of both code and data; therefore, both code and data must be considered during the task profiling process. In current uses of heterogeneous computing, tasks are typically decomposed by the application programmer. The complete automation of *task profiling* is a difficult problem, and represents a long-term goal in the field of heterogeneous computing.

task script a formal task documentation used in requirements. Specified in the form of an SVDPI sentence expressed as Subject-Verb-Direct. Object-[Preposition-Indirect.Object].

taxonomy any schema that describes a piece of knowledge defining relationships among simple entities. It is used for depicting classifications and the body of knowledge. *See* Flynn's taxonomy.

TCB *See* task control block, trusted computing base.

TCP *See* transmission control protocol.

teaching-by-doing *See* teaching-by-showing programming.

teaching-by-showing programming a programming technique in which the operator guides the manipulator manually or by means of a teach pendant along the desired motion path. During this movement the data read by joint position sensors (all robots are equipped with joint position sensors) are stored. During the execution of the motion (playing back) these data are utilized by the joint drive servos. Typical applications of this kind of programming are spot welding, spray painting, and simple palletizing. Teaching-by-doing does not require special programming skill and can be done by a plant technician. Each industrial robot is equipped with these capabilities. Also called teaching-by-doing.

team decision decision taken independently by several decision makers in charge of a given process (or a decision problem) and forming a team; i.e., contributing to a commonly shared goal.

team member any personnel involved in the building of a project.

technical metric refers to the software engineering aspects of system specification such as size, complexity, number of attributes, number of methods, etc.

technical reference a document describing the technical aspects of a system. The system presents both hardware and software aspects.

technical review a development stage, often occurring many different times during the life of a project, in which on-line support systems are examined for performance issues and by subject-matter experts.

technical risk any one of the set of potential technical problems or occurrences that may cause the project to fail.

technical writer a specialist responsible for documenting hardware and software or entire systems.

technological risk the risk related to the technology chosen in the project. The technology includes: tools, languages, operating systems, platform, communication media, support media, methodology, modeling paradigm, etc.

technology a collective term for a group of practices, tools, techniques, and/or methods.

template (**1**) a pattern, often in the form of a mask, which can be used to locate objects and features in an image. For large objects that might appear in many orientations, this procedure is very computation intensive, and it is normal to use small templates to search just for features and then infer the presence of the objects. Template matching is commonly performed for tasks such as edge detection and corner detection.

(**2**) a prototype from which specific instances can be generated.

(**3**) in the C++ language, a keyword which allows the specification of code without binding specific data types to the values. Instantiations of templates must bind the types. *See also* generic.

(**4**) in a dataflow architecture a way of organizing data into tokens. Also called an activity packet.

template mask a mask that forms a pixellated template of an object or feature, and which may then be used for template matching. *See also* template.

template matching a technique in which a model and an optimization method are used to deform a template to a study in order to find the best match for the purpose of detection or recognition.

templating the action of making a template. It is sometimes used as a synonym for cloning.

temporal averaging averaging a signal in the time domain. For discrete signals, *temporal averaging* by a finite impulse response filter is a way to smooth out the signal.

temporal cohesion a cohesion in which the tasks performed by a software module/class are all used in a specific phase of the program execution. *See* cohesion, logical cohesion, procedural cohesion.

temporal constraint a constraint on the system behavior regarding the time used to do something or the order in which things have to be done.

temporal determinism in a deterministic system when the response time for each set of outputs is known.

temporal interval logic extension of an interval logic to express timing constraints, generally are introduced operators to constrain the time duration of an interval or to translate an interval by a constant time value.

temporal logic (**1**) logic with a notion of time included. The formulas can express facts about past, present, and future states. The formulas are interpreted over Kripke structures, which can model computation; hence, *temporal logic* is very useful in formal verification.

(**2**) an extension of propositional or first order logic with operators to state when statements are true. Time can be continuous or discrete and extend indefinitely into the future and/or in the past. Three prefix operators are usually present, represented by a circle, a square, and a diamond; they mean "is true at the next time instant", "is true from now on", and "is eventually true". Usually the until operator is also present: "A until B" means A is true until B is true. "A until B" is the "strong until" and implies that there is a time when B is true. "A unless B" is the "weak until" in which it is not necessary that B holds eventually. There are two types of *temporal logic* used: branching time and linear time. Linear time considers only one possible future; in branching time there are several alternative futures. In branching time temporal logic temporal operators are duplicated; one version is used to state that the temporal condition holds for all the possible futures and the other one is used to state that there exists at least one future where the temporal condition is true.

temporal resolution the ability to resolve two closely spaced targets in the time domain. *See also* resolution.

temporary table a table whose existence is restricted to a session.

temporary update problem problem that arises when a transaction reads data from an uncommitted transaction which is subsequently rolled back.

term over a language $\mathcal{L}$, either a simple term or a complex term; a simple term is a variable symbol or a constant symbol from $\mathcal{L}$, and a complex term $f(t_1, \ldots, t_n)$ consists of a function symbol f from $\mathcal{L}$ of an arity $n \in I^+$ together with n terms $t_1, \ldots, t_n$. Example: $f(a, Y)$, where a is a constant symbol and Y is a variable symbol.

terminal a position within a component where a wire attaches. Usually a *terminal* is a single point on the boundary of a component, but a terminal can be on the interior of a component and may consist of a set of points, any of which may be used for the connection. A typical set of points is an interval along the component boundary.

terminal device an asynchronous input/output port to which a user's terminal is connected.

terminal driver a systems program that manages asynchronous input/output ports, that is, terminal devices.

terminal symbol one of a set of distinguished symbols used to define a grammar. A sentence generated by the grammar consists solely of *terminal symbols*. *Contrast with* nonterminal symbol. For any specific grammar, the set of terminal symbols and the set of nonterminal symbols are disjoint. Also known as a terminal vocabulary. Sometimes designated by the metasymbol Vt. A set of terminal symbols is sometimes called an alphabet. *See also* vocabulary.

terminal vocabulary *See* terminal symbol.

termination protocol a protocol by which individual sites can decide how to terminate a particular transaction when they cannot communicate with other sites where the transaction executes.

terminator a syntactic mark that is used to terminate a grammatical construct. *Contrast with* separator. For example, in most languages, a semicolon is a statement terminator, but in a few languages it is a statement separator.

tesselation in the Euclidean plane, a subdivision of that plane into polygonal cells that cover the whole of it, such that two neighboring cells have disjoint interiors (in other words they have in common either a vertex or a side with its two end-vertices). When the cells are isometric regular polygons, one of the following three cases occurs:

1. Each cell is a regular hexagon and has 6 neighboring cells each having a side in common with it.

2. Each cell is a square and has 8 neighboring cells, of which 4 have a side in common with it, and 4 have a vertex in common with it.

3. Each cell is an equilateral triangle and has 12 neighboring cells, of which 3 have a side in common with it, 3 have a vertex in common with it in such a way that the two neighboring cells are symmetric with respect to the common vertex, and 6 have a vertex in common with it but without symmetry with respect to the common vertex. The *tesselation* of the plane into regular cells is a mathematical model of the subdivision of an image into pixels, and the corresponding digital space is made of the centers of all the cells. In both the hexagonal and square tesselations, there is a vector basis such that the cell centers coincide with points with integer coordinates. Modern technology accords with Cartesian tradition in favoring the square tesselation, but the hexagonal tesselation has a simpler topology (with fewer neighboring cells, and all of the same type), which simplifies certain types of algorithm, such as thinning algorithms. *See also* pixel adjacency.

test technical operation that consists of verifying functionalities for a given product, process, or service according to a specified procedure.

testability (**1**) the capability of the software product to enable modified software to be validated.

(**2**) the measure of the ease with which a circuit can be tested. It is defined by the circuit controllability and observability features.

test access port a finite state machine used to control the boundary scan interface.

test-and-set instruction an atomic instruction that tests a Boolean location and if FALSE, resets it to TRUE. *See* atomic instruction.

test bed the environment containing the hardware, software, tools, and other supports for conducting the test.

test case (**1**) a set of test inputs, conditions for execution, and expected outputs with the aim of testing a particular part of the system, particular functionalities or particular requirements.

(**2**) documentation describing test cases as described in (**1**).

test case generator *See* test generator.

test checklist an outline of usability test procedures. It helps remind usability testers of their planned procedures and of the proper order of these procedures.

test coverage the degree to which a given test presents with respect to the whole functionalities and modules of the system.

test criteria the criteria that have to be satisfied by a system or component to pass a test.

test generator a tool to generate test cases on the basis of: test criteria, requirements, specification, and data structures. The test case generator has to produce the expected results and output for each corresponding test case.

testing operating a system (possibly under unrealistic conditions of use) in order to detect faults. *See* debug, software testing.

test library the entire collection of test cases that is available in an enterprise.

test pattern (**1**) input vector such that the faulty output is different from the fault-free output.

(**2**) a trace describing the evolution of the inputs of a system used for stimulating the system during testing.

test phase the period of time in the system life-cycle during which the test is performed.

Typically the testing phase is performed by a different team.

test plan and procedure a description of testing strategy and tactics.

test point (**1**) a physical contact for a hardware device that can be monitored with an external test device.

(**2**) a data element within a software module that is accessible to an external test module.

test register a register used in the processor to ease testing of some functional blocks (e.g., cache memory) by simplifying accesses to their internal states.

test repeatability an indicator associated to a test that gives a measure about the possibility to obtain the same results repeating the test several times.

test report a document reporting the test performed and the results obtained.

test response compaction the process of reducing the test response to a signature. Common compaction techniques use signal transition counting, accumulated addition, CRC codes, etc.

test vectors a test scheme that consists of pairs of input and output. Each input vector is a unique set of 1s and 0s applied to the chip inputs and the corresponding output vector is the set of 1s and 0s produced at each of the chip's output.

texel a bitmap used to texture a 3D polygon model, including adjustments for perspective correction, where vertices of the object model are mapped onto the 2D texture bitmap. In addition to color and brightness, textures may also be encoded with the properties of transparency and specular reflectivity.

This kind of texture may also be procedural in nature. A possible side-effect of texture mapping occurs unless the renderer can apply texture maps with correct perspective. Perspective-corrected texture mapping involves an algorithm that translates *texels,* or pixels from the bitmap texture image, into display pixels in accordance with the spatial orientation of the surface.

text a list of characters, usually thought of as a list of words separated by spaces. *See also* string.

text categorization *See* text classification.

text classification a task that involves labeling texts as either relevant or irrelevant to a particular category. For example, binary classification tasks require each text to be labeled as either relevant or irrelevant to a single category, while multiclass tasks (sometimes called text categorization problems) require each text to be assigned one or more different category labels.

text corpora text databases. A single text database is called a text corpus.

text filtering a task that involves filtering a set of texts so that only relevant texts that satisfy the filter are passed on to a user.

text representation the form that is used to represent a document. For example, a document might be represented as a set of words or as a set of linguistic structures.

text routing a task that involves categorizing texts with respect to a set of user profiles and sending relevant texts to the appropriate user.

text segment a synonym for code segment. So named in the Unix operating system because it contains the executable "text". *See* data segment.

textual equivalence a form of type equivalence in which two objects are considered to be of the same type if their external representations are the same. Note that this generally applies only to languages that do not have programmer-defined types. *Compare with* structural equivalence, name equivalence.

textural edgedness a measure of the mean edge contrast at every position in an image, where the average is taken over a significant region so as to smooth out small scale variations, thereby providing an indication of the type of texture present.

textural energy a measure of the amount of statistical, periodic, or structural variation at a location in a texture, "energy" being a suitable square-law unit corresponding to the variance imposed on the mean intensity at that location in the texture.

texture quantitative measure of the variation of the intensity of a surface that can be described in terms of properties such as regularity, directionality, smoothness/coarseness, etc.

texture analysis the process of analyzing textures that appear at various positions in images. The term also includes the process of demarcating the boundaries between different textural regions, and leads on to the interpretation of visual scenes.

texture modeling the process of modeling a texture with a view to

1. later recognition
2. generating a similar visual pattern in a graphics or virtual reality display.

Textures are usually partly random in nature, and texture models usually involve statistical measures of the intensity variations.

TGT *See* Kerberos ticket-granting ticket.

theorem solver program that aids in building proofs of properties, it can be fully automatic or semi-automatic.

theoretical maximum a maximum value for a random variable that can never be exceeded in a system. An example is the theoretical maximum of an end-to-end delay in a system without priorities or feedback and with bounded queue sizes and service times.

thermomagnetic recording recording method used with magneto-optical disks. It involves first using a focused laser beam to heat the disk surface and then forming or annihilating magnetized domains.

thermometer coding a method of coding real numbers in which the range of interest is divided into nonoverlapping intervals. To code a given real number, say x, the interval in which x lies is assigned the value $+1$, as are all intervals containing numbers less than x. All other intervals are assigned the value 0 (in the binary case) or -1 (in the bipolar case).

Θ a theoretical measure of the execution of an algorithm, usually the time or memory needed, given the problem size n, which is usually the number of items. Informally saying some equation $f(n) = \Theta(g(n))$ means it is within a constant multiple of $g(n)$. More formally, it means there are some c_1, c_2, and k, such that $0 \leq c_1 g(n) \leq f(n) \leq c_2 g(n)$ for all $n \geq k$. The values of c_1, c_2, and k must be fixed for the function f and must not depend on n. *See also* big-O notation, asymptotically tight.

think time the time that the user is not executing a process or job on a device in a system. Think time is measured as the interval of time that elapses from when the user receives the result of one command until the user issues a new command.

thinning image operator that clears, somehow symmetrically, all the interior border pixels of a region without disconnecting the region. Successively applying a thinning operator results in a set of arcs termed "skeleton".

third generation language (3GL) a language designed to be easier for a human to understand, including things like named variables. A fragment might be let c = c + 2 * d. FORTRAN, ALGOL, and COBOL are early examples of this sort of language. Most "modern" languages (BASIC, C, C++) are third generation. Most 3GLs support structured programming.

third normal form (3NF) a relation that is in second normal form and no non-prime attribute is transitively dependent on the key.

this in C++, a designation of the object being operated upon. In other object-oriented languages, known as self.

Thomas' write rule a rule that specifies that write operations superseded by later write operations on the same object are ignored.

thrashing a state in which a computer system is expending most of its resources on overhead operations rather than on intended computing functions.

thread a unit of concurrency within a process. Sharing a single CPU between multiple tasks (or "threads") is a way designed to minimize the time required to switch threads. This is accomplished by sharing as much as possible of the program execution environment between the different threads so that very little state needs to be saved and restored when changing thread. Multithreading differs from multitasking in that threads share more of their environment with each other than do tasks under multitasking. Threads may be distinguished only by the value of their program counters and stack pointers while sharing a single address space and set of global variables. There is thus very little protection of one thread from another, in contrast to multitasking. Multithreading can thus be used for very fine-grain multitasking, at the level of a few instructions, and so can hide latency by keeping the processor busy after one thread issues a long-latency instruction on which subsequent instructions in that thread depend. A lightweight process is somewhere between a thread and a full process.

threaded code an implementation technique for programming languages in which the program is represented by a sequence of addresses to interpretive routines. The most common example of this is an implementation of the FORTH language. *Compare with* byte code.

thread of control the most important task in a process. It typically controls all the other tasks.

threshold coding a coding scheme in transform coding in which transform coefficients are coded only if they are larger than a selected threshold.

thresholding any technique involving decision making based on certain deliberately selected value(s), known as threshold(s). For an example, refer to threshold coding. These techniques are also often utilized in image segmentation. For example, the transformation of an image I with depth > 1 into a binary image I_b in which $I_b(\mathbf{x}) = 1$ if $I(\mathbf{x})$ is greater than a given threshold (possibly dependent on $\mathbf{x}$), $I_b(\mathbf{x}) = 0$ otherwise.

threshold method a technique used in conjunction with the Hadamard matrix for selecting the compressed transform vector where those components with values above some threshold are kept. *Compare with* zonal method.

threshold set a way of partitioning an image into subimages that stack on top of each other, with each being a subset of the one below it. *See also* stack filter.

throughput **(1)** a measure of the amount of work performed by a computer system over a given period of time.

(2) the quantity of transactions or the amount of data which can be processed or delivered through a network or computer system or component of such a system within unit time.

tiepoint a point in an input image whose corresponding point in a transformed image is known. *Tiepoints* are often used to specify transformations in which the locations of transformed pixels change, as in geometric transformations and morphing. *See also* geometric transformation.

TIFF *See* tagged image file format.

tightly coupled multiprocessors a system with multiple processors in which communication between the processors takes place by sharing data in memory that is accessible to all processors in the system.

time may denote either real time (also referred to as elapsed, calendar, or wall-clock time) or execution time.

time analysis the analysis of the evolution of specific aspects or systems considered against the time and the supposed temporal constraints, if any.

time behavior attribute of software that bears on response and processing times and on throughput rates in performing its function. The capability of the software product to provide appropriate response and processing times and throughput rates when performing its function, under stated conditions.

time boxing a technique by which the product is always delivered at the stated date. If there are problems and delays, the functionality may be less than planned but the deadline never slips.

time-box prototype a special type of prototyping where a prototype is built in a fixed amount of time. The deadline is hard and normally is 3 to 6 weeks.

time-complexity analysis a software analysis/design technique where the performance of individual algorithms can be precisely determined from a timing perspective. The best-case, average-case, and worst-case times of different algorithms can be compared against one another to assist a software engineer in making the correct choice of an algorithm for an application.

time consistent busy hour the busy hour of the mean daily traffic profile. In general, the busy hours of the single days contributing to this mean profile need not be the same and they need not be the same as the time consistent busy hour.

time-constructible function a function $t(n)$ that is the actual running time of some Turing machine on all inputs of length n, for all n.

time critical task a task having a behavior depending on temporal constraint and whose failure may lead to a great damage. *See* real-time system.

time-delay neural network a multilayer feedforward network in which the output is trained on a sequence of inputs of the form $x(t), x(t-D), \dots, x(t-mD)$, where $x(.)$ is, in general, a vector. By specifying the required output at sufficient times t, the network can be trained (using back propagation) to recognize sequences and predict time series.

time dependent variable variable of the system that may change with time.

time domain the specification of a signal as a function of time; time as the independent variable.

timed Petri net a Petri net in which there is some time delay from the moment that all of its input places are filled until a transition fires. These firing delays can be either deterministic or controlled by a random variable.

time-driven simulation a form of discrete-event simulation in which the program structure consists of a main loop that at each iteration advances the simulation clock by some fixed increment and then updates the system state to reflect all events that should have taken place during that time interval. *See also* event-driven simulation.

time event an event that occurs at a particular time. It may be expressed as a time expression.

time explicit in temporal logic, when there is a specific variable generally called a clock used to state timing constraints. For example, if T is the clock variable, formula

$$\Box \forall t. E \wedge T = t \rightarrow <> T <= t + 10 \wedge A$$

states that for all occurrences of event E within 10 time instants, action A will occur.

time expression an expression that resolves to an absolute or relative value of time.

time frequency analysis any signal analysis method that examines the frequency properties of a signal as they vary over time.

time implicit in temporal logic, when the meaning of formulas depends on the evaluation time and this is left implicit in the formula.

time independent variable variable of the system that does not change with time; it is a parameter of the system.

timeliness a measure of how fast the event already perceived can be processed.

time-loading *See* CPU utilization.

timeout (**1**) to terminate a transaction after a fixed period of time.

(**2**) a period of time after which an error condition is raised if some event has not occurred. A common example is sending a message. If the receiver does not acknowledge the message within some preset timeout period, a transmission error is assumed to have occurred.

time-overloaded system a software system that has a 100% or more utilization factor.

time sharing an operating technique of a computer system that provides for interleaving in time of two or more processes.

The effect is achieved by repeatedly preempting the execution of each task to provide time to execute the others. Since tasks are often subject to execution delays for other reasons (i.e., waiting for interactive users to provide some input or for read/write operations from secondary storage), overall system efficiency usually increases with the number of concurrently executing tasks (called the "multiprogramming level") up to a certain point, then drops quickly due to the overhead of task switching (a condition known as "thrashing"). *See also* round-robin system, processor sharing.

time shift for a signal $x(t)$ a displacement in time t_0. The time shift is given by $x(t - t_0)$.

time slicing a mode of operation in which two or more processes are assigned quanta of time for execution on the same processor.

time/space complexity a function describing the maximum time/space required by the machine on any input of length n.

time stamp a value representing the current time. It can be the number of seconds elapsed from a specific date or the number of milliseconds from the boot of the system.

In event-driven simulations, time stamps can be used to determine the next event. In network routing, packets can be time-stamped to determine priority or to decide which will be discarded in case of buffer overflow.

timestamp a date and time value obtained from the system clock.

timestamp ordering an optimistic locking technique for achieving serializability, where a transaction only updates a database object if no other transaction has left a timestamp signifying a conflicting update to the object.

time-to-market the time between project start-up and delivery of the final concrete system. This duration is affected by organization factors and non-software elements of the system.

time to recover time taken for a system to resume operation following a transient failure. The recovering of a service is not necessarily synonymous with its repair; in the case of software failure, the system may recover rapidly and corrective maintenance can be done offline.

time to restore service *See* time to recover.

timing constraint a constraint on the temporal behavior of a system oriented to quantitative constraints. Extensions of temporal logic that permit the specification of this type of constraint have been introduced for real-time systems. *See also* temporal constraint.

timing error an error in a system due to faulty time relationships between its constituents.

TLB *See* translation look aside buffer.

Toeplitz matrix a matrix with the property that it is symmetric and the i, jth element is a function of $(i - j)$. The Toeplitz nature of autocorrelation matrices of wide-sense stationary discrete time random processes is exploited extensively in minimum mean square error prediction/estimation algorithms.

token (1) device that generates or assists in generation of one-time security code/passwords.

(2) in dataflow architectures, data items employed to represent the dynamics of a dataflow system.

token bus a method of sharing a bus-type communications medium that uses a token to schedule access to the medium. When a particular station has completed its use of the token, it broadcasts the token on the bus, and the station to which it is addressed takes control of the medium. Also called token ring.

token ring *See* token bus.

tolerance the amount of error allowable in an approximation.

tomography the process of reconstructing an image from projection data.

tool space space of a 6×1 vector representing the positions and orientations of the tool or end effector of the robot.

top-down starting with a small number of poorly resolved items and breaking them down into smaller, more manageable pieces. It is a "divide and conquer" technique.

top-down analysis an analysis of a context-free grammar which uses leftmost derivation. *See also* LL(k). *Compare with* bottom-up analysis.

top-down reasoning *See* backward chaining.

top hat transform a transform used in mathematical morphology. Let A be a structuring element centered about the origin, and for every pixel p, let A_p be its translate by p. The *top hat transform* measures the extent by which in a given gray-level image I the gray-levels of pixels in A_p are higher than those in the portion surrounding A_p. One way is to take the arithmetical difference $I - (I \circ A)$ between I and the opening $I \circ A$ of I by A; in a dual form, one takes the arithmetic difference $(I \bullet A) - I$ between the closing $I \bullet A$ of I by A and I. Another method considers a second structuring element B which forms a ring surrounding A, and at each pixel one computes the arithmetic difference between a "representative" gray-level in A_p (either minimum, maximum, median, or average) and a "representative" gray-level in B_p (either minimum, maximum, median, or average). *See* closing, opening, structuring element.

top node a synonym for the root node when the graph represents a lattice or semi-lattice. *See also* bottom node.

topological map an organization of nodes in which the similarity of any two nodes is a function of their distance from each other on the map. For example, with a 2-dimensional grid, Euclidean distance can be used as the distance between two nodes. Used in the self-organizing map.

topological order a numbering of the vertices of a DAG such that every edge in the graph that goes from a vertex numbered i to a vertex numbered j satisfies $i < j$.

topology for a neural network, topology refers to the number of layers and the number of nodes in each layer.

topology preserving skeleton result of an operation transforming a digital figure into a one-pixel-wide skeleton having the same "topology", in other words whose connected components and holes correspond in a one-to-one way with those of the figure. This is generally achieved by homotopic thinning. The medial axis transform or distance skeleton preserves the topology of a figure (binary image) in the Euclidean case, but not in the digital case; the same defect arises with the morphological skeleton. *See also* thinning.

topology tree tree that describes a balanced decomposition of another tree, according to its topology.

Torrance-Cook shading a shading model that incorporates an ambient lighting component, a diffuse component and a specular component. *See* Lambert's law.

total correctness during the proof of correctness, an indication about a program or system stating if the output assertions (post conditions) follow logically from input assertions (preconditions) and the corresponding processing steps; and if the program terminates as expected. *See* partial correctness.

total function a function that is defined for all of its domain type.

totally undecidable problem a problem that cannot be solved by a GOTO program. Equivalently, one for which the set of yes-instance strings is not a type-0 language.

total participation given an entity type E in a relationship R, total participation of E in R specifies that every entity in E must be involved in at least one relationship instance.

total quality management a company commitment to develop a process that achieves high quality product and customer satisfaction. It is similar to quality assurance but it is on every level in the organization impacting every aspect of it, including software.

touch input a means for selecting a location on the surface of the display unit using a variety of technologies that can respond to the placing of a finger or other pointing device on the surface. These are essentially data panels placed either on the display surface or between the user and the display surface.

tournament selection a probabilistic selection process adopted in EP, in which the fitness of each individual is compared with those of s randomly chosen individuals in the population, where s is usually referred to as the tournament size. The number of instances where the fitness of the current individual is greater is then recorded as the score for the individual. This is repeated for all parent and offspring individuals and the resulting scores are arranged in descending order. Those individuals corresponding to the top of the list are then incorporated into the next generation. *See also* evolutionary programming.

tower configuration a PC that is designed to sit sideways, making it easier to stack more peripherals inside the unit.

towers of Hanoi given three posts (towers) and n disks of decreasing sizes, move the disks from one post to another one at a time without putting a larger disk on a smaller one. The minimum is $2^n - 1$ moves. The "ancient legend" was written in 1884.

TP *See* transaction processing.

TPM *See* transaction processing monitor.

tps benchmark rating unit for the number of transactions per second a system can provide.

trace the record about the execution of a program showing the sequence of events, executed instruction, messages received, errors, etc.

traceability the property that allows the user to trace a features or an error along a system or process. For example, it can be useful to trace the requirements along the development life-cycle, the errors along the system structure, the messages along the communication channel, or the email along the network.

tracing in software engineering, the process of capturing the stream of instructions, referred to as the trace, for later analysis.

track concentric circle on which data is organized sequentially on the disk.

trackball the earliest version of an input device using a roller ball, differing from the mouse in that the ball is contained in a unit that can remain in a fixed position while the ball is rotated. It is sometimes referred to as an upside-down mouse, but the reverse is more appropriate as the *trackball* came first.

track buffer a memory buffer embedded in the disk drive. It can hold the contents of the current disk track.

traffic channel a channel in a communication network that is used to carry the main information or service, which is typically voice, data, video, etc. *Compare with* control channel.

traffic management the act of controlling the volume and statistical properties of network traffic using some combination of access restrictions and rate control or adaptation of routing.

traffic matrix a matrix containing the traffic between source and destination pairs. This can be the traffic volume, connection set-up rate, mean bandwidth, peak bandwidth, etc.

traffic model a stochastic model for arrival and service processes.

trainability the property of an algorithm or process by which it can be trained on sample data and thus rendered adaptable to different situations. *See also* training procedure, supervised learning, unsupervised learning.

training algorithm of self-generating neural network given as follows:

1. initially, the network contains nothing. When the first training example comes in, the system creates a neuron for it.

2. When a new training example comes in, the neurons in the hierarchy are examined for similarity to the training example, hierarchically from the root(s). Each time, the last winner and all the neurons in its child networks are examined to find a new winner. If the new winner is the old winner, the examination stops and the old winner is the final winner; otherwise, the new winner and its child network are examined until a leaf node is found to be the winner. During this process, the weights of all the winners and their neighbors should be updated according to the same rule as that of a self-organizing network. After the final winner is found, the network structure should be updated according to the following rules:

3. If the winner is a terminal node (i.e., it has no child network(s)), generate two new neurons, and copy the training vector and the original winner into them, respectively. Make them as two neurons in a new child network of the updated winner.

   If the winner is a non-terminal node (i.e., it has a child network), generate only one neuron, copy the training vector into it, and put it in the child network of the updated winner.

4. Repeat the above process until all the training examples are exhausted. One training epoch has been completed at this point, and more than one epoch may be required for the network of neural networks to reach an equilibrium. *See* self-generating neural network.

training procedure the method of calculating the set of free parameters of a function given a set of training data.

training set set of data used as the basis for determining the best set of free parameters. Typically used in iterative training algorithms.

trajectory planning a trajectory planning algorithm is the path description, the path constraints imposed by manipulator dynamics as inputs and position, velocity, and accelerations of the joint (end-effector) trajectories as the outputs. *See* path.

transaction (**1**) one or more database updates or actions that together are atomic, consistent, isolated, and durable.

(**2**) a computation which can be undone. Once a *transaction* is initiated, any computations it performs do not make a permanent change in the state being manipulated until a commit operation is performed. If there is some reason to not make the changes permanent, a rollback operation is performed. Many programming languages directly support transactions, and in those that do not, but support exceptions, the exception mechanism can be used to provide rollback or commit, although more programming is required. Transactions may also be nested transactions.

transaction isolation separation of transactions, such that each appears to run independently, to the extent defined by the level, or degree, of transaction isolation in use. The SQL92 standard defines four levels of transaction isolation (read uncommitted, read committed, repeatable read, and serializable). The ODMG2 standard uses the same transaction isolation levels.

transaction level read consistency where a transaction always reads data, as at the start of the transaction, and does not see changes made by the transaction itself.

transaction log a record of transactions that have been processed, often used to automate reprocessing of transactions.

transaction processing (TP) the exchange of transactions in a client-server system to achieve the same ends as would be performed by the equivalent single complex application.

transaction processing monitor (TPM) for mission-critical applications, it is vital to manage the programs that operate on the data. TPMs achieve this by breaking complex applications down into transactions. TPMs were invented for applications that serve thousands of clients. A TPM can manage transaction resources on a single server or across multiple servers.

transaction scope the extent of a transaction. What is included between the start and end of a transaction.

transaction throughput the number of transactions that can be processed in a given time period, often expressed as the number of transactions per second.

transfer delay *See* end-to-end delay.

transfer function a function that converts information of one data type to information of another data type. Operations such as rounding and truncation are *transfer functions* from floating point to integer. In some cases, the transfer function may not actually change representation, but simply be a mechanism for informing the compiler that the value is to be treated as a different type. *See also* type cast, coercion.

transfer rate a measure of the number of bits that can be transferred between devices in a unit of time.

transfer time in a hierarchical memory system, the time required to move a block between two levels.

transform the process of converting data from one form into another. Often used to signify a system that rotates the coordinate axes. Examples of *transforms* include the Fourier transform and the discrete Fourier transform. A discrete linear transform can be described as a product of the input vector with a transform matrix. *See also* transform kernel.

transformation rule a rule that can be applied to a query tree to result in a logically equivalent tree.

transform coding a method for source coding similar to subband coding. The input signal is transformed into an alternative representation, using an invertible transform (e.g., the Fourier transform), and the quantization is then performed in the transform domain. The method utilizes the fact that enhanced compression can then be obtained by focusing (only) on "important" transform parameters.

transform kernel a function that is multiplied with an input function, the result of which is integrated or summed to form a transformed output. For example, in the definition of the continuous Fourier transform the kernel is $e^{-j\omega t}$ and in the definition of the discrete Fourier transform the kernel is $e^{-j2\pi nm/N}$.

transform vector quantization the generalization of scalar transform coding to vector coding. *See* transform coding.

transform VQ *See* transform vector quantization.

transient class a class whose instances are not persistent.

transient fault *See* intermittent fault.

transient state (1) in reliability analysis, a fault or error resulting from temporary environmental conditions.

(2) in a stochastic process, a state that is not part of a recurrent chain and hence has probability zero of occurrence in steady state.

(3) in a computing system or a communication network, a system state under non-stationary load (e.g., after a load change) until a new steady state is reached.

transition the evolution from a state to the next. The term is represented by an arrow between two states in state diagrams.

transition function a function of the current state and input giving the next state of a Turing machine.

transitive closure an extension or superset of a binary relation such that whenever (a, b) and (b, c) are in the extension, (a, c) is also in the extension.

transitive dependency a property of a database relation. Given attributes A, B, and C, if $A \rightarrow B$ and $B \rightarrow C$, then $A \rightarrow C$ is a transitive dependency.

transitive rule a rule pertaining to databases. Given sets of attributes A, B, and C, if $A \rightarrow B$ and $B \rightarrow C$, $\models A \rightarrow C$.

translation a geometric transformation which simply adds an offset to the pixel coordinates of an image. Point **M** can be moved, or translated, to a new location **M'** by adding a vector **T**. More concisely: **M'** = **M** + **T**. Translation and many other simple transformations can be done simultaneously if positions and directions are represented in homogeneous coordinates.

translation look aside buffer (TLB) essentially a small fully associative address-cache used to provide fast address translation for the most used virtual addresses. The TLB is associatively searched on a virtual address and in the event of a hit, it returns the corresponding real address. In the event of a miss, if the addressed page is in main memory, then a TLB entry is made for it; otherwise the page is first brought in after a page fault and then the TLB entry is made; in either case, the TLB eventually returns a real address. The TLB may be fully associative, set associative, or hashed.

translucent a characteristic of a material allowing light to pass through partially or diffusely.

transmission control protocol (TCP) the basic transport-level protocol of the Internet. It provides for reliable, flow-controlled, error-corrected virtual circuits.

transmission loss refers to requirements that were originally requested for but that were lost during the development process. It is not uncommon to have a loss of 10% in large software systems.

transparent code a code in which the complement of every codeword also is a codeword.

transparent latch essentially, a flip-flop which continuously passes the input to the output when the clock is high but holds the last output during any interval when the clock is low. The circuit is said to have latched when it is holding its output constant regardless of the value of the input.

transput a term used to refer collectively to input and output. Synonym: input/output, I/O.

transputer a fully self-sufficient, multiple instruction set von Neumann processor designed to be connected to other transputers. Designed and manufactured by Inmos Corporation. The *transputer* was specifically designed to be used in arrays for parallel processing.

trap a conditional jump to an exception or interrupt handling routine, often automatically activated, with the location from which the jump occurred recorded.

trapped error an execution error that immediately results in a fault.

traveling salesman problem the problem of determining the optimal route through a set of cities, given the intercity travel costs.

traversal (**1**) a specification of movement along the arcs of a graph. May include reverse movements (backtracking) along the arcs.

(**2**) the act of traversing a graph.

traversal algorithm an implementation of a graph traversal. *See also* tree walk, prefix traversal, postfix traversal.

tree (**1**) a data structure represented as a directed graph of nodes and arcs, such that each node has at most one arc leading into it, and can have zero or more arcs leading from it.

See also tree node, leaf node.

(**2**) an abstract representation of a program, represented as a tree whose nodes correspond directly to terminal symbols and nonterminal symbols of the grammar used to parse the input. *See also* parse tree.

tree code a code produced by a coder that has memory.

tree coding old name for convolutional coding.

tree editing problem the problem of transforming one of two given trees into the other by an edit scripts of minimum cost.

tree network limited connection of subscriber nodes to a central control or distribution unit via other subscriber nodes in the network.

tree node (**1**) a node in a tree. The designation is independent of whether or not there are any arcs leading from the node.

(**2**) a node in a tree that has the property that it has one or more arcs leading from it to other nodes, also called an internal node. *Compare with* leaf node.

tree-search algorithms for searching through a tree-structured problem based on a certain cost function, or metric, increment associated with each branch of the tree.

tree structure diagram *See* hierarchical diagram.

tree structured vector quantization scheme to reduce the search processes for finding the minimum distortion codevector using a tree structured codebook, where each node has m branches and there are $p = \log_m N_c$ levels of the tree. Abbreviated tree structured VQ.

tree structured VQ *See* tree structured vector quantization.

tree structure robot any of a set of rigid bodies connected by joints forming a topological tree.

tree walk a traversal algorithm that specifies how to traverse a tree. *See* prefix traversal, postfix traversal.

trellis code a (channel) coding scheme in which the relation between information symbols and coded symbols is determined by a finite-state machine. The current block of information symbols and the state (in the finite-state machine) uniquely determine the block of coded output symbols as well as the next state. Thus, trellis coding can be viewed as generalized block coding for which the encoder function depends on the current as well as previous blocks of non-overlapping information symbols. In the class of trellis codes we, for example, find convolutional codes, trellis-coded modulation, and continuous-phase modulation. *See also* convolutional code.

trellis coded modulation a forward error control technique in which redundancy is introduced into the source stream through an increase in the number of symbol values rather than an increase in the number of symbols. Developed by G. Ungerboeck in the late 1970s, this approach has found widespread use in systems with limited bandwidth.

trellis coding *See* trellis code.

trellis diagram in convolutional codes and trellis coded modulation, a graphical depiction of all valid encoded symbol sequences, and the basis for the Viterbi decoding algorithm.

trellis search an algorithm for searching through a trellis-structured problem based on a certain cost function, or metric, increment associated with each branch of the trellis.

trellis vector quantization (trellis VQ) a method for structured vector quantization, where the input signal is classified and coded in a manner described by a mathematical structure known as a "trellis".

trellis VQ *See* trellis vector quantization.

trial phase in which a software product is tested in real-conditions by the end-user. It can be performed in-house or directly in the plant of the end-user.

triangle inequality a complete weighted graph satisfies the triangle inequality if $\text{weight}(u, v) \leq \text{weight}(u, w) + \text{weight}(w, v)$ for all vertices u, v, w. This will hold for any graph representing points in a metric space. Many problems involving edge-weighted graphs have better approximation algorithms if the problem is restricted to weights satisfying the triangle inequality.

triangular co-norm denoted $\dot{+}$, is a two place function from $[0, 1] \times [0, 1]$ to $[0, 1]$. The most used triangular co-norm is the union $\vee$ defined as

$$\mu_{A \vee B} = \max\{\mu_A(x), \mu_B(x)\},$$

where A and B are two fuzzy sets in the universe of discourse X.

The *triangular co-norm* is used as disjunction in the approximate reasoning, and reduces to the classical OR, when applied to two crisp (nonfuzzy) sets.

See also approximate reasoning, fuzzy set.

triangular loop a nested loop where the inner loop limits depend on the outer loop index.

triangular norm (1) in fuzzy systems, denoted $\star$, a two place function from $[0, 1] \times [0, 1]$ to $[0, 1]$. The most used *triangular norms* are the intersection $\wedge$ and the algebric product $\cdot$ defined as

$$\begin{aligned} \mu_{A \wedge B} &= \min\{\mu_A(x), \mu_B(x)\}, \text{ and} \\ \mu_{A \cdot B} &= \mu_A(x)\mu_B(x), \end{aligned}$$

where A and B are two fuzzy sets in the universe of discourse X.

The triangular norm is used as conjunction in the approximate reasoning, and reduces to the classical AND. when applied to two crisp (nonfuzzy) sets.

See also approximate reasoning, fuzzy set.

(2) in control, a function T from the closed unit square into the closed unit interval endowed with the following properties:

$$\begin{aligned} T(a, b) &= T(b, a); T(a, b) \leq T(c, d) \\ &\Leftarrow a \leq c, b \leq d; \\ T(a, 1) &= a; T(T(a, b), c) \\ &= T(a, T(b, c)); \end{aligned}$$

for all $a, b, c, d \in [0, 1]$. The most important triangular norms (t-norms) are defined by product of two numbers, operation of taking the smaller one of two numbers and a family of T_s given by

$$T_s(a, b) = \log_s\left(1 + \frac{(s^a - 1)(s^b - 1)}{s - 1}\right)$$

t-norms are used to define a triangle inequality in the family of probabilistic metric spaces called statistical metric spaces, which may be considered as uncertain counterparts of metric spaces. Different triangular norms are used as composition operators in fuzzy systems, although the standard operation is defined by minimum.

triangulation (1) the transformation of a model into a mesh of triangles to facilitate speedy rendering or other computational geometry algorithms. The initial model might be a planar graph, free-form surface, polygonal model, point cloud data, or volumetric data.

(2) an example of faceting. *See* faceting, facet.

trie digital tree, tree in which edges are labeled by symbols or strings.

trigger (1) the combination of circumstances which activates a latent fault. The circumstances include selection of particular inputs together with a certain internal state of the system.

(**2**) specifies a condition and an associated action. If the condition is satisfied, the rule is *triggered.*

trilinear filtering a level of detail blending technique used in MIP texture mapping. Pixels are taken from multiple MIP maps and blended to produce the final color. The purpose is to remove the bands between adjacent pixels taken from different MIP maps.

trinary function a function with three arguments. *See also* constant function, unary function, binary function, N-ary function.

tri-state circuit a circuit element or branch that is in a high impedance state and effectively isolated from the rest of the circuit from a current flow standpoint.

tristimulus coordinates *See* chromaticity coordinates.

tristimulus value one value in tristimulus color theory. Tristimulus color theory stems from the hypothesis that the human eye has three types of color receptors (cones) that have peak sensitivity in the red, green, and blue visible light wavelengths, respectively. The tristimulus color values are a set of three values X, Y, and Z which replace the red, green, and blue intensities with a weighted integral which calculates a spectral energy over the range of visible wavelengths of light for each value; the integrals allow for colors to be represented purely additively, while representing colors via red, green, and blue intensities often requires a subtractive interaction between the "primaries".

Trojan horse a method for penetrating a system's security measures by launching a program that masquerades as a common command program, such as the login program, in order to collect information to invade a user's account or the system.

tropism an external directional influence on the branching patterns of trees.

true a value that represents the "true" value in two-valued logic. Programming languages are, in the abstract, not constrained to a particular representation of the notion of "true" and traditional choices have been "the constant 1", "any nonzero value", "any odd value", "the constant -1", or "any positive value". *Compare with* false.

true complement in a system that uses binary (base 2) data, a representation of negative numbers as 1's complement of the positive number. This is also called a radix complement.

true concurrency *See* concurrency. *Contrast with* apparent concurrency.

true data dependency the situation between two sequential instructions in a program when the first instruction produces a result that is used as an input operand by the second instruction. To obtain the desired result, the second instruction must not read the location that will hold the result until the first has written its results to the location. Also called a read-after-write dependency and a flow dependency.

truncation (**1**) loss of information as a consequence of the representation used by the compiler, language, or underlying computer. In the case of floating point numbers, this usually happens on the least significant bit of the representation.

(**2**) loss of information caused by a need to store a value in a location that does not have the capacity to store the entire value. Unlike mathematical truncation, this usually involves the loss of high-order information, storing a modulus result and potentially losing sign information. *See* chopping, rounding.

truncation to minus infinity loss of low-order information of a value such that the result is always less than or equal to the untruncated value. Example: -2.7 becomes -3; -3.1 becomes -4. *Compare with* truncation to zero. *See also* rounding.

truncation to zero loss of low-order information of a value such that the absolute value of the result is less than or equal to the absolute value of the untruncated value. Important when negative numbers are involved. Example: −2.7 becomes −3, −3.1 becomes −3. *Compare with* truncation to minus infinity. *See also* rounding.

trust the willingness to believe messages, especially access control messages, without further authentication.

trusted computing base (TCB) the collection of all mechanisms in a system that enforces security policy. The TCB boundary represents the security perimeter of the system.

truth model a very detailed mathematical description of a process to be controlled. The truth model is also called the *simulation model*, since it is used in simulation studies of the process. *See also* design model.

truth table a listing of the relationship of a circuit's output that is produced for various combinations of logic levels at the inputs.

TS fuzzy model *See* Takagi-Sugeno-Kang fuzzy model.

TSK fuzzy model *See* Takagi-Sugeno-Kang fuzzy model.

tty driver *See* terminal driver.

tty port *See* terminal device.

tuple (1) an ordered set of values. Can be viewed as a row of values in the table.

(2) a representation of an intermediate language in the compilation process in which operations are represented by an ordered sequence of tuples, each tuple specifying the operation and the operands.

tuple relational calculus a calculus based on the predicate calculus that uses tuple variables that range over relations.

turbo code the parallel concatenated convolutional coding technique introduced by Berrou, Glavieux, and Thitimajshima in 1993. These codes achieve astonishing performance through parallel encoding of the source symbol sequence and iterative serial decoding of the demodulated symbol sequence.

turbomouse an input device that consists of a trackball and two buttons on a stationary pedestal.

turbulence a chaotic system condition characterized by disorder on all scales, with backward eddy currents and circular waves.

Turing machine the simplest formal model of computation consisting of a finite-state control and a semi-finite sequential tape with a read-write head. Depending on the current state and symbol read on the tape, the machine can change its state and move the head to the left or right. Unless otherwise specified, a *Turing machine* is deterministic.

Turing reduction a reduction computed by an oracle Turing machine that halts for all inputs with the oracle used in the reduction.

Turing test a test conceived by Alan Turing in 1950 designed to reveal whether a machine demonstrated "artificial intelligence".

In essence, *Turing's Test* says that if a computer cannot be distinguished from a human (by asking it a series of questions), then it must be concluded that the computer is "thinking". Turing imagined a man asking questions via a teletype of an entity behind some closed door. The man did not know whether the entity was another human or a computer answering the questions. If a computer program could be written such that the man could not determine the identity of the being through questioning, then it must be concluded that the computer had passed the test, and hence exhibited "artificial intelligence". To date, no machine has passed any form of Turing Test.

turnaround time the elapsed time between submission of a job and the return of the complete output.

tutorial a type of on-line support system that provides a supportative and safe educational environment for learning processes or products. They usually include interactivity, and the content and pace are controlled, to some degree, by users. *Tutorials* have a broader pedagogical scope than both documentation and help.

twenty-six connected *See* voxel adjacency.

two-address computer one of a class of computers using two or less address instructions.

two-address instruction a class of assembly language ALU instruction in which the two operands are located in memory by their memory addresses. One of the two addresses is also used to store the result of the ALU operation.

two-and-a-half-D sketch a representation of the input image which is augmented at every position by information relating to a 3D structure and which is deemed to constitute a significant step on the way to human image interpretation. The name arises as the basic representation is still 2D, whereas it is tagged with all available 3D information; it is important in forming a bridge between early (i.e., low level) visual processes and high level vision. It is strongly associated with the name of its developer, the late David Marr.

two-channel coding a coding scheme in which a signal is decomposed into two parts: low frequency and high frequency components. The low frequency component is undersampled and the high frequency component is coarsely quantized, thus saving data. It can be viewed as a special example of subband coding.

two-channel filter bank a filter bank that has one high frequency band and one low frequency band in both analysis and synthesis filters.

two-dimensional cellular automaton a cellular automaton where a cell's contents at time t are based on its own and the contents of all its immediate neighbors at time $t - 1$.

two-dimensional memory organization memory organization in which the arrangement on a single chip reflects the logical arrangement of memory. In the most straightforward case, each address is presented at once in its entirety and decoded by a single decoder. However, to reduce the number of pins required for addressing, the address may be split into two parts that are then sent in sequence on the same lines. Then during the row access strobe, one part is used to select a "row" of the memory, and during the column access strobe the other part is used to select a "column" of the selected "row". The "row" output may be held in a buffer and the "column" access then applied to the contents of the buffer.

In a two-and-a-half dimensional organization, the bits of each word are spread across several chips — one bit per chip in the most extreme case. Each chip is then equipped with two decoders, each of which decodes part of a split address in order to carry out a selection on the chip.

two-phase locking in database systems, a situation where all locks are acquired by a transaction in a growing phase, prior to release of all locks in a shrinking phase. *Two-phase locking* guarantees serializability, except when wormholes exist.

two-port memory a memory system that has two access paths. One path is usually used by the CPU and the other by I/O devices. This is also called dual port memory.

two-way interleaved in memory technology, a technique that provides faster access to memory values by interleaving memory values in two separate modules.

type the concept of *type* and class are in general distinguished. In most of the classical object-oriented programming languages this distinction is not performed.

type cast an operation that tells a compiler to treat a representation specified as one type as being a representation of another type. In some languages (such as C), the notion of *type cast* may be confused with coercion and transfer function, using the same syntax for each of these.

type checker the part of a compiler or interpreter that performs type checking.

type checking a form of semantic analysis in which rules not specified as part of the syntax of the language are used to determine if the data types of the variables of the program conform to those rules. For some languages, *type checking* can be performed at compile time, and is a form of static semantics. In other languages, type checking cannot be performed until the program executes, and is a form of execution semantics.

typed language a language with an associated (static) type system, whether or not types are part of the syntax.

type equivalence a specification in a language by which it is determined if the type of a specific value is equivalent to a specified type. *See* name equivalence, structural equivalence, textual equivalence.

type lattice a common misnomer for the hierarchy of types in a language. The type system in most languages is actually a semi-lattice, having a designated top node which is usually a type that can represent any other type in the language. In some languages, the types do not form a semi-lattice but references to the types do form a semi-lattice. If the language does not support multiple inheritance, the semi-lattice is a tree.

type-safe an operation that cannot violate the type semantics of the language. Some languages, such as C and C++, have type cast operations that allow the user to perform type-unsafe conversions.

type system a collection of type rules for a typed programming language. Same as static type system.

typing error an error reported by a type checker to warn against possible execution errors.

U

UART *See* universal asynchronous receiver transmitter.

UDP *See* user datagram protocol.

UEP code *See* unequal error protection code.

UID acronym for unique identifier. A value that can uniquely identify an object. An implication is that this is a value permanently assigned to the object during its entire existence, including on secondary storage, during information transfers between machines, and the like. This is distinguished from, for example, a memory address, which is a UID of the object if there is only one copy of the object in memory, but which can change if the object is removed from memory and later brought back in. A UID can be essential in a programming language that supports objects that can be stored long-term in a persistent store. *See also* GUID.

ULA *See* uniform linear array.

ULP acronym for unit in the last place. This is the value represented by the least significant digit in the sequence of digits of a fixed-point representation. That is, it is 1 for integers and r^{-n} for a radix-r representation with n digits to the right of the radix point.

ultimate goal quality the necessary and sufficient quality required to satisfy user stated and implied needs. This is a goal that developers must reach, but which cannot be fully met because the user is not aware of his real needs and needs may change after they are stated.

ultrasonic memory obsolete form of memory, in which data was stored as an ultrasonic sound wave recirculating through a thin column of mercury. Also called mercury delay line memory.

umask a value used to alter the default access mode of an ordinary file or directory that is subtracted from the default access mode to produce a new value for creating files and directories.

umbra the part of the shadow created by an extended light source that is entirely cut-off from the source. It is surrounded by the penumbra that receives some light from the light source.

umbra transform a morphological transform used for visualization of operations on gray-level images. Let the graylevel image F be a function $E \to T$, where E is the set of points and T the set of gray-levels. The umbra of F is the set

$$U(F) = \{(h, v) \in E \times T \mid v \leq F(h) \text{ and } v \neq \pm\infty\} .$$

The behavior of a morphological operator on gray-level functions can be visualized by applying the corresponding operator for sets to the umbras of the gray-level functions. *See* morphological operator.

UML *See* unified modeling language.

unary function a function that takes one argument. *See also* constant function, unary function, binary function, N-ary function.

unbiased estimate an estimate $\hat{\mathbf{x}}$ of $\mathbf{x}$ which is not subject to any systematic bias; that is,

$$E\left[\hat{\mathbf{x}} - \mathbf{x} = 0\right] .$$

See also bias.

unbound name a name whose meaning is not determinable from an analysis of the program fragment in which it is used. The meaning of such a name is only determined by some other specification, whether it is an enclosing lexical context which does not appear, or a dynamic context not determinable until execution time.

uncomputable function a function that cannot be computed by an algorithm or, equivalently, not by any Turing machine.

0-8493-2691-5/01/$0.00+$.50

unconditional branch an instruction that causes a transfer of control to another address without regard to the state of any condition flags.

undecidable *See* undecidable problem.

undecidable language a problem that cannot be decided by any algorithm; equivalently, whose associated language cannot be recognized by a Turing machine that halts for all inputs.

undecidable problem a problem that cannot be decided by any algorithm; equivalently, whose associated language cannot be recognized by a Turing machine that halts for all inputs. *See also* decidable problem, solvable problem, undecidable language, Post's correspondence problem.

underflow a condition in which a floating point computation results in a value too small to represent in the available floating point representation. *See also* sudden underflow, gradual underflow, overflow.

underflow detection a form of execution semantics in which an underflow produces an error condition. The exact nature of the condition and how it is handled are determined by factors such as the language, its implementation, and the platform on which it is running.

understandability the capability of the software product to enable the user to understand whether the software is suitable, and how it can be used for particular tasks and conditions of use.

undirected graph a graph whose edges are unordered pairs of vertices. That is, each edge connects two vertices. *See also* directed graph.

undo to remove a change made to data, by applying a before image of the data held in an undo, or before image log.

undo log a log of before images of data used to remove changes to a database.

undo record a group of before images of a single, atomic change to the database.

unequal error protection (UEP) code a code in which certain digits of a codeword are protected against a greater number of errors than other digits in the codeword.

UNF *See* unnormalized form.

unfinished work, residual work the amount of work that still has to be finished in a system regarded by the virtual waiting time approach.

unfolding the technique of transforming a program that describes one iteration of an algorithm to another equivalent program that describes iterations of the same algorithm.

Unibus bus standard used by Digital Equipment Corporation for its PDP and VAX computers.

unicode acronym for uniform code, a representation of a character set expressed as 16-bit characters. Allows representation of several international languages intermixed in the same data stream. *See also* ASCII, ANSI, multibyte character set.

unidirectional bus a group of signals that carries information in one direction. Example: The address bus of the microprocessor is unidirectional; it carries address information in one direction—from the microprocessor to memory or peripheral.

unified cache a cache that can hold both instructions and data. *See* cache.

unified modeling language (UML) a combination of a metamodel and a notation. It was endorsed by the Object Management Group as their standard approach to modeling object-oriented systems in 1997.

uniform circuit complexity the study of complexity classes defined by uniform circuit families.

uniform circuit family a sequence of circuits, one for each input length n, that can be efficiently generated by a Turing machine.

uniform distribution a probability distribution in which all events are equiprobable, i.e., $p(x) = k$ subject to $\int_{-\infty}^{\infty} p(s)ds = 1$.

uniform length coding a coding scheme that assigns the same number of bits to different messages no matter what probabilities the messages assume.

uniform linear array (ULA) in array processing, an array with evenly spaced sensors placed on a straight line.

uniform matrix a matrix having the same number of items in each row. *See also* ragged matrix.

uniform memory access refers to a class of shared memory multiprocessor systems in which accesses to all parts of the shared memory take the same time independently of which processor makes the access.

uniform resource locator (URL) in the World Wide Web, the Internet address of a file.

uniform sampling the sampling of a continuous signal at a constant sampling frequency.

unimplemented instruction (1) a numeric pattern in an instruction stream that does not correspond to any defined machine instruction.

(2) a type of trap operation executed by a processor when an unimplemented instruction is encountered.

uninitialized variable a variable whose value has not been explicitly assigned at the time the variable was declared. Some languages implicitly initialize all variables to well-specified values at their point of declaration; other languages will simply use whatever value was left in memory in the memory area now representing the variable. A common programming error is to use the value of a variable before the value has been established by the program.

uninterruptible power supply (UPS) a power supply designed to charge an energy storage medium, while providing conditioned output power, during the presence of input power and to continue providing output power for a limited time when the input to the supply is removed. These power supplies are typically used in critical applications to prevent shutdown of these systems during power failures, power surges, or brownouts.

union (1) in operations on mathematical sets, an operation that produces a new set such that for every element of the resulting set, the element is contained in one or the other of the two sets which formed the union, and furthermore the element is not duplicated in the resulting set.

(2) in an operation on two numeric values in the computer, the bitwise OR operation on the bits representing the values.

(3) in the C language, the term that designates a variant record.

(4) in a database system, performs the relational operation union. *See also* join, intersect.

union compatibility a property of two or more relations. Two relations are said to be *union compatible* if they have the same degree and the ith attributes of the two relations are of the same domain $\forall\ 1 \leq i \leq n$.

union of fuzzy sets the fuzzy analogy to set the set-theoretic union.

Let A and B be two fuzzy sets in the universe of discourse X with membership functions $\mu_A(x)$ and $\mu_B(x)$, $x \in X$. The membership function of the union $A \cup B$, for all $x \in X$, is

$$\mu_{A \cup B}(x) = \max\left\{\mu_A(x), \mu_B(x)\right\} .$$

See also fuzzy set, membership function.

union operator a logical OR operator.

In a crisp (non-fuzzy) system, the union of two sets contains elements that belong to either one or both of the sets. In fuzzy logic, the union of two fuzzy sets is the fuzzy set with a membership function that is the larger of the two. *See* union of fuzzy sets.

uniprocessor *See* single instruction stream, single data stream.

uniquely decodeable a channel code where the correct message sequence can always be recovered uniquely from the coded sequence as observed through the channel. Of particular interest for multiple access channels.

unitary transform a transform whose inverse is equal to the complex conjugate of its transpose.

unit delay in discrete time systems, the delay of a signal by a single sample interval, i.e., $x(n-1)$. Under the z-transform, $z^{-1}X(z)$.

unit of work *See* transaction.

unit test part of the testing strategy that focuses on tests to individual program components.

universal asynchronous receiver transmitter (UART) a standard interface often used in small computer systems, to buffer and translate between the parallel word format used by the CPU and the aynchronous serial format used by slow I/O devices.

universal coding coding procedure that does not require knowledge of the source statistics and yet is asymptotically optimal. A typical example is Lempel-Ziv coding.

universal fuzzy approximator a fuzzy system approximator in a sense that it can approximate any nonlinear function to any degree of accuracy on any compact set.

universal hashing a scheme that chooses randomly from a set of harsh functions.

universal quantification a first order logic operator used to quantify a variable over a finite or infinite set. It is used to state that a formula is true for every value of a variable. It is usually represented with ∀. *See* existential quantification.

universal source coding refers to methods for source coding that do not rely on explicit knowledge of the source statistics. One important method for universal lossless source coding is Ziv-Lempel coding.

universal synchronous/asynchronous receiver-transmitter (USART) a logic device that performs the data link layer functions, such as serializing, deserializing, parity generation and checking, error checking, and bit stuffing, of a serial transmission protocol for either synchronous or asynchronous transfer modes.

universal Turing machine a Turing machine that is capable of simulating any other Turing machine if the latter is properly encoded.

universe of discourse term associated with a particular variable or groups of variables, it is the total problem space encompassing the smallest to the largest allowable non-fuzzy value that each variable can take.

Unix the common name for a family of interactive, multiuser operating systems. Often capitalized as "UNIX". However, Unix is not an acronym, and therefore need not be capitalized unless referring to the registered trademark. Unix is available in one version or another for virtually any computer, ranging from desktop personal computers and workstations to the most powerful supercomputers.

Unix was originally developed on a Digital Equipment Corporation (DEC) PDP-7 in 1969 by Ken Thompson and Dennis Ritchie — employees of AT&T's Bell Laboratories. UNIX is a registered trademark of AT&T's UNIX System Laboratories.

(2) "unix" is the filename of the Unix executable.

unlock to release a semaphore or other type of lock.

unnamed pipe a pipe that does not have a directory entry, that is, a pipe which is accessible only to its creator and users.

unnormalized form (UNF) a relation is unnormalized if it does not satisfy any normal forms.

unordered file *See* heap file.

unpopulated card a card with no chips.

unrepeatable read a read of an object or data set producing different data the second or subsequent time it is performed, as the object or data set has been updated since the first read.

unrevealed failure *See* latent fault.

unsafe cast *See* unsafe conversion.

unsafe conversion a form of type cast or representation conversion in which the compiler is told that it must treat a value of a certain type (or reference to a value of that type) as if it were a value of a different type (or a reference to a value of the different type).

unsharp masking an edge enhancement technique that subtracts a blurred version of an image from the input image. *See also* edge enhancement.

unsigned integer an integer value in which the high-order bit is treated as a numeric bit rather than a sign bit. In most machines, 2's complement representation is used, so the interpretation of the sign bit is by convention. For example, a byte can represent a value in the range $-128 \ldots 127$ as a signed value but a value in the range $0 \ldots 255$ as an unsigned value.

unsigned shift a form of shift operation in a computer in which shifts towards the low-order bit use the constant 0-bit value to fill the new values required.

unsolvable problem a computational problem that is not solvable. The associated function is called an uncomputable function.

unspanned organization a file organization in which records do not span block boundaries.

unspecified in specifications of programming languages, conditions or results that cannot be known, are unpredictable, are implementation-specific, or which indicate an erroneous program whose results are not constrained by the language definition. *See also* indeterminate.

unstable equilibrium the state of a system that can easily be moved into a chaotic or unstable state.

unsupervised learning a training technique in statistical pattern recognition or artificial neural networks in which the training set does not include a predefined desired output. The learning system seeks to identify structure in the data by clustering similar input patterns.

unsupervised neural network neural networks that require predetermined output to form the interconnection weight matrix of a network. If no input-output pairs are known and a number of inputs are available, we can only memorize them in an organized order. Similar inputs are memorized in locations close to each other and different inputs are stored in locations far from each other. The network is able to recognize an input that has already appeared. If the input has never appeared, it will be stored in an appropriate location, which is close to similar inputs and far from different inputs. An example of an *unsupervised neural network* is the Kohonen self organizing map, which can be implemented using an adaptive correlator. A correlator has been long used in optical processing for measuring the similarity between two inputs.

untrapped error an execution error that does not immediately result in a fault.

untyped pointer a pointer whose type is unknown. In many languages, the untyped pointer represents the top node of the type lattice.

upcasting pointing to or referencing a derived object using a reference or a pointer to a base class. Another term for widening.

update anomaly in database systems, when an update command results in inconsistency in the data.

update authorization in database systems, permission to update a relation.

update lock a lock used to avoid deadlock, where the intention is to acquire an exclusive lock, but unlike an exclusive lock, an *update lock* may be downgraded to a shared lock.

update query in database systems, used to modify the values of attributes of one or more selected tuples in a relation.

up/down counter a register that is capable of operating like a counter and can be either incremented or decremented by applying the proper electronic signals.

up-level addressing a concept from programming languages. In languages that permit function definitions to be declared within the scope of existing function definitions, such that parameter values and values declared in all scopes enclosing the function definition being made, access to the enclosing values is effected by a technique called *up-level addressing.* This generally requires context to be maintained on the stack, and this historically is referred to as a display (from the Algol-60 implementation technology).

upper CASE a CASE tool available to support the initial tasks of software development: requirements, analysis, high level design.

upper triangular matrix a matrix that is only defined at (i, j) when $i \geq j$. *See also* lower triangular matrix, ragged matrix.

UPS *See* uninterruptible power supply.

upsampling a system that inserts $L - 1$ zeros between the samples of an input signal to form an output signal. An L-fold upsampler followed by an appropriate lowpass filter produces an output signal that is an interpolated form of the input signal, at L times the sampling rate. *Upsampling* also often refers to the operation of the upsampler and the lowpass filter together.

URL *See* uniform resource locator.

usability the capability of the software product to be understood, learned, used, and attractive to the user, when used under specified conditions. Some aspects of functionality, reliability, and efficiency will also affect usability, but for the purposes of ISO/IEC 9126 they are not classified as usability. Usability should address all of the different user environments that the software may affect, that may include preparation for usage and evaluation of results.

USAGE a specification in the COBOL language which determines the internal representation of a value.

usage the conditions of use of an item. A relationship that is a dynamic dependency. The actuation of an association link (which itself declares the architecture, not the dynamics).

USART *See* universal synchronous/asynchronous receiver-transmitter.

use a point in a program at which a name is used in a computation. *Contrast with* definition.

use bit in a paging system, a bit associated with a page entry in a lookup table which indicates that the page has been referenced since the last time the bit was reset. The bit is reset when it is read.

use case diagram a diagram showing the basic use cases of a software system together with the interactions that the user (expressed as an actor) has with these use cases. Each use case represents the functionality requested by a single actor in order to gain some benefit.

user an individual who uses the software product to perform a specific function. Users may include operators, recipients of the results of the software, or developers or maintainers of software.

user account consists of a login name, a password, a home directory, and some administrative files.

user CPU that component of CPU time spent supporting user application functions.

user datagram protocol (UDP) a datagram-level transport protocol for the Internet. There are no guarantees concerning order of delivery, or dropped or duplicated packets.

user documentation the information given with the software/hardware system to the user. It describes in an informal way what the software/hardware does and how the user can interact with it to achieve the wanted behavior. Often the user documentation is produced in an early stage of the development process to describe what the software/hardware system will do.

user friendly The typical definition for user interface presenting a set of appealing features that allow it to be perceived by users as an easy system to interact with.

user ID an integer that uniquely identifies a system user.

user interface (UI) the aspects of a computer system or program which can be seen (or heard or otherwise perceived) by the human user, and the commands and mechanisms the user uses to control its operation and input data. A graphical user interface emphasizes the use of pictures for output and a pointing device such as a mouse for input and control, whereas a command line interface requires the user to type textual commands and input at a keyboard and produces a single stream of text as output. *See* graphical user interface.

user manual a document describing all the needed information for the use of a software/hardware product, from the installation to the advanced details.

user mode in a multitasking processor, the mode in which user programs are executed. In *user mode,* the program is prevented from executing instructions that could possibly disrupt the system and also from accessing data outside the user's specified area.

user model a stochastic model for user behavior. *User models* are often used to define hierarchical traffic models as an input to a queueing system or a simulation.

username a sequence of characters identifying a user to the operating system.

user perceived quality quality that is perceived by the user when the software product is actually used in the users' environment, based on user attributes not visible to the developer.

user satisfaction a heuristic measure of whether users are happy with a bought/delivered software system.

user state a computer mode in which a user program is executing rather than a systems program.

Some computers have two modes of operation: the system state is used when parts of the computer's operating system are executing and the *user state* is used when the computer is executing application programs.

user-visible register an alternative name for general purpose registers, emphasizing the fact that these registers are accessible to the instructions in user programs. The counterpart to *user-visible registers* is registers that are reserved for use by privileged instructions, particularly within the operating system.

using declaration when a class inherits using private or protected inheritance level, one or more of the inherited members can be promoted again to a higher level (up to its original access level) by declaring the member in the derived class definition with the using keyword.

USL acronym for UNIX System Laboratories.

utility a software tool designated to perform some frequently used support function.

utilization (**1**) in computer performance evaluation, the fraction of the processor within a given time period spent supporting system and user functions.

(**2**) a ratio representing the amount of time a system or component is busy divided by the time it is available.

utilization law in operational analysis, the *utilization law* states that the utilization of a device is equal to the product of the mean service time between completions at that device and the throughput of that device.

UVPROM *See* erasable programmable read-only memory.

V

V&V *See* verification and validation.

vagueness a property indicating the lack of specifics and clarity and which is allied to imprecision and fuzziness.

validation (**1**) confirmation by examination and provision of objective evidence that the particular requirements for a specific intended use are fulfilled. In design and development, validation concerns the process of examining a product to determine conformity with user needs. Validation is normally performed on the final product under defined operating conditions. It may be necessary in earlier stages. Multiple validations may be carried out if there are different intended uses. It answers the question "Are we building the right system"?

(**2**) the performance or capacity planning study step in which the model solution is compared with actual system measurements. If the model output parameters are determined to be close enough to the corresponding system measurements, then the model is said to be "validated".

(**3**) the comparison of results obtained from different models for the same system, e.g., simulation and analytic models.

(**4**) in electronic active and passive device modeling, the pass/fail process in which a completed, ready to use model is used in a simulation, then compared to an intended application, and is determined to suitably predict reality.

validation of a measure the process of ensuring that the measure is a proper numerical characterization of the claimed attribute; this means showing that the representation condition is satisfied.

validation of a prediction system the process of establishing the accuracy of the prediction system by empirical means, i.e., by comparing model performance with known data points in a given environment.

validation phase the second phase of an optimistic concurrency control protocol where a transaction validates that its updates meet serializability requirements.

validation set the set of data to evaluate the performance of a system which was trained on a separate set of data.

valid unique defect the sum of the real defects, but not the duplicate defects. So when an error is mentioned more than once, it is not counted. Used for measuring software quality.

vanishing point a point in a perspective projection where parallel lines not parallel to the projection plane converge. A finite 2D projection of a point at infinity in 3D.

Van Wijngaarden grammar *See* W-grammar.

variable (**1**) a name in a programming language which holds a value. Typically a variable is a value which can be changed by operations of the programming language. Note that a name that is used as a formal parameter may be referred to as a "variable", even in some languages in which the formal parameter values may not be changed within the body of the function with which they are associated.

(**2**) a form of parallel job scheduling in which the size of the partition allocated to a job is determined at the time a job is submitted and is specified by the user.

variable bit rate (VBR) describes a traffic pattern in which the rate at which bits are transmitted varies over time; such patterns are also referred to as bursty. VBR sources often result from compressing CBR sources, for example, a 64 kbps voice source in its raw form has a constant bit rate; after compression by removing the silence intervals, the source becomes VBR.

variable length code to exploit redundancy in statistical data, and to reduce average number of bits per word, luminance levels having high

0-8493-2691-5/01/$0.00+$.50

probability are assigned short code words and those having low probability are assigned longer code words. This is called *variable length* coding or entropy coding. *See also* entropy coding.

variable length field in database systems, situation in which not all records in the file will have the same length for this field.

variable-length instruction the fact that the machine language instructions for a computer have different numbers of bits with the length dependent on the type of instruction.

variable resolution hierarchy an approach where images corresponding to the levels of the hierarchy vary in spatial resolution. This results in a pyramid structure where the base of the pyramid represents the full resolution and the upper levels have lower resolution.

variance the mean-squared variability of a random variable about its mean:

$$\sigma^2 = \int_{-\infty}^{\infty} (x - \mu)^2 p(x)\, dx \ ,$$

where σ^2 is the variance, μ is the mean, and $p(x)$ is the probability density function. *See also* covariance, correlation.

variant record a designation of a syntactic construct supported in many languages that allows a program to interpret the contents of the same area of storage in two or more ways. *See also* union.

variant selector within a variant record, the value that determines which of the variants is to be used for interpretation. Depending on the language, this value may be assignable by the programmer (thus allowing the interpretation to be changed at any time), or, once established, fixed for the lifetime of the object. It may be an explicit programmer-named value of the variant record, or implicit (such as in run-time type information). Not all languages that support variant records require a variant selector to exist. Some languages require the *variant selector* be used by providing syntactic constructs that require its use before operating on members of the variant; other languages impose no such limitations.

VBA acronym for Visual Basic for Applications.

VBR *See* variable bit rate.

VBScript a language based on Visual Basic, which is intended to execute in a limited environment. *See* sandbox.

VDL *See* Vienna Definition Language.

VDU *See* visual display unit.

VE *See* virtual environment.

vector **(1)** a list of numbers, typically Cartesian coordinates or a direction in 2D or 3D. For example, $\mathbf{d}_i = (d_x, d_y, d_y)^T = (1, 1.5, 0)^T$.

(2) a term often used to denote a one-dimensional array. Also used to represent a one-dimensional slice of a larger array.

vectored interrupt an interrupt request whereby the processor is directed to a predetermined memory location, depending on the source of the interrupt, by the built-in internal hardware. In the X86 processors, the addresses are stored in an array in memory (a mathematical vector) and indexed by the interrupt number. In the 8080 and Z80, the interrupt number becomes part of a CALL instruction with an implied address that is executed on an interrupt cycle.

vector graphics a type of computer graphic. The earliest computer graphics displays were drawn on so-called vector displays, because the electron beam which produced the image was under software control. The beam followed a chain of vectors (i.e., a polyline) from one point to another. *Vector graphics* are sometimes referred to as line-drawing graphics.

vector image an image consisting of mathematical descriptions of the objects in the scene, e.g., equations for lines and curves. The image is independent of resolution so it can be stretched, rotated, and skewed with no degradation. *Vec-*

tor images are often used in CAD applications. *See also* bitmapped image, CAD, image.

vector processor a computer architecture with specialized function units designed to operate very efficiently on vectors represented as streams of data. *See* linear array processor.

vector quantization (VQ) quantization applied to vectors or blocks of outputs of a continuous sources.

Each possible source block is represented by a reproduction vector chosen from a finite set (the "codebook"). According to rate-distortion theory, *vector quantization (VQ)* is able to perform arbitrarily close to the theoretical optimum if the lengths of the input blocks are permitted to grow without limit. The method was suggested by Claude Shannon in his theoretical work on source coding (during the late 1940s and the 1950s), but has found practical importance first in recent years (during the 1980s and 1990s) because of the relatively high complexity of implementation and design compared to scalar methods. Also referred to as "block source coding with a fidelity criterion".

vector quantization encoding an encoding scheme whereby an image is decomposed into n dimentional image vectors. Each image vector is compared with a collection of representative template or code vectors from a previously generated codebook. The best match code vector is chosen using a minimum distortion rule. Then the index of the code vector is transmitted. At the receiver this is used with a duplicate codebook to reconstruct the image. Usually called VQ encoding.

vector quantizer (VQ) a device that performs vector quantization.

vector space model a model used in information retrieval. The documents in the document set are represented as a set of vectors of weighted terms. The terms are weighted according to the frequency within that document and across the whole document set. The user's information need is also represented as a vector of weighted terms. The similarity or relevancy of a document to a given query is obtained by finding the cosine of the angle between the vectors.

vector stride the number of consecutive memory addresses from the beginning of one element to the next of a vector stored in memory. Also used to refer to the difference in vector index between two consecutively accessed vector elements.

Veitch diagram *See* Karnaugh map.

velocity field the three-dimensional vector attached to each image point that gives its current motion.

vendor the person or organization that sells a system or service to a set of customers.

Venn diagram a graphical notation from set theory that is useful for describing membership in sets according to binary properties, using overlapping ovals to divide the plane into regions. Regions inside an oval have the property the oval represents, while regions outside it do not have the property. Regions are shaded to show combinations of properties (or sets) of interest, or elements are placed in regions corresponding to their properties (or membership).

See also Karnaugh map.

verifiability a feature describing to which extent a system can be verified.

verification (1) steps taken to ensure that the output products of any development phase correctly implement the input products. Confirmation by examination and provision of objective evidence that specified requirements have been fulfilled. In design and development, *verification* concerns the process of examining the result of a given activity to determine conformity with the stated requirement for that activity. It answers the question: "Are we building the system right"? The verification includes activities such as inspection, proof of correctness, static analysis, etc. It may include the act of reviewing, inspecting, testing, checking, auditing, comparing, or otherwise establishing and documenting whether items, processes, services, or documents conform to specified requirements.

It is also the formal proof of program correctness.

(2) a formal technique to prove correctness, e.g., of a protocol specification. *See also* reachability graph.

verification and validation (V&V) the process of verification and validation. In some cases, with some techniques the distinction of the boundary between these formal activities is blurred.

Versa Module Europe bus (VME bus) a standardized processor backplane bus system originally developed by Motorola. The bus allows multiple processors to share memory and I/O devices.

version a concrete instance of an object. Multiple versions of one object can exist. A concrete configuration with concrete versions of the different objects belonging to this configuration. Also known as a release.

vertex the points in a model at which edges terminate, for example, the eight corners of a cube or the three corners of a triangle. Polyhedrons, polygonal surfaces, and triangulations are composed of vertices, edges, and faces.

vertex coloring an assignment of colors to the vertices of a graph such that no two adjacent vertices receive the same color.

vertex normal the direction vector pointing directly out of a polygonal/polyhedral model at a given vertex. This may be defined as the average of the surface normals of the faces adjacent to the vertex.

vertical fragmentation (1) of a relation in distributed databases, is the subdivision of the relation into subsets based on a set of projects. The key attributes of the relation are also stored with each fragment so that the original relation can be recreated.

(2) in object-oriented databases, fragmenting the instances of a class into more than one storage structure — one set for frequently accessed objects and another set for archives.

vertical microinstruction a field that specifies one microcommand via its op code. In practice, microinstructions that typically contain three or four fields are called vertical.

vertical propagation a type of privilege grant. A user granted privileges with the grant option with *vertical propagation* limited to N limits the length of the sequence of users awarded the grant to N.

very large scale integration (VLSI) a technology that allows the construction and interconnection of large numbers (millions) of transistors on a single integrated circuit.

very long instruction word (VLIW) a computer architecture that performs no dynamic analysis on the instruction stream and executes operations precisely as ordered in the instruction stream. It gets its name from the relatively longer microinstructions with respect to other architectures.

VGA *See* video graphics adapter.

via minimization given a set of trees in the plane, each interconnecting the terminals of a net, determining a layer assignment that minimizes the number of points (vias) at which a layer change occurs.

video (1) representation of moving images for storage and processing. Often used interchangeably with television. In particular, "video signal" and "television signal" are synonyms.

(2) a particular stored sequence of moving images, e.g., on a tape or within a database.

video card a special board that interfaces the computer monitor to the computer. Also called a video display adapter.

video coding compression of moving images. Coding can be done purely on an intraframe (within-frame) basis, using a still image coding algorithm, or by exploiting temporal correlations between frames (interframe coding). In the latter case, the encoder estimates motion between the current frame and a previously coded reference frame, encodes a field of motion vectors that describe the motion compactly, generates a motion-compensated prediction image and codes the difference between this and the actual frame with an intraframe residue coder — typically the 8×8 discrete cosine transform. The decoder receives the motion vectors and encoded residue, constructs the prediction picture from its stored reference frame, and adds back the difference information to recover the frame. *See also* MPEG.

video compression *See* video coding.

video graphics adapter (VGA) a video adapter proposed by IBM in 1987 as an evolution of EGA. It is capable of emulating EGA, CGA, and MDA. In graphic mode, it allows to reach 640×480 pixels (wide per high) with 16 colors selected from a pallet of 262144, or 320×240 with 256 colors selected from a pallet of 262144.

Vienna Definition Language (VDL) a semantic description language developed at the IBM Scientific Center in Vienna, Austria.

view (1) a software representation or a document about software. Examples are requirements and specification documents, hierarchy charts, flowcharts, petri nets, test data, etc. Each view is classified according to a particular view type: non-procedural, e.g., requirements documents; pseudo-procedural, e.g., software architecture documents; procedural, e.g., source code, data definition; analysis views which may accompany any other view.

(2) in databases, a single relation derived from other views or relations. Views are not explicitly stored in the database.

virtual address (1) an address that refers to a location in virtual memory.

(2) the address generated by the processor in a paging (virtual memory) system. *Compare with* real address.

virtual base class in the C++ language, multiple inheritance is allowed, and because of it a base class can be inherited indirectly more than once. In order to avoid the existence of two instances of the same base class in the derived object, the base class is inherited using the virtual keyword. This keyword forces the compiler to store one instance of the common base class in the memory space of the derived object.

virtual camera a set of parameters defining a 2D view of a 3D model. These might include: camera location, direction, or camera twist — defining the upwards direction in the rendered image.

virtual child record type a virtual record type that acts as a child record.

virtual cut-through a method for relaying packets through an intermediate switching node in a network similar to *store-and-forward* with the following difference: if the desired output link is free when a given packet arrives and there are no other packets queued, the transmission of that packet over the next hop will begin immediately, instead of waiting for the rest of the packet to arrive. This pipelining of the packet through the nodes of the network gives better performance than store-and-forward under light load, but when the load becomes high enough for queues to form, the performance is equivalent to store-and-forward. *See also* packet switching, store-and-forward.

virtual DMA DMA in which virtual addresses are translated into real addresses during the I/O operation.

virtual environment an artificial environment maintained by a computer which a user may interact with or view.

virtual function in the object-oriented paradigm when a base class demands the implementation of one member function, this function is a virtual function. Moreover, a function whose implementation is redefined by the derived class is a *virtual function.* The polymorphism exploitation is obtained by virtual function declaration; the declaration specifies the function name and interface of the member function for all derived classes.

virtual instrument an instrument created through computer control of a collection of instrument resources with analysis and display of the data collected.

virtually addressed cache a cache memory in which the placement of data is determined by virtual addresses rather than physical addresses. This scheme has the advantage of decreasing memory access times by avoiding virtual address translation for most accesses. The disadvantage is that data stored in the cache may have different virtual addresses in different processes (aliasing).

virtual machine (**1**) a process on a multitasking computer that behaves as if it were a stand-alone computer and not part of a larger system.

(**2**) an abstraction hiding the physical processor and its memory with one or more layers of software.

(**3**) a model of a computer instruction set that is typically implemented by an interpreter. Examples of virtual machines include p-Code and the Java Virtual Machine.

virtual memory main memory as seen by the processor, i.e., as defined by the processor-generated addresses, in contrast with real memory, which is the memory actually installed or that is immediately addressable.

The *virtual memory* corresponds to the secondary storage, and data is automatically transferred to and from real memory as needed. In paged virtual memory, secondary memory is divided into fixed-size pages that are automatically moved to and from page frames of real memory; the division is not logical and is usually invisible to the programmer. In segmentation, the divisions (known as segments) are logical and of variable-sized units that are much larger than pages. Segments are generally much larger than pages: 16K to 64K bytes vs. 0.5K to 4K bytes. Many machines combine both paging and segmentation.

Since secondary memory is much larger than main memory, virtual memory presents the programmer with the view of a main memory that appears to be larger than it actually is. Virtual memory also facilitates automatic transfer of data, protection, accommodation of growing structures, efficient management of main-memory, and long-term storage.

virtual memory interrupt interrupt that occurs when an attempt is made to access an item of virtual memory that is not loaded into main memory.

virtual method a feature of an object-oriented language by which a method that is invoked by a superclass will actually execute the appropriate method in a subclass that has defined it. In some object-oriented languages, all methods are implicitly virtual methods; in others (such as C++) the virtual methods must be explicitly indicated by the programmer. *See* vtable.

virtual page number in a paged virtual memory system, this is the part of the memory address that points to the page that is accessed, while the rest of the address points to a particular part of that page.

virtual parent child relationship (VPCR) a relationship between two record types, one of which is a virtual record type. These are used to model M:N relationships.

virtual parent record type a virtual record type that acts as a parent record.

virtual path a concept used to describe the unidirectional transport of virtual channels that are associated by a common identifier value.

virtual reality (VR) a simulation of a virtual environment which according to some must have an "immersive" quality encouraging the feeling of being present in the environment. Technology used with virtual reality includes stereo image helmets and 360° screens, but may be as simple as a standard monitor display.

virtual reality modeling language (VRML) a 3D model description format suited to transfer on the WWW.

virtual record a record that contains no data values. It contains only a logical pointer to a physical record.

virtual register one of a bank of registers used as general purpose registers to hold the results of speculative instruction execution until instruction completion. *Virtual registers* are used to prevent conflicts between instructions that would normally use the same registers. *See also* speculative execution.

virtual sensor a sensor used as a basis for implementing everyday human behavior such as visually directed locomotion, handling objects and responding to sounds and utterances. Virtual humans should be equipped with visual, tactile, and auditory sensors.

virtual split the process of dividing a hard disk into more than one logical DOS drive.

virtual time *See* simulation time.

virtual waiting time amount of unfinished work at a certain instant of time. This is equal to the waiting time which an imaginary arrival at the same regarded instant of time would experience. In certain queueing systems otherwise hardly analytically tractable (like n*D/D/1), the *virtual waiting time* can be used as a system state variable.

virus self-replicating program that is transmitted via email, newsgroups, disks of uncertain origin, and other means, that deliberately cause harm to a system's files.

visibility level of accessibility or privilege assigned to an entity/object/variable in the code.

visibility map a planar subdivision that encodes the visibility information. Two points p and q are visible if the straight line segment $\overline{p, q}$ does not intersect any other object. A horizontal (or vertical) visibility map of a planar straight-line graph is a partition of the plane into regions by drawing a horizontal (or vertical) straight line through each vertex p until it intersects an edge e of the graph or extends to infinity. The edge e is said to be horizontally (or vertically) visible from p.

visible surface determination during rendering of 3D scenes, it is necessary to determine which objects occlude others in order that the scene looks correct, and time wastage may be prevented by not drawing shapes that will be overdrawn. Techniques include: culling back-facing polygons, Z-buffering, and Warnock's algorithm.

visit ratio the number of visits a customer makes to a device each time it enters the system.

Visual Basic a language developed by Microsoft, loosely based on the syntax of the BASIC language, but containing so many extensions that the resemblance of the current language to the primitive BASIC language is almost undetectable. Contains significant features that support the Microsoft Windows Graphical User Interface, hence the name.

Visual Basic for Applications a dialect of Visual Basic designed as an extension language for applications.

visual display unit (VDU) a common means of input/output to/from a computer. Consists of a CRT and a keyboard.

visual interface a computer user interface based on visual/graphical components.

visual language a language based on visual/graphical tokens and on a visual syntax defined in terms of spatial relationships among visual tokens.

visual notation a notation based on a visual formalism, typically based on graphic symbols.

visual perception the perception of a scene as observed by the human visual system; it may differ considerably from the actual intensity image because of the non-linear response of the human visual system to light stimuli.

visual programming environment (VPE) software that allows the use of visual expressions (such as graphics, drawings, animation, or icons) in the process of programming. These visual expressions may be used as graphical interfaces for textual programming languages. They may be used to form the syntax of new visual programming languages leading to new paradigms such as programming by demonstration, or they may be used in graphical presentations of the behavior or structure of a program.

visual programming language (VPL) any programming language that allows the user to specify a program in a two- (or more) dimensional way. Conventional textual languages are not considered two-dimensional since the compiler or interpreter processes them as one-dimensional streams of characters. A VPL allows programming with visual expressions — spatial arrangements of textual and graphical symbols. VPLs may be further classified, according to the type and extent of visual expression used, into icon-based languages, form-based languages, and diagram languages. Visual programming environments provide graphical or iconic elements that can be manipulated by the user in an interactive way according to some specific spatial grammar for program construction. A visually transformed language is a non-visual language with a superimposed visual representation. Naturally visual languages have an inherent visual expression for which there is no obvious textual equivalent.

visual space the complete set of all possible images on a specific set of sampling and quantization parameters. Any specific image would be a member of this large space. For a 2×2 bi-level image, the space contains 16 members. Allowing all 3×3 bi-level images increases the size of the space to 512 (number of quantized levels raised to the power M, where M is the total number of pixels in the image).

VLIW *See* very long instruction word.

VLSI *See* very large scale integration.

VME bus *See* Versa Module Europe bus.

vocabulary in defining a formal grammar, the union of nonterminal symbols and terminal symbols.

voice means for enabling a computer or data processing system to recognize spoken commands and input data and convert them into electrical signals that can be used to cause the system to carry out these commands or accept the data. Various types of algorithms and stored templates are used to achieve this recognition.

void (1) possessing no value.

(2) in C and C++, a keyword that indicates a function returns no value.

(3) in C and C++, a keyword which when used with a pointer indicator, "void *", indicates an untyped pointer.

volatile (1) a value that can change in some way exogenous to the program or some section of the program.

(2) in the C language, a qualifier to a declaration that states that the variable being declared contains a volatile value. The implication of a volatile value is that certain optimizations that depend upon referential transparency cannot be made because of changes not visible to the compiler at the time the optimization analysis is made.

volatile data data that may change without the processor executing an instruction that changes it; registers on a variety of different hardware devices may act as volatile data. A volatile declaration on a data object in a program is designed to keep the compiler from optimizing away reads to that data object, since the compiler cannot know whether it has been changed since the last time it was read. Such data also cannot be cached.

volatile database the portion of the database that is stored in main memory buffers.

volatile device a memory or storage device that loses its storage capability when power is removed.

volatile fault *See* intermittent fault.

volatile memory memory that loses its contents when the power supply is removed. Examples include most types of RAM.

volatile storage *See* volatile memory.

voltage *See* potential difference.

volume rendering the visualization of 3D volume data, e.g., data sets such as MRI scans consisting of a volume of density samples or voxels.

volumetric rendering refers to techniques that model objects as three-dimensional volumes of material, instead of being defined by surfaces.

von Neumann architecture a stored program computer design in which data and instructions are stored in the same memory device and accessed and executed serially. *See also* Princeton architecture, single instruction stream,
single data stream.

von Neumann bottleneck in a von Neumann architecture, the fundamental performance limitation imposed by the fact that accessing of instructions and data from main memory must occur in a serial fashion.

Von Neumann, John (1903–1957) Born: Budapest, Hungary.

Von Neumann is best know for his role in the development of the theory of stored program flexible computers. He is honored by the reference to Von Neumann machines as a theoretical class of computers. Von Neumann also invented the idea of game theory. As a mathematician Von Neumann published significant work on logic, the theory of rings, operators, and set theory. His work, The Mathematical Foundations of Quantum Mechanics, was significant in the mathematical justification of that field. Von Neumann was a brilliant mathematician and physicist whose theoretical contributions are fundamental to modern physics and electrical engineering. He was the youngest member of the Institute of Advanced Studies at Princeton, and did important work on the Manhattan Project.

von Neumann rounding *See* jamming.

voting protocol any distributed concurrency control technique that does not use a distinguished copy; with voting protocols transactions send a lock request to all sites containing a copy of the item. If the majority of sites grant the lock, then the transaction may access the database item.

voxel the 3D analog of a pixel; abbreviation of volumetric picture element. Mathematically it is a point in 3D space having integer coordinates; concretely, it can also be interpreted as a cube of unit size centered about that point. *See* pixel.

voxel adjacency one of three types of adjacency relations defined on voxels:

1. 6-adjacency: Two voxels are 6-adjacent iff they differ by 1 in one coordinate, the other two coordinates being equal; equivalently, the two unit cubes centered about these voxels have one face in common.

2. 18-adjacency: Two voxels are 18-adjacent iff they differ by 1 in one or two coordinates, the remaining coordinates being equal; equivalently, the two unit cubes centered about these voxels have one face or one edge in common.

3. 26-adjacency: Two voxels are 26-adjacent iff they differ by 1 in one, two, or three coordinates, the remaining coordinates being equal; equivalently, the two unit cubes centered about these voxels have one face, one edge, or one vertex in common.

   In these definitions, the numbers 6, 18, and 26 refer to the number of voxels that are adjacent to a given voxel. *See* pixel adjacency, voxel.

VPCR *See* virtual parent child relationship.

VPE *See* visual programming environment.

VPL *See* visual programming language.

VQ *See* vector quantization, vector quantizer.

VQ encoding *See* vector quantization encoding.

VR *See* virtual reality.

VRML *See* virtual reality modeling language.

vtable a name applied to the virtual method table used in the implementation of some object-oriented languages. A virtual method table is a table of function pointers referenced by a pointer from each object in the class and is used to dispatch the virtual method calls to the correct method in a subclass needed to support an override of the virtual method by the subclass.

wait where a transaction stops pending the availability of an object it requires.

waiting *See* blocked.

waiting line *See* queue, buffer.

waiting probability the probability that a randomly chosen arrival finds that the server is already occupied with another task, and hence must wait before receiving service. *See also* Erlang C function.

waiting state in a dataflow architecture when the function cannot be executed since not all input lines contain a token.

waiting time the time between the arrival of a request in a queueing system and the beginning of service for this request. In most cases, the time a request spends in a queueing system is the sum of its waiting time and its service time.

wait state a bus cycle during which a CPU waits for a response from a memory or input-output device.

wait time the time a process spends waiting in the ready state before its first transition to the running state.

walk a path in which edges may be repeated.

Wallace tree multiplier a high-speed parallel multiplier that consists of levels of carry-save adders arranged in the form of a tree with a carry-propagate adder at the root. The operands are fed in at the top (leaves) of the tree. Partial carries and partial sums corresponding to various partial products then flow down the tree and are assimilated at the root to yield the final product.

wall-clock time total time taken by an application, including any delays due to the operating system or other processes and activities. This is in distinction to user time, as measured by an operating system, the time the system was actually busy with the work of that particular application.

Walsh transform *See* Walsh–Hadamard transform.

Walsh–Hadamard transform (WHT) a transform that uses a set of basis functions containing values that are either +1 or −1, and are determined from the rows of the Hadamard matrices. This has a modest decorrelation capability and is simple to implement.

WAM *See* Warren abstract machine.

warm boot *See* warm start.

warm start (**1**) reassumption, without loss, of some processes of the system from the point of detected fault.

(**2**) the restart of a computer operating system without going through the power-on (cold) boot process.

Warnock's algorithm a spatial partitioning technique for depth sorting a list of polygons so that they may be rendered correctly. The algorithm subdivides the screen rectangle until it may be painted entirely in the color of the foremost polygon or the background color.

warping the manipulation of 2D images by arbitrary geometric (i.e., position) transformations of the pixels of some or all of an image. Some simple types of warping are stretching, scaling, rotating, skewing, shearing, or perspective transform (perspective projection). This may be used to draw texture maps. Many simple transformations can be done simultaneously if positions and directions are represented in homogeneous coordinates.

Warren abstract machine (WAM) an intermediate (low-level) language that is often used as an object language for compiling PROLOG programs. Its objective is to allow the compilation of efficient PROLOG code.

0-8493-2691-5/01/$0.00+$.50

watchdog processor a processor that observes some process and signals an alert if a certain event happens or fails to happen.

watchdog timer a simple timer circuitry which keeps track of proper system functioning on the basis of time analysis. If the timer is not reset before it expires a fault is signaled, e.g., with an interrupt.

watercourse a line on a surface $f(x, y)$ which represents a watershed of the inverted surface $-f(x, y)$. The line of steepest descent from a saddle point to a minimum is a watercourse. Watercourses meet watersheds at saddle points.

waterfall life-cycle a life-cycle model dating back many decades in which phases were identified (analysis, design, code, etc.) such that these phases occurred sequentially with well-defined transitions. These change-over points between phases permitted easy project management, evaluation of progress, and so forth. Later modifications permitted some backtracking, but this was usually limited to between contiguous phases. It is highly appropriate for software developments in known domains when the current project, its requirements, environment, etc., are almost identical to past projects so that the risk factor (of "surprises") is very low.

watershed a line on a surface $f(x, y)$, typically an image, which divides it into "catchment areas". Within a catchment area, lines of descent all connect to the same minimum point. The line of steepest ascent from a saddle point to a maximum is a *watershed.* Watersheds often correspond to ridges.

watershed segmentation a morphological gradient edge detection scheme.

Watson, Thomas J., Jr. Watson is best known as the president of IBM who led the company into a dominant position in the computer industry. Watson took over his father's company, changed the structure, and moved the company away from the card tabulating business in which they held a dominant position. Watson Jr. oversaw the development of the IBM System/360 machines which were to give the company a dominant position in computing.

Watson, Thomas J., Sr. (1874–1956) Born: Cambell, New York.

Watson is best known as the president of IBM (International Business Machines). While Watson was not a technical person, his position as head of IBM put him in a position of supporting the development of a number of devices, both electronic and mechanical, leading to the development of the modern computer industry.

wavefront array processor similar to a systolic processor except that there is no external clock.

wavelet a basis function that is obtained by translating and dilating a mother wavelet; it has such properties as smoothness, time-frequency localization, orthogonality, and/or symmetry.

wavelet coding coding a signal by coding the coefficients of the wavelet transform of the signal. The discrete wavelet transform is often used in image compression.

wavelet packet a family of scaling functions and wavelets by translation and dilation of a mother wavelet and a scaling function following a binary tree structure.

wavelet shrinkage a non-parametric estimation method in order to remove noise from a signal by shrinking wavelet coefficients of a signal towards zero.

wavelet transform a computational procedure which, in order to represent a given function $x(t)$ by basis function ϕ, calculates

$$x(a, b) = \frac{1}{\sqrt{|a|}} \int_{-\infty}^{\infty} x(t)\phi\left(\frac{t-b}{a}\right) dt \, ,$$

where a and b are real numbers. *See also* inverse wavelet transform.

WBS *See* work breakdown structure.

weak coherency the property of a distributed file or memory system that does not necessarily provide sequential coherency by itself but that can provide sequential coherency if certain additional instructions are executed by any program running on the system.

weak entity type an entity type that does not have a key attribute. Weak entity types are related to a strong entity type (known as the identifying owner) by an identifying relationship. Weak entities are uniquely identified by the combination of the key of the strong entity and a set of attributes of the weak entity attribute. This set of attributes is known as the partial key.

weakest precondition a specification of state that must be satisfied for a segment of code to execute correctly and produce a well-defined result. Characterized by a predicate that evaluates to "true" if all the conditions are well-defined and meet the constraints for correct execution. In particular, the *weakest precondition* is the minimum specification of state that must be met, such that omitting any component of the state specified will result in undefined results of the computation. *See also* precondition, post condition.

weakest precondition semantics a variant of axiomatic semantics where a program and an output property are mapped to the weakest proposition that is necessary of the program's input to make the output property hold true.

weak fairness the kind of fairness whereby, when a process continuously makes a request, it will eventually be granted.

weak head normal form a lambda-expression in which all redexes are inside the bodies of lambda-abstractions.

weakly checked language a language that is statically checked but provides no clear guarantee of absence of execution errors.

weighted directed graph a directed graph which has a weight, or numeric value, associated with each edge. *See also* directed graph, weighted graph, adjacency-matrix representation.

weighted Euclidean distance for two real valued vectors $x = (x_1, x_2, \ldots, x_n)$ and $y = (y_1, y_2, \ldots, y_n)$ is defined as

$$D_\psi(x, y) = \sqrt{(x - y)^T \psi (x - y)}\ ,$$

where ψ is the inverse of the covariance matrix of x and y, and T denotes the transpose.

weighted graph a graph that has a weight, or numeric value, associated with each edge. Some applications require all weights to be nonnegative or to be positive. *See also* directed graph.

weighted mean squared error (WMSE) a generalization of the mean squared error.

weighted median a type of moving median filter using integer weights $a_1, a_2, \ldots, a_m$. It is found by repeating a_i times the observation x_i, ordering the new set of $a_1 + a_2 + \cdots + a_m$ values, and then choosing the middle value as the output. *See also* center weighted median.

weighted number system a number system in which each digit in a sequence of digits has an associated weight that corresponds to its position. For example, in conventional decimal notation the weights are 1, 10, 100, etc.

weight initialization the choosing of initial values for the weights in a neural network prior to training. Most commonly small random values are employed so as to avoid symmetries and saturated sigmoids.

weight sharing a scheme under which two or more weights in a network are constrained to maintain the same value throughout the training process.

Weiler-Atherton algorithm a technique for clipping one generalized polygon with the boundary of another.

well behaved a program fragment that will not produce forbidden errors at run time.

well-formed formula (WFF) in first order predicate logic, a formula built from atomic formulas using the connectives "$\wedge$" (and), "$\vee$" (or), "$\neg$" (not), and the quantifiers "$\forall$" (for all) and "∃" (there exists). All atomic formulas are WFF's, and if ϕ and ψ are *well-formed formulas,* then so are the conjunction $(\phi \wedge \psi)$, the disjunction $(\phi \vee \psi)$, the negation $\neg\phi$, and the quantified formulas $\forall X\phi$, $\exists X\phi$, where "X" is a variable symbol.

well-formed transaction a transaction where each read operation is preceded by a shared lock instruction, and each update preceded by an exclusive lock instruction.

WFM *See* workflow management.

W-grammar a context-sensitive grammar style created by Aad Van Wijngaarden [48, 49], which is a grammar consisting of two context-free grammars. The first level grammar creates model production rules; the second grammar generates the language. The language can then generate sentences. The two grammars are finite; the resulting set of productions is potentially infinite. *See also* metanotion, metaproduction rules, hypernotion, hyper rules.

Whetstone the speed of a processor as measured by the Whetstone Benchmark.

Whetstone benchmark one of several well-known synthetic benchmark programs, designed to measure the execution speed of a variety of floating point instructions, on scalar and vector data, but also on some integer code.

A benchmark test program for scientific computers originally written Whetstone was the first major synthetic benchmark, designed in the 1960s in the small town of Whetstone, England.

while|do a form of loop in which a predicate is evaluated, and if it evaluates to "true", causes the loop body to be executed, and repeats this process until the predicate becomes "false". If the predicate is initially "false", the loop body will not be executed at all. *See also* do|while, do|until.

while-loop a loop construct characterized by a control predicate evaluated before each iteration of the loop. *See also* while|do. *Compare with* do-loop.

white box reuse a style of reuse based on class inheritance. A subclass reuses the interface and implementation of its parent class, but it may have access.

white box testing testing the system by concentrating on the internal algorithmic correctness inside the subroutine or object class.

whitening filter a filter that whitens noise, i.e., one which brings noise whose power spectrum is not white into this condition, e.g., by means of a frequency dependent filter. Noise whitening is a vital precursor to matched filtering.

whitespace non-printing characters including the space and tab characters.

WHT *See* Walsh-Hadamard transform.

widening pointing to or referencing a derived object using a reference or a pointer to a base class. Another term for upcasting.

widget window gadget. A component of the interface that is used to represent, contain, and manipulate objects. Examples are icons, push buttons, check boxes, text fields, etc.

Widrow-Hoff learning rule a gradient descent learning rule for calculating the weight vector $\vec{w}$ for a linear discriminant which minimizes the squared error objective function. The vector $\vec{w}$ is modified as

$$\vec{w}(n+1) = \vec{w}(n) + \alpha(n)\left(d - \vec{w}^T(n)\vec{x}\right)\vec{x}\ ,$$

where $\vec{x}$ is an input vector, d is the desired output, and $\vec{w}(n)$ is the weight vector at iteration n.

Wiener, Norbert (1894–1964) mathematician whose contributions include: Brownian motion, stochastic processes, generalized functions, harmonic analysis, control theory, and optimal filtering. Established the field of cybernetics, author of "Cybernetics: or Control and Communication in the Animal and the Machine".

wildcard a symbol that matches any character, or in some cases any sequence of characters. Although most often used in the context of specifying file names in command lines typed to operating system shells, it is also used in some languages to designate pattern characters in string-matching patterns in programming languages that support such constructs.

wild failure *See* catastrophic failure.

Williams tube memory a memory device based on electric charges being stored on the screen of a cathode ray tube. Now obsolete.

Winchester disk a type of magnetic disk for data storage. Its characteristic property is that the disk and the read-write head are placed in a hermetically sealed box. This allows higher recording density as the read-write head can be moved closer to the disk surface. *See also* disk head.

window factor of the text that is aligned with the pattern.

windowed convolution a mathematical operation where a mask of numerical weights defined over some window is translated across a digital image pixel by pixel, and at each pixel the arithmetic sum of products between the mask weights and the corresponding image pixels in the translated window is taken.

windowing the process of opening a window. In signal processing, it is common to open only a certain restricted portion of the available data for processing at any one time; such a portion is called a window or sometimes a mask or neighborhood. For instance, in *FIR* filter design, a technique known as *windowing* is used for truncation in order to design an FIR filter. The design of window becomes crucial in the design.

In image processing, it is a common practice that a square window of (for example) 3×3 pixels is opened centered at a pixel under consideration. In this window operation, the gray level of the pixel is replaced by a function of its original gray level and the gray levels of other pixels in the window. Different functions represent different operations: in particular, they will be suitable for different filtering or shape analysis tasks. *See also* median filter, thinning.

window manager the software layer of a window-based operating system devoted to managing the windowing mechanisms.

window operation an image processing operation in which the new value assigned to a given pixel depends on all the pixels within a window centered at that pixel location.

window system system software that provides a virtual model of a physical terminal to the application programmer.

winner-take-all network a network in which learning is competitive in some sense; for each input a particular neuron is declared the "winner" and allowed to adjust its weights. After learning, for any given input, only one neuron turns on.

wired OR a circuit that performs an OR operation by the interconnection of gate outputs without using an explicit gate device. An open collector bus performs a wired OR function on active-low signals.

wireless local area network (WLAN) a computer network that allows the transfer of data without wired connections.

witness **(1)** a structure providing an easily verified bound on the optimal value of an optimization problem. Typically used in the analysis of an approximation algorithm to prove the performance guarantee.

(2) a mismatch of two symbols of string y at a distance of d is a "witness" to the fact that no subject y could occur twice at a distance of exactly d positions (equivalently, that d cannot be a period of y).

Wizard-of-Oz paradigm a procedure for collecting speech data to be used for training a conversational system in which a human wizard aids the system in answering the subjects' queries. The wizard may simply enter user queries verbatim to the system, eliminating the recognition errors, or may play a more active role by extracting appropriate information from the database and formulating canned responses. As the system becomes more fully developed it can play an ever-increasing role in the data collection process, eventually standing alone in a wizardless mode.

WLAN *See* wireless local area network.

WMSE *See* weighted mean squared error.

wobble in computer arithmetic, the maximum relative error corresponding to $1/2$ ulp.

word **(1)** a unit of information in a computer. This term is interpreted based upon cultural usage, and may represent 8, 16, or 32 bits of information, depending on the tradition of the computer and language. Many machines support word lengths other than these two common values. Some machines support only one word length. The word length is frequently directly reflected in a programming language as a specific data type. *See also* byte, halfword, longword, quadword.

(2) with respect to a text editor, any sequence of alphanumeric characters terminated by whitespace, a punctuation mark, or other non-alphanumeric character; any string of adjacent punctuation marks or non-alphanumeric characters is also treated as a word.

word parallel processing of multiple words in the same clock cycle.

work the total number of operations taken by a computation.

work bench a collection of software tools that are available to the software developers.

work breakdown structure (WBS) the set of work tasks required to build the software; defined as part of the process model.

work conserving a system in which the server is only idle when there is no request waiting. Polling systems with non-zero switchover times and systems with a dormant server are generally not work conserving.

work depth model a model of parallel computation in which one keeps track of the total work and depth of a computation without worrying about how it maps onto a machine.

work-efficient a parallel algorithm is work-efficient as asymptotically (as the problem size grows) it requires at most a constant factor more work than the best known sequential algorithm (or the optimal work).

workflow multiple tasks / steps / activities, of which there are two types: simple, representing indivisible activities, and compound, representing those which can be decomposed into sub-activities. An entire workflow can be regarded as a large compound task. The workflow defines the relationships among the activities in a project, from start to finish. Activities are related by different types of trigger relation. Activities may be triggered by external events or by other activities. Also the movement of documents around an organization for purposes including sign-off, evaluation, performing activities in a process and co-writing.

workflow management (WFM) workflow management is a technology that supports the reengineering and automation of business and information processes. It involves:

1. defining workflows, i.e., those aspects of process that are relevant to control and coordinate the execution of its tasks, and
2. providing for fast (re)design and (re)implementation of the processes as business/information needs change.

working set the collection of pages, $w(t, T)$, referenced by a process during the time interval $(t - T, t)$. These pages are most likely to be resident at any given point of program execution.

working space the portion of main memory that is assigned to a computer program for temporary storage of data.

workload characterization the process of identifying workload components, or customer classes, that are composed of transactions or jobs that have similar characteristics. *See* clustering (2).

workpackage a component/task of a project/workflow. The workpackage has: starting and ending date, objectives, milestones, results, and resources. The workpackage may have sub-workpackages.

work-preserving a translation of an algorithm from one model to another is work-preserving if the work is the same in both models, to within a constant factor.

World Wide Web (www) a network service that allows users to share multimedia information.

WORM *See* write once read many.

wormhole when a transaction completes after a second transaction, that depends on the result of the first transaction (the wormhole transaction). The presence of a wormhole precludes serialization.

worst-case cost the cost of an algorithm in the most pessimistic input possibility.

wraparound a phenomenon in signal processing that occurs in the discrete case when signals are not properly manipulated. For instance, in circular convolution, if the length of signals is not properly chosen, i.e., there are not sufficient zeros appended at the end of the signals, the so-called wraparound error will take place. That is, the contributions from different periods will overlap.

wrapper class a class having poor functionalities of its own. This kind of class is used to mask a non-object-oriented implementation, to hide software components provided by a third-party, to encapsulate objects with an interface that is not compatible.

wrist for a manipulator refers to the joints in the kinematic linkage between the arm and hand (or end-effector). Usually a wrist allows an orienting of the manipulator. Therefore the main role of the wrist is to change the orientation of the hand (or end-effector).

write to change the contents of a storage device or data medium.

write-after-read hazard *See* antidependency.

write ahead logging where a log record for an update is written to disk prior to the update itself being written.

write allocate part of a write policy that stipulates that if a copy of data being updated is not found in one level of the memory hierarchy, space for a copy of the updated data will be allocated in that level. Most frequently used in conjunction with a write-back policy.

write-back *See* copy-back.

write-back cache *See* copy-back.

write broadcast a protocol for maintaining cache coherence in multiprocessor systems. Each time a shared block in one cache is updated, the modification is broadcast to all other caches. Also referred to as write update.

write buffer a buffer that stores memory write requests from a CPU. The write requests in the buffer are then served by the memory system as soon as possible. Reduces the number of processor wait cycles due to long latency write operations.

write instruction a processor instruction that stores information into memory from a processor register or a higher level cache.

write invalidate a protocol for maintaining cache coherence in multiprocessor systems. Each time a shared block in one cache is updated, a message is sent that invalidates (removes) copies of the same block in other cache memories. This is a more common alternative than write broadcast protocols.

write lock a database synchronization mechanism. If a transaction T holds a write lock on a datum X, then no other transaction can obtain any lock on x.

write once read many (WORM) used to refer to memory devices that allow data to be written once after device fabrication, and to be read any number of times. A typical example is PROM.

write permission grants or denies the associated class of users the right to modify a file.

write phase The third and last phase in an optimistic concurrency control protocol, where the transaction updates the database.

write policy determines when copies of data are updated in a memory hierarchy. The two most common write policies are write through and write back (copy back).

write through a write policy that stipulates that when a copy of data is updated at one level of a memory hierarchy, the same data are also updated in the next outer level. Write through is usually only used in low level caches. Its advantages are that it is fast and simple to implement, and that it always guarantees that the next level of the memory hierarchy has a valid copy of all data. Its main disadvantage is that it generates much data traffic to the next level.

write-through cache when a location in the cache memory is changed the corresponding location in main memory is also changed.

write update *See* write broadcast.

WWW *See* World Wide Web.

XENIX a Unix version produced and distributed explicitly for personal computers; originally developed by Microsoft and released in 1986 and currently distributed by Santa Cruz Operation (SCO) and Interactive Corporation.

XOR an operator on two Boolean values whose result is true if exactly one of the values is true and the other false, and false if both values are true or both values are false. *See also* AND, OR, NOT, implies.

XOR gate a logic gate that performs the exclusive-OR function. Exclusive OR is defined for two inputs as one or the other being true but not both.

X-ray image a digital image whose pixels represent intensities of X-rays. The X-rays may come from artificial sources (medical images) or arise naturally (astronomy). Also important in modern inspection systems. *See also* imaging modalities, medical imaging.

xref an abbreviation for cross-reference.

XYZ color space a color model in which X specifies the red component and Y the green component. The blue component is $1 - X - Y$ (the color components are scaled so that $R+G+B = 1$). Z specifies the brightness.

0-8493-2691-5/01/$0.00+$.50

Y

Y2K acronym for the year 2000 problem. *See* millennium problem.

yacc acronym for yet another compilercompiler. A parser generator system originally developed under Unix, and now widely available. The program takes a description of a restricted context-free grammar defining a language and generates a parser for that language that can be integrated into an application. *See also* lex.

$\boldsymbol{YC_rC_b}$ *See* YUV color palette.

year 2000 problem (Y2K) *See* millennium problem.

yield to leave a processor voluntarily.

YIQ color space a chrominance/luminance color space model used in the American NTSC television standard, Y specifies luminance, I and Q specify chrominance. I specifies the red-orange/cyan (or blue-green) component, and Q specifies the green/magenta (or purple) component (called the quadrature component). In this standard the image is coded RGB values as follows: $Y = 0.30R + 0.59G + 0.11B$, $I = 0.28G + 0.59R - 0.32B$ and $Q = -0.53G + 0.21R + 0.31B$.

YUV color palette a chrominance/luminance color space model used in the British PAL television standard, Y specifies luminance, U and V specify chrominance. U specifies the blue/yellow component, and U specifies the red/cyan (or blue-green) component. YUV is also called Y C_rC_b.

0-8493-2691-5/01/$0.00+$.50

Z

Z-buffer a method for solving the visible (or hidden) surface problem using two aligned pixel buffers or images. The first buffer stores the current color of the pixel and the second buffer stores the distance from the viewer to the surface. When rendering a point a on a scene surface, if the distance from the observer to a is greater than that of a previous point b that projects to the same image pixel, then point a can be ignored (as it cannot be seen). If the distance to a is less than the stored distance to b, then distance and color of a replace the color and distance buffer entry of b.

A Z-buffer is often efficiently implemented as a hardware buffer with entries aligned with pixels. Unfortunately, these Z-buffers suffer a lot from aliasing effects and A-buffers are much better at dealing with visibility problems at subpixel accuracy. Also called depth buffer.

Z-buffering a technique for speeding up depth sorting while rendering. As each primitive in the frustrum is drawn, the distance from the viewpoint is recorded in the Z-buffer or depth buffer. If a pixel has already been drawn with a closer Z value the new pixel value is not recorded. *See* visible surface determination.

zero-address computer a class of computer based on zero-address instructions. Stack-based calculators are zero-address computers and can be programmed using postfix notation.

zero-address instruction a class of assembly language ALU instruction in which the operands are kept on a first-in-first out stack in the CPU, and thus require no explicit addresses.

zero-address machine a machine in which the operands for an instruction are implicit in the nature of the instruction. A representative zero-address machine is one in which all operations are performed on a stack. *See* zero-address computer.

zero divide *See* divide-by-zero.

zero flag a bit in the condition code register which indicates whether the result of the last arithmetic or logic instruction is zero (1 for zero, 0 for not zero).

zero insertion force (ZIF) a type of chip socket which does not require special tools for insertion or removal. A special locking/unlocking lever is incorporated into the socket package. Many newer motherboards accept the CPU.

zero order hold (ZOH) a procedure that samples a signal $x(t)$ at a given sampling instant and holds that value until the succeeding sampling instant.

zero padding technique where a discrete finite length signal is padded by adding some number of zeros at the end of the signal. The discrete Fourier transform of a zero padded signal has more frequency samples or components than that of a nonzero padded signal, although the frequency resolution is not increased. Zero padding is also sometimes used with the discrete Fourier transform to perform a convolution between two signals.

zero-sum game one of a wide class of noncooperative two-person games in which the sum of the cost functions of the decision makers is identically zero. In the zero-sum games cooperation between players is impossible because the gain of one player is a loss of the other one. Thus the game is characterized by only one cost function which is minimized by the first player and maximized by the second one. To the zero-sum game one could also transform a constant-sum game in which the sum of the cost functions is constant. The solution in the zero-sum games has a form of saddle-point equilibrium and roughly speaking it exists for problems in which max and min operations on the cost function commute. In zero-sum games without equilibrium in pure strategies it is possible to find a saddle point in mixed strategies if the game is played many times in the same conditions. The resulting outcomes are average gains or losses of the players.

0-8493-2691-5/01/$0.00+$.50

ZIF *See* zero insertion force.

zip a file format and a set of data compression algorithms used to store one or more files in a single file. Originally devised by Phil Katz and placed in the public domain.

Zipfian distribution a distribution of probabilities of occurrence which follow Zipf's law.

Zipf's law the probability of occurrence of words or other items starts high and tapers off exponentially. Thus, a few occur very often while many others occur rarely. *See also* Zipfian distribution.

Ziv-Lempel coding *See* LZ77, LZ78, and Lempel-Ziv-Welch coding.

ZOH *See* zero order hold.

zombie a process whose data structures still remain in the operating system, but which cannot be accessed.

zonal coding a coding scheme in transform coding in which only those transform coefficients located in a specified zone in the transform domain are coded. For its counterpart, refer to threshold coding.

zonal method a technique used in conjunction with the Hadamard matrix for selecting the compressed transform vector where one simply fixes the set of components to be kept. *Compare with* threshold method.

zone recording a technique that allows the number of sectors per track on a magnetic disk to vary with the radius of the track. The tracks are divided into several zones, such that the number of sectors per track is determined by the maximum possible bit density on the innermost track in each zone.

zooming viewing an image at different sizes. Zooming in creates an enlarged view of a portion of the scene in the image frame. Zooming out does the reverse.

zoom pyramid a data structure which stores an image at multiple sizes/resolutions. The zoom pyramid for a 640×320 image would include versions with sizes 240×160, 160×80, 80×40, etc. Zooming in to the image quickly is therefore possible.

REFERENCES

The following are the cited references for the definitions found in this dictionary.

[1] Banks, J., *Handbook of Simulation,* John Wiley & Sons, New York, 1998.

[2] Barfield, W. and Furness, T.A., Eds., *Virtual Environments and Advanced Interface Design,* Oxford University Press, New York, 1995.

[3] Boff, K.R. and Lincoln, J.E., Eds., *Engineering Data Compendium: Human Perception and Performance,* (3 volumes). Harry G. Armstrong Aerospace Medical Research Laboratory, Wright-Patterson AFB, OH, 1988.

[4] Burdea, G. and Coiffet, P., *Virtual Reality Technology,* Wiley-Interscience, New York, 1993.

[5] Chen, W. -K., Ed., *The Circuits and Filters Handbook,* CRC Press, Boca Raton, FL, 1995.

[6] Cleaveland, J.C. and Uzgalis, R.C., *Grammars for Programming Languages,* Elsevier, 1977.

[7] Dorf, R.C., Ed., *The Electrical Engineering Handbook,* CRC Press, Boca Raton, FL, 1997.

[8] Craig, J.J., *Introduction to Robotics: Mechanics and Control,* Addison-Wesley, Reading, MA, 1986.

[9] Diner, D.B. and Fender, D.H., *Human Engineering in Stereoscopic Viewing Devices,* Plenum Press, New York, 1990.

[10] Dougherty, E.R. and Laplante, P.A., *Introduction to Real-Time Image Processing,* SPIE Press/IEEE Press, January 1995.

[11] Draft proposed ANS FORTRAN X3J3/76, in SIGPLAN Notices (11,3), March, 1976.

[12] Durlach, N.I. and Mavor, A.S., Eds., *Virtual Reality: Scientific and Technical Challenges,* National Academy Press, Washington, DC, 1995.

[13] Durand, F., http://www-imagis.imag.fr/Membres/Fredo.Durand/Book/ index.html:\\ Computer graphics bookmarks.

[14] Enslow, H., Multiprocessor organization, *Computing Surveys,* 9(1), 103–129, 1977.

[15] Foley, J.D., et al., *Computer Graphics: Principles and Practice,* 2nd ed. in C, Addison-Wesley, Reading, MA, 1996.

[16] http://www.acm.org/tog/GraphicsGems/:Graphics Gems series, Academic Press Professional, London/Boston.

[17] Greenwood, D.T., *Principle of Dynamics,* Prentice-Hall, Englewood Cliffs, NJ, 1965.

[18] Hennessy, L. and Patterson, D.A., *Computer Architecture: A Quantitative Approach,* 2nd ed., Kaufmann, 1996.

[19] iDREAM^{TM} Software L.L.C., http://www.idreamsoftware.com/html/3dresource/3ddict.html: Online dictionary.

[20] *IEEE Standard Dictionary of Electrical Engineering,* 6th ed., 1996.

[21] Jang, H., http://www.scs.ryerson.ca/~h2jang/gfx_c.html: Some definitions and code samples.

[22] Jouppi, N.P., The nonuniform distribution of instruction-level and machine parallelism and its effect on performance, *IEEE Transactions on Computers,* 38(12), 1645–1658, Dec. 1989.

[23] Katevenis, G.H., *Reduced Instruction Set Computer Architectures for VLSI,* MIT Press, Cambridge, MA, 1985.

[24] Kumamoto, D., htpp://www.iluminati.com/~dnk/graphics.html: Graphics software, hardware and discussion compendium.

[25] Laplante, P.A., Ed., *Comprehensive Dictionary of Electrical Engineering,* CRC Press, Boca Raton, FL, 1998.

[26] Laplante, P.A., *Real-Time Systems Design and Analysis: An Engineer's Handbook,* 2nd ed., IEEE Press/IEEE CS Press, Piscataway, NJ, 1997.

[27] Laplante, P.A. and Stoyenko, A.D., Eds., *Real-Time Image Processing: Theory, Techniques, and Applications,* IEEE Press, 1996.

[28] Laplante, P.A., *Easy PC Maintenance and Repair,* 2nd ed., Windcrest/McGraw-Hill, Blue Ridge Summit, PA, 1995.

[29] Laplante, P.A., *Fractal Mania,* Windcrest/McGraw-Hill, Blue Ridge Summit, PA, 1993.

[30] Latham, R., *The Dictionary of Computer Graphics and Virtual Reality,* Springer-Verlag, 1995.

[31] Levine, W.S., Ed., *The Control Handbook,* CRC Press, Boca Raton, FL, 1995.

[32] Lucas, P., et al., Method and Notation for the Formal Definition of Programming Languages, Technical Report 25.087, IBM, Vienna Laboratory.

[33] Myers, J., *Advances in Computer Architecture,* 2nd ed., John Wiley & Sons, New York, 1982.

[34] Netmeg Internet, http://www.netmeg.net/faq/computers/graphics/: A collection of graphics FAQs.

[35] Nof, S.Y., Ed., *Handbook of Industrial Robotics,* John Wiley & Sons, New York, 1999.

[36] Norman, D.A. and Draper, S.W., *User-Centered System Design: New Perspectives on Human-Computer Interaction,* Lawrence Erlbaum Associates, Hillsdale, NJ, 1986.

[37] Omondi, A.R., *Computer Arithmetic Systems,* Prentice-Hall, Englewood Cliffs, NJ, 1994.

[38] Patterson, A. and Ditzel, D.R., The case for the RISC, *Computer Architecture News,* 8(6), 25–33, 1980.

[39] Patterson, A. and Sequin, C.H., A VLSI RISC, *IEEE Computer,* 15(9), 8–21, 1982.

[40] Radin, The 801 Minicomputer, *IBM Journal of Research and Development,* 21(3), 237–246, 1983.

[41] Reddy, M., http://www.dcs.ed.ac.uk/~mxr/gfx/index-hi.html: Graphics file formats and FAQs.

[42] Russ, J.C., Ed., *The Image Processing Handbook,* 2nd ed., CRC Press, Boca Raton, FL, 1994.

[43] Satava, R.M., Ed., *Cybersurgery: Advanced Technologies for Surgical Practice,* Wiley-Liss, New York, 1998.

[44] Shaw, M. and Garlan, D., *Software Architecture Perspectives on an Emerging Discipline,* Prentice-Hall, Englewood Cliffs, NJ, 1996.

[45] Sheridan, T.B., *Telerobotics, Automation, and Human Supervisory Control,* MIT Press, Cambridge, MA, 1992.

[46] South, D.W., *The Computer and Information Science and Technology Abbreviations and Acronyms Dictionary,* CRC Press, Boca Raton, FL, 1994.

[47] Tabak, *RISC Systems and Applications,* Research Studies Press and Wiley, 1996.

[48] Tucker, A.B., Ed., *The Computer Science and Engineering Handbook,* CRC Press, Boca Raton, FL, 1996.

[49] Van Wijngaarden, A., et al., Report on the Algorithmic Language Algol-68, Technical Report 101, Mathemastiche Centrum, Amsterdam.

[50] Van Wijngaarden, A., et al., Revised Report on the Algorithmic Language Algol-68, Acta Informatica, 5(1), 236.

[51] Waite, W. and Goos, G., *Compiler Construction,* Springer-Verlag, 1984.

[52] Whittaker, J.C., Ed., *The Electronics Handbook,* CRC Press, Boca Raton, FL, 1996.

[53] Yoshikawa, T., *Foundations of Robotics: Analysis and Control,* MIT Press, Cambridge, MA, 1990.